# The Mollusca

## VOLUME 1

Metabolic Biochemistry and Molecular Biomechanics

# The Mollusca

Editor-in-Chief
**KARL M. WILBUR**

Department of Zoology
Duke University
Durham, North Carolina

# The Mollusca

## VOLUME 1
## Metabolic Biochemistry and Molecular Biomechanics

Edited by

## PETER W. HOCHACHKA

*Zoology Department*
*University of British Columbia*
*Vancouver, British Columbia, Canada*

1983

## ACADEMIC PRESS
*A Subsidiary of Harcourt Brace Jovanovich, Publishers*
New York   London
Paris   San Diego   San Francisco   São Paulo   Sydney   Tokyo   Toronto

ACADEMIC PRESS, INC.
111 Fifth Avenue, New York, New York 10003

*United Kingdom Edition published by*
ACADEMIC PRESS, INC. (LONDON) LTD.
24/28 Oval Road, London NW1 7DX

Library of Congress Cataloging in Publication Data

Main entry under title:

The Mollusca.

Includes index.
Contents: v. 1. Metabolic biochemistry and
molecular biomechanics / edited by Peter W.
Hochachka -- v. 2. Environmental biochemistry
and physiology / edited by Peter W. Hochachka --
v. 3. Development / edited by N.H. Verdonk &
J.A.M. van den Biggelaar & A.S. Tompa -- v. 4-5.
Physiology / edited by A.S.M. Saleuddin & Karl M.
Wilbur.
1. Mollusks--Collected works. I. Wilbur, Karl M.
QL402.M6  1983    594              82-24442
ISBN 0-12-751401-5 (v. 1)

PRINTED IN THE UNITED STATES OF AMERICA

83 84 85 86      9 8 7 6 5 4 3 2 1

# Contents

## 3. Carbohydrate Metabolism in Cephalopod Molluscs

KENNETH B. STOREY AND JANET M. STOREY

## 4. Carbohydrate Catabolism in Bivalves

ALBERTUS DE ZWAAN

## 5. Carbohydrate Metabolism of Gastropods

DAVID R. LIVINGSTONE AND ALBERTUS DE ZWAAN

## 6. Amino Acid Metabolism in Molluscs

STEPHEN H. BISHOP, LEHMAN L. ELLIS,
AND JAMES M. BURCHAM

## 7. Lipids: Their Distribution and Metabolism

PETER A. VOOGT

## 8. Molluscan Collagen and Its Mechanical Organization in Squid Mantle

JOHN M. GOSLINE AND ROBERT E. SHADWICK

# 9. Molecular Biomechanics of Protein Rubbers in Molluscs

ROBERT E. SHADWICK AND JOHN M. GOSLINE

# 10. Molecular Biomechanics of Molluscan Mucous Secretions

MARK DENNY

# 11. Quinone-Tanned Scleroproteins

J. H. WAITE

# Contributors

Numbers in parentheses indicate the pages on which the authors' contributions begin.

**Stephen H. Bishop** (243), Department of Zoology, Iowa State University, Ames, Iowa 50011

**James M. Burcham** (243), Department of Zoology, Iowa State University, Ames, Iowa 50011

**Mark Denny**[1] (431), Department of Zoology, University of Washington, Seattle, Washington 98195

**Lehman L. Ellis** (243), Department of Zoology, Iowa State University, Ames, Iowa 50011

**J. H. A. Fields**[2] (55), Department of Zoology, University of British Columbia, Vancouver, British Columbia V6T 2A9, Canada.

**John M. Gosline** (371, 399), Department of Zoology, University of British Columbia, Vancouver, British Columbia V6T 2A9, Canada

**P. W. Hochachka** (55), Department of Zoology, University of British Columbia, Vancouver, British Columbia V6T 2A9, Canada

**David R. Livingstone** (177), Institute for Marine Environmental Research, Plymouth PL1 3DH, Great Britain

**T. P. Mommsen**[3] (55), Department of Zoology, University of British Columbia, Vancouver, British Columbia V6T 2A9, Canada

**R. Seed** (1), Department of Zoology, University College of North Wales, Bangor, Gwynedd LL57 2UW, Wales

[1]Present address: Department of Biology, Stanford University, Hopkins Marine Station, Pacific Grove, California 93950.
[2]Present address: Department of Zoology, University of Washington, Seattle, Washington 98105.
[3]Present address: Department of Biology, Dalhousie University, Halifax, Nova Scotia, Canada.

**Robert E. Shadwick** (371, 399), Department of Zoology, University of British Columbia, Vancouver, British Columbia V6T 2A9, Canada

**Kenneth B. Storey** (91), Institute of Biochemistry, Carleton University, Ottawa, Ontario K1S 5B6, Canada

**Janet M. Storey** (91), Institute of Biochemistry, Carleton University, Ottawa, Ontario K1S 5B6, Canada

**Peter A. Voogt** (329), Laboratory of Chemical Animal Physiology, The State University of Utrecht, 3508 TB Utrecht, The Netherlands

**J. H. Waite**[4] (467), Department of Biochemistry, University of Connecticut, Farmington, Connecticut 06032

**Albertus de Zwaan** (137, 177), Laboratory of Chemical Animal Physiology, State University of Utrecht, 3508 TB Utrecht, The Netherlands

[4]Present address: Orthopaedics Research Laboratory, University of Connecticut, Farmington, Connecticut 06032.

# General Preface

This multivolume treatise, *The Mollusca*, had its origins in the mid 1960s with the publication of *Physiology of Mollusca*, a two-volume work edited by Wilbur and Yonge. In those volumes, 27 authors collaborated to summarize the status of the conventional topics of physiology as well as the related areas of biochemistry, reproduction and development, and ecology. Within the past two decades, there has been a remarkable expansion of molluscan research and a burgeoning of fields of investigation. During the same period several excellent books on molluscs have been published. However, those volumes do not individually or collectively provide an adequate perspective of our current knowledge of the phylum in all its phases. Clearly, there is need for a comprehensive treatise broader in concept and scope than had been previously produced, one that gives full treatment to all major fields of recent research. *The Mollusca* fulfills this objective.

The major fields covered are biochemistry, physiology, neurobiology, reproduction and development, evolution, ecology, medical aspects, and structure. In addition to these long-established subject areas, others that have emerged recently and expanded rapidly within the past decade are included.

*The Mollusca* is intended to serve a range of disciplines: biological, paleontological, and medical. As a source of information on the current status of molluscan research, it should prove useful to researchers of the Mollusca and other phyla, as well as to teachers and qualified graduate students.

Karl M. Wilbur

# Preface

There is little that impresses classical biochemists and physiologists more than those principles to which all organisms conform. It is these common principles which are the necessary and sufficient focus for future studies of biochemistry and physiology. Parts of biology differ from these fields in that the important aspect of nature for many biologists is the immense diversity of organisms, which is insufficiently explained by the disciplines of biochemistry and physiology. For such biologists, the critical concern is to give attention to common mechanisms relating to the diversity of organisms and to environmental influences. The molluscs are an unusually favorable group for the investigation of such mechanisms in that they comprise one of the most successful invertebrate phyla in their species diversity and in the diversity of the environments they have exploited.

A principal goal of the first two volumes on the Mollusca is to begin bridging the gap between the points of view of biochemistry and physiology on the one hand and the rest of biology on the other. The chapters have been written within two kinds of frameworks: first, illustrating general principles, and second, exposing principles of design that fit molecular, metabolic, and mechanical mechanisms to life-style and the environment.

The volumes bring together long-needed summaries of advances in the traditional areas of biochemistry and in recently developed areas that have become a part of molluscan biochemistry. These more recent areas are molecular biomechanics and environmental biochemistry. Topics in molecular biomechanics are discussed in Volume 1, Chapters 8, 9, 10, and 11. Various aspects of environmental biochemistry are covered in Volume 2, Chapters 3, 4, 5, 6, 7, 9, and 10. For the most part, the topics presented in such chapters have not been reviewed in previously published volumes on molluscs. With their inclusion, the integration of biochemistry with research in physiology and in ecology becomes more evident than in the past.

The first chapter of Volume 1 introduces the phylum Mollusca and is intended as a reference chapter for all volumes of the treatise. It provides information about the general features of the major classes and their evolution, the anatomical organization of molluscs, and an abbreviated classification of the major taxonomic groups of molluscs.

The antecedents of these volumes are *Physiology of Mollusca* (edited by K. M. Wilbur and C. M. Yonge, Academic Press, 1964 and 1966) and *Chemical Zoology* (edited by M. Florkin and B. J. Scheer, Academic Press, 1972). Like them, the present volumes are addressed to researchers in molluscan studies and to others in the fields of biochemistry and physiology.

Peter W. Hochachka

# Contents of Other Volumes

## Volume 2: Environmental Biochemistry and Physiology

# Volume 3: Development

# Volume 4: Physiology, Part 1

# Volume 5: Physiology, Part 2

# Volume 6: Ecology

# 1

# Structural Organization, Adaptive Radiation, and Classification of Molluscs

## R. SEED

Department of Zoology,
University College of North Wales,
Bangor, Gwynedd LL57 2UW, Wales, United Kingdom

## I. Introduction

The phylum Mollusca comprises one of the largest and most successful phyla within the animal kingdom, being exceeded in terms of number of species only by the Arthropoda and perhaps also by the Protozoa and Ne-

THE MOLLUSCA, VOL. 1
Metabolic Biochemistry
and Molecular Biomechanics

matoda. Over 100,000 species of living molluscs have so far been described (more than twice the number of vertebrates), and a rich fossil record extends back to the Cambrian. Molluscs are remarkably diverse with regard to their external morphology, and in size they range from microscopic clams and snails (<1 mm) to giant oceanic squid and the massive tridacnid clams of Indo-Pacific coral reefs. They are successful in both terrestrial and aquatic environments and have penetrated perhaps a wider range of habitats than virtually any other animal group.

Molluscs have long been of interest to human beings because they include species of considerable socioeconomic importance, for example as pests, as exploitable food resources, and as intermediate hosts for a variety of disease-bearing parasitic helminths.

Seven classes are generally recognized (see Section VIII for classification), but their body forms vary so much that any attempt at a definition that satisfactorily embraces all of the variation within these groups proves quite elusive. Indeed molluscs are generally identified as such by any one of an array of traits rather than by any single diagnostic character.

The major objective of this introductory chapter is to provide an overview of the structural organization and classification of the Mollusca as a preface to the volumes on physiology and biochemistry. However, the molluscan body form is so incredibly plastic (compare, for example, a snail with a mussel or an octopus) that any introductory account would be substantially incomplete without some reference to the range of adaptive radiation that has occurred within this phylum. Accordingly a section of this chapter is devoted to the diversity and main evolutionary trends found within the major molluscan classes. Spatial constraints have clearly restricted the depth of coverage of certain topics. Some sections are intentionally brief because the material covered is documented in substantially greater detail elsewhere in this book.

## II. The Ancestral Mollusc

Despite their morphological plasticity, all molluscs broadly conform to the same relatively simple basic organizational plan. This has resulted in the now familiar concept of the "archetype" mollusc in which a theoretical ancestral form exhibiting those features generally considered to be the primitive bases of several molluscan traits is constructed as a model against which the diversity of the different classes can be contrasted. Whether such a hypothetical form ever existed is of course entirely speculative; it does, nonetheless, provide a convenient framework for describing some of the basic molluscan characteristics. Although there are clearly many intellectual dangers inherent in constructing such an artificial archetype, many malacologists have evidently preferred this approach to the alternative of describing "type species" because these latter are almost inevitably highly specialized forms specifically attuned to their own peculiar life-styles.

In Fig. 1, a possible archetypic mollusc has been reconstructed. This was probably a rather flat, sluggish animal with a low shieldlike shell and a broad ventral sole used for crawling over rocky surfaces. As with virtually all molluscs, the body can be broadly divided into two regions: the head–foot and the visceral mass (visceropallium). Primitively, the foot was probably a broad, flat structure whereas the poorly developed head carried simple sense organs. The mouth was located at the tip of a short snout, and particulate material was rasped by means of a filelike protrusible radula—a structure unique to the molluscs. The visceral mass is the main metabolic region of the molluscan body. It is situated dorsal to the foot totally enclosed by the protective shell. Unlike the head–foot, the visceral mass is soft and nonmuscular, and it relies largely on cilia and mucus for efficient functioning. Enveloping the visceral mass and hanging from it like a skirt is the mantle epithelium, which is responsible for shell secretion, particularly around its free margin. The space contained between the mantle skirt and the sides of the body is termed the mantle or pallial cavity, a feature of particular significance in molluscan organization. Primitively, this space may simply have been a marginal groove between the mantle and foot housing the paired series of respiratory organs and receiving waste material from the anus and excretory ducts (Fretter and Graham, 1962). This condition still largely persists in present-day chitons and monoplacophorans. Very early in molluscan evolution, however, posterior expansion of this space, possibly brought about through the forward movement of the shell to protect the vulnerable head region, produced a spacious cavity beneath the shell into which the head–foot could be withdrawn. Because this posterior chamber probably afforded considerable protection against excessive turbulence and silting, the concentration of the gills into this cavity may have offered some selective advantage.

Perhaps the most prominent feature of the primitive molluscan mantle cavity are the paired gills or ctenidia. These effectively divide the mantle cavity into two functional chambers: a ventral inhalent chamber and a dorsal exhalent chamber. The anus and urogenital ducts discharge into the exhalent chamber whereas a pair of chemosensory osphradia taste the inhalent water before it impinges on the gills. Each gill is supported by dorsal and ventral membranes and consists of a central axis with alternating rows of wedge-shaped filaments along either side (i.e., is bipectinate).

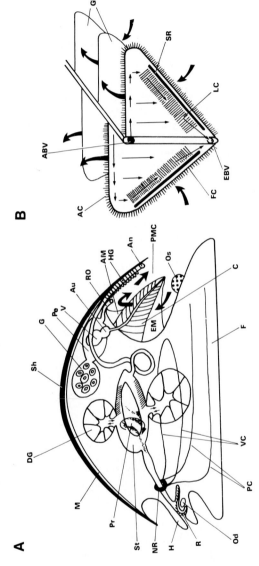

**Fig. 1.** (A) Schematic view of an archetypic mollusc. Heavy arrows indicate path of water circulation through the posterior mantle cavity. (B) Stereogram of the archetypic molluscan gill (aspidobranch). Water flow between adjacent gill plates indicated by heavy arrows. Small arrows show the countercurrent flow of blood within each gill plate. ABV, afferent blood vessel; AC, abfrontal cilia; AM, afferent membrane; An, anus; Au, auricle; C, ctenidium; DG, digestive gland; EBV, efferent blood vessel; EM, efferent membrane; F, foot; FC, frontal cilia; G, gonad; GP, gill plates; H, head; HG, hypobranchial gland; LC, lateral cilia; M, mantle; NR, nerve ring; Od, odontophore; Os, osphradium; PC, pedal cords; Pe, pericardium; PMC, posterior mantle cavity; Pr, protostyle; R, radula; RO, renal organ; Sh, shell; SR, skeletal rod; St, stomach; V, ventricle; VC, visceral cords. (Modified from various sources.)

A dorsal afferent blood vessel carrying deoxygenated blood to the gill and a ventral efferent vessel carrying oxygenated blood back to the heart run within the axis. The gill filaments are richly ciliated. Powerful lateral cilia on opposing faces of the filaments create the respiratory current, whereas frontal and abfrontal cilia, fringing the inhalent and exhalent margins of the gills, respectively, cleanse the gills by throwing off waste material into the exhalent stream. Chitinous skeletal rods running immediately inside the frontal margins of each filament prevent the gills from collapsing. Blood diffuses through the filaments from the afferent to the efferent vessels, thus constituting a countercurrent to the water flowing across the gill surfaces (see Fig. 1B)—an arrangement that makes for greater efficiency. The rectum is attached to the roof of the mantle cavity along its midline. Flanking the rectum above the gills lie the inappropriately named hypobranchial or mucous glands. These consolidate waste material and thus prevent the mantle cavity from becoming clogged with sediment. The organs of the mantle cavity thus comprise a group of functionally integrated structures.

The heart of our archetypic mollusc consists of a single median ventricle partially surrounding the rectum. Discharging into the ventricle are a pair of lateral auricles that in turn receive oxygenated blood from the gills. The ventricle pumps blood around the body via anterior and posterior aortas, which eventually discharge into an extensive system of hemocoelic sinuses. The viscera are therefore bathed directly by blood. Molluscs utilize the large, fixed volume of blood present in these hemal spaces as an effective hydraulic skeleton against which muscles can operate during locomotion. By channeling blood from one part of the body to another, remarkable changes in shape are possible. Backflow of blood is prevented by valves. From the extensive system of hemal sinuses, blood is returned to the gills, usually via the kidneys.

The true coelom is restricted to the cavities of the gonads and the pericardium surrounding the heart. Primitively, the gonads discharge into the pericardial cavity. Also associated with the pericardium and originating from it are a pair of mesodermal coelomoducts whose walls form the renal organs or kidneys. These also serve to convey gametes from the pericardium to the mantle cavity in the more primitive molluscs. Initially, sexes were probably separate and fertilization external. Spiral cleavage first produced a trochophore larva, and then the more typical molluscan veliger with its characteristic ciliated velum.

The ancestral mollusc, like many present-day molluscs, was probably a microphagous grazer, and the organization of the gut was thus adapted for processing a slow but more or less continuous stream of particulate material. Situated within the buccal cavity is the odontophore with its associated ribbon-like radula. This can be moved rhythmically in and out of the mouth by means of a complex system of muscles. The rows of chitinous teeth distributed over the surface of the radula are recurved posteriorly so that the effective stroke is a forward and upward licking movement as the radula is repeatedly withdrawn. New teeth are added posteriorly in the radula sac as those at the front of the ribbon are lost through excessive wear. In many present-day molluscs, the radula works in combination with biting jaws. These are often united into a single structure on the upper surface of the buccal cavity. Particulate material is mixed with mucus from the salivary glands before being carried by cilia down the esophagus and into the stomach as a mucous string. In the stomach, the food string is wound onto a stiff fecal rod (protostyle) that projects into the stomach from the intestine and is rotated by cilia. As the food string rotates, it continually sweeps over the ciliated stomach wall. During this process those particles that become detached are graded by a complex ciliary sorting mechanism. Anteriorly, a chitinous shield protects the stomach wall from abrasion by sharp particles that may have entered the gut along with the food. Smaller particles enter the digestive tubules whereas larger particles either become incorporated into the protostyle or pass directly into the intestine, where they are compacted into fecal pellets. The stomach of even the early mollusc is thus quite a complex structure.

The nervous system of the ancestral mollusc probably resembled that of the free-living flatworms. A concentration of nervous tissue around the esophagus served mainly to operate the odontophore–radula complex. Extending from this esophageal nerve ring were two pairs of longitudinal cords. Of these the pleurovisceral cords innervated the mantle and visceral mass whereas the pedal cords innervated the foot. Cross connectives between these cords are a feature of the primitive molluscan nervous system.

The earliest molluscan shell was probably little more than a tough dorsal cuticle of horny conchiolin. Calcification at one, two, or several centers gave rise, respectively, to the univalve, bivalve, or multivalve condition. This must have occurred very early in molluscan evolution because all the major existing groups occur together in the earliest fossil-bearing rocks. The archetypic shell provided support for the mantle cavity and also served as a skeleton for the multiply paired pedal muscles that ran down into the substance of the foot. The primitive shell could thus be effectively pulled down over the body for protection. Pedal muscles also served for locomotion. The molluscan shell is complex and typically consists of three layers. The outer periostracum and middle prismatic layers are secreted only at the mantle edge and are thus responsible for shell growth. An inner laminated crystalline layer thickens and strengthens the shell and is deposited by cells over the general outer face of the mantle. The molluscan shell has been extensively studied (e.g., Grégoire, 1972; Rhoads and Lutz, 1980).

Whereas the common organizational plan of the archetype mollusc can be recognized throughout virtually the whole subsequent evolution of the phylum, extensive adaptive radiation has produced a wide range of body forms that superficially appear to have little in common. Indeed, it is this very plasticity in their basic design that has been such a key factor in the enormous success of the molluscs. During the evolution of the various classes, specialized features have been added whereas some of the more primitive features have been suppressed.

## III. General Features of the Major Classes

### A. Gastropods

The class Gastropoda is the largest and most varied class, comprising over 80% of all living molluscs. Gastropods are widespread throughout the marine environment and have successfully invaded a vast array of freshwater habitats. They are the only molluscs to have colonized land; this they achieved by eliminating the gills and converting the mantle cavity into a highly vascularized lung. They have exploited virtually every conceivable mode of life as herbivores, deposit feeders, plankton feeders, scavengers, parasites, and predators.

However, despite their great diversity gastropods have much in common with the ancestral mollusc, for they typically possess a dorsal univalve shell and crawl around on a broad, flattened ventral foot. In the archetypic mollusc, however, the mantle cavity was posterior; in contrast, in all modern gastropods a process known as torsion brings the mantle cavity and its associated organs into an anterior forward-facing position immediately above the head. During this process the viscera as well as the mantle cavity itself are rotated counterclockwise through 180° such that the organs that were originally on the left-hand side now come to lie on the right side of the body, and vice versa. This rotation takes place around the narrow neck of tissue that connects the head–foot and the visceral mass. Internally the gut and nerve cords become twisted. Torsion is a unique feature of gastropod evolution and should not be confused with the process of shell coiling; these appear to have been two quite independent evolutionary events, and available evidence strongly suggests that planospirally coiled shells probably predate torsion (but see Solem, 1974). Torsion occurs as a result of the asymmetric growth of the two retractor muscles in the veliger larva. Although the right retractor is attached to the right side of the shell, some fibers pass over the gut to be inserted in the left side of the head –foot (Fig. 2A). When this muscle contracts, the visceral mass is pulled over to the left while the mantle cavity moves forward along the right. The entire rotation can be completed in a matter of minutes. However, whereas 180° rotation is sometimes achieved solely by muscular contraction, in most gastropods in which torsion has been carefully documented, complete rotation is achieved in two stages: an initial 90° rotation by larval retractor muscle contraction, followed by a slower rotation through the remaining 90° by a process of differential growth (Thompson, 1958).

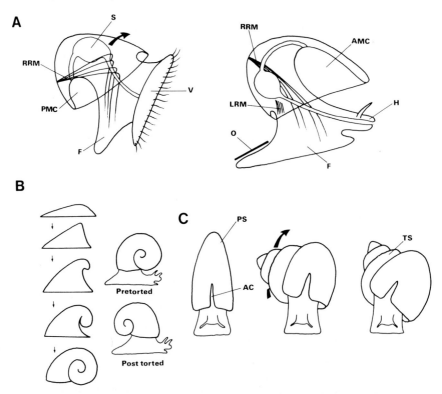

**Fig. 2.** (A) Torsion in a prosobranch veliger larva (*Patella*), pretorted (left) and posttorted (right) condition. (After Morton and Yonge, 1964.) (B) Evolution of a planospiral shell. (After Barnes, 1980.) (C) Evolution of the asymmetrical coiled gastropod shell. (After Hunter, 1979.) AC, anterior cleft; AMC, anterior mantle cavity; F, foot; H, head; LRM, left retractor muscle; O, osphradium; PMC, posterior mantle cavity; PS, planospiral shell; RRM, right retractor muscle; S, stomach; TS, turbinate spiral shell; V, velum.

Because torsion appears to be such an important event in the development and evolution of the gastropods, one is led to conclude, perhaps mistakenly, that it must confer certain adaptive advantages. No single totally satisfactory explanation of the evolutionary significance of torsion has yet been advanced despite several contributions. Garstang (1928) considered that torsion probably arose as a mutation that benefited the larva by enabling the vulnerable head and velum to be withdrawn into the anterior mantle cavity ahead of the tougher foot with its protective opercular seal. In the pretorted condition the head and velum were left exposed to micropredators until the foot had been withdrawn into the posterior cavity. Torsion was therefore seen as a larval adaptation with little or no direct value to the adult. Not all malacologists, however, share this view, and many prefer to think in terms of the eventual advantages to the adult gastropod (see, for example, Morton, 1979). The mantle cavity is clearly a dominant feature of most adult gastropods, and its anterior position, where the sensory osphradia can continually monitor the environment ahead of the advancing snail and where the gills are more easily flushed with clear undisturbed water, must be of considerable benefit. Stasek (1972) cogently argues that torsion permitted the withdrawal of the larger and more highly developed head that probably evolved on the molluscan line leading to the gastropods. In a reexamination of earlier work, Ghiselin (1966) concludes that the adaptive value of torsion may reside primarily with the newly settled spat. Considerable controversy therefore exists regarding the precise functional significance of gastropod torsion. Regrettably, much of this extensive theoretical debate has not been accompanied by a comparable acquisition of more empirical embryological data.

Regardless of the evolutionary significance of torsion, an anterior mantle cavity is not without its problems. Of these, perhaps the most serious is that waste materials are now voided immediately above the head where they threaten to foul the gills and major sense organs. Consequently many of the modifications exhibited by gastropods represent attempts to adjust to the problems imposed by torsion. In primitive gastropods, sanitation problems have been solved by the appearance of clefts or slits in the mantle and shell. These allow exhalent water to leave the mantle cavity well behind the head. Rollins and Batten (1968) suggest that similar marginal clefts may have already been present in the gastropod ancestors and, through these, the adaptive advantages of torsion were guaranteed.

Whereas the ancestral mollusc is thought to have had a low shield-like shell, most gastropods have asymmetric spirally coiled shells into which the soft parts can be withdrawn for protection. This change in morphology initially involved an increase in shell height. However, the resulting dorsally extended cone-shaped shell would not only have been a particularly unmanageable structure to carry around but would also have effectively prevented the snail from entering holes or crevices in search of food and shelter. This problem was partially solved by coiling the shell over the head in the form of a flat planospiral (Fig. 2B). Bilaterally symmetrical planospiral shells were characteristic of most early gastropods such as the fossil Bellerophontacea. Such shells, however, still suffered from one major disadvantage: They were exceedingly bulky because each shell whorl lay completely outside that of the preceding whorl. By suitable changes in the growth gradients at the mantle margin the planospiral shell was easily converted into a more compact asymmetric helical shell in which successive whorls are laid down one below another around a central axis, the columella. Although planospirals have never completely disappeared, the helical spire appeared early in gastropod evolution and has remained the predominant type in present-day forms. In order to distribute weight equally over the head–foot, the asymmetric shell had to be displaced obliquely to the right (Fig. 2C). Solem (1974) argues that this tilting of an asymmetrically coiled shell may in fact have provided the initial impetus for torsion. Shell displacement resulted in the partial occlusion of the right side of the mantle cavity, and the subsequent reduction and ultimate loss of the posttorsional right pallial organs. In most present-day gastropods there is therefore considerable asymmetry, especially in the organs associated with the mantle cavity. Even in those gastropods that subsequently lose their shells and show a secondary return to bilateral symmetry, marked asymmetries in their internal anatomy frequently persist. Gastropods have resolved the problems associated with torsion and pallial asymmetry in several unique ways. Consequently they exhibit a degree of morphological diversity unsurpassed elsewhere in the animal kingdom.

The class Gastropoda is divided into three subclasses: Prosobranchia, Opisthobranchia, and Pulmonata. Prosobranchs retain an anterior mantle cavity and exhibit the full effects of torsion including the streptoneurous (twisted) condition of the nervous system. Primitive prosobranchs belong to the order Archaeogastropoda. These have bipectinate feather-like gills (aspidobranchs) similar to the archetype mollusc. Most prosobranchs belong to the order Mesogastropoda. Here a single monopectinate comblike gill (pectinibranch) is present. The order Neogastropoda comprise the third and perhaps most highly specialized group of prosobranchs. Neogastropods are marine carnivores and like the mesogastropods have a single pectinibranch gill. In the opisthobranchs and pulmonates the effects of torsion are reduced or obscured by subsequent growth and development. Opisthobranchs are essentially marine gastropods in which there is a reduction or even loss of the shell and mantle cavity and a concomitant emphasis on alternative respiratory structures. Some opisthobranchs have assumed a secondary bilateral symmetry. In the pulmonates the mantle cavity has been converted into a highly vascularized lung for gaseous exchange in air or secondarily in water. Of the two main groups of pulmonates, the order Basommatophora consists mainly of freshwater snails, the order Stylommatophora of terrestrial snails, slugs, and semislugs.

## B. Bivalves

The Bivalvia comprise a group of highly specialized laterally compressed molluscs in which the head, radula, and associated buccal mass are absent and in which the sensory functions have largely been assumed by the mantle margin. As a group they are considerably less diverse than the gastropods. The shell consists of two valves hinged dorsally (Fig. 3). These are secreted by a pair of mantle flaps that totally enclose the foot and visceral mass. Most present-day bivalves have enormously hypertrophied gills housed in a capacious mantle cavity. In addition to their respiratory role, these also serve as the main food-gathering organs in this

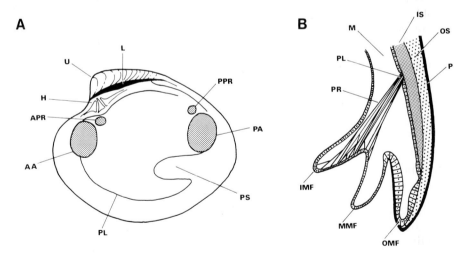

**Fig. 3.** (A) Inner surface of the right valve of a generalized bivalve. (B) Transverse section across the shell and mantle skirt. AA, anterior adductor; APR, anterior pedal retractor; H, hinge; IMF, inner mantle fold; IS, inner shell; L, ligament; M, mantle; MMF, middle mantle fold; OMF, outer mantle fold; OS, outer shell; P, periostracum; PA, posterior adductor; PL, pallial line; PPR, posterior pedal retractor; PR, pallial retractor; PS, pallial sinus; U, umbo. (After Fretter and Graham, 1976.)

predominantly filter-feeding group of molluscs. Although they are wide-spread in a variety of aquatic habitats, their methods of feeding, locomotion, and reproduction effectively exclude them from the terrestrial environment. Most bivalves are relatively sedentary. Many use their straplike foot for burrowing into soft sediments; others attach to or bore into hard surfaces such as rock or wood, whereas a few have become free swimmers.

Typically the bivalve shell consists of two similar valves that articulate middorsally and that can be tightly closed by a pair of powerful adductor muscles. These muscles, and the elastic ligament whose action they oppose, are peculiar to bivalves. The ligament consists mainly of a horny uncalcified material (conchiolin) and is secreted by part of the mantle known as the isthmus. It is normally situated immediately behind the oldest parts of the shell (the umbones) and can be internal or external to the shell. Functioning like a spring, the ligament opens the valves when the adductor muscles relax. Dorsally, on the inner surface of the valves immediately beneath the umbones, lies the hinge. This is usually furnished with teeth, the number and arrangement of which are important taxonomic features. Whereas the valve surfaces closely interlock at this juncture, they are in fact always separated by a narrow neck of tissue that continually enlarges the teeth and deepens the opposing troughs as the valves grow. Hinge teeth ensure that the valves are correctly aligned when the shell closes; they also prevent rotational and shearing movements of one valve over the other.

The mantle edge is marked by three parallel folds (Fig. 3B). The outer fold secretes the external horny periostracal layer as well as the outer calcareous part of the shell. The middle fold is principally sensory and may be furnished with tentacles and even eye spots. The innermost fold is muscular and controls the flow of water into the mantle cavity. The inner calcareous shell layer is deposited over the general outer surface of the mantle and thus effectively strengthens and thickens the shell. Primitively, the two margins of the mantle skirt are wholly separate from one another but among boring and deep-burrowing forms there has been a trend toward fusion of the mantle edges in order to reduce the risk of contaminating the mantle cavity. However, at least three openings must be retained: two for the inhalent and exhalent water streams and one through which the foot can emerge. Mantle fusion is frequently accompanied by the development of siphons. These exhibit extensive variation according to the degree to which the various mantle folds are involved in their formation. They may be naked and muscular, covered in periostracum, or even partially encased in shell.

Extrapallial fluid, into which the materials for shell secretion are first liberated, separates the mantle from the shell except in the regions of muscle attachment. These areas therefore lack the inner shell layer and are thus left as characteristic muscle scars. These scars provide valuable evidence concerning the structure and life habits of bivalves, and as they are frequently well preserved in fossil material they are of considerable paleoecological value. Extending around the inside of the valves near the shell margin is the pallial line representing the point of attachment of the pallial muscle. In taxa with posterior siphons this is deeply indented to form a pallial sinus (Fig. 3A). The largest and most prominent scars are those associated with the adductor muscles. These muscles probably originated as cross fusions of the pallial muscles at the ends of the mantle embayments in the ancestral bivalve (see Yonge and Thompson, 1976). The number and position of the adductor scars have been extensively used in bivalve classification. Typically there are two: one anterior and one posterior (dimyaria). These may be similar (isomyaria) or dissimilar (anisomyaria or heteromyaria) in size. Taxa in which only a single centrally placed posterior adductor is present comprise the monomyaria. Other prominent scars include those associated with the various pedal muscles. The presence of multiple pedal scars in some early bivalves such as the Ordovician *Babinka* recalls the archetypic condition.

The foot of most bivalves is laterally compressed, although in the most primitive taxa it still retains a flattened ventral sole. It is well developed in burrowing bivalves, but in attached forms it is reduced or absent. Behind or at the base of the foot is the byssal gland, which produces the proteinaceous byssus threads. Although present in the postlarval stages of most bivalves, where it is used for temporary attachment during metamorphosis, the byssus complex generally diminishes in importance in adult life except for those taxa that have successfully invaded firm substrata.

Gills are a predominent feature of the anatomy of most bivalves. These are paired structures suspended from the roof of the mantle cavity on either side of the foot. Most present-day species belong to the subclass Lamellibranchia. Here individual gill filaments are greatly elongated and folded so that each gill effectively consists of four broad filtering surfaces (lamellae). In the primitive subclass Protobranchia, the unfolded filaments are much simpler. The subclass Septibranchia comprise a small but highly specialized group of bivalves in which the gills are modified as a muscular pumping septum. Although this system of classification is still widely encountered, in view of its simplicity many malacologists now favor a rather more complex system (Section VIII). However, these older subclasses are readily accommodated in the newer scheme.

## C. Cephalopods

Cephalopods are unquestionably at the apex of molluscan evolution. Although ecologically perhaps not as successful as either the gastropods or bivalves—there are only a few hundred living species—cephalopods encompass the largest and most highly organized of all invertebrates. Their elaborate neural centers and complex behavioral patterns are surpassed only by the higher vertebrates. As a group they are exclusively marine and almost all are streamlined, active carnivores. Like the bivalves, the cephalopods are a comparatively uniform group having concentrated on just one of the many successful patterns of molluscan organization. They are particularly important as index fossils.

Two new molluscan organs appear in the cephalopods: the prehensile cephalic tentacles surrounding the mouth, and the muscular funnel. The latter is derived from part of the foot and directs the flow of water leaving the mantle cavity. During cephalopod evolution the body has become greatly extended along the dorsoventral axis. With their newly acquired method of jet propulsion, this axis has effectively become the functional anteroposterior axis. On morphological grounds, however, the mouth with its associated tentacles is ventral and the funnel posterior in position.

The ancestral cephalopod is thought to have had a cone-shaped shell. Most extinct species, however, had large external planospiral shells divided into chambers (camerae) by transverse septa. Only the largest, most recently formed chamber was occupied by the living animal. From the apex of the visceral hump, a cord of tissue called the siphuncle ran through a central perforation on each septum and secreted gas into the empty chambers of the shell. This served to counterbalance the weight of the bulky shell and so helped the animal to swim and adjust its position in the water column. Sutures, which mark the points of contact between the septa and shell wall, are especially important taxonomic features in fossil cephalopods. This primitive condition is retained only in *Nautilus,* now restricted to the Indo-Pacific. In all other modern cephalopods, the shell is either considerably reduced and internal, or has been completely lost. The visceral mass is now covered by a thick muscular mantle that functions in both respiration and locomotion by pumping water through the mantle cavity. In cephalopods, the cilia that are so characteristic of most molluscan systems have thus been largely replaced by muscles.

Three subclasses of cephalopods can be recognized: Nautiloidea, Ammonoidea, and Coleoidea. Nautiloids have straight or coiled external shells with simple sutures. *Nautilus,* the only surviving representative of this group, is unusual among living cephalopods in possessing four gills (tetrabranchiate) and four kidneys. Ammonoids, known only as fossils, had coiled external shells with complex septa and sutures. In the coleoids the shell is either internal or absent. Only one pair of gills (dibranchiate) and one pair of kidneys are now present. Virtually all modern cephalopods belong to this group; octopods have a ring of eight arms whereas decapods (cuttlefish, squid) have two long tentacles in addition to the eight shorter arms. A veliger is never present in cephalopods.

## IV. The Minor Classes

Apart from the three major groups already discussed, there also exist four minor classes: Aplacophora, Polyplacophora, Monoplacophora, and Scaphopoda. Of these only the Polyplacophora (chitons) achieve any degree of numerical or ecological importance.

The Aplacophora were once grouped with the Polyplacophora within the class Amphineura. However, on embryological and anatomical grounds (Thompson, 1960; Salvini-Plawen, 1969) the two classes are quite distinct. Aplacophorans are aberrant wormlike molluscs with a spiculose inrolled mantle but no shell. Typically they possess a simple radula and a small posterior mantle cavity in which gills may be present. There are two distinct orders of which the Ventroplicida is the larger. These live suctorially on gorgonians and hydroids. They possess a longitudinal ventral groove containing a vestigial ridgelike foot. In the Caudofoveata, the ventral foot groove is absent. These burrow into soft sediments and are probably deposit feeders. Aplacophorans occur at moderate depths throughout the world's oceans. To what extent they can be considered specialized or primitive is uncertain, but there is no evidence that they ever possessed a shell. Scheltema (1978) has recently reviewed their position within the phylum.

The scaphopods or tusk shells comprise the smallest molluscan class and have much in common with the bivalves. *Dentalium* is a typical form having a tapering tubular shell. The conical head, which bears numerous adhesive tentacles (captaculae), and the flanged foot, which is used for burrowing, emerge from the broad end of the shell. Both inhalent and exhalent currents pass through the narrower posterior end that projects above the surface of the substratum. The ventral elongated mantle cavity lacks typical gills, and respiration occurs through the folded mantle wall. Scaphopods feed on interstitial organisms, especially foraminiferans (Bilyard, 1974), using the adhesive captaculae and a strong radula.

The Monoplacophora encompass several families of fossil molluscs

and one living genus, *Neopilina*. The latter was first discovered from a
deep ocean trench off the Pacific coast of Central America in 1952 and
subsequently described in detail by Lemche and Wingstrand (1959) (see
also Morton and Yonge, 1964; Purchon, 1977). *Neopilina* possesses a
broad disk-shaped foot that is attached to a flattened shieldlike shell by
eight pairs of retractor muscles. The embryonic shell (protoconch) shows
evidence of spiral coiling. There are five or six pairs of lamellated gills
present in the shallow mantle grooves on either side of the foot. These
prectenidial gills, however, are quite unlike true molluscan ctenidia. The
head bears a series of elaborate folds and tentacles and there is a well-developed radula. The most important feature of *Neopilina* is the serial replication of the various organ systems along the body in an apparently
metameric fashion. The primitive nervous system has a ladder-like construction and is scarcely more centralized than that of the flatworms.
After its discovery, *Neopilina* was heralded as the "missing link" between the molluscs and the metamerically segmented annelid–arthropod
stock. Organ replication in *Neopilina,* however, is very irregular (e.g.,
two pairs of auricles, six pairs of nephridia, eight pairs of pedal muscles),
and if molluscs are indeed descended from segmented ancestors, then all
traces of this have certainly been lost in the more successful classes. *Neopilina* is probably quite specialized, having survived by becoming adapted
for life in the relatively noncompetitive conditions of the deep sea. Monoplacophorans evidently evolved along two distinct lines. In the subclass
Cyclomya there was a reduction in the number of gills and pedal muscles
whereas an increase in body height led ultimately to planospiral coiling.
This group may have included the progenitors of the gastropods. Members of the subclass Tergomya retained the flattened shell and the replicated organ systems, and survived only within the genus *Neopilina.*

The Polyplacophora or chitons are the most successful of the minor
classes. Their eight articulated dorsal shell plates, which in some species
may be partly or totally obscured by the reflected mantle margin, clearly
adapt these molluscs to life on irregular surfaces. Each plate consists of
several layers. The upper layer or tegmentum is composed of conchiolin
impregnated with calcium carbonate and covered by a thin periostracum.
Beneath this lies the thicker hypostracum (articulamentum) composed entirely of calcium carbonate. Most chitons live in the rocky intertidal zone
where they graze on algal or diatomaceous mats using a well-developed
radula. The mantle cavity consists of two narrow pallial grooves running
between the sucker-like foot and the flexible, often highly ornamented
mantle girdle. This latter extends well beyond the shell plates and can be
closely applied to the substratum. The primitive gills (each a true ctenidium) are serially repeated (up to 80 pairs) and form a curtain that divides

the pallial groove into inhalent and exhalent gutters. During development, gills first appear posteriorly; however, these are added to irregularly and sometimes independently along either side of the body as the chiton grows, and they do not therefore always constitute a perfectly paired series (see Russell-Hunter and Brown, 1965; Johnson, 1969). Gill multiplication here appears to be a functional requirement perhaps imposed by the restricted nature of the mantle cavity rather than an archaic form of some ancestral metameric condition.

## V. Molluscan Organization

### A. The Mantle Cavity and Gills

Primitively, the mantle cavity may simply have consisted of a marginal groove between the foot and the mantle edge that housed the gills and received waste material from the anus and excretory ducts. This condition still persists in *Neopilina* and in chitons. Early in molluscan evolution, however, posterior expansion of this space produced a cavity beneath the shell into which the head–foot could withdraw. The archetypic mollusc is therefore generally presumed to have had a spacious posterior mantle cavity in which were situated the anus and the paired gills, osphradia, urogenital ducts, and hypobranchial glands.

In the gastropods, torsion brings this mantle cavity into an anterior position above the head. Although this may have had benefits for both larva and adult, an anterior mantle cavity was not without its problems, particularly with regard to waste disposal. The fossil bellerophontaceans were among the earliest archaeogastropods. These had deeply cleft planospiral shells and were most probably already fully torted, with a pair of symmetrical gills in an anterior mantle cavity. Most gastropods, however, evolved asymmetric spirally coiled shells for the more efficient accommodation of the bulky visceral mass. This resulted in the progressive suppression of the organs in the posttortional right side of the mantle cavity. In forms such as *Pleurotomaria* and the keyhole limpet *Diadora,* the gills are still approximately equal in size, but in the abalone *Haliotis* the right gill is already somewhat smaller than its counterpart on the left side (Fig. 4A, B, and C). In all these primitive modern archaeogastropods in which the gills are paired (zeugobranch), water currents sweep over the gills to the midline where the shell and mantle open by means of slits or perforations. Such an arrangement allows contaminated exhalent water to escape well behind the head. In all zeugobranchs the bipectinate gills are attached to the mantle wall by

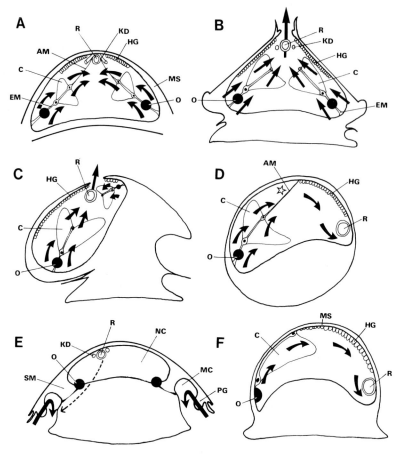

**Fig. 4.** Evolution of the gastropod mantle cavity. Cross sections of (**A**) a hypothetical primitive gastropod, (**B**) *Diadora* (keyhole limpet), (**C**) *Haliotis* (abalone), (**D**) *Trochus* (a higher archaeogastropod with a single bipectinate gill), (**E**) *Patella*, (**F**) *Buccinum* (a neogastropod with a single monopectinate gill. Solid arrows indicate the direction of water flow through the mantle cavity. ☆ Region liable to become clogged with silt. AM, afferent membrane; C, ctenidium; EM, efferent membrane; HG, hypobranchial gland; KD, kidney duct; MC, mantle cavity; MS, mantle skirt; NC, nuchal cavity; O, osphradium; PG, pallial gill; R, rectum; SM, shell muscle. (After Fretter and Graham, 1976, and Morton, 1979.)

membranes. Triangular filaments alternate along either side of the central gill axis. Each filament, which is supported by a chitinous skeletal rod, consists of a double fold of integument and encloses a blood space where gaseous exchange occurs. Tracts of lateral cilia lying immediately over the skeletal rods draw water over the gills. Frontal and abfrontal cilia convey mucus-bound particles around the free edges of the filaments where they are voided into the exhalent stream. Before passing over the gills, inhalent water is tested by paired osphradia. Apart from its anterior position, the organization of the mantle cavity in these primitive prosobranchs therefore departs little from that of the ancestral mollusc.

In the higher archaeogastropods (Trochacea, Neritacea) only the posttortional left gill remains (Fig. 4D). This, however, is still bipectinate as in the earlier zeugobranchs. The anus and kidney ducts (one in the neritaceans) open well to the right, and the oblique water current now sweeps through the mantle cavity from left to right. Among the patellacean limpets, *Acmaea* still retains an anterior mantle cavity with a single gill, but in *Patella* secondary gills formed from mantle folds completely encircle the body and the original mantle cavity is no longer a respiratory chamber (Fig. 4E). Limpets, like chitons, are highly specialized for life on surf-swept rocky surfaces. The condition found in the trochids and neritids foreshadows that of all the higher prosobranchs (meso- and neogastropods), in which only a single gill and auricle are present. Here, however, the gill is monopectinate with only a single series of gill filaments whereas the gill axis is fused to the wall of the mantle cavity (Fig. 4F). Inhalent water enters the mantle cavity on the left, passes between the gill filaments, and leaves on the right side of the head. A single osphradium lies at the base of the gill in the path of the inhalent stream. The hypobranchial gland is large and in some cases produces a purple dye. The shell in many of these higher prosobranchs has an anterior notch that houses an inhalent siphon. With the well-developed osphradium at its base, this siphon functions as an efficient nostril. The osphradium is particularly elaborate in carnivorous forms where it serves not only to determine the amount of sediment present in the water but also to locate prey. In some freshwater species, especially those inhabiting oxygen-poor water, the siphon can be effectively used as a breathing tube. A few prosobranchs like the slipper limpet *Crepidula* (Calyptraeacea) are modified for ciliary feeding. Here, as in the lamellibranchs, the gill filaments have been considerably lengthened to provide an increased surface for trapping particulate material.

In the opisthobranchs the mantle cavity moves back along the right side of the body. As it does so, the gill becomes reduced in importance until in the nudibranchs it is lost altogether and the naked dorsal body wall assumes a respiratory function. Primitive opisthobranchs such as the cephalaspids and anaspids (aplysioids) have a fleshy plicate gill, but this is quite different in structure from the ctenidial gill of the prosobranchs. Here, water currents are created not by gill cilia but by strips of cilia on the mantle wall. Many of these early opisthobranchs, such as *Scaphander,* are active burrowers, and the additional cilia required for cleansing and flushing the mantle cavity are housed in a special spiral cecum. The hypobranchial gland in these burrowers is still well developed. Aplysioids have a small triangular mantle cavity with a fleshy gill. On the roof of the mantle cavity, the hypobranchial gland produces a purple secretion whereas the opaline gland on the floor of the cavity produces a noxious secretion. Notaspids have lost the mantle cavity, but a naked gill remains on the right protected by the mantle skirt. By contrast, the thecosomatous pteropods have lost the gill but retain the hypobranchial gland in a spacious mantle cavity; in some species this is used in ciliary feeding.

Pulmonates retain an anterior mantle cavity that is lined with blood vessels and functions as a lung. The mantle cavity opens via the pneumostome on the right, which opens and closes rhythmically during ventilation. Hypobranchial glands have been lost and osphradia, if present, normally lie outside the mantle cavity. Lungs are equally suitable for both aquatic and aerial respiration, and many basommatophorans are amphibious. Colonization of land has probably occurred several times and this has rarely been unidirectional. In the many pulmonates that have subsequently become readapted to an aquatic life-style, the mantle cavity is often filled with water. *Lymnaea* can carry oxygen in the mantle cavity with the pneumostome closed. Where secondary extrapallial gills are developed, as in Planorbidae and Ancylidae, these are never true ctenidia. In the Anthoracophoridae, invaginations of the lung wall ramify through the adjacent blood spaces and probably function like tracheae.

The bivalve mantle cavity is particularly spacious, and the complex hypertrophied gills of the lamellibranchs form an efficient filtering device capable of sorting and straining exceedingly small particles. A single pair of bipectinate gills is present, and right and left gills are of equal size. In the primitive protobranchs such as *Nucula,* the gills lie obliquely at the back of the mantle cavity. These are small and their double row of triangular filaments recalls the primitive molluscan pattern (Fig. 5A$_1$). Patches of interlocking cilia hold the ctenidia in place and help support the individual filaments. Water is drawn over the gills by lateral cilia while frontal and abfrontal cilia remove sediment trapped on the gill surface. This is cast back into the inhalent chamber. A pair of large hypobranchial glands and simple osphradia are present in the exhalent chamber. Water flows through the mantle cavity in an anteroposterior direction and, though the gills are capable of trapping particles, they are not extensively used for feeding. Primitively, this is

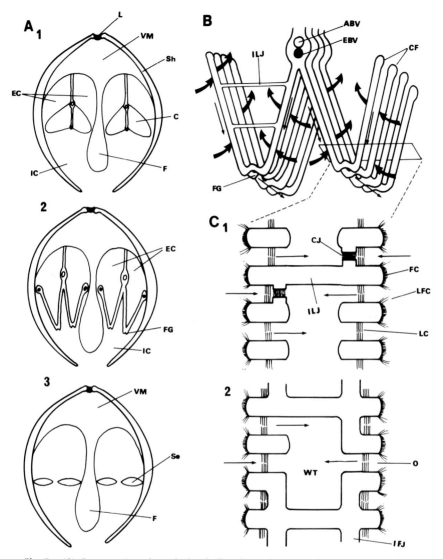

**Fig. 5.** (A) Cross sections through the shell and mantle cavity of (1) a protobranch, (2) a lamellibranch, (3) a septibranch. (B) Stereogram of four folded filaments in one lamellibranch ctenidium. Large arrows denote water currents created by lateral cilia. Small arrows denote the direction of "strained" food transported by frontal cilia. (C) Horizontal sections through a demibranch (half a ctenidium) in (1) a filibranch and (2) a eulamellibranch. Water currents between individual filaments indicated by arrows. ABV, afferent blood vessel; C, ctenidium; CF, ctenidial filaments; CJ, ciliary junction; EBV, efferent blood vessel; EC, exhalent chamber; F, foot; FC, frontal cilia; FG, food groove; IC, inhalent chamber; IFJ, interfilamentar junction; ILJ, interlamellar junction; L, ligament; LC, lateral cilia; LFC, laterofrontal cilia; O, ostium; Se, septum; Sh, shell; VM, visceral mass; WT, water tube. (Modified from various sources.)

achieved by the labial palps, two on each side of the mouth. Each outer palp carries a long tentacle-like structure that can be thrust into the sediment. These glandular and richly ciliated palp proboscoides collect particulate material, which is subsequently sorted by the ridged labial palps.

Not all protobranchs, however, are like *Nucula;* in the Malletiidae the gills form a muscular pumping membrane, whereas in the Solemyidae palp proboscoides are absent and the gills appear to be rather more important in food collection. In the lamellibranchs the gill filaments are greatly elongated and reflected. Each filament consists of two V-shaped demibranchs each with descending and ascending limbs (Fig. 5A$_2$,B). Cross connections at various points within the gill provide support for the elongated filaments. Tissue bridges (interlamellar junctions) run between the ascending and descending limbs of each filament. In the more primitive filibranch condition, adjacent filaments are simply held together by interlocking disks of cilia (interfilamentar junctions); such gills easily fray when dissected. In the more advanced bivalves, adjacent filaments become intimately united by vascular tissue bridges so that the interfilamentar spaces are now reduced to a series of holes or ostia through which the water percolates. These eulamellibranch gills have a much more cohesive structure. The pseudolamellibranch gill of the ostreids, pectinids, and pteriids can be regarded as an intermediate condition. In some bivalves the lamellae may also be folded, giving the gill an undulating appearance. Water is drawn over the lamellibranch gill by lateral cilia set along the sides of the filaments in filibranchs and in the ostia of eulamellibranchs. Frontal cilia are especially prominent, and lying between these and the lateral cilia are the laterofrontal cilia. These are compound cilia and their pinnate structure greatly enhances their efficiency in trapping particulate material (Owen, 1974). Particles retained by the laterofrontals are transported across the gills by the frontal cilia and delivered into ciliated food grooves that run to the labial palps.

The precise nature of the sorting mechanism on the gill filaments varies from species to species (see Barnes, 1980, pp. 399–401). Anteriorly the demibranchs terminate between the labial palps, whose surfaces bear a complex array of ciliary currents. Here heavy unsuitable particles are rejected from the tips of the palps whereas smaller particles are ingested. Waste material rejected by the gills is bound in mucus and voided as pseudofeces either via the pedal gape or along special rejection tracts in the inhalent siphon. In certain bivalves, like the deposit-feeding tellinaceans, the gills are considerably reduced in size and material collected by the elongate siphon is cast onto the palps for sorting. Here the gills are less important in food gathering.

In septibranchs the gills have become modified as a muscular septum (Fig. 5A$_3$). This divides the mantle cavity into dorsal and ventral chambers communicating with the exhalent and inhalent siphons, respectively. This septum is perforated by simple ciliated ostia (*Cuspidaria*) or by rather more complex sieves (*Poromya*). Septibranchs are rather unsuccessful deep-water scavengers or carnivores (Reid and Reid, 1974), and the gills and muscular labial palps no longer serve in sorting food.

The cephalopods are active molluscs, and water currents through the mantle are now produced by muscles rather than by cilia. Apart from producing a more efficient respiratory stream, this has also led to the evolution of jet propulsion. The mantle cavity houses a pair of gills (two pair in *Nautilus*) that are suspended from the roof of the mantle cavity by their afferent rather than by their efferent margins. These gills are homologous with and derivable from the aspidobranch condition. The gill filaments are not ciliated but their extensive folding greatly increases their surface area. They are also modified so as to resist the stress of the much stronger currents that now pass through the mantle cavity. The direction of water flow is reversed from the normal molluscan pattern. Water now enters the mantle cavity laterally and leaves midventrally via the funnel. Valves prevent backflow, and around the neck of the mantle there is a resisting device consisting of cartilaginous ridges and sockets. Hypobranchial glands are absent and an osphradium occurs only in *Nautilus*. Respiratory efficiency in *Nautilus* seems to have been increased by duplicating the ctenidia. With this duplication has gone a corresponding increase in the number of auricles and kidneys.

## B. The Alimentary Canal

The ancestral mollusc is generally presumed to have been a microphagous browser using a broad, toothed radula to scrape up particulate material from the rock surface. These particles were sorted and transported through the gut by a complex system of cilia. This primitive feeding method has largely been retained by many of the early gastropods as well as by the chitons and *Neopilina*.

Gastropods exhibit a wide variety of feeding habits and their radulae (beautifully illustrated by Solem, 1974) are accordingly quite diverse (Fig. 6). The rhipidoglossan condition found in many archaeogastropods is probably the most primitive type of radula. Here the broad straplike surface is covered with rows of numerous small teeth. The docoglossan radula of limpets and the taenioglossan radula of most mesogastropods are narrower and have fewer, more powerful teeth. In the rachiglossan radula

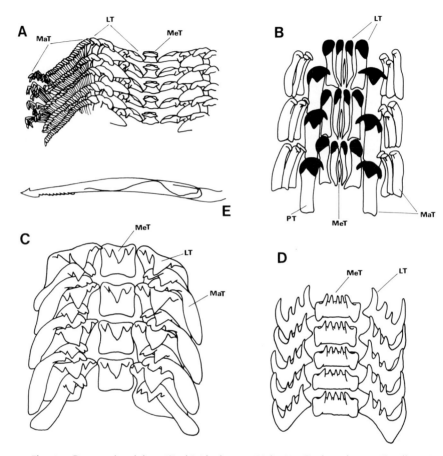

**Fig. 6.** Gastropod radulae. (**A**) rhipidoglossan (*Haliotis*), (**B**) docoglossan (*Patella*), (**C**) taenioglossan (*Littorina*), (**D**) rachiglossan (*Buccinum*), (**E**) single tooth from toxoglossan radula (*Conus*). LT, lateral teeth; MaT, marginal teeth; MeT, median tooth; PT, pleurocuspid tooth. (After Fretter and Graham, 1976.)

of the carnivorous neogastropods each row is reduced to three cutting teeth. The toxoglossan radula of the Conidae consists of a series of isolated harpoon-like teeth, each loaded with neurotoxins from the salivary glands. When discharged, these teeth are capable of subduing moving prey including fish. Among the gastropods, carnivorousness has been taken furthest in the higher prosobranchs and opisthobranchs. Pulmonates meanwhile have remained chiefly herbivorous, retaining a primitively broad many-toothed radula. Those that have become active carnivores (e.g., *Testacella*) have fewer, larger radula teeth. Opisthobranch radulae are highly variable.

The early molluscan stomach was probably quite a complex sorting area. Food strings entering the stomach were wound onto a rotating protostyle. This was housed in a special sac arising from the anterior part of the intestine and turned by cilia. Protostyles occur in *Neopilina* and in many primitive gastropods as well as the early bivalves. Chitons and limpets have lost this early mechanism whereas in noncarnivorous mesogastropods and the higher bivalves the protostyle has been superceded by the more complex crystalline style. In addition to winding-in the food string, this also acts as a source of enzymes. At the head of the rotating style the stomach wall is protected from abrasion by a cuticular shield.

In carnivorous prosobranchs, the style sac and ciliary sorting area have been lost and the stomach is reduced to a simple sac into which enzymes are secreted. The radula is now frequently associated with a proboscis, which allows the animal to probe deep into crevices or the carcasses of prey. In some cases (e.g., Muricacea, Naticacea), this proboscis can even be used to bore through calcareous shells.

Opisthobranchs are primarily grazing carnivores or suctorial feeders. However, as in the prosobranchs (and pulmonates), the more primitive forms are microphagous browsers. The cephalaspid *Actaeon,* for instance, has a broad radula, a style sac, and a sorting area. Most opisthobranchs and pulmonates, however, soon abandoned ciliary and mucus feeding and evolved muscular guts. The majority of cephalaspids have an esophageal gizzard with crushing plates whereas the stomach is comparatively simple. Some species (e.g., *Scaphander, Philine*) have evolved strong curved radula teeth for seizing prey, which is then crushed in the gizzard. Thecosomatous pteropods are primarily ciliary feeders but retain a gizzard for crushing diatoms. Aplysioids crop seaweed using paired jaws and a broad radula. Gymnosomatous pteropods have an impressive armory of jaws, pointed radula teeth, and prehensile hooks for feeding on thecosomatans. Suckered tentacles and adhesive oral papillae may also be present. Dorid nudibranchs exhibit an array of feeding mechanisms and their alimentary canals are specialized accordingly. Aeolids feed mainly on coelenterates using powerful jaws and a rasping radula. Their digestive diverticulae extend into the dorsal cerata in which undischarged nematocysts from their prey are stored, presumably as a defense mechanism. Saccoglossans are highly specialized suctorial feeders using blade-like radula teeth to lance the cell walls of green algae.

Some gastropods have become parasitic. The eulimids have lost their jaws and radula and the pharynx has become a muscular pump with glands to soften the host's tissues. Pyramidellids have elaborate mouth parts for sucking and piercing. The jaws form hollow needle-like stylets with upper suctorial and lower salivary canals. The radula is absent.

Scaphopods use captaculae around the head to pick up foraminiferans from the sand (Gainey, 1972). In *Dentalium*, the radula is well developed with five teeth per row. The stomach is a muscular bag. There is no style sac, and only vestiges of the gastric shield and ciliary sorting area remain.

The bivalves are ciliary feeders par excellence, and typically collect food using their hypertrophied gills and labial palps. There is no radula, buccal mass, or salivary glands, and the intestine is a relatively simple tube. By contrast, the stomach (Fig. 7A) forms a remarkably complex sorting system of ciliated ridges and furrows (Owen, 1955). Small particles are conveyed into the digestive diverticulae via the gastric cecae, whereas waste material enters the intestine where it is molded into feces. A crystalline style of mucoprotein occurs in the lamellibranchs. This rotates against the gastric shield, slowly releasing digestive enzymes. As particles on the style come into contact with the stomach wall, they are thrown across the sorting area. Their release from the style is evidently aided by the lower pH of the stomach, which reduces the viscosity of the mucus. In deposit-feeding tellinaceans, the stomach may serve to triturate the coarser food. Wood-boring bivalves (e.g., *Teredo*) have a large cecum attached to the stomach for storing wood fragments. Tridacnid clams farm symbiotic algae in their enormously expanded fleshy mantle lobes. Lens-

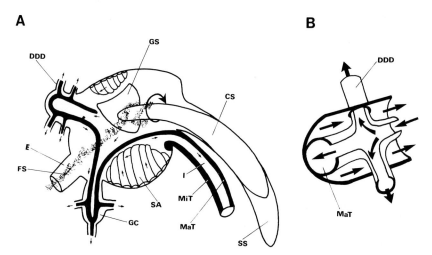

**A**                                                        **B**

**Fig. 7.** (A) Stomach of a generalized suspension-feeding lamellibranch. Small arrows indicate ciliary pathways of material entering stomach. (After Morton, 1979.) (B) Arrangement of major typhlosole within a gastric cecum showing extensions into the digestive diverticulae. (After Owen, 1955.) Ciliary currents indicated by solid arrows. CS, crystalline style; DDD, duct of digestive diverticulum; FS, food string; GC, gastric cecum; GS, gastric shield, I, intestine; MiT, minor typhlosole; MaT, major typhlosole; E, esophagus; SA, sorting area; SS, style sac.

like structures that function as light-collecting devices may also be present. Septibranchs are carnivores or scavengers, and the simplified stomach is modified for crushing prey.

In contrast to the bivalves, the cephalopods are essentially active predators and their gut is appropriately modified for a carnivorous life-style. Smooth muscles replace mucus and cilia, and there is considerable nervous coordination of peristalsis. In *Loligo,* food caught by the prehensile arms is masticated by the beaklike jaws. The radula is generally quite small but in some octopods it is used to drill prey. Salivary glands are especially prominent and produce neurotoxins. A spacious crop is present in octopods and *Nautilus,* but absent in squid and cuttlefish. The stomach is divided into proximal and distal chambers linked by a narrow sphinctered canal where the digestive gland opens. The proximal chamber forms a muscular gizzard in which food is triturated; the distal chamber is a thin-walled, often spirally coiled cecum. A ciliated region of the cecal wall separates out indigestible particles. The cephalopod digestive gland consists of ''liver'' and ''pancreas,'' the former secreting into the cecum, the latter into the gizzard. The whole cephalopod alimentary canal is specialized for the rapid processing of food.

## C. Circulatory and Excretory Systems

Although molluscs are coelomate, the true coelom is typically restricted to the cavities of the gonads and pericardium, and the main body cavity consists of a series of blood-filled lacunae that bathe the internal organs. Although the molluscan coelom was probably never a particularly large perivisceral space, it is perhaps worth noting that in the primitive molluscs such as the chitons and *Neopilina,* as well as in the more advanced cephalopods, the coelom is considerably more extensive. It is especially large in the cranchids where it may serve as a buoyancy chamber (Gilpin-Brown, 1972).

In view of the diversity that exists in the other organ systems, the molluscan circulatory system is surprisingly uniform. Blood circulates partly through lacunae and partly through a system of vessels. Typically the viscera are supplied with blood by way of aortas leading from the muscular ventricle. These discharge into a system of hemocoelic sinuses in which the tissues are directly bathed in blood. From these sinuses deoxygenated blood is conveyed to the gills, usually via the kidney. After percolating through the gill filaments, oxygenated blood is returned to the auricles through the efferent branchial vessels. In most molluscs a close relationship exists between the gills and auricles; in the chitons and monoplacophorans, however, common collecting ducts serve the replicated series of gills. Generally a single pair of auricles with their associated gills are present, but there is a tendency in many higher gastropods to suppress the posttortional right gill and auricle. In *Nautilus,* duplication of auricles and gills is probably a functional requirement imposed by the relative inefficiency of the nautiloid ctenidium and lack of accessory branchial hearts. Scaphopods lack a heart and pericardial cavity, and blood simply circulates between the organ systems by contractions of the body wall. Oxygenation occurs in folds in the mantle. Most cephalopods are highly active animals with high metabolic demands. Unlike most other molluscs they have a closed blood system, and the body cavity is now a spacious perivisceral coelom rather than a hemocoel. Blood is pumped around the body in an elaborate vascular system, with true capillaries linking arterial and venous blood. Oxygenated blood from the gills is received by the heart and dispatched around the body. Branchial hearts at the base of the gills receive deoxygenated blood, which they pump into the gills. As in other molluscs, blood passes through the kidneys on its way to the gills. The structure of the cephalopod vascular system, in which there is effectively a double circulation of blood (i.e., heart → tissues → branchial hearts → gills), is closely associated with the more active life-style of these molluscs. Capillaries, contractile arteries, and branchial hearts all serve to increase blood pressure and the rate of circulation. In octopods the reduced coelom forms a pair of capsules around the branchial hearts. These communicate with the gonocoel by aquiferous ducts that also open to the exterior. The precise function of these ducts is uncertain but they may serve to equalize coelomic and pallial pressures during locomotion.

The excretory system is inextricably linked with the vascular system. Two sites of excretory activity occur in the molluscs: pericardial glands, which represent the thickened pericardial wall; and coelomoducts. The latter discharge into the mantle cavity, whereas internally they are connected to the pericardium by the renopericardial canals. The coelomoducts are glandular and constitute the renal organs or kidneys. Molluscan kidneys vary quite extensively from one group to another. Only in the Monoplacophora and in *Nautilus* are more than a single pair present, whereas in some cases—as in the higher gastropods—one member of the original pair may even be lost except for a small portion that is incorporated into the genital duct. In prosobranchs the kidney opens into the back of the mantle cavity. In higher opisthobranchs it opens directly to the exterior on the right of the body, whereas in pulmonates an elongated duct discharges outside the mantle cavity close to the anus and pneumostome. Terrestrial gastropods must conserve water and in many species a reduced renopericardial aperture restricts the amount of water passing between the kidney and pericardium. The terrestrial Cyclophoridae and the

Ampullariidae (an essentially fresh-water group of snails, but capable of estivation during periods of drought) have two-chambered kidneys. The posterior chamber is concerned with excretion and the storage and resorption of water, whereas the anterior chamber is specialized for salt resorption. In terrestrial pulmonates the ureter functions in water and electrolyte regulation.

Bivalve kidneys are paired tubes that turn back on themselves. The lower section is glandular and opens from the pericardium; the upper section is thin-walled and discharges into the mantle cavity. In cephalopods the kidneys are relatively large thin-walled sacs. Within these sacs a mass of spongy glandular tissue (renal appendages) surrounds the afferent branchial veins. Apart from the kidneys and the pericardial glands, which in cephalopods form special appendages associated with the branchial hearts, the digestive gland may also serve an important excretory function.

## D. The Reproductive System

In the primitive molluscs, sexes were probably separate. Gonads discharged into the pericardium and from there the gametes were carried via the kidney ducts to the outside of the body for fertilization. This is still the condition in many of the more primitive present-day molluscs, such as *Neopilina* and some aplacophorans (though most Aplacophora are complex hermaphrodites). In primitive bivalves, gametes pass through both kidneys, whereas in scaphopods and archaeogastropods only the right kidney functions as a gonoduct. In chitons, gametes are shed through two separate gonoducts rather than through the kidneys. Chitons are rather unusual in that the trochophore develops directly into the adult.

Gastropods exhibit considerable diversity with respect to their reproductive systems. Most early gastropods are dioecious, with the gonad located in the visceral mass close to the digestive gland. The gonoduct varies in complexity from one group to another but it has developed in close association with the posttortional right kidney. In primitive archaeogastropods both kidneys are functional, but only the right kidney serves to convey gametes. These pass from the gonad into the kidney, its duct, or, as in the trochids, into the renopericardial duct (Fig. 8). The gametes then enter the back of the mantle cavity through the nephridiopore. In these early gastropods the gonoduct is thus relatively unspecialized and fertilization is external. In the higher prosobranchs, although the right kidney is lost, part of it survives as a section of the genital duct; a narrow genitopericardial duct may also persist. The gonoduct, which is now emancipated from the kidney, also becomes lengthened so that it opens at

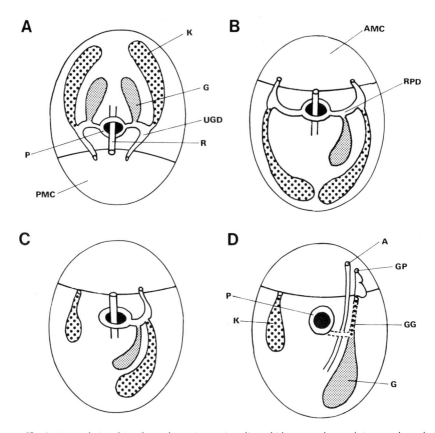

**Fig. 8.** Interrelationship of mantle cavity, pericardium, kidneys, and gonads in prosobranch gastropods. (**A**) Primitive mollusc before torsion, (**B**) ancestral prosobranch after torsion, (**C**) condition in many present-day archaeogastropods, (**D**) condition in higher prosobranchs (meso- and neogastropods). AMC, anterior mantle cavity; A, anus; G, gonad; GG, glandular gonoduct; GP, genital pore; K, kidney; P, pericardium; PMC, posterior mantle cavity; R, rectum; RPD, renopericardial duct; UGD, urogenital duct. (After Fretter and Graham, 1962.)

the mouth of the mantle cavity, thereby facilitating internal fertilization. This extension of the genital duct is termed the pallial duct. Initially this was probably little more than a ciliated gutter extending from the nephridiopore, but in most modern gastropods (in females at least) it has become closed over to form a distinct tube. It is this pallial section of the oviduct that receives and stores sperm for internal fertilization and that secretes the nutritive and protective layers around the eggs.

With this elaboration and differentiation of the pallial duct, new habitats in fresh water and on land were opened up to the gastropods. It is

significant, therefore, that among the archaeogastropods only the neritaceans have evolved a comparable system, and they too have successfully
exploited both freshwater and terrestrial habitats. In those gastropods
where tertiary egg membranes are secreted by the oviduct, a muscular
penis has evolved in the male so that fertilization can occur before the
eggs enter the secretory section of the oviduct. The penis develops from
the body wall on the right side of the head. In some gastropods sperm are
transported from the back of the mantle cavity to the penis along a ciliated
glandular pallial groove. In others this groove has closed over and a vas
deferens now opens on the tip of the penis. In meso- and neogastropods
the glandular pallial duct forms a ''prostate'' that secretes the seminal
fluids in which the sperm are conveyed. Sperm may be stored temporarily
in a copulatory bursa immediately inside the female aperture before being
transported along a ciliated ventral channel in the oviduct to the seminal
receptor. Fertilized eggs are transported from the female opening along a
ciliated groove to the foot where the pedal gland attaches them to the substratum. In higher prosobranchs fertilization is always internal but copulation need not always occur. In species with aphallic males, sperm enter
the female mantle cavity in the inhalent water. In certain families two
kinds of sperm occur: Normal eupyrenic sperm fertilize the eggs whereas
oligopyrenic sperm serve to transport or nourish the eupyrenes. Large
numbers of eupyrenes may become attached to oligopyrenes to form composite structures known as spermatozeugmata. After their liberation
these enter the female oviduct. Some prosobranchs are protandric hermaphrodites; a few parthenogenetic forms have also been described.

All opisthobranchs and pulmonates are complex hermaphrodites (Fig.
9). Early members of both groups are protandric, but in the more advanced forms, eggs and sperm mature simultaneously. The reproductive
systems are exceedingly complex and display endless variation. The various glands are often elaborately subdivided and there are tracts for the
separation of gametes. Early opisthobranchs such as *Aplysia* (Fig. 9C)
have a ciliated seminal groove that runs forward from the common genital
opening to the penis on the right side of the head. With the progressive
loss of the mantle cavity, however, the penis ultimately becomes invagi-
nated into the head. The arrangement of the female ducts in the opisthobranchs can be very complex indeed and in the dorid nudibranchs there
are three genital openings (Fig. 9D). Copulation with mutual sperm transfer is characteristic of opisthobranchs. In saccoglossans this occurs by
hypodermic impregnation. The hermaphroditic system of aquatic pulmonates recalls that of many early opisthobranchs. The presence of a veliger
larva in some species (e.g., *Siphonaria*) indicates that the pulmonates are
primitively marine. In land pulmonates such as *Helix*, mutual exchange of

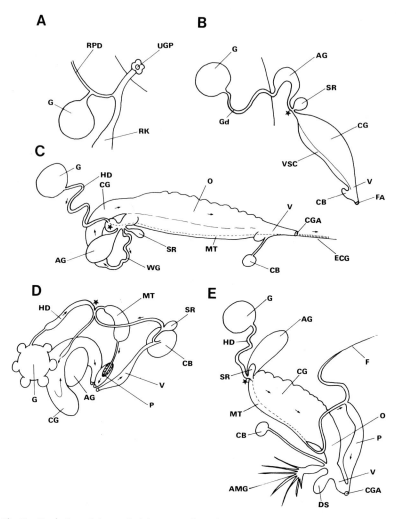

**Fig. 9.** Evolution of the genital ducts in selected gastropods. (**A**) A trochid (Archaeogastropoda), (**B**) female *Nucella* (Neogastropoda), (**C**) *Aplysia* (an early opisthobranch) (**D**) *Archidoris* (a nudibranch), (**E**) *Helix* (Pulmonata). AG, albumen gland; AMG, accessory mucous gland; CB, copulatory bursa; CG, capsule gland; CGA, common genital aperture; DS, dart sac; ECG, external ciliated seminal groove; F, flagellum; FA, female aperture; G, gonad; Gd, gonoduct; HD, hermaphrodite duct; O, oviduct; MT, male tract; P, penis; RK, right kidney; RPD, renopericardial duct; SR, seminal receptor; UGP, urogenital pore; V, vagina; VSC, ventral sperm channel; WG, winding gland; ★, site of fertilization. (After Morton, 1979, and Purchon, 1977.)

sperm usually involves a chitinous spermatophore secreted by the penile flagellum (Fig. 9E). Its transfer to the female is facilitated by mucous glands near the vaginal mouth. The vagina also contains a muscular dart sac that secretes a calcareous spicule. When driven into the body wall of the partner this "love dart" stimulates copulation. Copulating limacid slugs hang entwined in a string of mucus while pairing occurs.

Compared with gastropods, the bivalve reproductive system is remarkably simple. Most are dioecious and the short gonoducts are nonglandular. Primitively, these open into the mantle cavity via the kidneys but in higher forms they are separate (though they may still open on a common urogenital papilla). Fertilization is never strictly internal, though it may occur in the mantle cavity and some bivalves are known to brood their eggs. Some freshwater species release larvae (e.g., glochidia, lasidia) that are temporarily parasitic on fish. Relatively few bivalves are hermaphroditic, but those that are have been exhaustively studied.

Like the bivalves, most cephalopods are dioecious. The single median gonad is saclike with an internal coelomic cavity. Sperm pass into this lumen and then into the coiled male duct, which has specialized mucilaginous glands for the production of complex torpedo-shaped spermatophores. These complete their development in hardening and finishing glands before being stored in a spacious reservoir (Needham's sac) that opens into the left side of the mantle cavity. The female duct is also quite simple and terminates in an oviducal gland that coats the eggs with albumen. Nidamental glands may also be present; these secrete an additional protective covering around the eggs. Octopods are unusual in having two oviducts. Fertilization may occur inside or outside the mantle cavity. A penis is never present and the spermatophore is conveyed to the female by a specialized and often very elaborate arm, the hectocotylus. In the pelagic octopod *Argonauta* the hectocotylus is detachable and moves about freely within the female mantle cavity. The two most dorsal arms are here greatly expanded and secrete a papery bivalved shell (not homologous with the shell of other molluscs) in which the female incubates her eggs and into which she can retreat when disturbed. The reproductive system of *Nautilus* is broadly similar to that of the coleoids though here it is the right rather than the left gonoduct that is functional. Cephalopod reproduction typically involves complex courtship and mating rituals as well as a considerable degree of parental care.

## E. The Nervous System and Sense Organs

The molluscan nervous system varies in complexity from one class to another. Primitively, it resembles that found in free-living flatworms whereas in many cephalopods the nervous and sensory systems are surpassed only by the higher vertebrates. Typically the nervous system consists of a series of ganglia linked by long connectives. The relative simplicity of the system probably reflects the emphasis that molluscs (except the cephalopods) place on cilia and mucous glands rather than on fast-acting muscles.

In chitons the system closely approaches the primitive condition. Here ganglia are either absent or poorly developed. A nerve ring surrounds the esophagus and from it connectives run to the buccal ganglia, which control the odontophore complex. Running posteriorly from the nerve ring, a large pair of pedal cords innervate the foot and a lateral pair of visceropallial cords innervate the mantle, viscera, and shell sense organs. Both cords have numerous cross connectives.

In the gastropods, additional ganglia develop from the localization of nerve cells in the longitudinal cords (Fig. 10). In the pretorted condition a pair of cerebral ganglia innervate the sense organs of the head region as well as the paired buccal ganglia. From each cerebral ganglion, cords run to pedal and pleural ganglia. Of these the former innervate the foot, the latter the mantle and columella (retractor) muscle. Pleural and pedal ganglia are themselves interconnected, whereas a second pair of cords leaves the pleural ganglia and extends posteriorly as a visceral loop eventually terminating in a pair of visceral ganglia. These latter innervate the visceral mass. A pair of parietal (intestinal) ganglia, innervating the gills and osphradia, are located on the visceral loop. As a result of torsion the visceral cords become twisted so that the right parietal or supraintestinal ganglion, because it is located higher in the visceral mass, is now on the left, and the left parietal (infraintestinal) ganglion is on the right. This twisted condition, together with the separation of ganglia by nerve cords, is a primitive feature of the gastropod system and occurs with certain modifications in many prosobranchs. In most gastropods, however, this streptoneurous condition is obscured by two evolutionary events: a tendency for ganglia to become more concentrated or fused together with the attendant shortening of the connectives, and a tendency for the ganglia and cords to become secondarily bilaterally symmetrical. Considerable concentration of ganglia occurs in the higher prosobranchs such as *Busycon*, where all but the visceral ganglia have moved forward to surround the esophagus below the cerebral ganglia (Fig. 10B). All connectives except those between parietal and visceral ganglia have now been lost. In all but the most primitive opisthobranchs, detorsion untwists the visceral loop (euthyneury) as the mantle cavity moves back along the right side. All but the visceral ganglia are now concentrated in a nerve ring around the anterior gut (e.g., *Aplysia*, Fig. 10C). In the pulmonates even the vis-

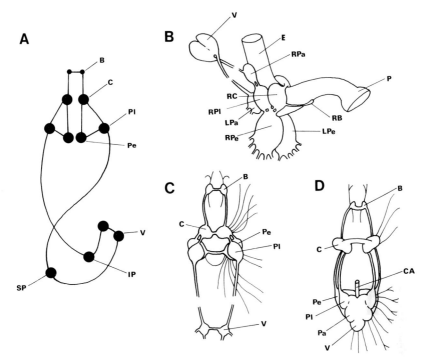

**Fig. 10.** (A) Schematic representation of a posttortional (streptoneurous) gastropod nervous system. (B) Concentrated nervous system of *Busycon* (Neogastropoda) (After Barnes, 1980.) (C) *Aplysia* (opisthobranch) and (D) *Helix* (pulmonate), secondarily symmetrical nervous systems. B, buccal ganglion; C, cerebral ganglion; CA, cephalic aorta; IP, Infraparietal ganglion; LPa, left parietal ganglion; LPe, left pedal ganglion; E, esophagus; P, proboscis; Pa, parietal ganglion; Pl, pleural ganglion; RB, right buccal ganglion; RC, right cerebral ganglion; RPA, right parietal ganglion; RPe, right pedal ganglion; RPl, right pleural ganglion; SP, superior parietal ganglion; V, visceral ganglion. (After Bullock and Horridge, 1965.)

ceral ganglia have been incorporated into the nerve ring. The resulting shortening of all connectives has resulted in a secondary bilateral symmetry in which torsion is no longer evident (Fig. 10D).

The nervous system in the essentially sedentary bivalves is comparatively simple. Above the esophagus are the cerebral ganglia with which the pleural ganglia have fused (except in protobranchs). From these cerebropleural ganglia a pair of connectives run to the pedal ganglia embedded in the base of the foot whereas another pair of long visceral connectives run to the visceral ganglia beneath the posterior adductor muscle. Foot movement and anterior adductor muscles are controlled by the pedal and cerebral ganglia whereas visceral ganglia control the posterior adductor muscle and the siphons. Cerebral ganglia are also responsible for coordinating pedal and valve movements. Some concentration of the nervous system occurs in the more active-swimming bivalves, such as *Pecten* and *Lima*. Here the cerebral ganglia move back to fuse with the visceral ganglia. These cerebrovisceral ganglia receive nerves from the mantle eyes.

The highly developed nervous system of the cephalopods is unrivaled among other invertebrates. This is clearly correlated with their active, carnivorous life-style. Although all the typical molluscan ganglia are present, they are no longer easily distinguished because they are largely fused together and greatly subdivided in the form of a complex brain encircling the esophagus. In *Nautilus* the brain consists of a cerebral band above, and pedal and pleurovisceral bands below the esophagus. The whole brain is protected by annuli of cartilage. Laterally paired optic lobes connect with the dorsal cerebral band. In the higher cephalopods (e.g., *Octopus*) the brain is more concentrated and more completely enclosed by a cartilaginous cranium. A large cerebral (supraesophageal) complex lies above the esophagus and at each side connects with the enormous stalked optic lobes. Anteriorly a pair of nerves run from the cerebral ganglia first to the superior buccal ganglia and then, by way of a circumesophageal commissure, to the inferior buccal ganglia. The subesophageal region consists of an anterior pedal center and a more posterior pleurovisceral center. These are interconnected and linked in a saddle-like fashion to the supraesophageal complex above. The pedal center innervates the arms and funnel whereas the pleurovisceral center gives rise to three pairs of nerves that innervate the visceral mass and gills (visceral nerves), the stomach region (sympathetic nerves), and the mantle wall (pallial nerves), respectively. Normal swimming and ventilatory contractions of the mantle muscles result from impulses conveyed through a system of numerous small motor neurones that radiate from the pair of stellate ganglia situated in the mantle wall. In fast-swimming cephalopods, where rapid movement is required for escape or attack, the pleurovisceral center has an additional median ventral or giant fiber lobe. This is responsible for the powerful and synchronized contractions of the mantle muscles. In the median ventral lobe, first-order giant neurones respond to impulses received from the sense organs. They run to another region in the pleurovisceral center where they synapse with second-order giant neurones. These send axons to the funnel muscles and, by way of the pallial nerves, to the stellate ganglia. Here they connect with third-order giant neurones that innervate the circular muscles in the mantle wall.

In general, whenever the subesophageal centers of the brain are stimulated they produce isolated responses rather than complete behavior patterns. They therefore constitute the lower motor centers. These are integrated by higher motor centers in the basal lobes of the supraesophageal complex. An anterior basal lobe coordinates the movement of the head and tentacles during feeding; a posterior basal lobe produces coordinated swimming movements involving the mantle and funnel, whereas a lateral basal lobe controls the chromatophores. In addition, buccal lobes coordinate the activity of the beak and radula. Above these higher motor centers there are two further integrative centers. Of these, the inferior frontal lobes integrate all tactile information received by the brain whereas the olfactory lobes, which are attached to the optic stalks, integrate the input from chemoreceptors. Finally, above all the other lobes are the frontal, vertical, and subvertical lobes. These contain vast numbers of nerve cells, but they do not receive any direct sensory input nor do they originate any motor axons directly to effector organs. These are the "silent areas" associated with memory and learning. For further details of the cephalopod nervous system see Wells (1962) and Young (1972).

In chitons the chief sense organs are the gustatory subradular organ and the unique system of aesthetes situated in vertical canals in the upper surfaces of the dorsal plates. The fine structure of these aesthetes is described by Boyle (1974). Megalaesthetes lie at the center of clusters of simpler micraesthetes and function as light receptors. In some species they may even form single eyes with a lens, a pigmented sheath, and a retina. The outer cuticle may serve as an elementary cornea. Paired osphradia are present in the mantle trough close to the posterior gills.

Early gastropods have a variety of sense organs including eyes and tentacles on the head, osphradia in the mantle cavity, and simple statocysts embedded in the foot. Additional tentacles may also occur around the mantle and foot. In higher gastropods these various sense organs are either emphasized or reduced according to life-style. Many gastropods use eyes for simple light orientation, but in the more active groups visual powers are especially pronounced. Heteropods, for instance, locate their food visually and have complex telescopic eyes with a large lens and folded retina. Strombids are also active carnivores with large eyes located at the ends of long peduncles. Land pulmonates explore largely by sight and have eyes at the tips of long cephalic tentacles. In burrowing gastropods, on the other hand, eyes are generally either lost or buried in the skin. Osphradia are essentially chemoreceptors but in prosobranchs, with their anterior mantle cavity, they may also serve as mechanoreceptors capable of detecting particles in the inhalent stream (Yonge, 1947). The evolution of the osphradium in gastropods closely parallels that of the gill; in primitive forms they are paired, but where only one gill survives there is only a single osphradium. In carnivorous forms the osphradia are particularly complex, and their association with an inhalent siphon provides an effective snout for sampling the environment and locating prey. With the lack of any marked water currents through the mantle cavity in pulmonates and opisthobranchs, the osphradia become less important and may even disappear. Typical olfactory organs in the opisthobranchs are the modified head tentacles (rhinophores). In the nudibranchs, tactile senses are especially important, and the whole naked dorsal surface becomes highly sensitive.

Bivalves lack a head, and the major sense organs therefore come to lie mainly around the mantle margin or at the tips of the siphons. Mantle tentacles are tactile and chemosensory. They are especially well developed in the free-swimming pectinids where they may be associated with quite complex eyes. Statocysts occur in the foot but these are reduced or absent in attached forms such as oysters. Osphradia retain their primitive position on the ctenidial axis and lie immediately beneath the posterior adductor muscle. Because they are now located in the exhalent chamber, their function as chemo- or mechanoreceptors is perhaps questionable.

Like the nervous system, the cephalopod sensory system is also highly developed (see Wells, 1966, for details). Food and enemies are recognized mainly by sight and the cephalopod eye is particularly well developed. Coleoid eyes are similar to those of vertebrates though their basis of construction is rather different. *Nautilus* relies mainly on its sense of smell, and its eyes, although large, are less complex than those of coleoids. Statocysts are present in nautiloids and coleoids. The particularly complex statocysts of *Octopus* are comparable with the vertebrate inner ear. Octopods depend more on tactile stimuli, and in this regard the arms and suckers are especially sensitive. Chemoreception is subordinate to the visual and tactile senses in cephalopods and only in *Nautilus* is the osphradium retained.

## VI. Evolution and Adaptive Radiation among the Major Classes

### A. Gastropods

The Gastropoda constitute by far the largest and most diverse molluscan class. That it is also the only class to have successfully colonized land may be at least partially due to torsion, the process peculiar to gastropods whereby the primitively posterior mantle cavity is brought into an anterior forward-facing position above the head. Most gastropods are prosobranchs that primitively exhibit the full effects of torsion. Opisthobranchs display detorsion and include some of the most highly specialized gastropods. Although structurally the most conservative group, the pulmonates have undergone extensive physiological adaptation.

The Archaeogastropoda are the most primitive and least specialized prosobranchs. The earliest archaeogastropods are those with cleft or perforate shells such as the abalone (Pleurotomariacea) and keyhole limpets (Fissurellacea). These have paired bipectinate gills similar to the ancestral mollusc and have solved the posttortional problem of sanitation by allowing exhalent water to escape much further back behind the head. The true limpets (Patellacea), with their low conical shell and powerful sucker-like foot, are especially well adpated for clinging to rocky surfaces on wave-exposed shores. Moreover, the absence of any breaks in the shells of these gastropods has enabled them to penetrate the high intertidal zone. Here the original mantle gills are either reduced (e.g., *Acmaea*) or completely replaced by a skirt of secondary gills (e.g., *Patella*). Limpets are secondarily evolved from coiled forms, and traces of their spiral ancestry are evident during their early development. The extensive gill surface of these early prosobranch groups is easily fouled by sediment, and consequently most species favor clean water habitats over rocky bottoms. The remaining archaeogastropods— the Neritacea and Trochacea (top shells and turbans)—retain the more characteristic spiral shell. Here, however, only the left gill (still bipectinate) is retained and the oblique ventilating system of these higher archaeogastropods is evidently a more effective solution to the problems imposed by torsion than that found in the more primitive groups. The neritaceans include freshwater (e.g., *Theodoxus*) and terrestrial forms as well as the more familiar intertidal species such as *Nerita*. In general, freshwater prosobranchs, as gill breathers, are rather more restricted to flowing, well-oxygenated water than are the aquatic pulmonates. The terrestrial helicinids are termed operculate land snails in order to distinguish them from pulmonate land snails, which lack an operculum. Here gaseous exchange occurs across the vascularized roof of the mantle cavity and the gill has consequently been lost. Like the pulmonates, their reproductive systems are adapted for internal fertilization and for the production of large eggs or viviparity.

Mesogastropods and neogastropods are much more varied than the archaeogastropods. Their facility for exploiting a wider range of habitats including soft sediments has largely been due to changes in gill structure and to the development of an inhalent siphon. The monopectinate gill is much less likely to clog with sediment than the more primitive bipectinate gill. Moreover, the development of a siphon by the inrolling of the mantle edge provided ready access to clearer water above the seabed. This siphon is accommodated in a notch that runs along the margin of the shell aperture. In carnivorous species this mobile siphon, with a chemosensory osphradium at its base, also serves as an efficient snout for locating prey. All the evolutionary developments subsequently encountered in the opisthobranchs and pulmonates are foreshadowed in the mesogastropods, and this largest prosobranch group contains a bewildering array of gastropods. It includes grazers such as the primitive periwinkles (Littorinacea), filter feeders such as the wormlike vermetids (Cerithiacea) and the curious slipperlimpets (Calyptraeacea), and deposit feeders such as the hydrobids (Rissoacea), as well as many higher carnivorous groups like the conches (Strombacea), cowries (Cypraeacea), moon shells (Naticacea), tuns, and helmet shells (Tonnacea). Although mainly a benthic group, a few mesogastropods have become pelagic, for example, *Ianthina* (Scalacea) and the heteropods, in which the shell is reduced as in *Carinaria* or finally lost as in *Pterotrachea*. The Lamellariidae (Cypraeacea) are sluglike mesogastropods with an internal shell. Like the neritaceans, the mesogastropods are also well represented in freshwater (e.g., *Valvata, Viviparus*) and terrestrial habitats. Terrestrial groups such as the Cyclophoridae and Pomatiasidae, like the archaeogastropod helicinids, are operculate land snails, and gaseous exchange occurs across the vascularized mantle wall. Special breathing tubes in the shell opening allow air to enter the mantle cavity even when the operculum is closed. The Ampullariidae are amphibious and their subdivided mantle cavity houses both a lung and a gill.

Three major groups arose from the mesogastropods: the neogastropods, the opisthobranchs, and the pulmonates. The neogastropods comprise a group of rather specialized carnivorous or scavenging marine prosobranchs that include the whelks (Buccinacea), the shell-boring muricids (Muricacea), the highly polished olives (Volutacea), and the poisonous cone shells (Conacea). The origin and evolution of the neogastropods is discussed by Pender (1973).

Although rather a small group compared with the prosobranchs and pulmonates, the opisthobranchs show a remarkable degree of adaptive radiation. Primitive opisthobranchs are similar in many respects to prosobranchs and early pulmonates, and in some cases an operculum—a predominantly prosobranch feature—may even be present. Throughout opisthobranch evolution, however, there has been a tendency toward detorsion, a process achieved during postlarval growth, in which the mantle cavity and its associated organs are pushed back along the right side of the body. With detorsion went the progressive reduction and eventual loss of the shell, mantle cavity, and gills (probably never true ctenidia in opisthobranchs), culminating in forms with a secondarily acquired bilateral symmetry. The precise reasons for detorsion are unknown, though it may conceivably have been initiated by the development of a particularly large cephalic shield in the early burrowing forms. Once the heavy shell had been abandoned, several new evolutionary pathways opened up to these newly streamlined opisthobranchs.

Opisthobranchs have radiated along numerous independent lines, though within each line recurring evolutionary themes are readily discerned. Typically the more primitive members of each lineage have spirally coiled shells and exhibit little evidence of detorsion. The earliest opisthobranchs are the cephalapids. These are essentially burrowers, and the lineage culminates in forms like *Philine* with flattened wedge-shaped bodies, an internal shell, and a calcified gizzard for crushing prey. The anaspids include the sea hares, in which a vestigial chitinous shell is overgrown by the mantle. A small triangular mantle cavity almost filled by the plicate gill is still present on the right side. General streamlining of the body for more efficient burrowing preadapted the opisthobranchs to an even more active life-style. Accordingly, most opishthobranch lineages have evolved their free-swimming representatives. In *Aplysia*, lateral extensions of the foot (parapodia) are used for swimming. Among the more active swimmers are the pelagic pteropods (sea butterflies) in which the parapodia are drawn out into ''wings.'' Pteropods may be shelled (Thecosomata) or naked (Gymnosomata). The former are ciliary feeders, the latter rapacious carnivores. Acochlidians are tiny naked interstitial forms of rather doubtful affinities. Sacoglossans (e.g., *Elysia*) are specially adapted for extracting fluids from green algae. They include the remarkable family Juliidae (e.g., *Berthelinia*), with bivalved shells. Notaspids display varying degrees of shell reduction and foreshadow the true nudibranchs to which they almost certainly gave rise. They lack a mantle cavity, but a naked gill still persists on the right (e.g., *Pleurobranchus*). The true sea slugs or nudibranchs are arguably the most remarkable and most beautiful of all molluscs. Here detorsion is complete; the shell, mantle cavity, and gills have all been lost (although a shell is present in the larva); and the body now assumes a secondary bilateral symmetry. The visceral mass is absorbed within the flattened sluglike body. Each group of nudibranchs has its own distinctive features. The naked dorsal body surface, which is generally greatly increased by complex and often brilliantly colored outgrowths (cerata), takes on sensory, respiratory, and protective functions. Aeolids store nematocysts from their coelenterate prey in club-shaped cerata (see Kaelker and Schmekel, 1976). The branched dendritic processes of *Dendronotus* are respiratory. Dorids have a circlet of secondary pinnate gills around the centrally placed posterior anus. The absence of a protective shell largely confines the nudibranchs to sublittoral or low intertidal habitats. Opisthobranchs are extensively reviewed by Thompson (1976).

Pulmonates are structurally the most conservative gastropods and evolutionary advances have here consisted mainly of physiological innovations to overcome problems of terrestrial life. Like the opisthobranchs, they probably arose from a single-gilled prosobranch. However, they do not exhibit detorsion and the mantle cavity forms a vascularized lung that communicates to the outside by a restricted opening, the pneumostome. Aquatic pulmonates belong to the order Basommatophora, in which the eyes are set near the base of a single pair of tentacles as in most prosobranchs. In the much larger group of terrestrial pulmonates, the order Stylommatophora, the eyes are at the tips of a second pair of tentacles. The Systellommatophora are a small group of unusual sluglike forms that lie apart from the mainstream of pulmonate evolution. Air breathing probably arose in the oxygen-poor waters of estuarine marshes and muds. Under such conditions the ability to breath air was probably an adaptive advantage. The evolution of the lung thus predisposed the pulmonates to a terrestrial existence. Limnitic basommatophorans (e.g., *Lymnaea, Planorbis*) are secondarily aquatic, having been clearly derived from air-breathing forms. The limpet form has been adopted by many aquatic pulmonates, which, like the archaeo- and mesogastropod limpets, show a partial approach to radial symmetry. Limpet-like pulmonates include both marine (e.g., *Siphonaria*) and freshwater (e.g., *Ancylus*) forms. *Amphibola* is the only operculate pulmonate, although the shells of many pulmonates, like those of several prosobranch neritids, have developed teeth and ridges in order to protect the shell aperture. Many land snails seal their shells by a temporary epiphragm of calcified slime. This prevents water loss but permits gaseous exchange. The Stylommatophora include the terrestrial snails (e.g., *Helix, Cypaea*) as well as the slugs (e.g., *Testacella, Limax, Arion*), in which the shell is either completely lost or reduced to a fragment and buried in the mantle. In these latter, as in the nudibranchs, bilateral symmetry has been secondarily regained. Shell reduction appears to have occurred several times among these higher pulmonates and most groups have their shell-less representatives. The evolution of the slug form is probably an adaptive response to low calcium levels. Semislugs retain a visible spiral shell but this is too small to function as an effective retreat. The land pulmonates have had a far more spectacular evolution and lead a much fuller terrestrial existence than the operculate land snails (prosobranchs). Although they are probably the most successful land invertebrates after the arthropods, slugs are unusual in that they have little obvious structural protection against desiccation. However, they are mainly nocturnal animals and are tolerant of extreme water loss. Moreover, the mantle cavity can be used as a reservoir to carry water, and water is readily absorbed across the body wall. Their streamlined compressible bodies allow them to slip easily into confined spaces or to burrow actively for prey. Although they have spread into numerous habitats throughout the world, stylommatophorans achieve their greatest diversity, as indeed do the operculate land snails, in hot, moist tropical forests. For detailed accounts of the structural and functional organization of pulmonates, see Runham and Hunter (1970), Solem (1974), and Fretter and Peake (1975).

Several gastropods have become successful parasites. Ectoparasitic forms (e.g., eulimids, pyramidellids) show the least amount of structural modification; endoparasitic forms (e.g., entoconchids), on the other hand, lose all typical molluscan features and are often recognized only by virtue of their larvae.

## B. Bivalves

Bivalves have concentrated on fewer structural patterns than the gastropods. The basic body plan is unmistakable, but superimposed upon this there has been a considerable degree of adaptive radiation. The bivalve shell is remarkably plastic and frequently reflects environmental demands to which it has been subjected. Similar life-styles in many unrelated lineages reveal numerous instances of parallel or convergent evolution. This has led to considerable problems in their classification. Bivalves probably arose from an ancestor with a domed shell, posterior mantle cavity, and one pair of gills. Lateral compression was made possible through a reduction in the zone of calcification along the dorsal midline, whereas near each end, cross fusion of mantle muscles produced the characteristic adductors. Although the protobranchs are generally believed to be the most primitive bivalves, the few surviving representatives of this once widespread group have become rather specialized. Most present-day protobranchs are surface deposit feeders, and this is thought to have been the original feeding method of the group [though Stasek (1961) and Allen and Sanders (1969) suggest that filter feeding may in fact be more primitive]. Organic material is generally collected from the sediment by elongate palp proboscoides whereas the small gills serve primarily or wholly for respiration. Septibranchs, by contrast, are a comparatively specialized group in which the gills form a muscular pumping septum. They have abandoned the more conventional bivalve method of filter feeding to become the carnivores and scavengers of deep-sea oozes. The vast majority of modern bivalves are lamellibranchs. These probably arose directly from the early protobranch stock through an extensive elaboration of the ctenidia. This development enabled them to collect food directly from the water column, thus freeing these bivalves from the constraints of deposit feeding, though some taxa have secondarily returned to this lifestyle. Although widespread in marine and freshwater habitats, their method of locomotion and their dependence on gills and pelagic development have effectively prevented the bivalves from becoming established on land.

Two major evolutionary events have been largely responsible for the successful radiation of bivalves: the neotenous retention of the postlarval byssus complex and the development of siphons often associated with varying degrees of mantle fusion. The byssus gland (a modified pedal mucous gland) originates as a postlarval organ used for temporary attachment during metamorphosis (Yonge, 1962). However, once metamorphosis is complete this organ is generally lost, but its retention into the adult stages in many bivalves was instrumental in the colonization of hard surfaces and allowed many diverse taxa to become independent of soft sediments. In these epifaunal bivalves the mantle remains unfused and siphons are generally absent. Byssal attachment on hard surfaces has had a profound influence on bivalve morphology, resulting in the gradual reduction (anisomyaria) and eventual loss (monomyaria) of the anterior adductor muscle. It has evolved independently in many different groups (e.g., Mytilidae, Arcidae, and the unusual Anomiidae in which the byssus forms a calcified plug), and has led in turn to the appearance of cemented forms such as oysters (Ostreidae) and jewel shells (Chamidae), and, perhaps most surprisingly of all, to free-swimming groups such as the scallops and fileshells (Pectinidae). Boring into the substratum confers considerable protection from predators and excessive water movement, and is known to have evolved by at least two independent routes, one of which was byssal attachment. The shells of many boring forms are greatly reduced and serve merely as cutting tools. Some byssate taxa like the fanshells (Pinnidae) have reinvaded sediments after having evolved many of the shell characteristics more typically associated with life on hard surfaces (Yonge, 1953). Reversion to an infaunal life-style has probably occurred many times during bivalve evolution, and this too may have been via neoteny because the postlarval stages of byssate taxa still retain an active burrowing foot. Considerable evolutionary potential is thus stored in the bivalve larva. Many of the shell features associated with epibyssal attachment to hard substrata are considered to be adaptations for improving physical stability (see Stanley, 1972; Seed, 1980). A transitional period of infaunal or semi-infaunal byssal attachment probably preceded the ultimate colonization of hard surfaces. Although these endobyssate forms were abundant during the Paleozoic, they now persist in only a few recent taxa such as *Modiolus* and *Brachidontes* (Mytilidae).

By contrast with byssally attached forms, bivalves inhabiting soft sediments are much more uniform in their shape and tend to retain the more primitive isomyarian condition. Nevertheless, considerable radiation in life habits has occurred, principally through the development of siphons and mantle fusion. Infaunal taxa are especially successful and account for well over 75% of all living bivalves. Siphon development opened up many new habitats by permitting infaunal bivalves to penetrate deeper into the sediment. This led to improved stability, reduced the risk of disinternment, and provided additional protection against predators. It also prevented the mantle cavity from becoming fouled with sediment and provided for a more orderly flow of water. Most suspension-feeding bivalves draw food and water into the mantle cavity through a posterior inhalent siphon. They may live near the surface and possess short siphons (e.g., Cardiidae, Veneridae) or they may be deep burrowers with either massive fused siphons (e.g., Myidae, Mactridae) or elongate shells (e.g., Solenidae). They may be relatively sedentary or active burrowers in shifting environments. Each life-style imposes its own suite of morphological features on the shell (Stanley, 1970). The siphonate tellinaceans (e.g., *Scrobicularia, Tellina*) have reverted to the ancestral condition of deposit feeding, but unlike the more primitive protobranchs they collect food by long mobile inhalent siphons that sweep the surface of the sediment like miniature vacuum cleaners. These active deposit-feeding bivalves are extremely successful especially in rich organic muds. Just as byssal attachment provided a route for the evolution of the boring habit, so too did certain adaptations for deep infaunal burrowing, because many of the features that evolved for deep burrowing are equally adaptable to a life spent boring into hard substrata.

Whereas the bivalve shell is thus subject to considerable phenotypic expression, perhaps one of the most striking features concerning the evolution of such a diverse group has been the repeated appearance of a comparatively restricted number of extremely successful shell morphologies, often in otherwise quite unrelated groups.

Although numerous lamellibranchs have invaded fresh water, these show little of the adaptive radiation found in marine bivalves. However, special adaptations, such as ectoparasitic larvae and the modification of part of the gill as a brood chamber, testify to the great antiquity of certain lineages. The most ancient of these are cosmopolitan in their distribution (e.g., *Sphaerium, Pisidium*). The Unionidae is the largest family and with some 1000 species accounts for about 20% of all living lamellibranchs. A few bivalves, especially within the family Erycinidae, have become commensals (e.g., *Montacuta*) or even endoparasites (e.g., *Entovalva*). Many, but not all, are associated with echinoderm hosts.

## C. Cephalopods

The body plan of the cephalopods is even more stereotyped than that of the bivalves. However, whereas the bivalves are an essentially sedentary group of filter feeders, the cephalopods are predominantly free-swimming pelagic predators with all the concomitant neurological and locomotory sophistications that such an active life-style demands. Cephalopods appear to be a waning group, and of the surviving forms only *Nautilus* approaches the condition of its Paleozoic ancestors. Much can therefore be learned about early cephlopod evolution from the study of this unique living fossil. *Nautilus* differs from other living cephalopods in its retention of the large-chambered external shell. Compared with most cephalopods, *Nautilus* is a poor swimmer. It is mainly nocturnal and locates its food by smell rather than by sight. The body is accommodated in the outermost chamber of the bilaterally symmetrical planospiral shell. The head bears a crown of adhesive, suckerless tentacles that are used to grope for food. Above the head a fleshy hood closes over like an operculum when the animal withdraws. The chambers of the shell are filled with gas, thus enabling the animal to adjust its position in the water column. The ventral foot (hyponome) consists of two separate fleshy lobes that overlap in the midline. Water enters the mantle cavity around the sides of the body and passes over the osphradia (two pair) and gills (two pair) before finally being forced out through the mobile funnel, not by pallial contractions as in the modern naked cephalopods, but by contractions of the intrinsic muscles of the funnel itself. The chambered flotation shell, jet propulsion, and prehensile tentacles are features that have contributed enormously to the success of the cephalopods.

The nautiloids and ammonoids were exceedingly successful groups and some were extremely large. Both groups evolved along similar lines, though the smooth, simply sculptured shells of the nautiloids contrast with the heavily sculptured and intricately sutured shells of the ammonoids. Whereas the most primitive nautiloids are believed to have had curved cone-shaped shells (Donovan, 1964) and probably swam in an upright position, the earliest fast-swimming pelagic forms appear to have been straight-shelled orthocones (e.g., *Orthoceras*). Subsequent planospiral coiling, however, produced more compact and easily maneuverable shells. These appeared early in nautiloid evolution and were soon to become the dominant form. Ammonoids arose from coiled nautiloids during the Silurian and persisted until the late Mesozoic. Many of the adaptations initiated by the nautiloids were taken a step further by the ammonoids. Some developed light, streamlined shells and became active swimmers; some became passive drifters and yet others reinvaded the seabed, pulling themselves along by means of their arms in an octopus-like fashion. Some ammonoids became secondarily uncoiled and resembled the earlier orthoconic nautiloids. During the Mesozoic, straight-shelled nautiloids gave rise to the belemnoids and it is from this group that all modern cephalopods (except *Nautilus*) have ultimately evolved (see Shrock and Twenhofel, 1953).

During the evolution of the coleoids the shell became completely enclosed by the muscular mantle. Coleoids are subdivided into two main groups, the Decapoda and the Octopoda. Decapods, which include the belemnoids, teuthoids (squid), and sepioids (cuttlefish), have 10 arms (2 of which are long and retractile), body fins, and an internal shell. Octopods are shell-less, have 8 arms, and generally lack fins. The belemnoids which are distinguished from straight-shelled nautiloids by virtue of their internal shell, are now totally extinct. The basic coleoid shell is well illustrated by the belemnoids. It consisted of three parts: a chambered phragmocone for buoyancy, a calcified torpedo-shaped rostrum for rigidity, and a chitinous pro-ostracum for attachment of the mantle muscles. Evidence indicates that the belemnoids were probably proficient swimmers. As a group, however, they showed little adaptive radiation and had largely disappeared by the Cretaceous.

In all living coleoids the shell is much reduced in importance. Only the open-coiled phragmocone of *Spirula*, a cosmopolitan deep-water cuttlefish, exhibits the primitive chambered condition to any marked degree. Traces of chambering may still persist in the familiar cuttlebone (phragmocone plus part of rostrum), but frequently all calcareous material is lost, leaving only a plate of horny material as in the pen of *Loligo* (pro-ostracum).

Whereas sepioids are generally small, benthic, and coastal in their distribution, the teuthoids are essentially fast oceanic swimmers. Some (e.g., *Architeuthis*) are exceedingly large; others (e.g., *Dosidiscus*) are capable of limited "flight." Cranchids and chiroteuthids are highly specialized bathypelagic forms. The virtual emancipation of these modern cephalopods from the clumsy external shell has led to a vastly improved system of locomotion involving powerful mantle wall muscles and a funnel that is now a complete tube. Most modern cephalopods have an ink sac alongside the rectum. This produces a melanoid cloud that, together with their often well-developed powers of color change, provides an effective means of evading potential predators.

Octopods have completely lost the shell and the two long tentacular arms that characterize the decapods. The eight remaining arms form a webbed circlet of tentacles around the mouth. In *Argonauta* two of these arms secrete a papery boat-shaped shell of calcified conchiolin in which the female carries the eggs. Although octopods can swim, they are generally more benthic than the decapods, using their arms to pull themselves along or anchor themselves to the seabed. Their bodies are much less streamlined than the decapods and there are no fins. The mantle edges are extensively fused to the body wall, resulting in a rather more restricted opening to the mantle cavity. Octopods are mainly coastal. Some offshore species are pelagic, others are deep-water forms with deeply webbed arms that may be used not only for swimming but also for gathering food. *Cirrothauma* is eyeless and its transparent body contains a thick jelly-like connective tissue. The Vampyromorpha or vampire squid consitute a small but distinct group of deep-water coleoids. These appear to be intermediate between octopods and squid.

The breakthrough in cephalopod evolution was unquestionably the appearance of the chambered shell. This was instrumental in the evolution of jet propulsion, which is so characteristic of this molluscan group. Packard (1972) considers that the streamlined coleoid design may in fact have evolved as a direct result of competition with fish. This could explain the many convergent features that these two groups appear to share. Despite their efficiency and their unquestionable success, the cephalopods also have their limitations particularly with regard to their respiratory and excretory physiology. These limitations have doubtless been instrumental in confining the cephalopods to the clearer waters of the open sea.

## VII. Molluscan Phylogeny

We may never know for certain what the characteristics of the ancestral mollusc might have been, and any statement concerning phylogeny and the interrelationships between present-day classes can be little more than hypothesis. However, the structural and functional plan throughout the molluscs is surprisingly uniform, and homologies established for the major classes have enabled a convincing working archetype to be constructed (Section II). Evidence suggests that this was probably an unsegmented, bilaterally symmetrical organism with a posterior mantle cavity that enclosed a set of paired organ systems. Whereas such an archetype is still perhaps the most likely ancestor of at least the two major classes, it has been suggested (see Russell-Hunter, 1979, p. 476) that a more plausible arche-type might be one with a fourfold basic organization similar to that in present-day *Nautilus*. From this condition there could then be either a reduction to a single-pair system (bivalves, gastropods) or replication to a many-pair system (chitons).

The discovery of *Neopilina* has led certain malacologists (e.g., Lemche and Wingstrand, 1959; Fretter and Graham, 1962) to conclude that the ancestral mollusc was perhaps metamerically segmented. This in turn has led to the view that molluscs are descended from annelids. If this view is correct then all traces of metamerism, even during early development, have been lost throughout the subsequent course of molluscan evolution. Organ repetition in chitons and *Nautilus* is probably a functional requirement rather than archaic metamerism. Even in *Neopilina* replication is very irregular as between individual organ systems, and it would be extremely difficult to determine what precisely constitutes a segment. Moreover, "segmentation" does not involve the coelom in the same way that it does in annelids. Unfortunately, no embryological evidence is yet available to establish the precise significance of organ replication in *Neopilina*. At present it seems doubtful whether metameric segmentation in the classical sense ever existed in the molluscs, and the weight of evidence currently suggests that the ancestors of this phylum (as well as the Annelida) are probably to be found among the free-living flatworm–nemertine stock (Vagvolgyi, 1967; Stasek, 1972). The precise nature of this ancestor is debatable, but a through gut, a trochophore-like larva, and some replication of structures (i.e., pseudometamerism) were probably already present. Annelids subsequently became truly metameric possibly as a locomotory adaptation for burrowing, but in the molluscs organ replication was largely abandoned as newly evolved life-styles began to emerge. Here fewer, presumably more efficient organs tended to replace the earlier, less efficient multiple systems.

Within the phylum itself relationships are equally obscure. All the major classes were already distinct by the time they first appeared in the fossil record, and at present they are not readily assembled into definite adaptive lines. Some interesting similarities exist between the scaphopods and bivalves, but whereas the planospiral shell of early gastropods recalls that of primitive cephalopods, this may simply reflect convergent evolution. The extensive development of a head seems to have been a gastropod–cephalopod innovation. A head may never have been present in the scaphopods and bivalves, and in the latter the adoption of ciliary feeding involved the loss of the radula. Valuable reviews of molluscan phylogeny are presented by Stasek (1972), Runnegar and Pojeta (1974), and Yochelson (1978). They argue that the Monoplacophora probably gave rise to the gastropods, bivalves, scaphopods, and cephalopods (subphylum Conchifera). The Aplacophora and Polyplacophora (subphylum Aculifera), on the other hand, appear to have diverged much earlier in molluscan evolution.

## VIII. Classification

A complete classification of the phylum Mollusca is not within the scope of this introductory chapter. Several such schemes, often based on various alternative diagnostic features, are already available in the literature. In the abbreviated classification that follows, only the major taxonomic groups together with a *selection* of families have been included.

---

Subphylum Aculifera
  Class Aplacophora
    Subclass Caudofoveata (Chaetodermomorpha)
    Subclass Ventroplicida (Neomeniomorpha)
  Class Polyplacophora (placed in separate subphylum Placophora by Stasek 1972)

Subphylum Conchifera

  Class Monoplacophora
    Subclass Cyclomya (extinct)
    Subclass Tergomya          Family Neopilinidae
  Class Gastropoda
    Subclass Prosobranchia
      Order Archaeogastropoda (Aspidobranchia, Diotocardia)
        Superfamily Bellerophontacea (extinct)
        Superfamily Pleurotomariacea[1]    Families Pleurotomariidae, Haliotidae
        Superfamily Fissurellacea[1]    Family Fissurellidae
        Superfamily Patellacea    Families Acmaeidae, Patellidea
        Superfamily Trochacea    Families Trochidae, Turbinidae
        Superfamily Neritacea[2]    Families Neritidae, Helicinidae
      Order Mesogastropoda (Taenioglossa. With order Neogastropoda these comprise
    the Pectinibranchia or Monotocardia.)
        Superfamily Cyclophoracea    Families Cyclophoridae, Viviparidae, Ampullariidae
        Superfamily Valvatacea    Family Valvatidea
        Superfamily Littorinacea    Families Littorinidae, Pomatiasidae
        Superfamily Rissoacea    Families Rissoidae, Hydrobiidae
        Superfamily Cerithiacea    Families Turritellidae, Vermetidae

---

| | |
|---|---|
| Superfamily Strombacea | Families Strombidae, Aporrhaidae |
| Superfamily Calyptraeacea | Families Calyptraeidae, Capulidae |
| Superfamily Cypraeacea | Families Cypraeidae, Lamellariidae |
| Superfamily Naticacea | Family Naticidae |
| Superfamily Scalacea (Epito- neacea) | Families Scalidae, Ianthinidae |
| Superfamily Aglossa | Families Eulimidae, Entoconchidae |
| Superfamily Heteropoda | Families Carinariidae, Pterotracheidae |
| Superfamily Tonnacea | Families Tonnidae, Cassidae, Cymatiidae |
| Order Neogastropoda (Stenoglossa) | |
| Superfamily Muricacea | Families Muricidae, Thaididae |
| Superfamily Buccinacea | Families Buccinidae, Nassariidae, Fascio- lariidae |
| Superfamily Volutacea | Families Volutidae, Olividae, Mitridae |
| Superfamily Conacea (Toxo- glossa) | Families Conidae, Terebridae |
| Subclass Opisthobranchia | |
| Order Cephalaspidea (Bullomorpha) | Families Actaeonidae, Scaphandridae, Phi- linidae |
| Order Pyramidellacea | Family Pyramidellidae |
| Order Anaspidia (Aplysiomorpha) | Families Aplysiidae, Akeratidae |
| Order Thecosomata[3] | Families Limacinidae, Cavoliniidae |
| Order Gymnosomata[3] | Families Clionidae, Cliopsidae |
| Order Acochlidiacea | Family Acochlidiidae |
| Order Notaspidea | Families Pleurobranchidae, Umbraculidae |
| Order Sacoglossa | Families Elysiidae, Juliidae, Oxynoidae |
| Order Nudibranchia (Acoela) | Families Dendronotidae, Aeolidiidae, Doridae |
| Subclass Pulmonata | |
| Order Systellommatophora | Families Onchidiidae, Veronicellidae |
| Order Basommatophora | Families Lymnaeidae, Planorbidae, An- cylidae |
| Order Stylommatophora | Families Achatinidae, Helicidae, Limaci- dae, Arionidae, Testacellidae |
| Class Scaphopoda | |
| Class Bivalvia (Pelecypoda)[4] | Taxa containing former subclasses Proto- branchia and Septibranchia are indi- cated; the rest contain only lamelli- branchs |
| Subclass Palaeotaxodonta (subclass Protobranchia in part) | |
| Order Nuculoida | Families Nuculidae, Malletiidae |
| Subclass Cryptodonta (subclass Protobranchia in part) | |
| Order Solemyoida | Family Solemyidae |
| Subclass Pteriomorpha | |
| Order Arcoida | Families Arcidae, Glycymerididae |
| Order Mytiloida | Families Mytilidae, Pinnidae |
| Order Pterioida | Families Pteriidae, Pectinidae, Anomiidae Ostreidae, Limidae |

*(Continued)*

**CLASSIFICATION** (Continued)

| | |
|---|---|
| Subclass Palaeoheterodonta | |
| Order Unionida | Families Unionidae, Mutelidae, Etheriidae |
| Order Trigonioida | Family Trigoniidae |
| Subclass Heterodonta | |
| Order Veneroida | Families Lucinidae, Chamidae, Erycinidae, Cardiidae, Tridacnidae, Mactridae, Solenidae, Tellinidae, Pisidiidae Veneridae, Dreissenidae |
| Order Myoida | Families Myidae, Pholadidae, Teredinidae |
| Order Hippuritoida (extinct rudists) | |
| Subclass Anomalodesmata (includes former subclass Septibranchia) | |
| Order Pholadomyoida | Families Poromyidae, Cuspidariidae, Clavagellidae |
| Class Cephalopoda (Siphonopoda) | |
| Subclass Nautiloidea | Family Nautilidae |
| Subclass Ammonoidea (extinct) | |
| Subclass Coleoidea | |
| Order Decapoda | |
| Suborder Belemnoidea (extinct) | |
| Suborder Sepioidea | Families Sepiidae, Sepiolidae, Spirulidae |
| Suborder Teuthoidea | Families Loliginidae, Architeuthidae, Cranchiidae, Chiroteuthidae |
| Order Vampyromorpha | Family Vampyroteuthidae |
| Order Octopoda | Families Octopodidae, Argonautidae, Cirroteuthidae |

[1] Together these comprise the superfamily Zeugobranchia.
[2] Sometimes placed in a separate order.
[3] Together these comprise the order Pteropoda.
[4] This system follows that in the Treatise of Invertebrate Palaeontology (Moore, 1957–1971). The older subdivision of the lamellibranchs into Filibranchia, Pseudolamellibranchia, and Eulamellibranchia has little or no systematic significance.

# References

Because much of this chapter is concerned with material more fully documented in standard works, only a selection of references is provided. Some of these references, however (e.g., Morton, 1979; Purchon, 1977; Stasek, 1972), contain extensive bibliographies.

Allen, J. A., and Sanders, H. L. (1969). *Nucinella serrei* Lamy (Bivalvia: Protobranchia) a monomyarian solemyid and possible living actinodont. *Malacologia* **7**, 381–396.

Barnes, R. D. (1980). "Invertebrate Zoology," 4th ed. Saunders, Philadelphia, Pennsylvania.

Bilyard, G. R. (1974). The feeding habits and ecology of *Dentalium entale stimsoni*. *Veliger* **17**, 126–138.

Boyle, P. R. (1974). The aesthetes of chitons. 2. Fine structure in *Lepidochitona cinereus*. *Cell. Tissue Res.* **153**, 383–398.

Bullock, T. H. and Horridge, G. A. (1965). "Structure and Function of the Nervous Systems of Invertebrates." Freeman, San Francisco.

Donovan, D. T. (1964). Cephalopod phylogeny and classification. *Biol. Rev.* **39**, 259–287.

Fretter, V., and Graham, A. (1962). "British Prosobranch Molluscs." Ray Society, London.

Fretter, V., and Graham, A. (1976). "A Functional Anatomy of Invertebrates." Academic Press, London.

Fretter, V., and Peake, J., eds. (1975). "Functional Anatomy and Physiology," Pulmonates, Vol. 1. Academic Press, New York, London.

Gainey, L. F. (1972). The use of the foot and captacula in the feeding of *Dentalium*. *Veliger* **15**, 29–34.

Garstang, W. (1928). Origin and evolution of larval forms. *Rep. Br. Ass. Advmt. Sci.,* Sec. D, 77 pp.

Ghiselin, M. T. (1966). The adaptive significance of gastropod torsion. *Evolution* **20**, 337–348.

Gilpin-Brown, J. B. (1972). Buoyancy mechanisms of cephalopods in relation to pressure. *Symp. Soc. Exp. Biol.* **26**, 251–259.

Grégoire, C. (1972). Structure of the molluscan shell. *In* "Chemical Zoology" (M. Florkin and B. T. Scheer, eds.), Vol. 3, pp. 45–102. Academic Press, New York.

Johnson, K. M. (1969). Quantitative relationship between gill number, respiratory surface and cavity shape in chitons. *Veliger* **11**, 272–276.

Kaelker, H., and Schmekel, L. (1976). Structure and function of the cnidosac of the Aeolidoidea. *Zoomorphologie* **86**, 41–60.

Lemche, H., and Wingstrand, K. G. (1959). The anatomy of *Neopilina galatheae*. *Galathea Rep.* **3**, 9–71.

Moore, R. C., ed. (1957–1971). "Treatise of Invertebrate Paleontology." Mollusca, Vol. I–N. Geological Society of America and University of Kansas Press, Lawrence.

Morton, J. E. (1979). "Molluscs," 5th ed. Hutchinson, London.

Morton, J. E., and Yonge, C. M. (1964). Classification and structure of the Mollusca. *In* "Physiology of Mollusca" (K. M. Wilbur and C. M. Yonge, eds.), Vol. 1, pp. 1–58. Academic Press, New York.

Owen, G. (1955). Observations of the stomach and digestive diverticula of the Lamellibranchia. 1. The Anisomyaria and Eulamellibranchia *Q. J. Microsc. Sci.* **96**, 517–537.

Owen, G. (1974). Studies on the gill of *Mytilus edulis:* the eu-laterofrontal cirri. *Proc. R. Soc. London, Ser. B.* **187**, 83–91.

Packard, A. (1972). Cephalopods and fish: the limits of convergence. *Biol. Rev.* **47**, 241–307.

Pender, W. F. (1973). The origin and evolution of the Neogastropoda. *Malacologia* **12**, 295–338.

Purchon, R. D. (1977). "The Biology of the Mollusca," 2nd ed. Pergamon, New York.

Reid, G. B., and Reid, A. M. (1974). The carnivorous habit of members of the septibranch genus *Cuspidaria*. *Sarsia* **56**, 47–56.

Rhoads, D. C., and Lutz, R. A., eds. (1980) "Skeletal Growth of Aquatic Organisms." Plenum, New York.

Rollins, H. B., and Batten, R. L. (1968). A sinus-bearing monoplacophoran and its role in the classification of primitive molluscs. *Palaeontology* **11**, 132–140.

Runham, N. W. and Hunter, P. J. (1970). "Terrestrial Slugs." Hutchinson, London.

Runnegar, B., and Pojeta, J. (1974). Molluscan phylogeny: the paleontological viewpoint. *Science (Washington, D. C.)* **186**, 311–317.

Russell-Hunter, W. D. (1979). "A Life of Invertebrates." Macmillan, New York.

Russell-Hunter, W. D., and Brown, S. C. (1965). Ctenidial number in relation to size in certain chitons, with a discussion of its phyletic significance. *Biol. Bull. Mar. Biol. Lab., Woods Hole, Mass.* **128**, 508–521.

Salvini-Plawen, L. (1969). Solenogastres und Caudofoveata (Mollusca, Aculifera): Organisation and phylogenetische Bedentung. *Malacologia* **9**, 191–216.

Scheltema, A. H. (1978). Position of the class Aplacophora in the phylum Mollusca. *Malacologia* **17**, 99–109.

Schrock, R. R., and Twenhofel, W. H. (1953), "Principles of Invertebrate Paleontology." McGraw-Hill, New York.

Seed, R. (1980). Shell growth and form in the Bivalvia. *In* "Skeletal Growth of Aquatic Organisms" (D. C. Rhoads and R. A. Lutz, eds.), pp. 23–67. Plenum, New York.

Solem, A. (1974), "The Shell Makers: Introducing Mollusks." Wiley, New York.

Stanley, S. M. (1970). Relation of shell form to life habits of the Bivalvia (Mollusca). *Mem. Geol. Soc. Am.* **125**, 296 pp.

Stanley, S. M. (1972). Functional morphology and evolution of byssally attached bivalve molluscs. *J. Paleontol.* **46**, 165–216.

Stasek, C. R. (1961). The ciliation and function of the labial palps of *Acila castrensis* (Protobranchia, Nuculidae) with an evaluation of the role of protobranch organs of feeding in the evolution of the Bivalvia. *Proc. Zool. Soc. London* **137**, 511–538.

Stasek, C. R. (1972). The molluscan framework. *In* "Chemical Zoology" (M. Florkin and B. T. Scheer, eds.), Vol. 3, pp. 1–44.

Thompson, T. E. (1958). The natural history, embryology, larval biology and post larval development of *Adalaria proxima* (Alder and Hancock) (Gastropoda, Opisthobranchia). *Philos. Trans. R. Soc. London, Ser. B* **242**, 1–58.

Thompson, T. E. (1960). The development of *Neomenia carinata* Tullberg (Mollusca: Aplacophora). *Proc. R. Soc. London, Ser. B* **153**, 263–278.

Thompson, T. E. (1976), "Biology of Opisthobranch Mollusca," Vol. 1. Ray Society, London.

Vagvolgyi, J. (1967). On the origin of molluscs, the coelom and coelomic segmentation. *Syst. Zool.* **16**, 153–168.

Wells, M. J. (1962), "Brain and Behaviour in Cephalopods." Stanford University Press, Stanford, California.

Wells, M. J. (1966). Cephalopod sense organs. *In* "Physiology of Mollusca" (K. M. Wilbur and C. M. Yonge, eds.), Vol. 2, pp. 523–545. Academic Press, New York.

Yochelson, E. L. (1978). An alternative approach to the interpretation of the phylogeny of ancient mollusks. *Malacologia* **17**, 165–191.

Yonge, C. M. (1947). The pallial organs in the aspidobranch Gastropoda and their evolution throughout the Mollusca. *Philos. Trans. R. Soc. London, Ser. B* **232**, 443–518.

Yonge, C. M. (1953). The monomyarian condition in the Lamellibranchia. *Trans. R. Soc. Edinburgh* **62**, 443–478.

Yonge, C. M. (1962). On the primitive significance of the byssus in the Bivalvia and its effects in evolution *J. Mar. Biol. Assoc. U. K.* **42**, 113–125.

Yonge, C. M., and Thompson, T. E. (1976), "Living Marine Molluscs." Collins, London.

Young, J. Z. (1972). "The Anatomy of the Nervous System of *Octopus vulgaris.*" Oxford Univ. Press, New York.

# 2

# Metabolic and Enzyme Regulation during Rest-to-Work Transition: A Mammal versus Mollusc Comparison

**P. W. HOCHACHKA**

**J. H. A. FIELDS**[1]

**T. P. MOMMSEN**[2]

Department of Zoology
University of British Columbia
Vancouver, British Columbia V6T 2A9, Canada

[1] Present address: Department of Zoology, University of Washington, Seattle, Washington 98105.

[2] Present Address: Department of Biology, Dalhousie University, Halifax, Nova Scotia, Canada.

THE MOLLUSCA, VOL. 1
Metabolic Biochemistry
and Molecular Biomechanics

# I. Introduction

Any attempt to describe quantitatively the interwoven pathways and reactions that comprise the whole of metabolism must include the chemical steps involved, the cellular activities of enzymes, and the concentrations of metabolic intermediates and modifiers, as well as interconversion rates under various conditions. Such a holistic view of molluscan metabolism is neither a possible nor a desirable goal for this chapter; rather, our aim is to provide an overview of mechanisms and principles of metabolic control that may be generally applicable. Because control characterizes every level of biological function, including the cellular metabolic one, we could approach this task by a simple and abstract treatment of current concepts in regulation. However, this has been done in other contexts (Atkinson, 1977; Fersht, 1977); and instead of recasting and updating these earlier reviews in a biochemical setting, we propose to analyze metabolic and enzymatic regulatory functions in a specific molluscan and in a general biological setting: the transition from rest to work at varying intensities. By borrowing heavily from the biochemical literature on vertebrate exercise for our overall framework, we hope to expose not only organization and control principles that are generally valid for molluscs as well as other animals, but also to expose a great deal of uncharted territory in molluscan biochemistry.

# II. Fundamental Strategies

A consequence of work often, although not always, is locomotion or at least movement of body parts. Physiologists tell us that in general the power needs of moving are a function of body size, velocity, and mode of movement. Large mass and fast movement demand high power output; small size and slow movement are energy efficient. High-velocity (burst-type) performance generally requires the most effort, much more than is required in slower steady-state movement. These concepts arise from vertebrate studies and one wonders if they are applicable to the molluscs. For some cephalopods, squid in particular, which are capable of short-du-ration bursts of swimming, medium-term steady-state swimming, and much longer term migrations over thousands of kilometers, these consid-erations seem quite appropriate. They may be less appropriate in the case of bivalves and that of most gastropods, where locomotor abilities are ei-ther greatly modified or hardly exist at all. In these, the costs of laying down a slime track or of moving a bivalve shell into a locked position are an intrinsic part of the work load. In the vertebrates, three fundamental biochemical adaptations give animals the necessary flexibility in terms of how fast, how hard, and how long they can work. These involve the fol-lowing:

1. The specialization of muscle ultrastructure and contractile ma-chinery into two or more classes of muscle cells, each with differ-ent functions (different twitch times, different strengths, etc.)
2. The development of high anaerobic and high aerobic metabolic po-tentials in different fiber types, each typically being fired by differ-ent substrate sources (carbohydrate in anaerobic fibers; fat or fat plus carbohydrate in others)
3. The development of finely tuned mechanisms for control of meta-bolic output

Metabolic output is controlled (a) on a short-term basis (stimuli calling for essentially instantaneous adjustments in muscle metabolism and work), (b) on a medium-term basis (stimuli such as training effects calling for more profound and stable adjustments), and (c) on a long-term basis (stimuli involving phylogenetic time and leading to genetically fixed differ-ences between species in the aforementioned characteristics, and thus in performance style and capacity). These principles are equally applicable to molluscs and vertebrates; data for the former, however, are sparse [none in (b) in the list just given, for example], whereas they are abundant for the latter. Hence we shall organize our discussion in part as a contrast between the two groups.

### III. Muscle Work and Contribution to Metabolism

Muscle in most animals is one of the largest tissues in the body, in ver-tebrates constituting up to about three-quarters of total body mass. In humans and mammals in general, muscle $O_2$ uptake accounts for only about 30% of basal metabolic rate; however, during heavy work, muscle contribution to the metabolic rate of the whole organism is more closely proportional to its relative mass. In the extreme case, muscle $O_2$ consumption can account for up to 90% of the total whole-organism metabolic rate (Weibel, 1979).

Similar data are not so easily obtainable for molluscan groups. In the octopus, mantle muscle alone constitutes only about 10% of the animal's mass, but if tentacles are included the value rises to well over 60%. In squid, the mantle itself may constitute 35–40% of the animal's mass, and of course if one includes the tentacles, the fractional mass of muscle in the body increases well above this value (Hochachka and Fields, 1982). Data on fractional contribution to metabolism are not available for many species, but in *Octopus macropus* the rate of $^{14}CO_2$ release from [$^{14}C$]proline by mantle muscle alone exceeds that of the kidney, branchial hearts, systemic heart, gills, and liver (Hochachka and Fields, 1983). So it is a reasonable assumption that, as in vertebrates so in molluscs, muscle is the dominant contributor to overall energy demands, particularly under working conditions.

In mammals, the rate at which muscles demand energy (and can generate it) varies with the length of time involved and the kind of work required. Highest rates of energy output can only be sustained for 5–10 sec, but, typically, higher rates of energy production can be sustained for the subsequent 2–3 min than for the next hour or so. And finally, any long-distance runner will acknowledge that the rate of energy generation that can be sustained for the first 2–3 h cannot be sustained indefinitely. This decline with time in the maximum power that can be generated by working muscle arises from the use of different energy sources for contraction (Howald et al., 1978); and, because these are preferentially stored in different kinds of muscle fibers, it also follows that different fiber types are used for different phases of work. Before considering the mobilization of different sources of energy, it is necessary to define what we mean by different fiber types.

## A. Skeletal Muscle Fiber Types in Vertebrates

In virtually all vertebrate animals that have been examined in depth, at least two—and usually three—fiber types are always distinguishable: they can be termed white, intermediate, and red according to their appearance (which derives mainly from their myoglobin content). In phylogenetically old fish groups (elasmobranchs, holosteans, chondrosteans, and some teleosts with primitive taxonomic features), the different energetic requirements of steady-state and burst-type swimming have led to a complete anatomical and functional division between fast and slow motor systems: Typically, steady-state swimming is supported by thin lateral strips of tonically active red muscle fibers, whereas burst swimming is supported by phasically active white fibers, which constitute the bulk of the musculature. Unlike the situation in teleosts, however, where little if any mixing of fibers occurs, in mammals skeletal muscle typically is a mosaic mixture of all three fiber types (Johnston and Moon, 1980a,b; Boyes and Johnston, 1979).

The standard explanation for the occurrence of these different fiber types is that it allows some separation and specialization of labor. Thus in fishes and mammals, red fibers are preferentially recruited during work of light and moderate intensity, whereas white fibers are recruited either when the excitatory input into the motor neurons increases to high levels during very strenuous work or when the red fibers become fatigued (Holloszy and Booth, 1976).

## B. Fiber Type Differentiation in Mollusc Muscles

Interestingly, a similar specialization of muscle fiber types appears to be expressed in the molluscs (see Volume 4). Although this has not been as extensively examined, it is an old observation that the adductor muscle of bivalves is differentiated into at least two types. One, the catch muscle, is specialized both in contractile and in metabolic properties for sustained tension development and work, whereas the other, the phasic muscle, is more effective in actually closing the shell (de Zwaan et al., 1980). In *Nautilus* retractor and funnel muscles, there is some evidence for specialization of muscles into forms varying in the ratio of oxidative: glycolytic capacities (Hochachka et al., 1978). However, the best example of this kind of phylogenetic parallelism is to be found in the differentiation of mantle musculature in squid. In at least six species of squid, it is now known that the mantle muscle is differentiated into bands of muscles biochemically specialized for either oxidative or anaerobic energy generation. The evidence is based on light microscopy, electron microscopy, histochemistry, and enzyme biochemistry (Mommsen et al., 1981). The pattern seems to vary somewhat between different species, depending on their locomotor specialization (Mommsen et al., 1981; Bone et al., 1981). The situation thus is analogous to that in vertebrates, where white muscle is physiologically specialized for short bursts of intense work (separated by long periods of recovery) whereas the slow red fibers are designed for prolonged work of intermediate intensity. In both groups, these higher level structural adjustments correlate with, and indeed critically depend on, underlying enzymatic and metabolic adaptations.

## C. Energy Stores and Utilization Sequence during Different Kinds of Work

One of the most important levels of exercise adaptation involves the kind of substrate that is stored and utilized. The available energy stores in human skeletal muscle are high-energy phosphate compounds, glycogen, and triglycerides. (Amino acids and proteins are not usually considered to be mobilized in most kinds of exercise in humans.) In terms of total useful high-energy phosphate that can be generated, ATP and CrP (creatine phosphate) constitute only a minor reserve, whereas glycogen and triglycerides provide ample amounts of energy at the local level, two to three orders of magnitude more than from ATP and CrP, respectively.

In molluscs, the available energy sources in muscle are arginine phosphate, glycogen, and amino acids, whereas triglyceride reserves and enzymatic potentials for $\beta$ oxidation can, for practical purposes, be ignored (Mommsen and Hochachka, 1981). Although a variety of guanidino compounds have been shown to occur in molluscs, including (methylagmantine, $\gamma$-guanidinobutyrate, $\gamma$-guanidinoacetate, arcaine, octopine, and arginine, only arginine phosphate has been established to act as the tissue phosphagen (Florkin and Bricteux-Grégoire, 1972; Stephens et al., 1965).

## IV. Phosphagen-Based High-Intensity Muscle Exercise

In nature, the needs for high-velocity short bursts of muscle work probably exceed all other locomotor requirements. The signal for activation of muscle work at its highest possible rate is neuronal membrane depolarization, leading to $Ca^{2+}$ activation of myosin ATPase and thus contraction. The immediate source of energy for contraction is ATP, which is stored at highest levels in white muscles specialized for burst work. But even at these high levels, there is only enough ATP to support a few seconds of heavy work. In few species does that seem to be an adequate burst-work mechanism, so it is usually bolstered by two additional processes: phosphagen buffering of the ATP supplies and anaerobic glycolysis (Holloszy and Booth, 1976).

In vertebrates, CrP is utilized as a high-energy phosphate reservoir. It is most abundant in white muscles, occurring at average levels of about 30 $\mu$mol/g wet weight. Under high-intensity burst-work conditions, creatine phosphate hydrolysis is catalyzed by creatine phosphokinase (CPK):

$$\text{CrP} \xrightarrow[\text{ADP} \quad \text{ATP}]{\text{Mg}^{2+}} \text{creatine}$$

CPK occurs in two main isozyme forms in mammalian muscles, a so-called soluble or cytosolic CPK (termed CPK-I) and a mitochondrial CPK-II. Because the fractional volume of white muscle fiber occupied by mitochondria is very small in most vertebrates, from less than 0.2 to about 2% (Hulbert et al., 1979), most of the CPK activity in white fibers is represented by CPK-I.

Functionally, it will be evident that CPK-I occupies a pivotal position in white muscle function in that it serves as a direct means of channeling ATP to sites of utilization: myosin ATPase of the contractile machinery. This strategic metabolic role is greatly facilitated by some CPK-I being attached to the M band of the sarcomere, close to the ATPase region of myosin (Bessman and Geiger, 1981).

Whereas the amount of creatine phosphate stored in muscle determines the total *amount* of work that can be sustained by this mechanism, the *rate* of energy production (in terms of $\mu$moles of ATP hydrolyzed per minute) depends on the amount of CPK-I available, which is the reason white fiber types typically have higher levels of this enzyme than red muscle fibers (Holloszy and Booth, 1976).

## A. The Phosphagen System in Molluscs

Exactly the same considerations as just described apply for arginine phosphate and phosphoarginine kinase (PAK) in mollusc muscles. For instance, arginine phosphate content is higher in the phasic muscle than in the catch muscle of the scallop *Placopecten megallanicus*. The same is true for the activity of arginine phosphokinase and for the rate of energy production in these two muscles (de Zwaan et al., 1980). In addition, phosphoarginine kinase seems to occur in different forms: a 40,000-molecular weight enzyme that is associated with tonic functions and an 80,000-molecular weight enzyme that has been described for burst-type muscles in eulamellibranchs (Moreland and Watts, 1967), although these results have been disputed by Moré et al. (1971). Generally, as in the analogous vertebrate system, the activity of cytosolic phosphoarginine kinase is highest in the anaerobic-type fibers, lower in the oxidative ones.

The arginine phosphate system in molluscs also differs in a number of minor yet fascinating ways from the mammalian creatine phosphate system. First, at this time, there is no evidence for phosphoarginine kinase association with the myofibrils (something that should be checked in future work); second, as all mollusc phosphoarginine kinases described to date have been localized within the cytosol, and not within or associated with mitochondrial fractions, the existence of an analogous phosphate shuttle system (see later), as in the vertebrates, is questionable. A third and perhaps more critical difference is that the phosphagen in molluscan metabolism is closely coupled to glycolysis: arginine, released by phosphoarginine kinase-catalyzed hydrolysis of arginine phosphate, is a cosubstrate with pyruvate for the terminal dehydrogenase reaction of glycolysis, leading to the formation of octopine, an imino acid (rather than lactate as found in vertebrates). Because in a number of instances (and perhaps as a rule) arginine phosphate hydrolysis precedes glycolytic activation, a competitive situation could in principle develop at such times for arginine as substrate for both octopine dehydrogenase (ODH) in the octopine-forming direction and for phosphoarginine kinase in the arginine phosphate-replenishing direction:

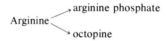

Kinetic studies (Vollmer et al., 1981) of squid mantle PAK and ODH separately, and in simultaneous two-enzyme experiments, indicate that they are in fact set up to avoid this situation; that is, they can either work in a linear forward sequence

$$\text{Arginine-P} \rightarrow \text{arginine} \rightarrow \text{octopine}$$

or a linear backward sequence

$$\text{Octopine} \rightarrow \text{arginine} \rightarrow \text{arginine-P}$$

which indeed seems to occur *in vivo* at least in scallops (de Zwaan et al., 1980).

A fourth difference is more biological than biochemical, for whereas in vertebrates there are no known species that seem to rely solely on phosphagen for supporting work (although this may often occur in quick-strike predators where action extends over only a few seconds), in molluscs there are probably many species that utilize this mechanism almost solely to power muscle work and locomotion. This is dramatically illustrated in the slipping and sliding locomotion of the slug—where $O_2$ uptake is elevated only after moving, not during movement, presumably to recharge phosphagen reserves (Denny, 1980)—and in at least two bivalve species where flapping locomotion is driven mainly by arginine phosphate hydrolysis (Gäde et al., 1978; Grieshaber, 1978). In these cases, anaerobic glycolysis seems to be activated mainly in recovery, again for purposes of recharging carbon and energy reserves.

Although phosphagen-driven muscle work may be adequate for quick-strike predators or sluggish bivalves and slugs, any animal that needs to work at elevated rates for longer than 5–20 sec must utilize an endogenously driven backup system to augment the ATP initially being produced by phosphagen hydrolysis. In vertebrates and at least in cephalopods, this provision is met by anaerobic glycolysis.

## V. Elevated Glycogen Stores and Elevated Enzyme Levels in Anaerobic Muscles

Because glycolysis behaves as a closed system during $O_2$-limited work, the total amount of work it can support depends on the available glycogen, whereas the rate of energy generation depends on the amount of glycolytic enzyme machinery. In mammals both parameters apparently are closely modulated; in consequence, glycogen is stored at higher levels in white muscle than in red, and the levels of all glycolytic enzymes are higher in white muscle than in any other tissue or organ in the vertebrate body (Holloszy and Booth, 1976).

In molluscs, similar information on glycogen levels is not available for the white and red muscle analogs of squid mantle. However, in bivalves, the adductor phasic muscle is known to be particularly rich in glycogen, whereas the catch muscle, being more oxidative, stores less. Enzyme activities in these "anaerobic" muscles, as in the vertebrate analogs, show higher capacities for glycolysis, particularly in some species (e.g., *Illex*; see Mommsen et al., 1981; see also Baldwin, 1983).

Although effective, such wholesale adjustments in enzyme level cannot account for a most dramatic features of muscle glycolysis: its capacity for generating very large changes in flux on transition from rest to maximum burst work. Newsholme (1982) has argued that in trained sprint runners, leg muscle glycolysis may be activated by as much as 2000-fold; that is, an increase in absolute flux of carbon (glycogen → lactate) of over three orders of magnitude! Because levels of octopine dehydrogenase in mantle muscle of some active octopus species may surpass 500 units/g (Fields et al., 1976a), and in pelagic squid, 1500 units/g (Baldwin, 1983), a similar degree of glycolytic activation may be anticipated at least in the most vigorous of molluscan swimmers. How this may be achieved is not well understood and is not a simple matter. Most, and sometimes all, intermediates along the pathway have been monitored in mammals (Holloszy and Booth, 1976), fishes (Guppy et al., 1979; Driedzic et al., 1981), and molluscs (de Zwaan et al., 1980). Although the pathway intermediates usually increase in concentration during activated flux, none change nearly as much as can the actual flux through the system. Because the typical 2- to 10-fold increase in concentrations of glycolytic intermediates cannot account for over a $10^3$-fold increase in flux, it is widely accepted that amplification mechanisms must be harnessed in order to achieve the very large absolute activation of catalytic machinery in muscle during burst work. The most plausible single mechanism for turning on muscle glycolysis in mammals and probably in molluscs as well is the glycogen phosphorylase control cascade. [Other mechanisms that have been proposed may be functional (Newsholme, 1982), but we believe to a lesser extent.]

## A. Turning on Muscle Glycogen Phosphorylase

In mammals, glycogen phosphorylase occurs in at least three isozyme forms, one of which is characteristic of skeletal and cardiac muscle (Cohen, 1978). Two classes of signals, endocrine and neuronal, act indirectly to activate glycogen phosphorolysis. Various endocrines (epinephrine being one of the most potent) serve to activate membrane-bound adenylate cyclase, which initiates the cascade control system. The neuronal mechanism for activating glycogen breakdown involves membrane depolarization and $Ca^{2+}$ release from sarcoplasmic reticulum; $Ca^{2+}$ control cuts in at the level of inactive phosphorylase $b$ kinase by binding to its smallest subunit, calmodulin. Calmodulin in muscle, unlike other tissues, is incorporated into the integral oligomeric structure of phosphorylase $b$ kinase; as in calmodulin from other tissues, it has two $Ca^{2+}$-binding sites. The binding of $Ca^{2+}$ leads to a large (about 15-fold) increase in affinity for substrate, which under physiological conditions leads to an ATP-dependent phosphorylation of $b$ kinase and thus its activation. Phosphorylase $b$ kinase in turn leads to the polymerization of phosphorylase $b$ to phosphorylase $a$. In this cascade system in white muscles, the molar ratios of protein kinase : $b$ kinase : phosphorylase $a$ are about $1 : 10 : 240$, whereas in red muscle they are $1 : 0.25 : 14$; the latter are considered to be similar to those in the liver (Cohen, 1978). In liver, the degree of amplification of an incoming epinephrine signal is estimated at about $10^6$ (Lehninger, 1975). Because the aforementioned molar ratios indicate that in white muscle the amplification potential is even higher, it seems probable that this control cascade is adequate to account for all potential flux changes occurring in glycolysis during burst work.

In molluscs, although the regulatory properties of glycogen phosphorylase have been investigated (Ebberink, 1980; Ebberink and Salimans, 1982), this kind of critical information is thus far lacking; however, the cephalopod phosphorylase system is under close investigation at this time (P. Cohen, personal communication), so a more precise comparison between the mammalian and the molluscan systems will soon be possible. From already available data, however, it seems safe to assume that a cascade system similar to that in mammals is operative in molluscs (see Volume 2, Chapter 5).

This is not a trivial assumption, for we must emphasize that in its capacity to function over such large absolute ranges of flux, the pathway of glycogenolysis is unique in metabolism; no other pathway ever does, or even potentially can, achieve the same degree of activation. At least at the level of glycogen phosphorylase, achieving such large changes in flux appears to depend on two key characteristics: (a) the occurrence of phosphorylase in two different (low and high activity) forms and (b) the maintenance of glycogen at levels that are always saturating for phosphorylase $a$.

As a result, the flow of carbon through this step in metabolism is directly proportional to the concentration of enzyme in the active state. Because this most essential feature is mechanistically dependent on the aforementioned control cascade, it is evident that the control cascade plays two critical metabolic functions: First, the cascade per se functions to amplify incoming signals greatly, making possible very large glycolytic activations typical of burst working muscle; and second, the "on–off" control mechanisms allow phosphorylase to serve as a flux generator, initiating flux of carbon into the glycolytic pathway. In these processes, the activity of glycogen phosphorylase is integrated with at least one additional control site in the pathway: the phosphofructokinase-catalyzed formation of fructose 1,6-bisphosphate (FBP).

## B. Regulatory Functions of Phosphofructokinase

Phosphofructokinase (PFK) catalyzes the first unidirectional step in glycolysis and as such has long been recognized as an important locus of control (Hue, 1981). As in the case of phosphorylase, muscle PFK occurs as a tissue-specific isozyme form, displaying unique catalytic and regulatory properties that tailor it for burst-type work in muscle. In all mammals and most other animals that have been studied in this regard (Mansour, 1972), PFK control is based on its interactions with both substrates, fructose 6-phosphate (F6P) and ATP: The F6P saturation curve is typically sigmoidal (at physiological pH) with the $S_{(0.5)}$ in the 0.1 m$M$ range, similar to physiological levels of this substrate. The saturation curve for ATP, in sharp contrast, is hyperbolic and through physiological concentrations of ATP (above 1 m$M$), the enzyme displays a kinetic order less than one.

In principle, these two key functional characteristics could go a long way toward appropriate control of PFK: During burst work, for example, F6P levels rise as flux is activated, and increasing F6P availability would serve to increase its rate of utilization by PFK (positive cooperativity) while reversing ATP substrate inhibition. At the same time, falling ATP levels simultaneously lead to reduced substrate (ATP) inhibition and to an increased affinity for F6P, in effect increasing reaction velocity. These are elegant and synergistic regulatory properties that help in integrating PFK function with that of glycogen phosphorylase and indeed with myosin ATPase. But built upon this control foundation are other inputs that add even more versatility to the control of this locus in glycolysis: The most important of these (Hochachka and Somero, 1983) are activation by AMP, ADP, FBP, $P_i$, and $NH_4^+$ as well as inhibition by citrate and fatty acyl-CoA derivatives.

AMP and FBP activations are considered particularly important because they also serve as inhibitors of fructose 1,6-bisphosphatase (FBPase), catalyzing the hydrolysis of FBP back to F6P. In addition, because they are inhibitory to a physiologically reverse reaction while being stimulatory to PFK, their net effects are to amplify flux through the FBP → F6P locus in metabolism, a function stressed particularly strongly by Newsholme (1982).

All of the foregoing mechanisms can be viewed as mainly *internal* to the glycolytic pathway: They integrate PFK activity with internal pathway functions. $NH_4^+$ plays a somewhat different role, however. During activated muscle metabolism, $NH_4^+$ levels rise by over 10-fold, reaching values in the 1 $\mu$mol/g range, which are far higher than necessary to lead to a significant PFK activation (Tornheim and Lowenstein, 1976). $NH_4^+$ is not an intermediate in glycolysis, being formed by the AMP deaminase reaction (itself under very tight regulation in skeletal muscle). Because AMP deaminase may function in the purine nucleotide cycle (Lowenstein, 1972), $NH_4^+$ appears to be serving an interpathway function: It contributes to PFK activation when its own rate of production has been elevated preventing its own further accumulation, as $NH_4^+$ is potentially highly disruptive to cell metabolism.

Because citrate also is not a direct intermediate in the glycolytic pathway, it is not surprising that it also is thought to play an important interpathway role: In this case, it serves to integrate anaerobic glycolysis with the Krebs cycle per se and indirectly with fatty acid oxidation. During Krebs cycle activation, and particularly during activated fatty acid oxidation, the concentration of citrate in muscles rises to about 0.4 $\mu$mol/g. These levels are high enough to inhibit PFK activity significantly at a time when the need for anaerobic glycolysis is reduced (Guppy et al., 1979).

This inhibition does not override all other modulator effects, which is the reason mammalian muscles can to some extent utilize *both* carbohydrate and fat during sustained work (Howald et al., 1978; Holloszy et al., 1978).

## C. Metabolite Control of Molluscan Phosphofructokinase

Although some of the regulatory properties just discussed are also displayed by PFK homologs from molluscan sources, there are also a number of key differences. Some of these will be more fully developed in companion chapters (see Chapter 4, this volume). At this time, let it suffice to mention that in the mantle muscle of squid, PFK control differs from that in mammalian muscle in at least one critical way: an absence of ATP inhibition. Under physiological conditions, its role in negative modulation of PFK is probably replaced by NADH (see Chapter 3, this volume). In bivalve molluscs, PFK from adductor muscle displays a characteristic phosphoenolpyruvate (PEP) sensitivity; because PEP accumulates to quite high levels (1 $\mu$mol/g) after prolonged anoxia, and contributes to redirecting carbon flow toward oxaloacetate (OXA) (via PEP-carboxykinase), this regulatory interaction is readily appreciated (Ebberink, 1980; Ebberink and de Zwaan, 1980).

## D. pH Effects on Molluscan Phosphofructokinases

The catalytic and regulatory properties of PFK from at least one mollusc (*Mytilus*) are also notably sensitive to pH. Among other effects, modestly decreasing pH (from 6.9 to 6.7) leads to a pronounced fall in affinity for F6P and general inhibition of catalytic rate at all substrate levels (Ebberink, 1980). During extended anoxia, *Mytilus* tissues do not express a Pasteur effect; instead the organism appears to lower its energy demands and thus metabolic rate down to about $\frac{1}{20}$ of normoxic levels (see Hochachka, 1980, for further discussion of this strategy). The drop in pH that is known to occur during anaerobiosis may well be a signal mechanism for "turning down" the activity of PFK and thus the gradual if wholesale reduction in overall metabolic rate. The aforementioned inhibitory effect of PEP may act in concert with these actions of pH.

## E. pH Effects on Muscle Phosphofructokinase in Mammals

In addition to the metabolite modulation just discussed, mammalian muscle PFK is also extremely sensitive to other microenvironmental factors, particularly monovalent cations ($K^+$ is stimulatory), divalent cations ($Mg^{2+}$ or $Mn^{2+}$ fulfill an absolute requirement for catalytic function), and pH. Of these, the latter is by far the most important (Hochachka and Mommsen, 1983) because pH change profoundly influences the effectiveness of metabolite modulators. In addition, pH plays a role in integrating PFK function with that of pyruvate kinase (PK) and with lactate dehydrogenase(LDH)-catalyzed formation of lactate. In fishes and other lower vertebrates, decreasing pH automatically activates PK, because in these groups PK typically displays an acidic pH optimum. (Rat muscle PK, for a contrasting example, displays optimal activity at about pH 8.1.) This effective $H^+$ activation is potentiated by FBP (Guppy and Hochachka, 1979; Sutton et al., 1981).

In lower vertebrates (fishes and reptiles) and in molluscs (see Chapter 4, this volume) PK is an allosteric enzyme under close metabolite regulation. FBP, a product of the PFK reaction, serves as a potent feed-forward activator of muscle PK, assuring closely coupled activation of both enzymes. In most mammals, in contrast, the major integration mechanism merely involves adenylate coupling; that is, ADP, the product of the phosphofructokinase reaction, is a substrate for pyruvate kinase, and this in itself serves to automatically coordinate the activities of these two enzymes. With the activation of PK, one final control function (that of redox balance) is necessary for completing the process of anaerobic glycolysis.

## F. Maintenance of Redox Balance during Anaerobic Glycolysis

In muscles of mammals, cephalopods (Hochachka and Fields, 1982), and bivalves (de Zwaan et al., 1980), the concentration of glycogen is high (in the range of 100 $\mu$mol/glucosyl glycogen/g or higher), much if not all of which can be converted to glycolytic end products during anaerobic glycolysis. The pool size of [NADH] + [NAD$^+$], however, is small, about 1 $\mu$mol/g. So for anaerobic glycolysis to be prolonged there is the need for continuously reoxidizing NADH formed at the glyceraldehyde 3-phosphate dehydrogenase (GAPDH) step. Usually, it is explained that this is achieved by a 1:1 functional coupling between the GAPDH-oxidative step and a terminal reductive step, some end product accumulating as the process continues. However, recent studies indicate that the situation is not quite so simple, particularly during early stages of glycolytic activation. This is nicely illustrated in tuna white muscle, which contains one of the highest glycolytic potentials thus far described. During early stages of glycolytic activation, conditions are not suitable for NAD$^+$ regeneration by LDH for two main reasons: (a) The pH is relatively high, making the $K_m$ for pyruvate very high and (b) pyruvate availability is low (levels below 0.1 $\mu$mol/g when the $K_m$ is well over 1 m$M$). In order to keep glycolysis supplied with a continuous source of $NAD^+$, a high amount of cytosolic malate dehydrogenase (MDH) is maintained. Unlike LDH, MDH is kinetically well suited for function during these early stages of glycolytic activation (a very high affinity for both NADH and oxaloacetate). Its key carbon substrate (oxaloacetate) is derived from aspartate via transamination. Only modest amounts of aspartate are stored, so this system cannot sustain glycolytic redox for very long; long enough, however, to build up pyruvate levels and make LDH more competitive. It is presumed that as LDH begins to take on its usually assigned role in redox regulation, the accumulation of $H^+$ concomitant with lactate serves to activate further lactate formation by two mechanisms: a direct activation of LDH catalysis (because LDH in the forward direction displays an acidic pH optimum) and lowered $K_m$ for pyruvate (Guppy and Hochachka, 1978 a,b). Thus the effects of decreases in pH may be utilized not only locally (to facilitate LDH activation) but also to integrate the terminal function of glycolysis with that at other control sites such as the PK reaction.

## G. Redox Balance during Anaerobic Glycolysis in Molluscs

In molluscs, LDH may be present in visceral tissues, particularly in cephalopods (J. H. A. Fields, unpublished data), but it is often deleted from muscle metabolism. In cephalopod muscles, it is replaced by ODH and thus octopine becomes the anaerobic end product. In other molluscs, it may be replaced by alanopine, or strombine dehydrogenase; the latter is notable, for example, in some gastropods (Baldwin and England, 1983), whereas alanopine and strombine dehydrogenases may co-occur in many bivalves (Dando et al., 1981). The way these may contribute to overall redox regulation under a variety of conditions is explored elsewhere in this treatise (Chapters 4 and 5, this volume). Let it suffice to mention at this time that a part of the problem mentioned earlier for tuna white muscle LDH also seems to be displayed by squid and octopus mantle ODH. Like the LDH just discussed, at pH 7–7.5 these ODHs have a very high $K_m$ for pyruvate as well as for arginine. Thus during early stages of glycolysis, they would not be suitably geared for reoxidation of NADH. As in the vertebrate case, the idea has been advanced that MDH, which also occurs at very high activities relative to ODH in these muscles, is primed by aspartate-derived OXA and serves to balance redox. As pyruvate and arginine levels rise, they lead to concomitantly lowered $K_m$ values (each influencing the $K_m$ for cosubstrate), making ODH more competitive and capable of serving in redox regulation (Hochachka, 1980). How and if concomitant pH changes are involved in this process currently remain unanswered questions.

## VI. Fermentation and Acidification

A correlation is often observed between end-product accumulation and acidification, but there is considerable misunderstanding of why this occurs. The confusion probably arises mainly because several factors, in particular $Mg^{2+}$, pH, and substrate source, influence the proton stoichiometry of any fermentative pathway. In fact, because these parameters vary in different tissues, it is not possible to write a single equation for glycolysis in all tissues or even for the same tissue under differing conditions. This can be easily illustrated by considering anoxic muscle (see Hochachka and Mommsen, 1983, on literature in this area). From recent evidence, free $Mg^{2+}$ levels in muscle are known to be about 4.4 m$M$, so that through *in vivo* pH ranges, practically all ATP is in the complexed form; ADP is only partially complexed whereas $P_i$ binds $Mg^{2+}$ so weakly that it can be assumed to be mainly uncomplexed. Because the true adenylate reactants in glycolysis are necessarily determined by $Mg^{2+}$ availability (Table IA,B), and because free $Mg^{2+}$ levels vary among different species and among different tissues, the overall equations for glycolysis must be very cell specific. Because the pK values of Mg•ATP and free ATP are different, the amount of $H^+$ formed per mole of glucose fermented differs, and at any given pH is greater if the ATP is fully complexed than if ATP is uncomplexed (Table IA,B). Whereas a potential role for $Mg^{2+}$ in the control of glycolytic $H^+$ production may be implied, its potential has not been experimentally assessed. The situation is somewhat clearer, however, with respect to pH, which undergoes fairly large changes during aerobic–anaerobic transitions (Sutton et al., 1981).

The effect of pH change on glycolytic $H^+$ production can be analyzed by extending the example of anoxic vertebrate muscle fermenting glucose at varying pH but unchanging $Mg^{2+}$ levels. When this is done, an obvious relationship is uncovered between pH and the amount of $H^+$ formed per mole of glucose fermented to lactate: *the lower the pH, the greater the amount of $H^+$ formed* (Table IA,B). Overall, the pattern is the same if glycogen is the starting fermentable substrate, but because a $H^+$-producing step, that catalyzed by hexokinase, is not utilized, glycogen fermentation to lactate, at pH values only slightly above neutrality, proceeds with the consumption (not the production) of $H^+$ (Table IC).

A close examination of molluscan metabolic pathways leading to alanopine, strombine, or octopine indicates that, in terms of proton production, they are essentially equivalent to mammalian glycolysis (Table 1D). Where bivalve anaerobiosis greatly differs from that in vertebrates is in its capacity for, and utilization of, alternate fermentation pathways. Of these, the glucose fermentations either to succinate or to propionate are

**TABLE I**

**Proton Stoichiometry for Different Fermentation Pathways** [a]

---

(A) Glucose → lactate, assuming no free $Mg^{2+}$:

pH 6.8

Glucose + 1.10 $ADP^{2-}$ + 0.90 $ADP^{3-}$ + 1.00 $P_i^{2-}$ + 1.00 $P_i^-$
→ 2 lactate⁻ + 1.17 $ATP^{3-}$ + 0.83 $ATP^{4-}$ + 0.93 $H^+$

pH 7.4

Glucose + 0.46 $ADP^{2-}$ + 1.54 $ADP^{3-}$ + 0.39 $P_i^-$ + 1.61 $P_i^{2-}$
→ 2 lactate⁻ + 0.52 $ATP^{3-}$ + 1.48 $ATP^{4-}$ + 0.33 $H^+$

pH 8.0

Glucose + 0.14 $ADP^{2-}$ + 1.86 $ADP^{3-}$ + 0.11 $P_i^-$ + 1.89 $P_i^{2-}$
→ 2 lactate⁻ + 0.16 $ATP^{3-}$ + 1.84 $ATP^{4-}$ + 0.09 $H^+$

(B) Glucose → lactate, assuming 4.4 mM free $Mg^{2+}$

pH 6.8

Glucose + [1.36 MgADP⁻ + 0.30 $ADP^{3-}$ + 0.30 $ADP^{2-}$ + 0.04 MgADP
+ 0.86 $P_i^{2-}$ + 0.86 $P_i^-$ + 0.28 $MgP_i$ + 0.25 $Mg^{2+}$]
→ 2 lactate⁻ + [1.91 $MgATP^{2-}$ + 0.04 $ATP^{4-}$ + 0.03 $ATP^{3-}$
+ 0.02 MgATP⁻] + 1.15 $H^+$

pH 7.40

Glucose + [1.56 MgADP⁻ + 0.34 $ADP^{3-}$ + 0.09 $ADP^{2-}$ + 0.01 MgADP
+ 1.26 $P_i^{2-}$ − 0.32 $P_i^-$ + 0.42 $MgP_i$]
→ 2 lactate⁻ [1.94 $MgATP^{2-}$ + 0.04 $ATP^{4-}$ + 0.01 $ATP^{3-}$
+ 0.01 MgATP⁻ + 0.04 $Mg^{2+}$] + 0.40 $H^+$

pH 8.00

Glucose + [1.62 MgADP⁻ + 0.36 $ADP^{3-}$ + 0.02 $ADP^{2-}$
+ 1.43 $P_i^{2-}$ + 0.09 $P_i^-$] + 0.48 $MgP_i$
→ 2 lactate⁻ + [1.96 $MgATP^{2-}$ + 0.04 $ATP^{4-}$ + 0.14 $Mg^{2+}$] + 0.11 $H^+$

(C) Glycogen → lactate, assuming 4.4 mM free $Mg^{2+}$ [b]

**pH 6.80**

**Glycogen + 3 [$a_{6.8}$] → 2 lactate⁻ + 3 [$b_{6.8}$] + 0.72 $H^+$**

pH 7.40

Glycogen +3 [$a_{7.4}$] + 0.40 $H^+$ → 2 lactate⁻ + 3 [$b_{7.4}$]

pH 8.00

Glycogen + 3 [$a_{8.0}$] + 0.84 $H^+$ → 2 lactate⁻ + 3 [$b_{8.0}$]

(D) Alanopine production, assuming 4.4 mM free $Mg^{2+}$

pH 6.80

Glucose + 2 alanine + 2 [$a_{6.8}$] → 2 alanopine⁻ + 2 [$b_{6.8}$] + 1.15 $H^+$

---

(Continued)

TABLE I (*Continued*)

pH 7.40

Glucose + 2 alanine + 2 $[a_{7.4}]$ → 2 alanopine$^-$ + 2 $[b_{7.4}]$ + 0.40 H$^+$

Glycogen + 2 alanine + 3 $[a_{7.4}]$ + 0.40 H$^+$ → 2 alanopine$^-$ + 3 $[b_{7.4}]$

pH 8.00

Glucose + 2 alanine + 2 $[a_{8.0}]$ → 2 alanopine$^-$ + 2 $[b_{8.0}]$ + 0.11 H$^+$

(E)  Succinate production, assuming 4.4 m$M$ free $Mg^{2+}$

pH 6.80

Glucose + 1.53 $HCO_3^-$ + 0.47 $H_2CO_3$ + 4 $[a_{6.8}]$ → 2 succinate$^{2-}$
$$+ 4 \, [b_{6.8}] + 0.77 \; H^+$$

pH 7.4

Glucose + 1.87 $HCO_3^-$ + 0.13 $H_2CO_3$ + 4 $[a_{7.4}]$ + 1.07 H$^+$ →
$$2 \; succinate^{2-} + 4 \, [b_{7.4}]$$

Glycogen + 1.87 $HCO_3^-$ + 0.13 $H_2CO_3$ + 5 $[a_{7.4}]$ + 1.87 H$^+$ →
$$2 \; succinate^{2-} + 5 \, [b_{7.4}]$$

pH 8.0

Glucose + 1.97 $HCO_3^-$ + 0.03 $H_2CO_3$ + 4 $[a_{8.0}]$ + 1.75 H$^+$ →
$$2 \; succinate^{2-} + 4 \, [b_{8.0}]$$

(F)  Propionate production, assuming 4.4 m$M$ free $Mg^{2+}$

pH 6.80

Glucose + 6 $[a_{6.8}]$ + 0.55 H$^+$ → 2 propionate$^-$ + 6 $[b_{6.8}]$

pH 7.40

Glucose + 6 $[a_{7.4}]$ + 2.80 H$^+$ → 2 propionate$^-$ + 6 $[b_{7.4}]$

Glycogen + 7 $[a_{7.4}]$ + 3.60 H$^+$ → 2 propionate$^-$ + 7 $[b_{7.4}]$

pH 8.00

Glucose + 6 $[a_{8.0}]$ + 3.67 H$^+$ → 2 propionate$^-$ + 6 $[b_{8.0}]$

(G)  Ethanol production with and without $CO_2$ hydration

pH 7.4

Glucose   + 2 $[a_{7.4}]$ → 2 ethanol + $[b_{7.4}]$ + 1.87 $HCO_3^-$
+ 0.13 $H_2CO_3$ + 0.27 H$^+$

Glycogen + 3 $[a_{7.4}]$ + 0.53 H$^+$ → 2 ethanol + 3 $[b_{7.4}]$
+ 1.87 $HCO_3^-$ + 0.13 $H_2CO_3$

Glucose   + 2 $[a_{7.4}]$ + 1.60 H$^+$ → 2 ethanol + 2 $[b_{7.4}]$
+ 2 $CO_2$ (removed)

[a] For all calculations in Table II and in subsequent tables, the dissociation constants for all relevant species are taken from (5) given in Hochachka and Mommsen (1982).

[b] For all equations in the remainder of the table, [a] = sum of all species of ADP and $P_i$ at respective pH to unity; [b] = all species of ATP at respective pH to unity.

of particular importance (Table IE,F), so it is interesting to note that they too can proceed with net $H^+$ consumption. For propionate production, for example, the fermentation reaction proceeds with the consumption, at near neutral pH, of about 1 mol $H^+$/mol of glucose (Table IF). At higher pH values, which may be expected in ectotherms at low temperatures, the molar yield of $H^+$ is even higher. Glycogen fermentation to propionate also consumes $H^+$ but in substantially higher amounts than when glucose is the starting substrate. Only when the pH falls below about pH 6.5 does this pathway yield a net production of $H^+$ (Table IF).

These considerations make it clear that fermentation reaction pathways per se do not necessarily lead to net proton accumulation; indeed, sometimes they actually consume protons. Why then is the association between end-product accumulation and acidosis commonly noted? If the observed $H^+$ ions are not formed primarily in fermentation reactions, from where do they arise? The answer is from the coupled hydrolysis of ATP.

## A. ATP Hydrolysis and $H^+$ Production

During anaerobiosis, the fate of fermentatively generated ATP is hydrolysis for the support of various cell work functions. The hydrolysis of ATP of course is not determined by whether the ATP is derived from aerobic or anaerobic metabolism; in both cases it proceeds with the release of ADP, $P_i$, and $H^+$. The two main intracellular factors determining the amount of $H^+$ released during ATP hydrolysis again are $Mg^{2+}$ and pH. For muscle (at constant 4.4 m$M$ free $Mg^{2+}$ levels), the equations (Table II) show an interesting dependence on pH: over the pH 6.5–8.0 range, *the lower the pH, the fewer the moles of $H^+$ formed per mole of ATP hydrolyzed.* The effect of pH on $H^+$ production during ATP hydrolysis is almost exactly opposite to that observed for the glycolytic production of $H^+$,

**TABLE II**

**Stoichiometry of ATP Hydrolysis under Different pH Conditions**[a]

pH 6.80
$$[b_{6.8}] \rightarrow [a_{6.8}] + 0.425\ H^+$$

pH 7.40
$$[b_{7.4}] \rightarrow [a_{7.4}] + 0.80\ H^+$$

pH 8.00
$$[b_{8.0}] \rightarrow [a_{8.0}] + 0.945\ H^+$$

[a] For explanation of symbols, see footnote to Table IC.

which is important because *in vivo* the two processes (glycolytic generation of ATP and ATPase-catalyzed hydrolysis of ATP) are, to a lesser or greater extent, coupled.

## B. Stoichiometry of Net Proton Production during Anaerobiosis

It is widely appreciated that the coupling of ATP-forming and ATP-utilizing reactions is a centerpiece of aerobic and of anaerobic metabolism. The adenylates are the basis for the coupling because ATP is one of the products of glycolysis or of oxidative phosphorylation, while it is at the same time the key substrate for cell ATPases. However, this coupling in anaerobiosis is different from that observed during aerobic metabolism because in the latter, all the products of ATP hydrolysis (ADP, $P_i$, and $H^+$) are reutilized during oxidative phosphorylation. In the former, by contrast, although ADP and $P_i$ are stoichiometrically reutilized during glycolytic replenishment of ATP, $H^+$ ions are not reutilized if glucose is the fermentable substrate, and are only partially reutilized if glycogen is being fermented above pH 7.2 (see Table IB,C) or if the succinate–propionate pathways are used (Table IE,F). Because of the opposite pH dependencies of $H^+$ production by glycolysis and by ATP hydrolysis, however, the

TABLE III

Stoichiometry of Proton Production in Tightly Coupled Fermentations and ATPases[a]

| Fermentation process | pH | $H^+$ formed in fermentation: moles $H^+$/moles glucosyl unit | Moles $H^+$ obtained by hydrolysis of ATP formed in fermentation[b] | Sum: moles $H^+$/mole glucosyl unit plus moles $H^+$ from hydrolysis of ATP formed in fermentation |
|---|---|---|---|---|
| Mammalian tissues | | | | |
| Glucose → lactate | 6.8 | 1.15 | 0.85 (2) | 2.00 |
| | 7.4 | 0.40 | 1.60 (2) | 2.00 |
| | 8.0 | 0.11 | 1.89 (2) | 2.00 |
| Glycogen → lactate | 6.8 | 0.72 | 1.28 (3) | 2.00 |
| | 7.4 | −0.40 | 2.40 (3) | 2.00 |
| | 8.0 | −0.84 | 2.84 (3) | 2.00 |
| Molluscan tissues | | | | |
| Glucose → alanopine | 7.4 | 0.40 | 1.60 (2) | 2.00 |
| Glucose → propionate | 7.4 | −2.80 | 4.80 (6) | 2.00 |
| Glucose → succinate | 7.4 | −1.07 | 3.20 (4) | 2.13 |

[a] From Hochachka and Mommsen (1982), with modification.

[b] Numbers in parentheses refer to assumed moles of ATP formed per mole unit glucosyl during fermentation process.

total number of moles of $H^+$ generated (moles of $H^+$ per mole of glucose or glucosyl units fermented *plus* moles of $H^+$ released during ATPase-catalyzed hydrolysis of the ATP formed during the fermentation) is always the same: two. *For the overall system (glycolysis plus ATPase), the process can be described by one equation with the stoichiometry (of proton production) being independent of pH and $Mg^{2+}$ levels* (Table III).

An unexpected outcome of analyzing molluscan fermentation pathways is that *exactly the same stoichiometry of net $H^+$ production prevails*. That is, the fermentation of either glucose or glycogen (a) to alanopine, strombine, or octopine or (b) to succinate or propionate, proceeds with a variable utilization or production of protons associated with ATP replenishment (Table I), but when this process is coupled to the hydrolysis of the ATP formed in the fermentation (Table II), one equation always satisfactorily describes the net stoichiometry of proton production (Table III), and in all cases net proton production is independent of pH:

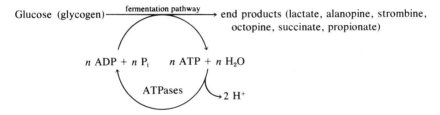

That is why, in mammals and molluscs, sustained anaerobiosis results in a net accumulation of $H^+$ ions, and why end-product accumulation and acidosis are usually correlated.

## C. Relationship between Intracellular Buffers and Anaerobic Glycolysis

To this point in our analysis we have argued that mammalian muscle enzyme machinery is adapted, in effect, to take advantage of pH shifts during activated anaerobic glycolysis. But there is another side to the coin. Although pH is allowed to drop during anaerobic metabolism supporting burst work, it clearly cannot be left totally unbridled; it too must be regulated. Muscles of vertebrates invariably rely to a large extent on the imidazole group on histidine for intracellular buffering. Anserine and carnosine are typically utilized in mammals; free histidine is often utilized in teleosts. Whichever form of buffer is utilized, its buffering capacity is closely matched with the muscle's capacity to generate lactate and $H^+$; that is why when muscle glycolytic machinery is tuned up, so also is its buffering system (Davey, 1960; Abe, 1981; Castellini, 1981).

In molluscs, on the other hand, the situation is less clear. Provocative data are available to be sure, implying important roles for pH in the regulation of both PFK and PK at least in bivalves (Holwerda and de Zwaan, 1972; de Zwaan et al., 1976). However, what remains missing in all molluscan studies to date is an assessment of what intracellular buffering components are present and an assessment of their properties. Indeed, in view of the release of arginine during earlier stages of anaerobic metabolism, there is some question as to direction of pH change at this time. It is possible that when arginine levels become high, intracellular pH is elevated; but in prolonged anoxia, this effect is counteracted by "dumping" this basic amino acid into octopine and by the general accumulation of other metabolic acids (e.g., succinate). That is, in short-term anaerobic metabolism (e.g., during anaerobic work) pH may temporarily rise, whereas in the long haul (prolonged anoxia) pH is known to fall quite drastically (Wijsman, 1975). Clearly this area represents a promising field of discovery for the future.

## D. Adaptative Strategies for Anaerobic Work: An Overview

From these considerations it will be evident that molluscan and mammalian organisms can call on a combination of three basic mechanisms during adaptations for sustaining anaerobic work: (a) adaptation of enzyme machinery, (b) adaptation of the enzyme milieu, and (c) adaptation of metabolic output. In vertebrates, enzyme machinery is adjusted both in amount and kind, with white muscle fibers containing the highest activities of glycolytic enzymes found in the organism. In addition, essentially every step in the pathway is catalyzed by tissue-specific isozymes that are catalytically suited for function during burst work.

The cellular milieu in which muscle glycolytic enzymes operate is adapted in at least three ways. First, large quantities of glycogen are stored, usually as $\beta$ granules, upon which glycogen phosphorylase and most of the enzymes of its control cascade are "plated out" (Cohen, 1978). Second, the elevated glycolytic potential of white muscle is matched by a similar elevation in imidazole-based buffering power. And third, modest changes in pH per se as well as in the ratio of $H^+$ to $OH^-$ necessarily develop during burst work, and even these may be harnessed to facilitate function of key enzymes in glycolysis (PK and LDH, for example).

A conceptually different mechanism, involving adaptation of metabolic output of the system, must be viewed as more than just the sum of enzyme adjustments plus adjustments in cellular milieu. This is because many of the control features of anaerobic metabolism in muscle are not formally a part of the pathway. The most critical of these is the control cascade initiated by hormone activation of adenylate cyclase, but others include the calmodulin-based $Ca^{2+}$ activation, as well as bypass steps (such as FBPase), and facilitatory functions (such as those of MDH in early redox regulation). Their summed effects, however, raise the potential for a controlled 1000- to 2000-fold activation of anaerobic glycolysis during burst work in humans (Newsholme, 1982). This unique feature, in no way representative of metabolic control and metabolic pathways in general, has developed in vertebrate muscles as a backup to CrP priming of ATP supplies for burst work.

Insofar as molluscan systems have been analyzed, many of the same considerations hold, although they are modified in a number of critical ways. The main gaps in our knowledge here relate to (a) the nature of amplification mechanisms and a quantification of their need during glycolytic activation, (b) the role of pH in glycolytic control, and (c) the role and nature of intracellular buffers. Nevertheless, enough information is already available to justify the parallel that we have drawn between the anaerobic machinery and its adaptation in the two groups. In both, because the potential metabolic output and its control vary between muscles in a way that correlates with their anaerobic work capacities, we can confidently conclude that these characteristics are products of selection and adaptation. In bivalves, these mechanisms may be able to sustain reasonable work rates for hours to days. But even at their best in more active organisms, anaerobic mechanisms can sustain high work rates for only a short time: 3–5 min in humans (Howald et al., 1978), 5–10 min in the tuna (Guppy et al., 1979), and maybe even less in squid (Grieshaber and Gäde, 1976) or scallops (de Zwaan et al., 1980), to mention four examples. For exercise that must be sustained for longer periods, these organisms must harness oxidative energy metabolism, and the question now arises of how aerobic metabolism is adapted for this purpose.

## VII. Elevated Oxidative Capacity in Red and Intermediate Muscles

Just as in the case of glycolysis, where anaerobic power output depends on the catalytic potential of glycolytic enzymes, so in the case of aerobic sustained performance the maximum power output depends on the catalytic capacity of oxidative enzymes, largely in red and intermediate fiber types. Across essentially all vertebrate phylogenetic groups, red muscle typically contains substantially higher levels of oxidative enzymes than does white muscle. This difference between red and white muscle of course goes hand in hand with differences in mitochondrial abundance and ultrastructure. Insofar as our data go, the same is true at least for cephalopods (Mommsen et al., 1981; Bone, et al., 1981). The biochemical adaptation strategy, therefore, is one we have seen before, merely maintaining higher amounts of what otherwise might be termed standard metabolic machinery. An important exception to this involves creatine phosphate, but it is not known how the role of arginine phosphate is modified in red muscle analogs of active molluscs such as squid.

## A. The Role of Creatine Phosphate in Oxidative Muscle Metabolism

Because the storage function of creatine phosphate is obviously unnecessary in aerobic work, one might well wonder why creatine phosphate and enzymes for its metabolism are retained at all in red muscle. One possible answer is that creatine phosphate is put to work in a different role: Thus, during sustained aerobic work, creatine phosphate is thought to play a role in shuttling high-energy phosphate between sites of ATP formation and sites of ATP utilization. This idea arises from the findings that both substrates (creatine phosphate and creatine) and creatine phosphokinase (CPK) isozymes *are to some extent partitioned in muscle cells* (Bessman and Geiger, 1981). Experimental tracing of the pathway of the γ-phosphate of ATP shows that *oxidative phosphorylation can supply ATP to CPK-II without first mixing with the extramitochondrial pool of ATP*. This result is in fact predictable if a shuttling mechanism is assumed, but otherwise it is not easily explained.

To what extent arginine phosphate may play analogous roles in oxidative fibers of molluscs is not known. However, in mammals it is clear that whereas creatine phosphate may be validly considered an energy source in white muscle, in aerobically working red muscle it is really not a fuel at all. It only indirectly influences the *rate* of energy generation but makes essentially insignificant contributions to the *total* amount of energy that can be generated. The latter is determined by the true carbon fuels utilized by oxidative metabolism.

## B. Determinants of Total Amount of Energy Produced in Aerobic Work

A critical difference between anaerobic and aerobic performance arises when we consider the total amount of power that can be generated. In anaerobic metabolism, because it is in effect a closed system, the total amount of power that can be released is determined by the endogenous stores of substrate (mainly glycogen). Aerobically working muscles, in contrast, remain open systems: They are continuously perfused with blood and are thus in communication with central depot supplies of carbon and energy. The total amount of work they can generate is influenced in part by endogenous substrate supply and in part by substrate supply in central depots. The two main endogenous substrates available for red muscle metabolism in vertebrates are glycogen and fat; both occur in red and intermediate-type muscles, but fat is especially abundant. In addition, however, the total amount of power that can be generated is also influenced by the availability of substrates in other depots, fat in adipose tissue and glycogen in liver. As already pointed out, a reciprocal relationship between glucose and fat metabolism is well known in mammals, and it is widely accepted that when fat metabolism is activated, glucose oxidation is reduced. Although the mechanism for this is being explored in mammalian systems, in cephalopods, fat is not a quantitatively important substrate, so we shall not develop current concepts involved in its mobilization and control.

In active molluscs, mainly cephalopods, aerobically working muscles also must remain in communication with the rest of the body, hence total amount of power that can be generated must also be determined by the *sum* of endogenous glycogen and a large free amino acid pool plus exogenous substrates deliverable from other parts of the body.

The interaction between glucose and amino acid metabolism in molluscs is poorly understood (see Chapter 3, this volume). However, at least under one set of conditions spawning migration of squid, it is clear that the bulk energy needs of the animal must be met by protein and hence amino acid catabolism. As in analogous vertebrate situations (e.g., the spawning migration of salmon), there simply is not enough storage glycogen "on board" to supply the energy needs of the migration, which, in both instances, proceeds with minimal feeding. In salmon, the nature of protein mobilization during spawning migration at least at one level of organization is now known (Mommsen et al., 1980), but to our knowledge, the same information is not available for migrating squid. Interestingly, it has been proposed that even in octopus, protein and amino acids are a major, if not the sole carbon and energy source (R. K. O'Dor, personal communication).

In mammals, particularly humans, amino acids contribute to exercise metabolism in only a minor way, at least under most experimental conditions thus far studied. When and where they are utilized, the catabolism of most amino acids feeds substrate into the Krebs cycle. That of course is the final pathway for the catabolism of carbohydrates and fatty acids as well, so it is imperative to consider how this process is regulated.

## C. Krebs Cycle Control Options

We have pointed out that in tuning up white muscle metabolism for support of burst work, vertebrates adjust both the amount and the isozyme type of enzyme present in muscle tissue. Tissue-specific isozyme arrays then are an important part of the adaptational arsenal that makes muscle glycolysis exceptionally suitable for supporting burst-type work. In principle, the same mechanism could underlie Krebs cycle adaptation for muscle work, so it is instructive that this qualitative strategy is rarely if ever utilized. Although as we have seen, the levels of Krebs cycle enzymes in oxidative muscles in mammals and molluscs are high, sometimes an order of magnitude higher than in anaerobic muscle (Gupply et al., 1979), these typically do not occur in tissue-specific isozymic form (see, e.g., Matsuoka and Srere, 1973). That is, the catalytic and regulatory properties of Krebs cycle enzymes in oxidative muscles are the same as they are in any other tissue, which means it is the unique compositional and content changes of muscle metabolites during transition from rest to exercise that must be at the heart of Krebs cycle regulation in muscle. This most fundamental assertion seems well confirmed for mammalian systems but is in need of further proof in molluscan ones. Currently available information on Krebs cycle enzymes such as citrate synthase, (CS), however, are certainly consistent with only one enzyme form (Fields and Hochachka, 1976, and unpublished data). If so, then in molluscs too, the transition from low to high Krebs cycle activity must be determined by tissue-specific, uniquely mediated, metabolite control mechanisms.

## D. Control of the Krebs Cycle

Singly, the most important feature of the Krebs cycle is that it is *cyclic and catalytic,* with no net accumulation or depletion of intermediates. [It may, however, take on different functions during anaerobiosis in molluscs (see Chapter 4, this volume).] If for a moment one views the Krebs cycle as a superenzyme, granted an oversimplification, it becomes evident that the simplest way of increasing the flux of carbon through it is merely to increase its spinning rate without any concomitant changes in cycle intermediates. This kind of control has two fundamental prerequisites. The first is that at least one enzyme in the process occurs in "on" and "off" (low versus high activity) states. Within the cycle, three such enzymes are well known: citrate synthase and NAD-linked isocitrate dehydrogenase in the first span of the cycle; 2-ketoglutarate dehydrogenase in the second span. Citrate synthase is under close regulation by the adenylates and CoASH. Isocitrate dehydrogenase (IDH) is regulated by the adenylates, ADP being an absolute allosteric requirement for catalysis; 2-ketoglutarate dehydrogenase (KGDH) is product inhibited by NADH (Atkinson, 1977). If substrate supply for any one of these enzymes were saturating under resting conditions, flux through the Krebs cycle could be activated by simply increasing spinning rate (i.e., increasing the amounts of catalytically active enzymes at these key control sites). This condition, that these enzymes not be limited by substrate supply, in fact is the second prerequisite for this simplest kind of control mechanism, and unfortunately it cannot be met in muscle at rest. From the best evidence available in both mammals and molluscs, at least one of these enzymes (CS) is undoubtedly limited by vanishingly low concentrations of oxaloacetate; moreover, it is likely that IDH and KGDH are also limited by isocitrate and 2-ketoglutarate availability. In order to increase the spinning rate of the Krebs cycle, it is thus also necessary to augment the pool size of the cycle intermediates.

## E. Augmentation of Krebs Cycle Intermediates

There are a variety of pathways by which intermediates can either be bled from, or fed into, the Krebs cycle. In the powerfully energetic flight muscles of hymenoptera, pyruvate carboxylase channels pyruvate from mainline aerobic glycolysis into the mitochondrial oxaloacetate pool from which carbon is "spread" throughout the cycle intermediates. In the flight muscle of the blowfly and the tsetse fly, and probably in the mantle muscle of squid, proline serves as both a substrate for complete oxidation and as a means for augmenting the Krebs cycle (Mommsen et al., 1983; Fields and Hochachka, 1983). In mammalian muscle, aspartate is mobilized through the coupled operation of two transaminases—aspartate aminotransferase (EC 2.6.1.1) and alanine aminotransferase (EC 2.6.1.2) —forming oxaloacetate for the Krebs cycle, with alanine accumulating as a result (see Evered, 1981). Skeletal muscle, in addition to this mechanism, contains significant levels of PEP-carboxykinase, which is capable of channeling carbon from mainline glycolysis for purposes of Krebs cycle augmentation during activated fat or amino acid catabolism. Indeed, when glycogen is fully depleted from mammalian muscle, the rate of ATP formation from fat catabolism alone drops by about half; in part, this could be the result of a gradual loss of Krebs cycle intermediates (because of a variety of side reactions,) and hence a fall in its rate of spinning.

## F. Adaptive Strategies for Sustained Aerobic Work: An Overview

To return to our theme, it will now be clear that the same three basic adaptational strategies just introduced (adaptation of enzyme machinery, enzyme milieu, and metabolic output) are utilized in adjusting mammalian and molluscan muscle for sustained performance. The main adjustments at the enzyme level appear to involve simple increases in the levels of enzyme activities per gram of red-type muscle. Thus the potential aerobic power output is elevated in oxidative-type muscles, but unlike the situation in glycolysis, where essentially every enzyme step in the pathway is catalyzed by a muscle-specific isozyme or isozyme array, in mammals at least the enzymes of the Krebs cycle do not occur in isozymic form. Thus transition from rest to maximum work rates requires adjustments at other levels, namely, in cellular milieu and in the way metabolic output is regulated.

In mammals, the milieu in which muscle enzymes operate during aerobic exercise is adjusted in at least two ways. First, large quantities of intracellular triglyceride are retained in red muscle. Second, creatine phosphate, although at lower levels than in white muscle, plays a rather critical role as a shuttle between sites of ATP formation and utilization. There is not enough information on molluscs to know whether similar milieu adjustments occur. However, in both groups a third and final level of adaptation of muscle metabolism for aerobic exercise occurs in the way in which fuel utilization is controlled. During early stages of highest intensity aerobic work, glycogen and fat are mobilized simultaneously in order to obtain highest rates of energy generation (Holloszy et al., 1978). As endogenous glycogen stores are depleted, the importance of fat as a fuel source rises, until in long-duration work it is the sole fuel being utilized. In cephalopod molluscs, glycogen and amino acids appear to be the key fuel sources utilized. In fact, in the greatly extended energy requirements of spawning migration, protein and amino acids become the dominant fuel. In either event, both processes feed into the hub of metabolism, the Krebs cycle, whose activity must pace either that of $\beta$ oxidation (in mammals) or of amino acid mobilization (in cephalopods). This appears to be achieved by mechanisms for augmenting the pool of Krebs cycle intermediates, which in concert with modulator effects on key control enzymes lead to increased rates of spinning of the cycle.

Summed together, these adjustments underpin sustained aerobic performance. Although the degree of activation of aerobic power output is less than that for phosphagen-based or glycolysis-based performance, it is nevertheless impressive. 5- to 20-fold in most vertebrates (Marsh, 1981). In squid, for the one species studied (*Loligo opalescens*), the situation is complex. Estimates of standard and routine metabolic rates are high, up to six times higher than for salmon of the same size; yet maximum metabolic rates are similar (O'Dor, 1982). This implies that either the scope for aerobic activity in *Loligo* is much less than in salmon—the interpretation assumed by O'Dor (1982)—or that the measurements of standard metabolic rates are not realistic representations of the situation in nature. We envisage the possibility of two artifact sources: First, squid in captivity may not reduce their metabolic rates down to the basal levels they may routinely reach in nature. Second, and probably more significant, the squid utilized in O'Dor's (1982) experiments had completed their spawning migration and were programmed for death. As this condition could affect both standard and maximum metabolic processes, it is imperative that O'Dor's provocative experiments be further extended, hopefully with animals at different stages in their life cycles. If the data are taken at face value, however, they indicate that the scope for activity (maximum—standard metabolic rate) is substantially less than with vertebrates, which may correlate with the lower efficiency of jet propulsion in the squid compared to swimming in fishes.

In summary, both aerobic and anaerobic metabolic systems are thus well suited for their particular purpose, with appropriate stimuli leading to nearly instantaneous adjustments accommodating new energy and locomotor demands. At a general level these adaptations are well illustrated by comparing and contrasting red- and white-type muscles, as we have done already. This kind of comparison fails, however, to yield insight into any finer tuning that may be possible. In vertebrates, two other levels of such tuning of energy metabolism are well known: medium-term training effects and long-term genetic adaptations specializing organisms to different modes of performance. In molluscs, there is no information on the former. On the other hand, enough information is available to indicate some interesting parallels in genetically based exercise adaptations in both mammals and molluscs.

### VIII. Sites of Specialization for Anaerobic Work

The foregoing analysis is useful not only for focusing on metabolism during work transitions, but also for allowing us to pinpoint those sites and functions that in principle might be modifiable during training (as in human athletes) or during long-term adaptation for anaerobic work. A list of such theoretically adjustable features in anaerobic metabolism in muscle would include:

1. Increasing size of muscles utilized for the specific kind of work (i.e., hypertrophy)
2. Fiber-type replacement
3. Increasing levels of endogenous substrates (phosphagen and glycogen)
4. Increasing levels of key enzymes in anaerobic metabolism and its control
5. Adjustments in isozyme type at specific loci in metabolism
6. Improving buffering capacity of working muscles

The contributions of these various factors to training in humans and other mammals have been reviewed elsewhere (Holloszy et al., 1978). On a phylogenetic scale, it turns out that all these factors may also be utilized. Among birds, for example, long-distance flyers (such as waterfowl), display predominantly red fibers in their flight muscles, whereas shortdistance flyers (such as grouse) typically retain mainly white fibers in their breast muscles; the former store intracellular fat, the latter, glycogen, with the expected reorganization of enzymatic machinery (Marsh, 1981). Similarly, among fishes, it is a standard (and old) observation that species specialized for short, intense bouts of burst swimming have a high proportion of white muscle, but relatively reduced amounts of red. Those that are sluggish, but still rely on this strategy of locomotion, have low levels of glycolytic enzymes; in contrast, tuna white muscle sustains over 1000 units of PK and over 5000 units of LDH, for sustaining very high velocity bursts of swimming (Guppy et al., 1979). In addition, the kinetic properties of tuna LDH have been tailored for the intense kind of glycolytic system that has developed. Finally, there also is a correlation between the anaerobic capacities of vertebrate muscles and their intracellular buffering capacities (Abe, 1981).

Similar strategies are evident among molluscs. For example, the adductor muscle of sessile bivalves, although a relatively anaerobic tissue, displays low absolute activities of enzymes in anaerobic metabolism; whereas the adductor muscle in scallops (utilizing a flapping pattern of swimming for predator escape) contains substantially higher levels (de Zwaan et al., 1980). These in turn are surpassed by glycolytic enzymes in the mantle muscles of octopus species and of squid (Fields et al., 1976 a; Ballantyne et al., 1981).

The parallelism between mollusc and mammal is even more evident in squid, where the metabolic organization of mantle muscle is closely analogous to the white versus red differentiation seen in vertebrates (Mommsen et al., 1981). Here there are some notable differences between squid adapted for different life-styles. The most vigorous swimmers studied (*Ommastrephes, Symplectoteuthis,* and *Illex*) all display expanded bands of oxidative fibers on either inner or outer sides of the mantle sandwich; until the biomechanical significance of this organization is clarified, it is not entirely clear what biochemical consequences, if any, arise from unequal thickening of the outer or inner oxidative bands. However, in metabolic terms, the correlation of expanded oxidative-type muscle mass with sustained swimming abilities is clear. In contrast, *Berryteuthis* (a slow-moving benthic squid) and *Octopoteuthis* (a watery, very sluggish midwater species) both display very narrow bands of oxidative muscle, which would be easily overlooked if one did not know how to look for them (Mommsen et al., 1981; Bone et al., 1981). Associated with these adjustments in fiber types are adjustments both in enzymatic composition and in the levels of key substrate sources; for example, arginine phosphate and proline both occur at substantially higher levels in the midmantle of *Illex* than in its outer oxidative bands of muscle.

## IX. Sites of Specialization for Aerobic Work

In considering sites and functions in aerobic work that might theoretically be adjustable during training (in mammals) or during genetic adaptation for aerobic work, many of the same considerations made already are fully applicable and have been discussed elsewhere (Hochachka and Somero, 1983). In addition, because aerobically working muscles remain open systems, important adaptations may occur at higher levels of organization as well, particularly in cooperative metabolic interactions between visceral tissues and muscle. In mammals, these adaptations may involve

1. Adjustments in glycogen and fat storage depots
2. Adjustments in the capacity of liver and adipose tissue to release glucose and fatty acids, respectively, during aerobic work
3. Adjustments in the capacity of muscle to take up blood glucose and blood free fatty acids, either for immediate catabolism or for deposition as endogenous storage form (as muscle glycogen or muscle triglyceride)
4. Adjustments in the capacity to deliver $O_2$ to working muscle and in its capacity to take up $O_2$ from blood

In the most aerobically active of molluscs, the cephalopods, parallel adaptations may be involved in

1. Storage and cycling of amino acids (particularly of proline, alanine, and arginine) and of octopine

2. Storage and cycling of carbohydrate
3. Adjustments in the capacity of red- versus white-type muscle to take up glucose or amino acids, either for utilization or storage
4. As in mammals, adjustments in the capacity to deliver $O_2$ to working muscle and its capacity to extract $O_2$ from blood

However, to date, the metabolic picture emerging in cephalopods still retains large, unclarified areas, and this surely represents an area in need of further research and development.

## References

Abe, H. (1981). Determination of L-histidine-related compounds in fish muscle using high-performance liquid chromatography. *Nippon Suisan Gakkaishi* **47**, 139.

Atkinson, D. E. (1977). "Cellular Energy Metabolism and Its Regulation." Academic Press, New York.

Baldwin, J. (1983). Correlations between enzyme profiles in cephalopod muscle and swimming behaviour. *Pac. Sci.* **36**, 349–356.

Baldwin, J., and England, W. R. (1983). Gastropod octopine dehydrogenase: distribution, properties and function of the pedal rectractor muscle enzymes. *Pac. Sci.* **36**, 381–394.

Ballantyne, J. S., Hochachka, P. W., and Mommsen, T. P. (1981). Studies on the metabolism of the migratory squid, *Loligo opalescens:* enzymes of tissues and heart mitochondria. *Mar. Biol. Lett.* **2**, 75–85.

Bessman, S. P., and Geiger, P. J. (1981). Transport of energy in muscle: the phosphoryl-creatine shuttle. *Science* **211**, 448–452.

Bone, Q., Pulsford, A., and Chubb, A. C. (1981). Squid mantle muscle. *J. Mar. Biol. Assoc. U.K.* **61**, 327–342.

Boyes, G., and Johnston, I. A. (1979). Muscle fibre composition of rat *vastus intermedius* following immobilization at different muscle lengths. *Pfluegers Arch.* **381**, 195–200.

Castellini, M. A. (1981). Biochemical adaptations for diving in marine mammals. Ph.D. Thesis, Univ. of California, San Diego.

Cohen, P. (1978). The role of cyclic-AMP-dependent protein kinase in the regulation of glycogen metabolism in mammalian skeletal muscle. *Curr. Top. Cell. Regul.* **14**, 117–196.

Dando, P. R., Storey, K. B., Hochachka, P. W., and Storey, J. M. (1981). Multiple dehydrogenases in marine molluscs: electrophoretic analysis of alanopine dehydrogenase, strombine dehydrogenase, octopine dehydrogenase and lactate dehydrogenase. *Mar. Biol. Lett.* **2**, 249–257.

Davey, C. L. (1960). The significance of carnosine and anserine in striated skeletal muscle. *Arch. Biochem. Biophys.* **89**, 303–308.

Denny, M. (1980). Locomotion: the cost of gastropod crawling. *Science* **208**, 1288–1290.

de Zwaan, A., Kluytmans, J. H. F. M., and Zandee, D. I. (1976). Facultative anaerobiosis in molluscs. *Biochem. Soc. Symp.* No. 41, 133–168.

de Zwaan, A., Thompson, R. J., and Livingstone, D. R. (1980). Physiological and biochemical aspects of the valve snap and valve closure responses in the giant scallop *Placopecten magellanicus.* II. Biochemistry. *J. Comp. Physiol.* **137**, 105–114.

Driedzic, W. R., McGuire, G., and Hatheway, M. (1981). Metabolic alterations associated with increased energy demand in fish white muscle. *J. Comp. Physiol.* **141**, 425–432.

Ebberink, R. H. M. (1980). Regulation of anaerobic carbohydrate degradation in the sea mussel *Mytilus edulis* L. Ph.D. Thesis, Univ. of Utrecht, Utrecht.

Ebberink, R. H. M., and de Zwaan, A. (1980). Control of glycolysis in the posterior adductor muscle of the sea mussel *Mytilus edulis*. *J. Comp. Physiol.* **137**, 165–171.

Ebberink, R. H. M., and Salimans, M. (1982). Control of glycogen phosphorylase activity in the posterior adductor muscle of the sea mussel *Mytilus edulis*. *J. Comp. Physiol.* **148**, 27–33.

Evered, D. F. (1981). Advances in amino acid metabolism in mammals. *Biochem. Soc. Trans.* **9**, 159–169.

Fersht, A. (1977). "Enzyme Structure and Mechanism." Freeman, San Francisco, California.

Fields, J. H. A., and Hochachka, P. W. (1976). Oyster adductor citrate synthase: control of carbon entry into the Krebs cycle of a facultative anaerobe. *Can. J. Zool.* **54**, 863–870.

Fields, J. H. A., and Hochachka, P. W. (1983). Glucose and proline metabolism in *Nautilus. Pac. Sci.* **36**.

Fields, J. H. A., Baldwin, J., and Hochachka, P. W. (1976a). On the role of octopine dehydrogenase in cephalopod mantle muscle metabolism. *Can. J. Zool.* **54**, 871–878.

Fields, J. H. A., Guderley, H., Storey, K. B., and Hochachka, P. W. (1976b). Octopus mantle citrate synthase. *Can. J. Zool.* **54**, 886–891.

Florkin, M., and Bricteux-Grégoire, S. (1972). Nitrogen metabolism in molluscs. *In* "Chemical Zoology" (M. Florkin, and B. T. Scheer, eds.) Vol. VII, 301–342. Academic Press, New York.

Gäde, G., Weeda, E., and Gabbott, P. A. (1978). Changes in the level of octopine during the escape responses of the scallop, *Pecten maximus* (L.). *J. Comp. Physiol.* **124**, 121–117.

Grieshaber, M. (1978). Breakdown and formation of high-energy phosphates and octopine in the adductor muscle of the scallop, *Chlamys opercularis* (L.), during escape swimming and recovery. *J. Comp. Physiol.* **126**, 269–276.

Grieshaber, M., and Gäde, G. (1976). The biological role of octopine in the squid, *Loligo vulgaris* (Lamarck). *J. Comp. Physiol.* **108**, 225–232.

Guppy, M., and Hochachka, P. W. (1978a). Role of dehydrogenase competition in metabolis regulation. *J. Biol. Chem.* **253**, 8465–8469.

Guppy, M., and Hochachka, P. W. (1978b). Controlling the highest lactate dehydrogenase activity known in nature. *Am. J. Physiol.* **234**, R136–R140.

Guppy, M., and Hochachka, P. W. (1979). Pyruvate kinase functions in hot and cold organs of tuna. *J. Comp. Physiol.* **129**, 185–191.

Guppy, M., Hulbert, W. C., and Hochachka, P. W. (1979). Metabolic sources of heat and power in tuna muscle. II. Enzyme and metabolite profiles. *J. Exp. Biol.* **82**, 303–320.

Hochachka, P. W. (1980). "Living Without Oxygen: Closed and Open Systems in Hypoxia Tolerance." Harvard Univ. Press, Cambridge, Massachusetts.

Hochachka, P. W., and Fields, J. H. A. (1983). Arginine, glutamate and proline as substrates for oxidation and for glycogenesis in cephalopod tissues. *Pac. Sci.* (in press).

Hochachka, P. W., and Mommsen, T. P. (1983). Protons and anaerobiosis. *Science* (in press).

Hochachka, P. W., and Somero, G. N. (1983). "Biochemical Adaptation." Princeton Univ. Press, Princeton, New Jersey. In press.

Hochachka, P. W., French, C. J., and Meredith, J. (1978). Metabolic and ultrastructural organization in *Nautilus* muscles. *J. Exp. Zool.* **205**, 51–62.

Holloszy, J. O., and Booth, F. W. (1976). Biochemical adaptations to endurance exercise in muscle. *Annu. Rev. Physiol.* **38**, 273–291.

Holloszy, J. O., Winder, W. W., Fitts, R. H., Rennie, M. J., Hickson, R. C., and Conlee, R. K. (1978). Energy production during exercise. In "Biochemistry of Exercise" (F. Landry and W. A. R. Orban, eds.), pp. 61–74. Symposia Specialists, Miami, Florida.

Holwerda, D. A., and de Zwaan, A. (1973). Kinetic characteristics of allosteric pyruvate kinase from muscle tissues of the sea mussel *Mytilus edulis* L. *Biochim. Biophys. Acta* **309**, 296–306.

Howald, H., von Glutz, G., and Billeter, R. (1978). Energy stores and substrates utilization in muscle during exercise. In "Biochemistry of Exercise" (F. Landry and W. A. R. Orban, eds.), pp. 75–86. Symposia Specialists, Miami, Florida.

Hue, L. (1981). The role of futile cycles in the regulation of carbohydrate metabolism in the liver. In *Adv. Enzymol. Relat. Areas Mol. Biol.* **52**, 249–331.

Hulbert, W. C., Guppy, M., Murphy, B., and Hochachka, P. W. (1979). Metabolic sources of heat and power in tuna muscles I. Muscle fine structure. *J. Exp. Biol.* **82**, 289–301.

Johnston, I. A., and Moon, T. W. (1980a). Exercise training in skeletal muscle of brook trout (*Salvelinus fontinalis*). *J. Exp. Biol.* **87**, 177–194.

Johnston, I. A., and Moon, T. W. (1980b). Endurance exercise training in the fast and slow muscles of a teleost fish (*Pollachius virens*). *J. Comp. Physiol.* **135**, 147–156.

Lehninger, A. L. (1975). "Biochemistry," 2nd ed. Worth, New York.

Lowenstein, J. M. (1972). Ammonia production in muscle and other tissues: The purine nucleotide cycle. *Physiol. Rev.* **52**, 382–414.

Mansour, T. E. (1972). Phosphofructokinase *Curr. Top. Cell. Regul.* **5**, 1–39.

Marsh, R. L. (1981). Catabolic enzyme activities in relation to premigratory fattening and muscle hypertrophy in the gray catbird (*Dumetella carolinensis*). *J. Comp. Physiol.* **141**, 417–423.

Matsuoka, Y., and Srere, P. A. (1973). Kinetic studies of citrate synthase from rat kidney and rat brain. *J. Biol. Chem.* **248**, 8022–8030.

Mommsen, T. P., and Hochachka, P. W. (1981). Respiratory and enzymatic properties of squid heart mitochondria. *Eur. J. Biochem.* **120**, 345–350.

Mommsen, T. P., French, C. J., and Hochachka, P. W. (1980). Sites and patterns of protein and amino acid utilization during the spawning migration of salmon. *Can. J. Zool.* **58**, 1785–1799.

Mommsen, T. P., Ballantyne, J., MacDonald, D., Gosline, J., and Hochachka, P. W. (1981). Analogues of red and white muscle in squid mantle. *Proc. Natl. Acad. Sci. U.S.A.* **78**, 3274–3278.

Mommsen, T. P., French, C. J., Emmett, B., and Hochachka, P. W. (1983). The fate of arginine and proline carbon in squid tissues. *Pac. Sci.* **36**.

Moré, P., Moré, M.T., Monnet, R., and Poisbeau, J. (1971). *Compt. Rend. Séanc. Soc. Biol. Fil.* **165**, 1987–1990.

Moré, P., Moré, M. T., Monnet, R., and Poisbeau, J. (1971). Sur l'arginine kinase des protéines solubles du muscle adducteur de l'huître portugèse (*Crassostrea angulata* Lmk.) et de l'huître japonaise (*Crassostrea gigas* Th.)

Moreland, B., and Watts, D. C. (1967). Molecular weight isoenzymes of arginine kinase in the mollusca and their association with muscle function. *Nature (London)* **215**, 1092–1094.

Newsholme, E. A. (1982). Control of fuel supply and fatigue in sprinting and endurance exercise. *Proc. Hypoxia Symp., Banff, Alberta, 1981* (in press).

O'Dor, R. K. (1982). Respiratory metabolism and swimming performance of the squid, *Loligo opalescens. Can. J. Fish. Aquat. Sci.* **39**, 580–587.

Stephens, G. C., van Pilsum, J. F., and Taylor, D. (1965). Phylogeny and the distribution of creatine in invertebrates. *Biol. Bull. (Woods Hole, Mass.)* **129**, 573–581.

Sutton, J. R., Jones, N. L., and Toews, C. J. (1981). The effect of pH on muscle glycolysis during exercise. *Clin. Sci.* **61**, 331–337.

Tornheim, K., and Lowenstein, J. M. (1976). Control of phosphofructokinase from rat skeletal muscle. *J. Biol. Chem.* **251**, 7322–7328.

Vollmer, M., Hochachka, P. W., and Mommsen, T. P. (1981). Octopine dehydrogenase and phosphoarginine kinase in squid mantle: cooperation of two enzymes at the arginine branchpoint in cephalopod muscle. *Can. J. Zool.* **59**, 1447–1453.

Weibel, E. R. (1979). Oxygen demand and the size of respiratory structures in mammals. *In* "Evolution of Respiratory Processes: A Comparative Approach" (S. C. Wood and C. Lenfant, eds.), Lung Biology in Health and Disease, Vol. 13, pp. 289–346. Dekker, New York.

Wijsman, T. C. M. (1975). pH fluctuations in *Mytilus edulis* L. in relation to shell movements under aerobic and anaerobic conditions. *Proc. Eur. Mar. Biol. Symp., 9th,* pp. 139–149.

# 3

# Carbohydrate Metabolism in Cephalopod Molluscs

### KENNETH B. STOREY
### JANET M. STOREY

Institute of Biochemistry
Carleton University
Ottawa, Ontario K1S 5B6, Canada

Our knowledge of carbohydrate metabolism in cephalopod molluscs is largely confined to studies involving muscle work in these fast-swimming invertebrates. This chapter then, will concentrate mainly on metabolism and metabolic regulation in cephalopod muscle while integrating the available information on carbohydrate metabolism in other tissues.

THE MOLLUSCA, VOL. 1
Metabolic Biochemistry
and Molecular Biomechanics

## I. An Overview of the Design of Cephalopod Mantle Muscle

Cephalopod molluscs have in common a jet propulsion mode of swimming powered by cyclic inhalant and exhalant movements of the mantle muscle (funnel muscle in *Nautilus*). Life-styles within the group, however, are quite diverse. The fast-swimming life-style of the pelagic squid make them probably the fastest aquatic invertebrates; speeds of up to 200 cm/sec have been recorded for *Loligo* (Packard, 1969). At the other end of the scale are the more sluggish groups such as octopus and *Nautilus*, which are rather slow moving, swimming only infrequently. This diversity of life-styles is reflected in the structural and biochemical makeup of the mantle musculature.

### A. Structural Design of Mantle Muscle

A more detailed review of mantle muscle structure can be found elsewhere within these volumes. Briefly, the mantle muscle of cephalopods performs two basic kinds of work: slow, rhythmic contractions that move water in and out of the mantle cavity to provide ventilation of the gills; and rapid, powerful contractions that provide high-speed bursts of jet propulsion. The mantle muscle is composed mainly of circular muscle fibers that are divided at intervals by partitions of radial fibers (Ward and Wainwright, 1972; Moon and Hulbert, 1975; Bone et al., 1981). The circular fibers provide the power stroke for mantle contractions whereas the radial fibers aid expansion of the mantle cavity during inhalation (Bone et al., 1981). Individual muscle fibers show a dense packing of obliquely striated myofilaments and have a central sarcoplasmic core packed with mitochondria (Ward and Wainwright, 1972; Moon and Hulbert, 1975; Bone et al., 1981).

Three zones of circular fibers have been identified within the mantle. The inner and outer zones of the mantle have a rich vascularization, contain fibers with a high content of mitochondria (about 47% of the cross-sectional area in *Alloteuthis subulata*) (Bone et al., 1981), show strong histochemical staining for succinate dehydrogenase (Bone et al., 1981; Mommsen et al., 1981) and cytochrome oxidase (Leray and Ladanyi, 1969), and contain high activities of mitochondrial oxidative enzymes, in particular citrate synthase (CS) (Mommsen et al., 1981). The central zone, by contrast, has a sparser vascular bed, contains fibers of lower mitochondrial content (about 6% of the cross-sectional area in *A. subulata*) (Bone et al., 1981), shows weaker histochemical staining (Bone et al., 1981; Mommsen et al., 1981), and relatively lower activities of oxidative enzymes but higher activities of "anaerobic" enzymes such as phosphoarginine kinase (PAK) (Mommsen et al., 1981). Radial fibers resemble the fibers of the central zone. The properties of the fibers of the central zone versus those of the inner and outer mantle zones bear strong resemblances to those of vertebrate white versus red muscle, respectively, and like the vertebrate muscle types, they are specialized for different types of work. Placement of electromyograph electrodes into the different zones of *Sepia officinalis* muscle demonstrated that only the inner and outer zones were active during respiratory contractions although during jetting, the central zone, which is innervated by the giant axon, is active (Bone et al., 1981). A similar dual fiber type, one of high and one of low mitochondrial content, is present in the funnel in *Nautilus pompilius* (Hochachka et al., 1978).

## B. Biochemical Design of Mantle Muscle

Metabolism in cephalopod muscles can be either aerobic (utilizing the Krebs cycle, oxidative phosphorylation, and $O_2$ as the terminal electron acceptor) or anaerobic (utilizing the reactions of glycolysis alone in the absence of sufficient $O_2$ supply to the muscle). Among cephalopods, both of these modes of metabolic energy production are apparently fueled by carbohydrate catabolism. This contrasts to the general situation among vertebrates, where aerobic muscle work is generally found to be lipid based (Drummond, 1971).

Although the capacity for both aerobic and anaerobic energy production is present in the mantle muscle of all cephalopod species, various species have emphasized one mode or the other depending on their lifestyle. The capacity for sustained aerobic work is most highly developed among a number of the pelagic squid. These species have a very high respiratory rate (Johansen et al., 1978), and a mantle muscle with a high content of mitochondria (Moon and Hulbert, 1975) and high activities of oxidative enzymes (Hochachka et al., 1975a; Mommsen et al., 1981). The muscle is also modified with an increased thickness of the inner mantle zone (Mommsen et al., 1981). Aerobic carbohdyrate catabolism using the $\alpha$-glycerophosphate ($\alpha$-GP) cycle for transport of cytoplasmic reducing equivalents into the mitochondria supports muscle work with a simultaneous oxidation of proline augmenting energy production (Hochachka et al., 1975a) (see Section III). More sluggish types of cephalopods such as *Octopus* lack a capability for sustained swimming but use short, powerful bursts of muscle work to catch prey or evade predators. In these species respiratory rates are lower, mitochondrial content is less, and the activities of oxidative enzymes are low (Johansen et al., 1978; Hochachka et al., 1978; Fields et al., 1976a). Whereas respiratory movements can be supported aerobically, burst muscular work is supported by anaerobic glycolysis coupled to the hydrolysis of arginine phosphate reserves, and culminates in the condensation of the products of these processes, pyruvate and arginine, to form the end product octopine (see Section IV).

## II. Metabolic Fuel Sources

### A. Carbohydrate

Carbohydrate appears to be the major fuel utilized to support muscle work in cephalopods. Blood glucose concentrations range from 0.8 to 2.0 $\mu$mol/ml in *S. officinalis* (Storey et al., 1979); similar values for total blood sugar are reported by Goddard and Martin (1966) for three species. Blood glucose levels in *S. officinalis* appear to be closely regulated and respond to physiological stress. During hypoxia blood glucose levels decline steadily, but upon return to normoxic conditions, blood glucose concentration is rapidly restored (after an initial overshoot) to control levels (Storey et al., 1979). Blood glucose is perhaps a key substrate for tissues such as brain; this is substantiated by activities of hexokinase in brain and heart which are ten to twenty times greater than those of mantle muscle (Ballantyne et al., 1981). However, as a muscle fuel for high-speed swimming, blood glucose may be of little value for two reasons: (a) The central zone of the mantle muscle, which provides the power for burst swimming, is poorly vascularized (Bone et al., 1981) and (b) the activity of hexokinase in mantle muscle is low (Hochachka et al., 1975a). The ratio of hexokinase:phosphorylase:phosphofructokinase (PFK) activity in *Loligo forbesi* mantle muscle is 1:109:98, suggesting that muscle work is more likely to depend on glycogen catabolism than on the oxidation of blood glucose (Zammit and Newsholme, 1976).

Various authors have commented on the low glycogen content of cephalopod muscles. A number of early reports, summarized by Goddard and Martin (1966), listed muscle glycogen contents at between 0.02 and 1.28%, whereas Suryanarayanan and Alexander (1971) found 0.2–0.7 g glycogen/100 g dry weight in *Sepia* and *Loligo*. Hochachka et al. (1975a) detected 0.3 g/100 g dry weight in the pelagic squid, *Symplectoteuthis oualaniensis,* a value that was taken as minimal because the animals struggled violently during capture. In addition, distinct glycogen granules are not apparent in electron micrographs of squid muscle (Moon and Hulbert, 1975).

The glycogen content of mantle muscle from resting *S. officinalis* was considerably higher, however, averaging 24 $\mu$mol (measured as glucose)/g wet weight (equivalent to 0.43 g/100 g wet weight), but glycogen was rapidly depleted to levels of 10–20% of resting content by hypoxic stress or burst swimming (Storey and Storey, 1979a). Thus it is apparent that determinations of muscle glycogen content in cephalopods must be made with great care because of the rapid catabolism of glycogen that occurs if animals struggle during handling.

Little information is available on the enzymes involved in glycogen mobilization in cephalopods, but it is assumed that the control of glycogen phosphorylase activity is similar to that found in vertebrate muscle systems (Cohen, 1980). Cohen and co-workers have purified *Octopus* muscle phosphorylase and found a subunit molecular weight of 97,500, similar to that of rabbit muscle phosphorylase (P. Cohen, personal communication). They demonstrated that *Octopus* phosphorylase *b* can be converted to phosphorylase *a* by rabbit muscle phosphorylase kinase and that phosphorylase kinase activity from *Octopus* extracts can convert rabbit muscle phosphorylase *b* to *a*. *Octopus* phosphorylase *b* was activated only slightly by AMP in comparison to the rabbit muscle enzyme, but was considerably activated by sulfate. These properties suggest that the *Octopus* muscle enzyme more closely resembles the mammalian liver enzyme than it does the muscle form. Calmodulin, the protein that confers $Ca^{2+}$ sensitivity on a number of enzyme systems, has recently been isolated from octopus tissues (Seamon and Moore, 1980; Molla et al., 1981). The protein closely resembles vertebrate calmodulin and, like vertebrate calmodulin, is likely intimately involved in the regulation of glycogen catabolism through its role as an active subunit of phosphorylase kinase (Cohen, 1980).

## B. Lipid

The lipid content of *Sepia* and *Loligo* muscles ranges between 0.56 and 1.78 g/100 g dry weight (Suryanarayanan and Alexander, 1971), whereas Hochachka et al., (1975a) found 1.5% lipid on a wet weight basis in *S. oualaniensis*. Lipid droplets are not visible on electron micrographs of squid muscle (Moon and Hulbert, 1975). These low contents of lipid in muscle, coupled with low tissue activities of lipid-oxidizing enzymes such as 3-hydroxyacyl-CoA dehydrogenase (Mommsen et al., 1981; Ballantyne et al., 1981; Mommsen and Hochachka, 1981) and thiolase (K. B. Storey, unpublished data), suggest that lipid is not an important fuel for metabolism in most cephalopod tissues. In addition, lipid substrates (palmitoyl-CoA, palmitoyl carnitine, acetyl-carnitine) are not oxidized by isolated mitochondria from squid heart (Ballantyne et al., 1981; Mommsen and Hochachka, 1981). However, lipid may be of greater importance in the metabolism of other tissues; hepatopancreas of *S. officinalis* contains 20% lipid (mostly triglycerides) by dry weight (Boucaud-Camou, 1971).

## C. Amino Acids and Protein

Intracellular free amino acid concentrations are high in cephalopods as in most marine invertebrate species. Apart from taurine, the major free amino acids in most species are arginine and proline (Robertson, 1965; Florkin, 1966; Hochachka et al., 1975a, 1978, 1982; Storey and Storey, 1978; O'Dor and Wells, 1978; Suyama and Kobayashi, 1980; Ballantyne et al., 1981). Muscle arginine concentrations can range up to 60 $\mu$mol/g wet weight whereas proline levels exceed 100 $\mu$mol/g wet weight in some species. In muscle, the arginine pool is interconvertible with the pool of muscle phosphagen, arginine phosphate, and with the glycolytic end product, octopine. Arginine is vigorously metabolized as a substrate for the Krebs cycle, the urea cycle, and proline synthesis by many of the soft tissues (particularly gill, kidney, and hepatopancreas) of *S. officinalis* (Hochachka et al., 1982). Little catabolism of arginine takes place in mantle muscle; however, proline is readily oxidized by muscle during aerobic work (Hochachka et al., 1975a, 1982; Storey and Storey, 1978) (see Section III). Mitochondria from squid heart readily oxidize proline at high rates (Ballantyne et al., 1981; Mommsen and Hochachka, 1981).

Although muscle work in cephalopods is carbohydrate based, O'Dor *et al.* (1978) have suggested that the direct mobilization of muscle protein provides metabolic energy during periods of starvation. Because wide variations in body size can be tolerated by aquatic animals, a direct storage of dietary amino acids as muscle protein is possible, avoiding the more inefficient conversion of excess amino acids into lipid or glycogen reserves. Such a direct use of protein as an energy reserve may account for the lack of major lipid or glycogen reserves in cephalopod tissues.

## III. Carbohydrate Catabolism and Aerobic Muscle Work

### A. Design Features: Coupled Carbohydrate and Proline Catabolism

The metabolic pathways of aerobic energy production in squid mantle muscle are shown in Fig. 1. The basic routes of glycogen degrada-

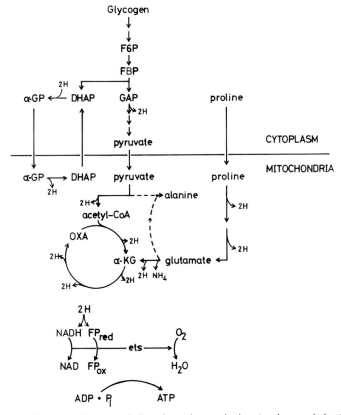

**Fig. 1.** Aerobic metabolism in cephalopod mantle muscle showing the coupled oxidation of glycogen and proline and the use of the α-glycerophosphate cycle for transporting cytoplasmic-reducing equivalents into the mitochondria. Two metabolic fates of glutamate are possible. During the initiation of muscle work, glutamate can be channeled into the glutamate–pyruvate transaminase reaction producing alanine to alleviate a temporary accumulation of pyruvate. Further catabolism of glutamate could utilize the glutamate dehydrogenase reaction. Abbreviations: F6P, fructose 6-phosphate; FBP, fructose 1,6-bisphosphate; GAP, glyceraldehyde 3-phosphate; DHAP, dihydroxyacetone-phosphate; α-GP, α-glycerophosphate; OXA, oxaloacetate; α-KG, α-ketoglutarate; FP, flavoprotein; ets, electron-transport system.

tion, glycolysis, and pyruvate oxidation via the Krebs cycle fit the classical pattern for animal tissues. However, in the design of its aerobic machinery for high power output (high ATP production per unit time) and high energetic efficiency (high ATP yield per mole substrate consumed), squid mantle muscle has incorporated two features that were first described for another highly aerobic invertebrate muscle, the insect flight muscle (Sacktor, 1970, 1976; Hochachka, 1976).

The first of these is the use of the $\alpha$-glycerophosphate cycle for the efficient transfer of cytoplasmic-reducing equivalents into the mitochondria. The combined actions of $\alpha$-glycerophosphate dehydrogenase ($\alpha$-GPDH) in the cytoplasm and $\alpha$-glycerophosphate oxidase in the mitochondrion drain only catalytic amounts of dihydroxyacetone-phosphate from mainline glycolysis during normal aerobic function but can instanteously respond to alterations in glycolytic flux by increasing or decreasing the rate of cycling. A system that uses the $\alpha$-glycerophosphate cycle can, however, be severely strained under anoxic conditions. In the absence of a terminal glycolytic dehydrogenase (e.g., lactate dehydrogenase, LDH), anaerobic glycolysis can remain in redox balance only by the production of equimolar amounts of $\alpha$-glycerophosphate and pyruvate, a relatively hopeless situation in which only 1 mol ATP can be produced per mol glucose oxidized. This situation has been well documented in insect flight muscle, which is incapable of anoxic work for this reason (Sacktor, 1970), and also appears to limit the ability for muscle work under anoxic conditions in the pelagic squid, *S. oualaniensis*. When held out of water, creating an anoxic stress, the mantle muscle of this species rapidly accumulates $\alpha$-glycerophosphate ($>10$ $\mu$mol/g wet weight) and pyruvate (Hochachka et al., 1975a).

The second key feature of the design of squid aerobic metabolism is the coupling of oxidation of the amino acid proline to carbohydrate catabolism. The utilization of proline provides an input of carbon into the Krebs cycle at the level of $\alpha$-ketoglutarate (Fig. 1). In conjunction with an increased flux through glycolysis, this provides an increased availability of $C_4$ (oxaloacetate, OXA) units for condensation with incoming $C_2$ (acetyl-CoA) units from glycolysis to achieve a coordinated activation of glycolysis and the Krebs cycle (Hochachka, 1976). The oxidation of proline also contributes significantly to overall aerobic energy production during metabolic activation: Per mole of proline oxidized to oxaloacetate, 1 mol of reduced flavoprotein and 4 mol NADH (5 mol when $\alpha$-ketoglutarate is produced by the glutamate dehydrogenase reaction) are produced, for a substantial yield of ATP.

## B. Metabolic Activation: The Rest-to-Work Transition in Mantle Muscle

The activation of aerobic energy production during the transition from rest to active swimming has been studied in the mantle muscle of the squid, *Loligo pealeii* (Storey and Storey, 1978) (Table I). Burst-swimming activity (for 10 sec) did not alter the muscle concentrations of four anaerobic products ($\alpha$-glycerophosphate, lactate, octopine, and succinate), sug-

TABLE I

Effect of Exercise on Metabolite Levels in Squid Mantle Muscle[a,b]

| Metabolite | At rest | After exercise |
|---|---|---|
| Glucose 6-phosphate | 0.03 ± 0.01 | 0.08 ± 0.03[c] |
| Fructose 6-phosphate | 0.008 ± 0.001 | 0.01 ± 0.002 |
| Fructose 1,6-bisphosphate | 0.07 ± 0.04 | 0.20 ± 0.05[c] |
| α-Glycerophosphate | 0.15 ± 0.03 | 0.15 ± 0.03 |
| Pyruvate | 0.05 ± 0.01 | 0.19 ± 0.01[c] |
| Lactate | 0.13 ± 0.05 | 0.12 ± 0.06 |
| Octopine | <0.1 | <0.1 |
| Isocitrate | 0.07 ± 0.001 | 0.04 ± 0.02[c] |
| α-Ketoglutarate | 0.06 ± 0.02 | 0.12 ± 0.02[c] |
| Malate | 0.41 ± 0.13 | 0.35 ± 0.10 |
| Succinate | 1.48 ± 0.13 | 1.30 ± 0.72 |
| Alanine | 3.80 ± 1.06 | 6.10 ± 1.25[d] |
| Proline | 3.15 ± 1.00 | 1.59 ± 0.53[c] |
| Arginine phosphate | 10.5 | 1.0 |
| Arginine | 15.2 ± 3.0 | 27.1 ± 5.5[c] |
| ATP | 5.3 ± 0.6 | 2.1 ± 0.6[c] |
| ADP | 0.7 ± 0.2 | 2.2 ± 0.4[c] |
| AMP | 0.3 ± 0.2 | 1.8 ± 0.7[c] |
| Energy charge | 0.90 | 0.52 |

[a] From Storey and Storey (1978).

[b] Metabolites are given as $\mu$mol/g wet weight, means ±SD, $n = 6$. Exercised squid were subjected to 10 sec of vigorous swimming.

[c] Significantly different from the level in the muscle at rest, $p < .01$;

[d] $p < .05$.

gesting that muscle work was supported aerobically. The initiation of muscle work was, however, accompanied by a rise in muscle glucose-6-phosphate levels, indicating an increased flow of carbon into the glycolytic pathway (Storey and Storey, 1978; Rowan and Newsholme, 1979; Edington et al., 1973). Facilitation of the two key regulatory enzymes of glycolysis, phosphofructokinase and pyruvate kinase, was also indicated from the observed changes in the concentrations of substrates and products of these reactions. Similar patterns of changes in the levels of glycolytic intermediates accompany the activation of glycolysis in other muscle systems (Sacktor and Wormser-Shavit; 1966; Williamson, 1966; Edington et al., 1973; Rowan and Newsholme, 1979). Alterations in the levels of various Krebs cycle intermediates similarly suggested an activation of the Krebs cycle, and the presence of a "cross-over" point at the NAD-dependent isocitrate dehydrogenase (IDH) reaction implied that this reaction could be a major control point of the cycle in mantle muscle. Muscle work also resulted in a decrease in muscle proline concentration that was accompanied by an almost equal increase in alanine levels (Table I). In this and other muscle systems, a temporary accumulation of pyruvate occurs at the initiation of muscle work as a result of the faster activation of glycolysis than of the Krebs cycle (Sacktor and Wormser-Shavit, 1966; Safer and Williamson, 1973). During this time, pyruvate is shunted into the production of alanine, the necessary amino groups being derived, via coupled transaminations, from proline in cephalopod and insect systems (Fig. 1) or from aspartate in vertebrate muscles. Once the Krebs cycle has been primed by the influx of $\alpha$-ketoglutarate derived from proline, and cycle activation is completed, alanine accumulation ceases and the full glycolytic output of pyruvate flows into the Krebs cycle. Thus a coordinated activation of aerobic metabolism is achieved (Hochachka, 1976; Sacktor, 1976; Storey and Storey, 1978).

The mobilization of arginine phosphate reserves also contributes to muscle energy output during the transition from resting to active muscle work in cephalopods. Muscle phosphagen typically acts as an energy reserve from which ATP pools can be buffered and replenished. Muscle arginine phosphate was rapidly depleted in working *Loligo* muscle, but despite the mobilization of the phosphagen, ATP was also considerably depleted (Table I). Most muscles accomplish the transition from resting to active states with little alteration in ATP concentrations (Sacktor and Hurlbert, 1966; Neely et al., 1973), but a number of studies have now indicated that this is not so in cephalopod muscles and that quite wide fluctuations in ATP concentration can be tolerated in mantle muscle (Storey and Storey, 1978, 1979a; Baldwin and England, 1980). The fall in muscle ATP levels during mantle muscle work resulted in a sharp rise in AMP concentrations. As will be discussed in the following section, AMP is a key activator of muscle phosphofructokinase and as such plays an important role in metabolic activation in squid muscle.

## C. Regulation of Aerobic Metabolism in Mantle Muscle

The transition from resting to active states in aerobic muscle requires a coordinated activation of glycogen breakdown, glycolysis, the Krebs cycle, the $\alpha$-glycerophosphate cycle, arginine phosphate hydrolysis, and proline oxidation. This is accomplished by specific regulation of key enzymatic sites via the actions of metabolite modulators. In part, glycolytic and Krebs cycle activation in squid mantle muscle is regulated by the adenylates plus several key intermediates within the pathways by effects similar to those documented for muscle enzymes from other sources. However, in a situation that is perhaps unique in metabolic regulation, a key regulatory role for NADH is apparent in squid muscle. In a system with an active $\alpha$-glycerophosphate cycle, NADH may well be the most immediate signal of the metabolic status of the cell, and in squid muscle direct allosteric effects by NADH appear to be the primary signals regulating the activities of key enzymes and coordinating the activities of catabolic pathways. Direct inhibitory effects of NADH contribute to the regulation of phosphofructokinase, glyceraldehyde 3-phosphate dehydrogenase (GAPDH), pyruvate kinase (PK), citrate synthase, adenylate kinase, arginine phosphokinase, and glutamate dehydrogenase (Storey and Hochachka, 1975a,b,c; Hochachka et al., 1975b; Storey, 1976, 1977a; Storey et al., 1978). In three instances—phosphofructokinase, pyruvate kinase, and citrate synthase—the normal regulatory effects of ATP on muscle enzymes appear to have been replaced by NADH inhibitions.

Before discussing details of the regulation of individual enzymes, an outline of some of the metabolic events that occur when squid muscle is stimulated to contract and that participate in enzyme activation can be given.

1. An influx of $Ca^{2+}$ into the cytoplasm stimulates myosin ATPase, activating muscle contraction. $Ca^{2+}$ also stimulates the activity of phosphorylase kinase (resulting in activation of glycogen phosphorylase) and mitochondrial $\alpha$-glycerophosphate oxidase.
2. A drop in cellular ATP levels due to myosin ATPase activity is rapidly translated and amplified into a sharp percentage increase in AMP concentration through the action of adenylate kinase.
3. Arginine phosphokinase hydrolyzes arginine phosphate reserves to help buffer ATP levels.
4. The fall in ATP levels relieves ATP inhibition of $\alpha$-glycerophosphate dehydrogenase; with both enzymes of the $\alpha$-glycerophosphate cycle now activated, cycling is rapidly increased, leading to a sudden fall in NADH levels.
5. Decreased NADH levels lead to a deinhibition of key enzyme activities.
6. Decreased arginine phosphate concentrations relieve inhibition of phosphofructokinase.
7. The sharp increase in AMP levels activates phosphofructokinase and citrate synthase directly and also overrides NADH inhibition of these enzymes.

This general scheme of metabolic regulation is derived from *in vitro* studies of the kinetic and regulatory properties of enzymes from the mantle muscle of the pelagic squid, *S. oualaniensis*. Enzyme control in this squid is elegantly integrated to produce a highly efficient aerobic metabolism. Regulatory metabolites such as NADH, arginine phosphate, AMP, and citrate each affect the activities of several key enzymes to produce coordinated alterations in carbon flux in response to muscle energy demand. Although many of the regulatory features to be discussed later are probably common to aerobic metabolism in all cephalopod muscles, certain features, in particular the regulatory effects of NADH, appear to be specialized adaptations in metabolic control in the highly aerobic pelagic squid only. Parenthetically it should be noted that the muscle enzymes of *S. oualaniensis* are also regulated and coordinated with respect to temperature and pressure to produce minimal disruptions to metabolism during the diurnal vertical migrations of this species (from surface waters of 1 atm and 25°C to deep waters of 140 atm and 12°C) (Storey and Hochachka, 1975a,b,c,d; Storey et al., 1975a,b; Hochachka et al., 1975b).

### 1. Phosphofructokinase

Phosphofructokinase (PFK) is generally held to be the key control site of glycolysis, and as such it displays complex control properties that allow enzyme activity to be attuned to a variety of metabolic signals (Mansour, 1972). Squid muscle PFK displays some unusual properties. Unlike most muscle PFKs, the squid enzyme shows hyperbolic fructose 6-phosphate saturation kinetics at low pH, is not inhibited by ATP, and does not show product activation by either ADP or fructose 1,6-bisphosphate (Storey and Hochachka, 1975a). The normal inhibitory control of the enzyme by ATP has been replaced by NADH inhibition, NADH being a mixed competitive inhibitor of the enzyme with respect to both substrates. Squid muscle PFK is also inhibited by arginine phosphate (providing control by the energy status of the cell) and by citrate (providing integration with the state of Krebs cycle activation) (Storey and Hochachka, 1975a; Storey, 1977a) (Fig. 2). The key positive modulator of squid muscle PFK is AMP, which provides activation on its own as well as deinhibition of NADH, citrate, and arginine phosphate effects. $P_i$ is

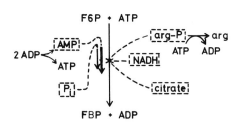

**Fig. 2.** Control of phosphofructokinase in squid mantle muscle. This key regulatory site in glycolysis is closely controlled by NADH, arginine phosphate, and citrate inhibition, and by AMP and $P_i$ activation. For abbreviations see Fig. 1.

also an activator of enzyme activity. At the initiation of muscle work, alterations in the levels of these metabolite modulators—a fall in tissue NADH levels due to an activation of the $\alpha$-glycerphosphate cycle, a decrease in arginine phosphate levels, and a sharp rise in AMP concentration—all combine to produce a very effective and rapid activation of PFK and also, therefore, of flux through glycolysis.

### 2. α-Glycerophosphate Dehydrogenase

$\alpha$-Glycerophosphate dehydrogenase ($\alpha$-GPDH), the cytoplasmic arm of the $\alpha$GP cycle, is intimately linked with flux through glycolysis by a redox link with glyceraldehyde 3-phosphate dehydrogenase (Fig. 1). The enzyme from squid muscle is strongly poised for function in the direction of $\alpha$-glycerophosphate production, with control of enzyme activity vested in ATP and NAD inhibition (Storey and Hochachka, 1975d). ATP is a particularly strong inhibitor of enzyme function, the $K_i$ for ATP being 0.7 m$M$ with respect to NADH compared to muscle ATP levels of 5–8 $\mu$mol/g wet weight in cephalopods (Storey and Storey, 1978, 1979a). These ATP effects allow the rate of $\alpha$-glycerophosphate cycling to be closely attuned to the energy requirements of the cell. Thus at the initiation of muscle work, a fall in cellular ATP levels would produce a deinhibition of $\alpha$-GPDH and an increase in cycling, leading to a fall in cellular NADH concentration. The fall in NADH levels has potent effects in activating PFK and other enzymes. Regulatory control of $\alpha$-GPDH by ATP appears to be a feature of the enzyme from tissues with active $\alpha$-GP cycles (e.g., squid mantle muscle, honeybee flight muscle) but is not a property of the enzyme from tissues with poorly developed $\alpha$-GP cycles (Storey and Hochachka, 1975e).

### 3. Glyceraldehyde 3-Phosphate Dehydrogenase

Squid muscle glyceraldehyde 3-phosphate dehydrogenase, (GAPDH), like the enzyme from other sources, is inhibited by the adenylates, but because these effects are easily reversed by levels of $P_i$ well below the levels of inorganic phosphate *in vivo*, it is likely that adenylate effects are of little consequence to enzyme regulation *in vivo* (Storey and Hochachka, 1975b). However, the enzyme is strongly inhibited by its product, NADH, a feature that integrates the activity of GAPDH with the activity of PFK and other enzymes of energy production in squid muscle.

### 4. Pyruvate Kinase

Pyruvate kinase (PK) is an important regulatory site in glycolysis. The enzyme from squid muscle appears to be primarily regulated by NADH and citrate inhibition. NADH effects produce a coordinated control of activity at this site with other regulatory sites in glycolysis and the Krebs cycle, whereas the effects of citrate provide an important feedback regulation of PK activity in response to the degree of Krebs cycle activation (Storey and Hochachka, 1975c). In common with other muscle PKs, squid muscle PK is inhibited by ATP, although inhibition by ATP appears to be relatively unimportant to overall enzyme regulation compared to NADH effects on the enzyme.

Guderley et al. (1976) have studied *Octopus cyanea* mantle muscle PK. Inhibitory control of this enzyme is provided by ATP, arginine phosphate, and citrate, although citrate effects on the *Octopus* enzyme are much weaker than those seen for the squid enzyme. In addition, NADH is only a weak inhibitor of *Octopus* PK. It appears, therefore, that only the squid with its highly aerobic metabolism, but not the more sluggish *Octopus*, has evolved a key regulatory function for NADH.

Cephalopod muscle PKs do not display inhibition by alanine or activation by fructose 1,6-bisphosphate (Storey and Hochachka, 1975c; Guderley et al., 1976), effects that play important roles in the control of muscle PKs from anoxia-tolerant bivalve molluscs (Mustafa and Hochachka, 1971; de Zwaan, 1972; Livingstone and Bayne, 1974).

## 5. Citrate Synthase

Control of this first enzyme of the Krebs cycle in squid muscle is provided through inhibitions by ATP, NADH, citrate, and $\alpha$-ketoglutarate (Hochachka et al., 1975b). Regulation of citrate synthase activity by citrate and $\alpha$-ketoglutarate is a common feature of many citrate synthases (Srere, 1974), including both the *Octopus* and the oyster muscle enzymes (Fields et al., 1976a; Fields and Hochachka, 1976). Regulation by $\alpha$-ketoglutarate, which is a competitive inhibitor with respect to oxaloacetate and a noncompetitive inhibitor with respect to acetyl-CoA, provides a feedback signal of the degree of activation of the second span (starting with $\alpha$-ketoglutarate dehydrogenase) of the Krebs cycle (Hochachka et al., 1975b). This is particularly important in squid muscle because the input of carbon from the oxidation of proline is at the level of $\alpha$-ketoglutarate. Citrate inhibition of the enzyme provides an integration of enzyme activity with the activities of PFK and PK, which are both citrate inhibited in squid muscle. ATP inhibition, which is competitive with respect to acetyl-CoA, provides a regulation by the energy status of the cell, whereas potent NADH inhibition ($K_i = 20 \mu M$) integrates the activity of citrate synthase with that of a number of glycolytic enzymes in squid muscle. However, increasing concentrations of oxaloacetate, or AMP at high levels, can reverse NADH inhibition of the enzyme. Thus at the initiation of muscle work, a sharp drop in mitochondrial NADH levels coupled with increased oxaloacetate availability (provided from the mobilization of proline) would allow an effective activation of citrate synthase activity.

## 6. Glutamate Dehydrogenase

Squid mantle muscle contains high activities of NAD-specific glutamate dehydrogenase (GDH) [12 units/g wet weight in *L. pealeii;* up to 26 units/g wet weight in other squid species (Storey et al., 1978; Mommsen et al., 1981)], which is strongly poised, as is the insect flight muscle enzyme, for function in the direction of the oxidative deamination of glutamate (Storey et al., 1978). GDH is distributed with highest activities in the inner zone of the mantle muscle, suggesting the importance of the enzyme in aerobic metabolism, specifically in the oxidation of muscle proline (Fig. 1). The enzyme is activated over 100-fold by optimal (5 m$M$) ADP concentration, indicating that ADP may be an almost obligatory cofactor for achieving significant enzyme activity *in vivo*. Like the enzyme from other sources (Goldin and Frieden, 1971; Smith et al., 1975), squid mantle muscle GDH is activated by ADP, AMP, leucine, and $P_i$ whereas GTP, GDP, ATP, NADH, glutamate, and $\alpha$-ketoglutarate are inhibitory. Control of GDH is key to regulating the mobilization of proline reserves during metabolic activation. An activation of the enzyme to provide increased influx of $\alpha$-ketoglutarate, thereby priming the Krebs cycle, is possibly achieved through a combination of the release of ATP, GTP, and NADH inhibition coupled to ADP activation.

## 7. Adenylate Kinase

During the initiation of muscle work, a fall in ATP levels is transformed via the action of adenylate kinase (2 ADP $\rightleftharpoons$ ATP + AMP) into a large percentage increase in cellular AMP levels. In a variety of muscle tissues, including squid mantle, this sharp rise in AMP has potent effects in activating energy metabolism (Sacktor, 1976; Hochachka et al., 1975a; Storey and Storey, 1978), in particular in activating and deinhibiting PFK (Fig. 2). Adenylate kinase from squid muscle is strongly inhibited by NADH and also by arginine phosphate (Storey, 1976). This, coupled with the nonequilibrium nature of adenylate kinase in resting muscle (Storey, 1976; Ballard, 1970; Yushok, 1971), suggests that enzyme activity is closely regulated under resting conditions. However, conditions that lead to the initiation of muscle work in squid (decreased NADH, arginine phosphate, and ATP levels) lead to a deinhibition of enzyme activity and a return to equilibrium function, a situation that would promote a rapid buildup of AMP concentrations.

## 8. Arginine Kinase

It is generally accepted that the phosphagen/phosphagen-kinase system in muscle functions primarily to "buffer" ATP levels during muscular work. Additionally, phosphagens have also been shown to have roles as modulators of enzyme activity; in particular arginine phosphate has inhibitory effects on a number of enzymes in cephalopod mantle muscle (Storey, 1977a, 1981; Guderley et al., 1976). In a study of squid mantle muscle arginine kinase (Storey, 1977a), it was found that the enzyme, like other enzymes of energy metabolism in squid muscle, was inhibited by NADH. NADH was a noncompetitive inhibitor (in the direction of ATP synthesis) with a ($K_i$ of 0.15 m$M$). During metabolic activation, then, falling NADH levels would result in a deinhibition of enzyme activity, leading to both ATP production from arginine phosphate reserves and a release of arginine phosphate inhibitory effects on enzymes.

To summarize, then, we have seen how aerobic metabolism in the mantle muscle of fast-swimming pelagic squid is closely regulated to bring about a rapid and efficient activation of energy production during the transition from rest to sustained work. The control properties of a number of key enzymes have been modified to bring them all under NADH control. In several cases this has involved replacing what is normally an ATP effect on enzyme activity with an NADH effect, and we can especially appreciate this specialized mode of metabolic regulation in the pelagic squid when we consider that although NADH is a key modulator of enzyme activities in squid muscle, it has little or no effect on the homologous enzymes from octopus muscle.

However, as highly specialized as the pelagic squid is for performing sustained aerobic work, the mantle muscle of these animals, like that of all cephalopods, still maintains the capacity for anaerobic glycolysis, that potential being based on the presence of high activities of the terminal glycolytic enzyme, octopine dehydrogenase (ODH).

## D. Aerobic Metabolism in Squid Heart Mitochondria

Two recent studies have examined oxidative metabolism in isolated mitochondria from the systemic hearts of *Loligo opalescens* (Ballantyne et al., 1981) and *Illex illecebrosus* (Mommsen and Hochachka, 1981) with results supportive of the scheme of aerobic muscle metabolism outlined in the previous section. Mitochondria from both species oxidized pyruvate, proline, glutamate, malate, and succinate at high rates while Mommsen and Hochachka (1981) also reported the oxidation of aspartate, alanine, pyrroline carboxylate, and ornithine. State 3 rates of substrate oxidation by *I. illecebrosus* mitochondria were 71.8 ± 14.3, 81.0 ± 11.2, 71.2 ± 15.6,

$33.4 \pm 8.5$, $47.0 \pm 7.5$ and $34.7 \pm 8.1$ nmol $O_2$ consumed/min mg mitochondrial protein for the oxidation of malate (5 m$M$), succinate (5 m$M$), proline (5 m$M$) + pyruvate (0.05 m$M$), glutamate (5 m$M$) + pyruvate (0.1 m$M$), ornithine (5 m$M$) + pyruvate (0.05 m$M$), and pyruvate (5 m$M$) + proline (0.1 m$M$), respectively (Mommsen and Hochachka, 1981). Mitochondrial studies confirm, therefore, that coupled carbohydrate (pyruvate) and amino acid (proline) catabolism is the basis of aerobic energy production in cephalopods.

Neither lipid substrates nor $\alpha$-glycerophosphate were oxidized by squid heart mitochondria. The failure of heart mitochondria to oxidize $\alpha$-glycerophosphate is at odds with the previous data suggesting the presence of an active $\alpha$-glycerophosphate cycle in cephalopod muscle (Hochachka et al., 1975 a); this may however be a difference between heart and mantle muscle metabolisms. Heart tissue showed substantial activities of malate dehydrogenase and aspartate aminotransferase in both mitochondria and cytoplasm (Ballantyne et al., 1981; Mommsen and Hochachka, 1981) suggesting that the malate–aspartate shuttle may be the preferred mode of hydrogen ion transfer in this tissue.

Mommsen and Hochachka (1981) examined the effects of the aminotransferase inhibitor, aminooxyacetate, on amino acid oxidation by heart mitochondria. The inhibitor did not affect proline oxidation or the oxidation of pyruvate when "sparked" by low amounts of proline. However glutamate and ornithine oxidation were inhibited. Two routes for the catabolism of these amino acids and the delivery of $\alpha$-ketoglutarate to the Krebs cycle are therefore indicated. Proline oxidation may proceed largely through the action of glutamate dehydrogenase (3.6 units/g in *I. illecebrosus* heart) with the catabolism of glutamate or ornithine using the glutamate–pyruvate transaminase (7 units/g) route. Proline oxidation by insect flight muscle mitochondria has one of two uses: (a) in some species proline is used only as a source of $C_4$ units for augmentation of the pool size of Krebs cycle intermediates during the activation of metabolism, and (b) in other species proline is completely oxidized to $CO_2$ and $H_2O$ as an aerobic substrate (Sacktor, 1976). The first use for proline apparently occurs in cephalopod tissues (Storey and Storey, 1978). The second use is indicated in squid heart by the presence of substantial activities of malic enzyme which would allow the conversion of malate to pyruvate and the complete oxidation of the carbon skeleton of proline (Mommsen and Hochachka, 1981).

The high rates of ornithine oxidation by squid heart mitochondria (Mommsen and Hochachka, 1981) suggest that arginine could be a potentially important aerobic substrate in squid tissues. The presence of a cytosolic arginase (2.5 units/g) in *I. illecebrosus* heart would allow the conversion of arginine to ornithine with the subsequent oxidation of ornithine. Radiotracer studies substantiate this showing a ready catabolism of [$^{14}$C]-arginine by squid tissues (Hochachka et al., 1982).

## IV. Carbohydrate Catabolism and Anaerobic Muscle Work

### A. Design of Anaerobic Metabolism

The anaerobic design of muscle metabolism in cephalopods couples the fermentation of carbohydrate via the reactions of glycolysis with the hydrolysis of phosphagen reserves (Fig. 3). This mode of cytoplasmic energy production results in a high ATP output per unit time but only at the expense of a low ATP yield per fuel equivalent. Only short-term bursts of intense muscle work are possible as a result. In more sluggish species of cephalopods such as *Octopus*, this mode of energy

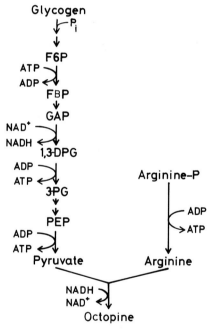

**Fig. 3.** Anaerobic metabolism in cephalopod mantle muscle showing the coupling of glyco-genolysis to arginine phosphate hydrolysis and the production of octopine. Four moles ATP are generated per mole octopine produced from glycogen (glucose 1-phosphate) and arginine phosphate. Abbreviations: 1,3-DPG, 1,3-diphosphoglycerate; 3-PG, 3-phosphoglycerate; PEP, phosphoenol-pyruvate; and for others see Fig. 1.

production may represent the sole power source for infrequent bursts of swimming, whereas in many of the squid species, with their well-developed capacities for sustained aerobic swimmung, the anaerobic fermentation of carbohydrate represents an "overdrive" mechanism for use during intense bursts of jetting. A key requirement for sustained energy production using glycolysis is the need to maintain redox balance, regenerating the NAD that is utilized at the glyceraldehyde 3-phosphate dehydrogenase reaction. In vertebrate and many invertebrate species, glycolytic redox balance is maintained through the action of lactate dehydrogenase. Among the Mollusca and a number of other invertebrate phyla, however, lactate dehydrogenase has often been replaced by or coexists with one or more of a family of imino acid dehydrogenases (Fig. 4). These enzymes catalyze a reductive condensation between a keto acid (generally pyruvic acid) and an amino acid to produce a conjugated secondary amino or imino acid end product:

$$\begin{array}{c} R_1-C-COOH + R_2-CH-COOH + NADH + H^+ \rightleftharpoons R_1-CH-COOH + NAD^+ + H_2O \\ \quad\;\| \qquad\qquad\quad | \qquad\qquad\qquad\qquad\qquad\qquad\quad | \\ \quad\;O \qquad\qquad\quad NH_2 \qquad\qquad\qquad\qquad\qquad\qquad NH \\ \qquad\qquad\qquad\qquad\qquad\qquad\qquad\qquad\qquad\qquad\quad | \\ \qquad\qquad\qquad\qquad\qquad\qquad\qquad\qquad\qquad\quad R_2-CH-COOH \end{array}$$

Among cephalopods, octopine and octopine dehydrogenase have functionally replaced lactate and lactate dehydrogenase in anaerobic muscle metabolism. During burst muscular work the production of octopine ef-

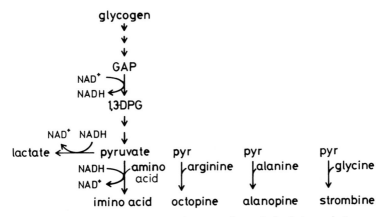

**Fig. 4.** Terminal dehydrogenases of glycolysis in molluscs. Redox balance during anaerobic glycolysis in mollusc tissues is maintained through the action of lactate dehydrogenase and/or one of a number of imino acid dehydrogenases such as octopine dehydrogenase, alanopine dehydrogenase, or strombine dehydrogenase.

fectively couples the products of glycolysis (pyruvate and NADH) with that of phosphagen breakdown (arginine) (Fig. 3).

## B. Glycolytic End Products and Terminal Dehydrogenases

### 1. Octopine and Octopine Dehydrogenase

Octopine dehydrogenase (ODH) (EC 1.5.1.11) catalyzes the reaction:

$$\text{Pyruvate} + \text{L-arginine} + \text{NADH} + \text{H}^+ \rightleftharpoons \text{D-octopine} + \text{NAD}^+ + \text{H}_2\text{O}$$

The enzyme is widely distributed among the Mollusca, including many bivalve and gastropod species and all cephalopod species tested to date (Thoai and Robin, 1959; Regnouf and Thoai, 1970; Gäde, 1980a).

Octopine was first identified as a natural product of *Octopus octopodia* muscle by Morizawa (1927). Subsequently the compound has been found in the muscles of other cephalopod species (Ackermann and Mohr, 1937; Irvin, 1938; Roche et al., 1952; Suyama and Kobayashi, 1980), as well as in a variety of other mollusc species (Roche et al., 1952; Moore and Wilson, 1937a,b; Regnouf and Thoai, 1970). Octopine and/or ODH have also been identified in nemerteans (Robin, 1964), sipunculids (Thoai and Robin, 1959; Haas et al., 1973), sea anemones (Zammit and Newsholme, 1976; Ellington, 1979) and crown gall tumors of plants (Ménagé and Morel, 1964). In the postmortem muscle of *Sepia* and the scallop, *Pecten,* octopine accumulates in inverse proportion to the disappearance of tissue arginine (Irvin and Wilson, 1939; Hiltz and Dyer, 1971; Sakaguchi et al., 1975). Using extracts of *S. officinalis* muscle, Thoai and Robin (1959) were the first workers to demonstrate that octopine was a product of muscle glycolysis and that its formation required the provision of reducing power, in the form of NADH.

The naturally occurring stereoisomer of octopine is D-octopine, which is $S(\text{L})$ at the arginine center and $R(\text{D})$ at the alanine center (Robin and Guillou, 1977; Biellmann et al., 1977). Of the three other possible stereoisomers, none is oxidized by *Pecten* ODH, but L-allooctopine is a competitive inhibitor of the enzyme (Biellmann et al., 1977).

Various authors have speculated on the choice of octopine over lactate for the glycolytic end product in molluscan muscles. Some of the possible advantages of octopine production include the following (a) Octopine couples the products of glycolysis and phosphagen breakdown, thus promoting a coordinated catabolism of these two energy sources; (b) removal of arginine via octopine formation could facilitate the action of the equilibrium enzyme arginine kinase in the further hydrolysis of arginine phosphate reserves (Zammit and Newsholme,

1976); (c) octopine formation results in no net change in muscle osmolarity as an accumulation of lactate would; (d) octopine is a more neutral product (with an isoelectric point of about 5.9) than is arginine, pyruvate, or lactate; octopine production would minimize alterations in intracellular pH during muscle work (Hochachka, 1980); (e) cephalopod hemocyanin has a pronounced Bohr shift; the presence of the weak acid octopine in the blood would not affect oxygen loading at the gills (Zammit, 1978); (f) arginine, but not octopine, has deleterious effects on the kinetics of various enzymes; production of octopine helps to regulate the arginine pool size (Bowlus and Somero, 1979); and (g) ODH may be capable of maintaining a lower cytoplasmic redox ratio (NADH:NAD) during anaerobic function than can LDH (Fields and Quinn, 1981).

ODH occurs in very high activities in the muscles of cephalopods, often at activities 100-fold or more greater than those of LDH (Table II). The ODH:$\alpha$-GPDH ratio provides a measure of the relative potentials of the muscle of different species for anaerobic glycolysis versus aerobic muscle work. As Table II shows, the ratio is highest in the more sluggish species, such as *Nautilus* and *Octopus*, which swim with short bursts of jet propulsion. The more "aerobic" squid, which have a high capacity for sustained swimming, exhibit a lower ODH:$\alpha$-GPDH ratio; in particular the pelagic squid, *S. oualaniensis*, shows a higher $\alpha$-GPDH than ODH activity in mantle muscle. Four muscle types in *Nautilus* (spadix, retractor, funnel, and heart) show a strong inverse correlation between ODH activity and aerobic capacity (measured by mitochondrial content and citrate synthase activity) (Hochachka et al., 1978). In addition, a difference in the ODH:$\alpha$-GPDH ratio occurs within the different mantle muscle zones of a single species (Mommsen et al., 1981) (Table II). The muscle of the inner mantle zone, which functions during breathing movements and/or sustained swimming, has a higher $\alpha$-GPDH content than does the central zone, which functions during burst jetting. The ODH:$\alpha$-GPDH ratio also varies between tissues of a single species. In *L. opalescens* the ratio was 22.4, 9.6, 4.6, 4.2, and 2.9 in mantle, fin, gill, heart and brain, respectively, reflecting the anaerobic versus aerobic potentials of metabolism in these tissues (Ballantyne et al., 1981).

ODH is also distributed throughout the soft tissues of cephalopods (Table III). In most tissues, ODH is the dominant dehydrogenase activity although in some tissues, LDH activity exceeds that of ODH. The presence of a dual dehydrogenase system in many of the soft tissues suggests the possibility of competition between the two enzymes for the pyruvate derived from glycolysis (see Section IV,E).

TABLE II

Activities of Octopine Dehydrogenase, Lactate Dehydrogenase, and $\alpha$-Glycerophosphate Dehydrogenase in Cephalopod Muscles

| Species[a] | Muscle | Enzyme activity ($\mu$mol/min/g wet weight) | | | Ratio ODH:$\alpha$-GPDH |
|---|---|---|---|---|---|
| | | ODH | LDH | $\alpha$-GPDH | |
| Nautiloidea | | | | | |
| Nautilus pompilius[1] | Spadix | 84.0 | 0.7 | 1.0 | 84.0 |
| | Retractor | 83.6 | 5.8 | 3.3 | 25.3 |
| | Funnel | 41.2 | 3.6 | — | — |
| Octopoda | | | | | |
| Octopus ornatus[2] | Mantle | 640.0 | — | 30.0 | 21.3 |
| Hapalochlaena maculosa[3] | Mantle | 458.0 | 11.0 | 24.5 | 18.7 |
| | Tentacle | 591.0 | 5.6 | 35.0 | 16.9 |
| Sepioidea | | | | | |
| Sepia officinalis[4] | Mantle | 99.8 | 1.6 | 5.7 | 17.5 |
| | Tentacle | 119.2 | 3.1 | — | — |
| Teuthoidea | | | | | |
| Symplectoteuthis oualaniensis[2,5] | Mantle | 110.0 | 0.6 | 240.0 | 0.46 |
| Loligo vulgaris[6] | Mantle | 130.0 | 0.4 | 11.0 | 11.8 |
| Alloteuthis subulata[7] | Mantle | 214.0 | 2.6 | — | — |
| Loligo forbesi[8] | Mantle | 146.5 | <0.1 | 10.8 | 13.6 |
| | Fin | 113.0 | <0.1 | 16.7 | 6.8 |
| Todarodes sagittatus[8] | Fin | 38.5 | <0.1 | 12.7 | 3.0 |
| Loligo opalescens[9] | Mantle | | | | |
| | Inner | 176.0 | — | 10.8 | 16.3 |
| | Middle | 182.0 | — | 7.7 | 23.6 |
| | Outer | 128.0 | — | 8.4 | 15.2 |
| Ommastrephes sp.[9] | Mantle | | | | |
| | Inner | 192.0 | — | 26.9 | 7.1 |
| | Middle | 277.0 | — | 20.8 | 13.3 |
| | Outer | 37.2 | — | 10.2 | 3.6 |

[a] References are as follows: [1]Hochachka et al. (1978); [2] Fields et al. (1976b); [3] Baldwin and England (1980); [4] Storey (1977b); $\alpha$-GPDH activity for mantle from J. M. Storey (unpublished data);[5] Hochachka et al. (1975a);[6] Gäde (1980a);[7] K. B. Storey (unpublished data); [8] Zammit and Newsholme (1976);[9] Mommsen et al. (1981).

## 2. Other Imino Acids and Imino Acid Dehydrogenases

In recent years, both alanopine dehydrogenase and strombine dehydrogenase (Fig. 4), along with their corresponding imino acids, have been identified in a variety of bivalve and gastropod tissues (Sangster et al., 1975; Collicutt and Hochachka, 1977; Fields, 1976, 1977; Fields et al., 1980; Fields and Hochachka, 1981; de Zwaan and Zurburg, 1981; Dando et al., 1981; Plaxton and Storey, 1982). Alanopine and strom-

## TABLE III
### Activities of Octopine Dehydrogenase and Lactate Dehydrogenase in Cephalopod Tissues

| Species[a] | Tissue | Enzyme activity ($\mu$mol/min/g wet weight) ODH | LDH | Ratio ODH:LDH |
|---|---|---|---|---|
| Nautilus pompilius[1] | Ventricle | 6.9 | 1.1 | 6.3 |
| | Hepatopancreas | 17.5 | — | — |
| | Gill | 5.4 | — | — |
| Sepia officinalis[2] | Ventricle | 103.0 | 17.1 | 6.0 |
| | Branchial heart | 43.5 | 6.4 | 6.8 |
| | Hepatopancreas | 18.5 | 2.9 | 6.4 |
| | Gill | 10.0 | 13.8 | 0.7 |
| | Brain | 32.2 | 4.0 | 8.1 |
| | Kidney[3] | 27.3 | — | — |
| Alloteuthis subulata[3] | Hepatopancreas | 3.5 | 1.4 | 2.5 |
| | Gill | 17.0 | 1.7 | 10.0 |
| | Brain | 60.0 | 3.2 | 18.8 |
| Symplectoteuthis oualaniensis[4] | Brain | 1.6 | 2.8 | 0.6 |
| Loligo vulgaris[5] | Gill | 22.0 | 4.0 | 5.5 |
| | Optic lobe | 17.0 | 5.0 | 3.4 |
| Loligo opalescens[6] | Ventricle | 64.0 | 0.2 | 320 |
| | Gill | 14.6 | 0.2 | 73 |
| | Brain | 28.0 | 0.7 | 40 |

[a] References are as follows: [1] Hochachka et al. (1978); [2] Storey (1977b); [3] K. B. Storey (unpublished data); [4] Fields et al. (1976c); [5] Gäde (1980a); [6] Ballantyne et al. (1981).

bine accumulate as products of anaerobic glycolysis in some species (Collicutt and Hochachka, 1977; Fields, 1977; de Zwaan and Zurburg, 1981). These enzymes, however, do not appear to be prominent among cephalopod molluscs. Neither alanopine dehydrogenase nor strombine dehydrogenase was detected in the tissues (mantle muscle, brain, gill, hepatopancreas) of S. officinalis or A. subulata (Dando et al., 1981). However, Sato et al. (1977) have reported the isolation of 2,2'-iminodi-propionic acid (alanopine) from the mantle muscle of the squid Todarodes pacificus, suggesting that alanopine dehydrogenase may be present in some cephalopod species.

### 3. Lactate and Lactate Dehydrogenase

LDH activities are low in the muscles of cephalopods, often negligible compared to ODH activities (Table II). It is not surprising, therefore, that lactate does not accumulate during glycolytic muscle work (Hochachka et al., 1975a; Grieshaber and Gäde, 1976a,b; Storey and

Storey, 1978, 1979a; Baldwin and England, 1980). However, LDH activities are significantly higher (sometimes exceeding ODH activities) in many of the soft tissues of cephalopods (Table III), suggesting that the enzyme may have an important function in these tissues.

Several authors have examined the stereospecificity of invertebrate LDHs and found that in almost all instances molluscan species have a D-lactate-specific enzyme (Scheid and Awapara, 1972; Long, 1976; Long et al., 1979). Among the Cephalopoda, D-lactate-specific LDH occurs in *N. pompilius* (J. H. A. Fields, personal communication), *Eledone cirrosa* (Hammen, 1969), *Octopus bimaculatus* (Long, 1976), *Octopus vulgaris* (Baccetti et al., 1975), *Octopus dolfeini martini, Loligo opalescens* (Martin et al., 1976), and *S. officinalis* (J. M. Storey, unpublished observations). However, Gäde (1979) found that the mantle muscle of *Loligo vulgaris* contained an L-lactate-specific LDH.

LDH in vertebrate tissues occurs in muscle-(M) and heart-(H) specific isozymic forms. Pyruvate substrate inhibition is the key kinetic difference distinguishing the two isozymes; M-type LDH is uninhibited by high pyruvate, H-type LDH shows pyruvate substrate inhibition with an $I_{50}$ of about 10 m$M$. Based on a study of the pyruvate-inhibition characteristics, the tissues of *S. officinalis* also appear to contain H- and M-type LDH complements. Brain LDH shows substrate inhibition by pyruvate ($I_{50}$ = 10 m$M$), whereas mantle muscle LDH does not (Storey and Storey, 1979b). Tentacle muscle, skin, and hepatopancreas also appear to have the M-type enzyme, whereas all other tissues show the H type (Storey, 1977b).

LDH from *L. vulgaris* mantle muscle and *S. oualaniensis* brain has been characterized (Gäde, 1979; Fields et al., 1976c). L-Lactate-specific LDH from *L. vulgaris* is a dimer of MW 70,000 (Gäde, 1979). The kinetic properties of the purified enzyme are, for the most part, those typical of a muscle-type LDH ($K_m$ values for pyruvate, NADH, L-lactate, and NAD$^+$ are 0.69, 0.036, 17, and 0.15 m$M$, respectively). However, the enzyme differs from a typical M-type LDH and from *S. officinalis* muscle LDH in showing substrate inhibition by pyruvate ($I_{50}$ = 20 m$M$). An interesting note is that the specimen of *L. vulgaris* used in Gade's study contained no ODH in its mantle muscle. This apparent genetic deficiency was compensated for by a greatly increased muscle LDH activity [~12 units/g wet weight versus 0.4 units/g found in muscle containing ODH (Gäde, 1979, 1980a)].

LDH and lactate metabolism in cephalopod sperm is worthy of a special note. Spermatozoa of *Octopus* use stored carbohydrate as their energy source; electron micrographs of the sperm tail show nine discrete glycogen bodies surrounding the central axoneme (Baccetti et al.,

1975). Mitochondria are present, although sperm can survive with full motility for at least 48 h in an anoxic environment (Mann et al., 1974, 1977). Anaerobic survival is based on the quantitative conversion of glycogen to D-lactate (Martin et al., 1976; Mann et al., 1977). D-LDH occurs in high activities in sperm and testis (316 units/g wet weight in testis) of *O. dolfeini martini*, but ODH is not present (Martin et al., 1976). The kinetic properties of *Octopus* sperm D-LDH are similar to those of an M-type LDH (Baccetti et al., 1975).

## C. Muscle ODH: Kinetic Properties and Enzyme Control

Octopine dehydrogenase is a monomer of MW 38,000–43,000 (Olomucki et al., 1972; Monneuse-Doublet et al., 1980; Fields et al., 1976b). ODH from the adductor muscle of *Pecten maximus* has been extensively studied by Olomucki and co-workers; Monneuse-Doublet et al. (1980) cite an 18-article bibliography by their group. Studies of the kinetic and physical properties of the enzyme have included enzyme kinetic constants (Thoai et al., 1969; Monneuse-Doublet et al., 1980), substrate specificity (Thoai et al., 1969; Pho et al., 1970; Biellmann et al., 1973), enzyme kinetic mechanism (Doublet and Olomucki, 1975; Monneuse-Doublet et al., 1978), and temperature effects on enzyme kinetic parameters (Luisi et al., 1975).

In speculating on the evolutionary origin of ODH, Hochachka (1980) summarizes some of the physical properties of ODH. ODH shares characteristics with both LDH and glutamate dehydrogenase (GDH) ODH and LDH are alike in sharing a similar subunit size, a blocked amino terminus, a catalytically important histidine residue at the active site, and essential histidine residues. In other features, however, ODH more closely resembles GDH. In both ODH and GDH a carbon–nitrogen bond is the reactive species whereas in LDH it is a carbon–oxygen bond. Both ODH and GDH show B-side stereospecificity for NADH, both exhibit similar spectral properties for the enzyme–NADH binary complex, and fluorometric data show similar conformational changes occurring on substrate binding (Hochachka, 1980).

Cephalopod muscle ODH has been characterized in four species: *Octopus ornatus, S. oualaniensis, N. pompilius,* and *S. officinalis* (Fields et al., 1976b; Hochachka et al., 1977; Storey and Storey, 1979b). Table IV summarizes the apparent $K_m$ values of the four enzymes. The $K_m$ values for arginine and NADH are likely well within the physiological concentration ranges of these compounds *in vivo* and are similar among all species. However, Hochachka (1980) identifies two types of ODH with respect to the $K_m$ for pyruvate. A low $K_m$ for pyruvate is found in *Nautilus, Sepia,*

**TABLE IV**

**Comparative Kinetics of ODHs from Muscle of Cephalopods and *Pecten maximus***

| Species[a] | Apparent $K_m$ (m$M$)[b] | | | | |
| --- | --- | --- | --- | --- | --- |
| | Pyruvate | Arginine | NADH | Octopine | NAD$^+$ |
| *Symplectoteuthis oualaniensis*[1] | 2.0 | 10.5 | 0.04 | 12.0 | 0.15 |
| *Octopus ornatus*[1] | 1.7 | 7.1 | 0.03 | 1.5 | 0.07 |
| *Nautilus pompilius*[2] | 0.3 | 5.3 | — | — | — |
| *Sepia officinalis*[3] | 0.3 | 3.2 | 0.03 | 0.85 | 0.25 |
| *Pecten maximus*[4] | 2.0 | 2.4 | 0.02 | 2.1 | — |

[a] References are as follows: [1] Fields et al (1976b); [2] Hochachka et al. (1977); [3] Storey and Storey (1979b); [4] Monneuse-Doublet et al. (1980) for the A form of muscle ODH.
[b] The $K_m$ for pyruvate or arginine is dependent on cosubstrate concentration. Cosubstrate concentrations were 30, 1.2, 0.2, 10, 0.1 m$M$ for *S. oualaniensis;* 30, 1.5, 0.2, 5, 0.1 m$M$ for *O. ornatus;* 20, 1, 0.15, 2, 0.2 m$M$ for *S. officinalis* for arginine, pyruvate, NADH, octopine, and NAD$^+$, respectively; and 30, 0.9, 0.2 m$M$ for *N. pompilius* for arginine, pyruvate, and NADH, respectively.

and the scallop, *Pecten,* suggesting that these species could readily accumulate octopine under conditions of anaerobic muscle work. Squid and octopus ODHs, however, have much higher $K_m$ values for pyruvate, which might suggest that octopine formation is favored only under more extreme conditions when muscle pyruvate concentrations become very high. Similarly, in the reverse direction species differences in the $K_m$ for octopine are apparent, suggesting perhaps a differing capability for the reversal of the ODH reaction in muscle *in vivo.* In the forward direction, the apparent $K_m$ values for pyruvate and arginine are strongly dependent on cosubstrate concentration, the $K_m$ decreasing in response to increasing cosubstrate concentrations (Fields et al., 1976b). For example, using *S. oualaniensis* ODH, the apparent $K_m$ for pyruvate decreases from 4 to 2 m$M$ when arginine concentration is raised from 3 to 30 m$M$; similarly, $K_m$ for arginine decreases from 13.5 to 6.7 m$M$ when pyruvate concentration is raised from 0.24 to 6 m$M$ (Fields et al., 1976b). During muscle work, increased glycogenolysis coupled with a breakdown of arginine phosphate reserves would result in the elevation of both pyruvate and arginine concentrations in muscle. This cooperative effect between $K_m$ values and cosubstrate concentrations could, therefore, be the key effect *in vivo* potentiating ODH activity and octopine synthesis during burst muscular work.

Cephalopod muscle ODH, like most dehydrogenases, is inhibited by the adenylates, which are competitive inhibitors with respect to NAD and NADH, although only ATP effects occur at concentrations within the physiological range (Fields et al., 1976b). Enzyme activity in both directions is affected by product inhibitions (Fields et al., 1976b; Storey and Storey, 1979b). In the forward direction, octopine gives a mixed pattern of inhibition with respect to both pyruvate and arginine, whereas in the reverse direction both pyruvate and arginine inhibit with respect to octopine. Arginine inhibition is quite strong ($K_i \sim 5$ m$M$) and might affect enzyme reversibility *in vivo* (Fields et al., 1976b). The enzyme exhibits strong redox control; both NAD and NADH are competitive inhibitors (with respect to NADH and NAD) with $K_i$ values well within the physiological range of these compounds *in vivo*.

## D. Isozymes of Octopine Dehydrogenase: Brain versus Muscle ODH

Multiple molecular forms of ODH have been reported in several bivalve species but no tissue-specific distribution seems to occur (Gäde, 1980a,b; Beaumont et al., 1980; Monneuse-Doublet et al., 1980). Two forms of ODH in *Pecten* adductor muscle display essentially identical kinetic properties and differ only in their charge and in their sensitivity to certain chemical treatments (Monneuse-Doublet et al., 1980).

In cephalopods, however, ODH occurs in distinct tissue-specific forms that are both electrophoretically and kinetically separable (Storey, 1977b; Storey and Storey, 1979b; Gäde, 1980b). Four isozymes of ODH were found in the tissues of *S. officinalis*, occurring in mantle muscle, ventricle, brain, and hepatopancreas (Storey, 1977b). The kinetic differences between the isozymes were largely in their apparent $K_m$ values for pyruvate and octopine. Brain ODH displayed the lowest apparent $K_m$ for both of these substrates whereas ventricle showed the highest $K_m$ for octopine and hepatopancreas the highest $K_m$ for pyruvate. Electrophoretic analysis showed that the mantle muscle form of ODH also occurred in tentacle muscle and skin whereas branchial heart, gill, ovary, and nidamental gland contained a mixture of the ventricle- and brain-specific enzymes, in varying proportions in each tissue (Storey, 1977b). Brain and mantle muscle isozymes of ODH have also been identified in *L. vulgaris* and *A. subulata* (Gäde, 1980b; Dando et al., 1981). In *N. pompilius*, a fast-moving isozyme of ODH characterizes muscle tissues whereas a slower migrating form is found in tissues such as gill and also in retractor muscle, which shows two bands of ODH activity (Hochachka et al., 1977, 1978). The gill form of the enzyme showed a lower apparent $K_m$ for octopine than did the muscle form.

The occurrence of tissue-specific forms of ODH in cephalopods is reminiscent of the M and H forms of LDH occurring in other animal groups and could imply a tissue-specific metabolism of octopine in cephalopods. The M isozyme of LDH functions in muscle in the maintenance of cytoplasmic redox balance during anaerobic glycolysis and displays kinetic properties (including a low $K_m$ for pyruvate, a high $K_m$ for lactate, and little or no substrate inhibition by pyruvate) that favor lactate formation (Everse and Kaplan, 1973). Compared to the M isozyme, however, H-type LDH has a much higher affinity for lactate and also exhibits strong substrate inhibition by elevated levels of pyruvate. H-Type LDH is found in more "aerobic" tissues and is believed to function largely in allowing the utilization of exogenous lactate as an aerobic fuel for tissues such as heart and brain or as a substrate for gluconeogenesis in tissues such as kidney cortex. Pyruvate substrate inhibition is the key kinetic criterion separating the M and H forms of LDH. When this test was applied to the ODH complement of *S. officinalis* tissues, only the brain-specific form of ODH was found to show pyruvate substrate inhibition (Storey and Storey, 1979b) (Fig. 5).

To investigate further the properties of ODH isozymes, brain and mantle muscle ODH from *S. officinalis* were purified and characterized using a series of criteria that have been used to differentiate between the M and H forms of LDH (Storey and Storey, 1979b). The study revealed that the muscle and brain forms of ODH do indeed display properties analogous to the M and H forms of LDH, respectively. Compared to muscle-type ODH, the brain form showed key kinetic differences including (Table V) (a) lower $K_m$ values for both octopine and $NAD^+$, properties that would favor octopine oxidation in tissues containing the brain-type isozyme; (b) stronger substrate inhibitions by both pyruvate (Fig. 5) and octopine, properties characteristic of an H-type LDH; and (c) stronger product inhibition by $NAD^+$ and octopine, effects that could limit the synthesis of octopine in tissues containing the brain isozyme (Storey and Storey, 1979b). Similar kinetic differences between the mantle muscle and brain forms of ODH have also been seen in *S. oualaniensis* and *L. vulgaris* (Fields et al., 1976b,c; Gäde, 1980b). Brain and muscle ODH also differ in their substrate specificities. *S. officinalis* muscle ODH utilizes only L-arginine as its amino acid substrate, whereas the brain isozyme also shows activity with L-lysine or L-ornithine (activity ratio arginine:lysine:ornithine = 100:7:6) (Storey and Dando, 1982).

Muscle- and brain-type ODH showed properties analogous to those of M- and H-type LDH in their responses to the hypoxanthine derivative of NADH. Brain-type ODH, like H-type LDH, shows a higher activity ratio NHxDH:NADH than does the muscle form (Storey and Storey, 1979b). In addition, brain ODH resembled H-type LDH in the formation of a

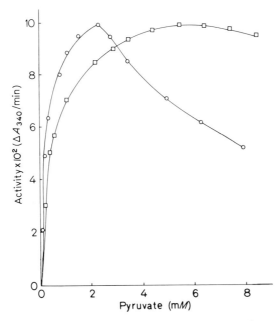

**Fig. 5.** Effect of pyruvate substrate concentrations on the activities of octopine dehydrogenase from *Sepia officinalis* mantle muscle (□) or brain (○). Brain ODH is strongly inhibited by elevated pyruvate substrate concentrations, the $I_{50}$ being about 10 m$M$. Similar pyruvate substrate inhibition of brain lactate dehydrogenase from *S. officinalis* occurs but the muscle forms of both enzymes show only slight inhibition by pyruvate at very high levels. (From Storey and Storey, 1979b.)

dead-end complex, enzyme–NAD⁺–substrate(s). Pyruvate substrate inhibition of H-type LDH is believed to derive from the formation of a dead-end ternary complex, enzyme–NAD⁺–pyruvate, which competes with NADH and which manifests itself in a slow decline with time in the apparent activity of the enzyme in an incubate containing enzyme, pyruvate, and NAD⁺ (Everse et al., 1971; Eichner and Kaplan, 1977). Incubation studies revealed that a similar dead-end complex was formed with brain ODH. With ODH, however, the complex was a quaternary one: enzyme–NAD⁺–pyruvate–arginine (Storey and Storey, 1979b) (Fig. 6). Mantle muscle ODH, like M-type LDH, showed no indication of inhibitory complex formation.

Thus in a number of features mantle muscle and brain isozymes of ODH bear strong resemblances to the M and H forms of LDH, respectively. This suggests that like M- and H-type LDH, these isozymes could

**TABLE V**

**A Comparison of the Kinetic Properties of the Mantle Muscle and Brain Isozymes of Octopine Dehydrogenase from *Sepia officinalis*[a]**

| | Apparent $K_m$ (mM) | | | | Substrate inhibition (%) | | | Product inhibition (%) | | | |
|---|---|---|---|---|---|---|---|---|---|---|---|
| | | | | | Octopine | | Pyruvate | Octopine | $NAD^+$ | Pyruvate | Arginine |
| | Pyruvate | Arginine | NADH | Octopine | $NAD^+$ | 15 mM | 40 mM | 10 mM | 2 mM | 0.4 mM | 1 mM | 10 mM |
| Mantle muscle | 0.3 | 3.2 | 0.03 | 0.85 | 0.25 | 9 | 50 | 0 | 20 | 5 | 10 | 45 |
| Brain | 0.25 | 0.4 | 0.02 | 0.09 | 0.11 | 50 | 85 | 50 | 55 | 42 | 12 | 55 |

[a] From Storey and Storey (1979b).

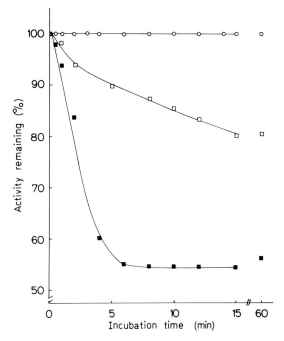

**Fig. 6.** Effect of preincubation in the presence of NAD$^+$, pyruvate, and arginine on the apparent activity of ODH isozymes from *Sepia officinalis*. The formation of the inhibitory quaternary complex, enzyme–NAD$^+$–pyruvate–arginine, which competes with NADH, manifests itself in a slow decline in apparent enzyme activity in an incubate containing brain ODH, NAD$^+$, pyruvate, and arginine. Quaternary complex formation is apparent when brain ODH is incubated with either high (■) (20 mM pyruvate, 100 mM arginine, 1 mM NAD$^+$) or low (□) (0.2 mM pyruvate, 30 mM arginine, 0.2 mM NAD$^+$) substrate concentrations but is not formed (○) when the muscle isozyme of ODH is used. (From Storey and Storey, 1979b.)

be involved in tissue-specific aspects of octopine metabolism, a proposal that has been tested and borne out in *in vivo* studies of the catabolism of octopine in cephalopod tissues (Storey and Storey, 1979a) (see Section IV,G).

## E. Brain ODH versus LDH: Competition at the Pyruvate Branchpoint

The presence of substantial activities of both ODH and LDH in many of the soft tissues of cephalopods (Table III) creates a situation in which the two enzymes could compete for the pyruvate derived from glycolysis. Such a situation has been studied in the brain of the squid *S. oualaniensis*, where the ODH:LDH activity ratio is 4:7 (Fields et al., 1976c). The study showed a number of kinetic differences between the two enzymes, including (a) a higher affinity by LDH for pyruvate, (b) a 10-fold higher affinity by ODH for $NAD^+$, and (c) a very low affinity by LDH for lactate ($K_m$ = 14 m$M$, similar to an M-type LDH) versus a very high affinity by ODH for octopine ($K_m$ = 0.2 m$M$). Based on these properties, Fields et al. (1976c) argued that brain ODH and LDH are unlikely to compete for pyruvate *in vivo* and that in fact they have separate and distinct functions. Squid brain, if it is like brain tissue of most other animals, likely depends on the aerobic oxidation of exogenous substrate (glucose) for most of its energy needs. Brain ODH, with its kinetic similarities to H-type LDH, would function largely in allowing blood octopine to be utilized as an alternate aerobic substrate for brain metabolism. Brain LDH, on the other hand, with its high affinity for pyruvate, provides the system for cytoplasmic redox balance during periods of anoxic stress. A similar division of function may apply to other tissues of the cephalopod that display dual dehydrogenase activities.

## F. Octopine Production and Anaerobic Glycolysis in Cephalopod Muscle

### 1. Octopine Formation in Muscle

The accumulation of octopine as the product of anaerobic glycolysis during muscle work in cephalopods has now been demonstrated in several species (Grieshaber and Gäde, 1976a,b; Hochachka et al., 1977, 1978; Storey and Storey, 1979a; Baldwin and England, 1980). Hochachka et al. (1977) demonstrated that isolated cirri of the spadix muscle of *N. pompilius* produced octopine when stimulated to contract against a constant work load; after 4 min of contraction, octopine concentration had reached 14 $\mu$mol/g wet weight in stimulated muscle versus 3.4 $\mu$mol/g in control muscle. Similarly, the isolated retractor muscle of *Nautilus* accumulated octopine in proportion to the duraton of contraction. Grieshaber and Gäde (1976a) demonstrated the stoichiometry between arginine phosphate breakdown and octopine accumulation in *L. vulgaris* muscle. In the mantle at rest, arginine phosphate, arginine, and octopine contents averaged 178, 170, and 12 $\mu$mol/g dry weight whereas after swimming to exhaustion muscle levels of these compounds were 14, 179, and 98 $\mu$mol/g dry weight, respectively. The decrease in the arginine phosphate pool during muscle work appeared as a direct increase in octopine levels. A similar relationship between arginine phosphate and octopine levels in muscle at rest versus after exhaustive swimming was found by Baldwin and England (1980) for mantle and tentacle muscle of the octopus, *Hapalochlaena maculosa*.

## 2. Anaerobic Glycolysis in S. officinalis Mantle Muscle

Anaerobic glycolysis in mantle muscle has been studied in detail in the cuttlefish, *S. officinalis* (Storey and Storey, 1979a). Both hypoxic stress and burst-swimming activity resulted in the accumulation of octopine as the major product of muscle metabolism (Table VI). In both situations, octopine production was inversely correlated with the depletion of muscle glycogen and arginine phosphate reserves, suggesting that muscle energy production followed the pathways outlined in Fig. 3. Burst swimming resulted in the coordinated mobilization of both glycogen and phosphagen reserves; indeed high-speed jet propulsion, which can only be sustained for a short time, is probably limited by the depletion of both these reserves. Of the energy expended by *S. officinalis* mantle muscle during burst swimming, about 33% (30 $\mu$mol ATP/g wet weight) was derived from the hydrolysis of arginine phosphate reserves, whereas about 60 $\mu$mol ATP/g wet weight was derived from the anaerobic fermentation of glycogen (assuming 3 mol ATP produced per mol glucose 1-phosphate converted to pyruvate). During hypoxic stress, energy production relies even more heavily on the catabolism of glycogen; 30 min exposure to a $P_{O_2}$ of 10 mm Hg almost totally exhausted glycogen reserves whereas more than half of the arginine phosphate pool remained. Muscle octopine levels were correspondingly lower under hypoxic conditions.

In addition to octopine, mantle muscle of *S. officinalis* produced low amounts of other anaerobic end products during exercise or hypoxic stress. Pyruvate, alanine, $\alpha$-glycerophosphate, and malate concentrations were all elevated in stressed muscle. $\alpha$-Glycerophosphate accumulation demonstrates the blockage of the $\alpha$-glycerophosphate cycle, a means for aerobic redox balance, thereby necessitating redox regulation via ODH. The accumulation of malate in *S. officinalis* muscle along with a similar low accumulation of succinate by working *L. vulgaris* muscle (Grieshaber and Gäde, 1976a) suggests that cephalopod muscle has a capacity for producing succinate as an anaerobic end product of carbohydrate catabolism, although this capacity is poorly developed compared to that found in bivalve molluscs (de Zwaan et al., 1976).

## 3. Control of Anaerobic Glycolysis

Activation of muscle glycolysis during burst swimming in *S. officinalis* probably utilizes many of the same principles as outlined in Section III,C for the activation of glycolysis during aerobic muscle work, although the relative importance of regulatory control by NADH is not known. Arginine phosphate, however, has important regulatory effects on *S. officinalis* glycolytic enzymes (Storey, 1981). Arginine phosphate is a strong inhibitor of mantle muscle phosphofructokinase ($K_i$ = 5 m$M$, competitive

## TABLE VI

### Effect of Exercise and Hypoxic Stress on the Levels of Some Metabolites in *Sepia Officinalis* Mantle Muscle[a,b,c]

| Metabolite | At rest | Mild hypoxia | Severe hypoxia | Exercise to exhaustion | Exercise + hypoxia |
|---|---|---|---|---|---|
| Arginine phosphate | 33.62 ± 0.66 | 21.40 ± 3.30 | 18.80 ± 3.00 | 3.76 ± 0.63 | 0.80 ± 0.40 |
| Arginine | 29.60 ± 0.60 | 35.90 ± 2.43 | 37.50 ± 2.47 | 45.30 ± 1.64 | 38.60 ± 1.35 |
| Octopine | 0.17 ± 0.12 | 3.72 ± 1.70 | 3.64 ± 1.13 | 8.59 ± 0.71 | 13.43 ± 0.42 |
| Total | 63.4 | 61.0 | 59.9 | 57.7 | 52.8 |
| Glycogen (as glucose) | 23.70 ± 1.61 | 13.80 ± 2.30 | 2.15 ± 0.85 | 4.50 ± 2.20 | 1.90 ± 1.20 |
| Pyruvate | 0.09 ± 0.01 | 0.30 ± 0.05 | 0.16 ± 0.02 | 0.76 ± 0.08 | 0.26 ± 0.02 |
| Alanine | 4.77 ± 0.88 | — | 8.63 ± 0.85[d] | 10.00 ± 1.50[d] | 9.56 ± 1.06[d] |
| α-Glycerophosphate | 0.33 ± 0.02 | — | 0.79 ± 0.18 | 1.36 ± 0.07 | 1.16 ± 0.15 |
| n | 7 | 3 | 4 | 4 | 3 |

[a] From Storey and Storey (1979a).

[b] Control animals were allowed to rest in aerated seawater. In mild hypoxia, animals were allowed to deplete seawater $P_{O_2}$ by respiration from an initial 130 mm Hg to a final 30 mm Hg. In severe hypoxia, animals were exposed for 30 min to seawater that had been bubbled with nitrogen gas to a $P_{O_2}$ of 10 mm Hg. Exercise was accomplished by forcing animals to swim to exhaustion in large, well-aerated seawater tanks. Exercise plus hypoxia combined the last two stresses; animals were exercised first and then placed for recovery in hypoxic seawater at 10 mm Hg.

[c] Metabolite levels are expressed as $\mu$mol/g wet weight ± SEM. Metabolite levels in stressed animals are all significantly different from those in control animals, $p < .01$.

[d] $p < .02$.

with respect to fructose 6-phosphate), and at physiological levels ($\sim 35$ $\mu$mol/g wet weight) the effects of phosphagen would severely dampen flux through this key site in the muscle at rest. Arginine phosphate effects are reversed by low concentrations of the activator AMP. At the initiation of muscle work, therefore, the combined effects of decreasing arginine phosphate levels in muscle and a sharp rise in cellular AMP concentration would effectively activate flux through the PFK locus and lead to a rapid increase in glycolytic energy production. Arginine phosphate also has inhibitory effects on *S. officinalis* mantle muscle hexokinase, aldolase, $\alpha$-glycerophosphate dehydrogenase, glyceraldehyde 3-phosphate dehydrogenase, phosphoglycerate kinase, and pyruvate kinase, although with the exception of hexokinase ($K_i = 5$ m$M$), the effects of phosphagen on these enzymes are relatively weak ($K_i$ values between 15 and 50 m$M$) (Storey, 1981). Possibly these inhibitions may be important in dampening glycolytic flux in the muscle at rest when arginine phosphate levels are high.

Octopine also has inhibitory effects on glycolytic enzymes from *S. officinalis* muscle. Like arginine phosphate, octopine is a strong inhibitor of hexokinase and PFK. The phosphagen also has inhibitory effects on glyceraldehyde 3-phosphate dehydrogenase and phosphoglycerate kinase as well as positive effects on the activities of phosphoglucomutase and pyruvate kinase (Storey, 1981). Octopine effects on PFK are reversed by AMP, so a buildup of octopine during glycolytic muscle work would not affect glycolytic flux but octopine might become an important effector on the return to the resting state. Octopine could act to inhibit glycolytic flux at a time when arginine phosphate and glycogen reserves are low and when accumulated octopine might be the preferred aerobic substrate during muscle recovery.

## G. Metabolic Fate of Muscle Octopine

The production of octopine in mantle muscle during anaerobic work is analogous to the accumulation of lactate during vertebrate white muscle work. In addition, ODH, like LDH, occurs as tissue-specific isozymes, one isozyme specialized as a pyruvate reductase and occurring predominantly in muscle tissue with another isozyme, geared to the oxidation of octopine, occurring mainly in the more aerobic tissues. Recent studies indicate a further analogy with the lactate–LDH system of vertebrates: Octopine produced during muscle work can be catabolized outside of the muscle mass by tissues that can make effective use of the compound as an aerobic substrate (Storey and Storey, 1979b; Storey et al., 1979; Hochachka et al., 1982). Octopine is released into the bloodstream at the cessation of muscle work and utilized by other tissues that are capable of oxidizing the pyruvate moiety as an aerobic substrate or channeling it into the resynthesis of glucose. A modified form of the Cori cycle, taking into account the necessary cycling of arginine, may be present in cephalopods (Fig. 7).

Considerable experimental evidence supports the proposition that some or all of the octopine produced in muscle is oxidized outside of the muscle mass. Some of this evidence follows:

1.   The total pool of arginine phosphate–arginine–octopine in mantle muscle decreases as a result of swimming to exhaustion or hypoxia (Storey and Storey, 1979a; Grieshaber and Gäde, 1976a) (Table VI). Total pool size was decreased by 17% in *S. officinalis* specimens that were first exercised and then allowed to recover in hypoxic water. The results suggest that octopine is flushed out of the muscle.

2.   Blood octopine (as well as pyruvate and alanine) levels were elevated during hypoxia and during recovery from hypoxia or exercise stress (Storey and Storey, 1979a; Gäde, 1980b). Although the levels of all blood metabolites are low in cephalopods, octopine concentration in the bloodstream increased up to eightfold during recovery from stress. This could allow significant movement of octopine out of the muscle mass, particularly in light of the rapid uptake of blood octopine by *S. officinalis* tissues.

3.   Blood octopine and glucose levels are inversely related over the

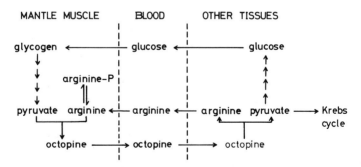

**Fig. 7.**  Intertissue cycling of metabolites in cephalopods: a modified Cori cycle. Octopine produced during muscle work is released into the bloodstream and utilized as an aerobic substrate in other tissues. The pyruvate moiety of octopine can be metabolized by two routes: (a) oxidation by the Krebs cycle or (b) reconversion to glucose via gluconeogenic reactions. A modified form of the Cori cycle appears to exist in which octopine leaving the muscle is replaced by arginine and glucose recycled to the muscle.

course of a hypoxic stress and recovery (Storey et al., 1979) (Fig. 8). Blood glucose levels in *S. officinalis* declined during hypoxia, indicating an increased uptake of glucose by tissues, whereas octopine levels in blood were elevated as energy production by anaerobic glycolysis increased. During recovery the relationship was reversed. Octopine was cleared from the blood and a rapid restoration of blood glucose levels occurred, possibly as a result of gluconeogenic reactions involving the use of octopine as the substrate.

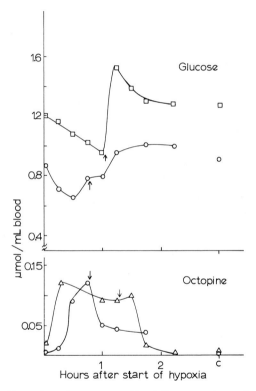

**Fig. 8.** Effect of hypoxia and recovery on blood glucose and octopine concentrations in *Sepia officinalis.* Hypoxic stress was begun at a seawater $P_{O_2}$ of 45 mm Hg and continued for 45–75 min to final seawater oxygen tensions of 19–31 mm Hg. Individuals were returned to normoxic water for recovery at the times indicated by the arrows. During hypoxia, blood glucose concentration declined steadily while blood octopine was elevated. During recovery, blood octopine was quickly cleared and glucose concentrations were restored to normal, after initial overshoots. Control concentrations (c) of blood glucose and octopine, taken before hypoxic stress, are given on the right of the figure ($\square$, $\bigcirc$, $\triangle$ represent data from three individuals.) (From Storey et al., 1979.)

4. An intravenously injected bolus of octopine (resulting in initial blood octopine levels 200 times normal) was rapidly cleared from the blood of *S. officinalis* with a half-time of less than 15 min and was accompanied by a transient elevation of blood glucose levels (Storey et al., 1979). This suggests a very high capacity by the tissues of *S. officinalis* for the uptake of blood octopine and also demonstrates the tight regulation of blood metabolite levels in the cephalopod.

5. [$^{14}$C]Octopine (uniformly labeled in the arginine moiety) was rapidly cleared from the blood and was concentrated by tissues such as ventricle, branchial heart, gill, and brain (Storey and Storey, 1979a). Mantle muscle was the least effective tissue in concentrating exogenous [$^{14}$C]octopine.

6. Tissues such as brain, gill, and ventricle not only showed a high uptake of blood [$^{14}$C]octopine but also readily oxidized the end product (Storey and Storey, 1979a) (Table VII). In particular, brain and ventricle showed a high capacity for the oxidation of octopine *in vivo*. This capacity in brain agrees well with the presence of the brain-specific isozyme of ODH with its low $K_m$ (0.085 m$M$) for octopine, and suggests that cephalopod brain can make effective use of octopine as an aerobic substrate just as vertebrate brain can utilize blood lactate. Mantle muscle, on the other hand, showed a very low capacity for the oxidation of [$^{14}$C]octopine *in vivo*.

TABLE VII

Catabolism of [$^{14}$C]Octopine by Tissues
of *Sepia officinalis* [a,b]

|  | Mode of injection | |
| --- | --- | --- |
| Tissue | Vena cava | Specific site |
| Mantle muscle | 0.0 | 5.0 |
| Gill | 5.8 | — |
| Ventricle | 32.0 | — |
| Brain | 20.0 | 40.0 |

[a] From Storey and Storey (1979a).

[b] [$^{14}$C]Octopine (uniformly labeled in the arginine moiety) was administered either via injection into the vena cava ($n = 2$) or by direct injection into tissues ($n = 3$). After 20 min rest under aerobic conditions, animals were killed, tissues excised, and the radioactivity in [$^{14}$C]octopine versus [$^{14}$C]arginine determined. Results show [$^{14}$C]arginine as a percentage of total tissue [$^{14}$C]content.

7. Both [$^{14}$C]octopine and [$^{14}$C]arginine are vigorously metabolized by the tissues of *S. officinalis*. Both were rapidly removed from the blood (as shown by vena cava–branchial artery differences) across three tissues—kidney, branchial heart, and gills—and catabolized by these tissues (Hochachka et al., 1982). In kidney, [$^{14}$C]arginine was catabolized by the urea cycle or fed, through the intermediary glutamate, into the carboxylic acid pool of the Krebs cycle or into the synthesis of proline. [$^{14}$C]octopine, however, was handled differently; only a limited amount of radioactivity was incorporated into intermediates of the urea cycle whereas most of the $^{14}$C label appeared in the proline pool.

The results demonstrate a rather complex metabolism of octopine in cephalopod tissues. Some portion of the octopine produced during muscle work may be reoxidized *in situ*. This would retain arginine within the muscle pool for the resynthesis of arginine phosphate reserves. The pyruvate moiety of octopine could be oxidized by the Krebs cycle in muscle as an aerobic substrate or could be channeled into gluconeogenic reactions. Although the activities of gluconeogenic enzymes in mantle muscle are very low (Hochachka et al., 1975a), recent studies on vertebrate skeletal muscle indicate that a significant fraction of the lactate produced in muscle is actually catabolized within the muscle. This may also be true in cephalopod muscle. Bone et al. (1981) point out that the blood supply to the mantle muscle passes first through the central zone of the mantle and then flows to the inner and outer zones, and they suggest that anaerobic metabolites produced in the central zone during high-speed jetting could be transferred to the more aerobic fibers of the inner and outer zones for reoxidation.

However, it is clear that some fraction, perhaps a high fraction, of the octopine produced in mantle muscle is shunted out of the muscle during recovery from exercise and delivered via the bloodstream to tissues that can utilize octopine as a substrate. A reversal of the ODH reaction in brain would provide pyruvate to be oxidized as an aerobic substrate by the Krebs cycle whereas arginine could be recycled to the muscle. Other tissues, such as hepatopancreas, may utilize octopine in a modified Cori cycle, directing the pyruvate moiety into gluconeogenesis and cycling both glucose and arginine back to the muscle (Fig. 7). And finally, other tissues, such as kidney, are capable of catabolizing the arginine moiety of octopine as well as the pyruvate moiety. The arginine moiety is channeled into proline synthesis with proline likely recycled to the muscle. This arginine–proline cycle may function to restore muscle proline reserves that are utilized during aerobic muscle work (Hochachka et al., 1982).

## Acknowledgments

The authors wish to thank Dr. Philip Cohen, University of Dundee, Scotland, for allowing us the use of unpublished data on octopus phosphorylase. We are also grateful to the director and staff of the Marine Biological Association of the United Kingdom for their facilities and goodwill during our many visits to study cephalopods. P. W. Hochachka and J. H. A. Fields provided us with many stimulating hours of discussion on mollusc metabolism.

## References

Ackermann, D., and Mohr, M. (1937). The occurrence of octopine, agmatine and arginine in the octopod, *Eledone maschata*. *Hoppe-Seyler's Z. Physiol. Chem.* **250**, 244–252.

Baccetti, B., Pallini, V., and Burrini, A. G. (1975). Localization and catalytic properties of lactate dehydrogenase in different sperm models. *Exp. Cell Res.* **90**, 183–190.

Baldwin, J., and England, W. R. (1980). A comparison of anaerobic energy metabolism in mantle and tentacle muscle of the blue-ringed octopus, *Hapalochlaena maculosa* during swimming. *Aust. J. Zool.* **28**, 407–412.

Ballantyne, J. S., Hochachka, P. W., and Mommsen, T. P. (1981). Studies on the metabolism of the migratory squid, *Loligo opalescens:* Enzymes of tissues and heart mitochondria. *Mar. Biol. Lett.* **2**, 75–85.

Ballard, F. J. (1970). Adenine nucleotides and the adenylate kinase equilibrium in livers of foetal and newborn rats. *Biochem. J.* **117**, 231–235.

Beaumont, A. R., Day, J. R., and Gäde, G. (1980). Genetic variation at the octopine dehydrogenase locus in the adductor muscle of *Cerastoderma edule* (L.) and six other bivalve species. *Mar. Biol. Lett.* **1**, 137–148.

Biellmann, J. F., Branlant, G., and Olomucki, A. (1973). Stereochemistry of the hydrogen transfer to the coenzyme by octopine dehydrogenase. *FEBS Lett.* **32**, 254–256.

Biellmann, J. F., Branlant, G., and Wallén, L. (1977). Stereochemistry of octopine and of its isomers and their enzymatic properties. *Bioorg. Chem.* **6**, 89–93.

Bone, Q., Pulsford, A., and Chubb, A. D. (1981). Squid mantle muscle. *J. Mar. Biol. Assoc. U.K.* **61**, 327–342.

Boucaud-Camou, E. (1971). Lipid constituents of the liver of *Sepia officinalis*. *Mar. Biol. (Berlin)* **8**, 66–69.

Bowlus, R. D., and Somero, G. N. (1979). Solute compatibility with enzyme function and structure: Rationales for the selection of osmotic agents and end products of anaerobic metabolism in marine invertebrates. *J. Exp. Zool.* **208**, 137–152.

Cohen, P. (1980). Well established systems of enzyme regulation by reversible phosphorylation. *In* "Recently Discovered Systems of Enzyme Regulation by Reversible Phosphorylation" (P. Cohen, ed.), pp. 1–10. Elsevier/North-Holland, New York.

Collicutt, J. M., and Hochachka, P. W. (1977). The anaerobic oyster heart. Coupling of glucose and aspartate fermentation. *J. Comp. Physiol.* **115**, 147–157.

Dando, P. R., Storey, K. B., Hochachka, P. W., and Storey, J. M. (1981). Multiple dehydrogenases in marine molluscs: Electrophoretic analysis of alanopine dehydrogenase, strombine dehydrogenase, octopine dehydrogenase and lactate dehydrogenase. *Mar. Biol. Lett.* **2**, 249–257.

de Zwaan, A. (1972). Pyruvate kinase in muscle extracts of the sea mussel, *Mytilus edulis* L. *Comp. Biochem. Physiol. B* **42B**, 7–14.

de Zwaan, A., and Zurburg, W. (1981). The formation of strombine in the adductor muscle of the sea mussel, *Mytilus edulis* L. *Mar. Biol. Lett.* **2**, 179–192.

de Zwaan, A., Kluytmans, J. H. F. M., and Zandee, D. I. (1976). Facultative anaerobiosis in molluscs. *Biochem. Soc. Symp.* No. 41, 133–168.

Doublet, M. O., and Olomucki, A. (1975). Investigations on the kinetic mechanism of octopine dehydrogenase. I. Steady-state kinetics. *Eur. J. Biochem.* **59**, 175–183.

Drummond, G. I. (1971). Microenvironment and enzyme function: control of energy metabolism during muscle work. *Am. Zool.* **11**, 83–97.

Edington, D. W., Ward, G. R., and Saville, W. A. (1973). Energy metabolism in working muscle: concentration profiles of selected metabolites. *Am. J. Physiol.* **224**, 1375–1380.

Eichner, R. D., and Kaplan, N. O. (1977). Catalytic properties of lactate dehydrogenase in *Homarus americanus*. *Arch. Biochem. Biophys.* **181**, 501–507.

Ellington, W. R. (1979). Octopine dehydrogenase in the basilar muscle of the sea anemone, *Metridium senile*. *Comp. Biochem. Physiol. B* **63B**, 349–354.

Everse, J., and Kaplan, N. O. (1973). Lactate dehydrogenase: Structure and function. *Adv. Enzymol. Relat. Areas Mol. Biol.* **37**, 61–148.

Everse, J., Barnett, R. E., Thorne, C. J. R., and Kaplan, N. O. (1971). The formation of ternary complexes by diphosphopyridine nucleotide-dependent dehydrogenases. *Arch. Biochem. Biophys.* **143**, 444–460.

Fields, J. H. A. (1976). A dehydrogenase requiring alanine and pyruvate as substrates from oyster adductor muscle. *Fed. Proc., Fed. Am. Soc. Exp. Biol.* **35**, 1687.

Fields, J. H. A. (1977). Anaerobic metabolism in the cockle, *Clinocardium nuttali*. *Am. Zool.* **17**, 943.

Fields, J. H. A., and Hochachka, P. W. (1976). Oyster citrate synthase: control of carbon entry into the Krebs cycle of a facultative anaerobe. *Can. J. Zool.* **54**, 892–895.

Fields, J. H. A., and Hochachka, P. W. (1981). Purification and properties of alanopine dehydrogenase from the adductor muscle of the oyster, *Crassostrea gigas*. *Eur. J. Biochem.* **114**, 615–621.

Fields, J. H. A., and Quinn, J. F. (1981). Some theoretical considerations on cytosolic redox balance during anaerobiosis in marine invertebrates. *J. Theor. Biol.* **88**, 35–45.

Fields, J. H. A., Guderley, H. E., Storey, K. B., and Hochachka, P. W. (1976a). Octopus mantle citrate cynthase. *Can. J. Zool.* **54**, 886–891.

Fields, J. H. A., Baldwin, J., and Hochachka, P. W. (1976b). On the role of octopine dehydrogenase in cephalopod mantle muscle metabolism. *Can. J. Zool.* **54**, 871–878.

Fields, J. H. A., Guderley, H., Storey, K. B., and Hochachka, P. W. (1976c). The pyruvate branchpoint in squid brain: competition between octopine dehydrogenase and lactate dehydrogenase. *Can. J. Zool.* **54**, 879–885.

Fields, J. H. A., Eng, A. K., Ramsden, W. D., Hochachka, P. W., and Weinstein, B. (1980). Alanopine and strombine are novel imino acids produced by a dehydrogenase found in the adductor muscle of the oyster, *Crassostrea gigas*. *Arch. Biochem. Biophys.* **201**, 110–114.

Florkin, M. (1966). Nitrogen metabolism. *In* "Physiology of Mollusca" (K. M. Wilbur and C. M. Yonge, eds.), Vol. 2, pp. 309–351. Academic Press, New York.

Gäde, G. (1979). L-Lactate specific, dimeric lactate dehydrogenase from the mantle muscle of the squid, *Loligo vulgaris*: purification and catalytic properties. *Comp. Biochem. Physiol. B* **63B**, 387–393.

Gäde, G. (1980a). Biological role of octopine formation in marine molluscs. *Mar. Biol. Lett.* **1**, 121–135.

Gäde, G. (1980b). A comparative study of ODH isozymes in gastropod, bivalve and cephalopod molluscs. *Comp. Biochem. Physiol. B* **67B**, 575–582.

Goddard, C. K., and Martin, A. W. (1966). Carbohydrate metabolism. *In* "Physiology of Mollusca" (K. M. Wilbur and C. M. Yonge, eds.), Vol. 2, pp. 275–308. Academic Press, New York.

Goldin, B. R., and Frieden, C. (1971). L-Glutamate dehydrogenases. *Curr. Top. Cell. Regul.* **4**, 77–117.

Grieshaber, M., and Gäde, G. (1976a). The biological role of octopine in the squid, *Loligo vulgaris* (Lamarck). *J. Comp. Physiol.* **108**, 225–232.

Grieshaber, M., and Gäde, G. (1976b). The biological role of octopine in molluscs. *Verh. Dtsch. Zool. Ges.* **69**, 222.

Guderley, H. E., Storey, K. B., Fields, J. H. A., and Hochachka, P. W. (1976). Catalytic and regulatory properties of pyruvate kinase isozymes from *Octopus* mantle muscle and liver. *Can. J. Zool.* **54**, 863–870.

Haas, S., Thome-Beau, F., Olomucki, A., and Thoai, N. van. (1973). Purification de l'octopine déshydrogénase de *Sipunculus nudus*. Etude comparative avec l'octopine déshydrogénase de *Pecten maximus*. *C. R. Seances Soc. Biol. Ses Fil.* **276**, 831–834.

Hammen, C. S. (1969). Substrate specificity of lactate dehydrogenase in some marine invertebrates. *Am. Zool.* **9**, 1105.

Hiltz, D. F., and Dyer, W. J. (1971). Octopine in the postmortem adductor muscles of the sea scallop (*Placopecten magellanicus*). *J. Fish. Res. Board Can.* **28**, 869–874.

Hochachka, P. W. (1976). Design of metabolic and enzymic machinery to fit lifestyle and environment. *Biochem. Soc. Symp.* No. 41, 3–31.

Hochachka, P. W. (1980). "Living Without Oxygen," pp. 42–59. Harvard Univ. Press, Cambridge, Massachusetts.

Hochachka, P. W., Moon, T. W., Mustafa, T., and Storey, K. B. (1975a). Metabolic sources of power for mantle muscle of a fast swimming squid. *Comp. Biochem. Physiol. B* **52B**, 151–158.

Hochachka, P. W., Storey, K. B., and Baldwin, J. (1975b). Squid muscle citrate synthase: control of carbon entry into the Krebs cycle. *Comp. Biochem. Physiol. B* **52B**, 193–199.

Hochachka, P. W., Hartline, P. H., and Fields, J. H. A. (1977). Octopine as an end product of anaerobic glycolysis in the chambered *Nautilus*. *Science* **195**, 72–74.

Hochachka, P. W., French, C. J., and Meredith, J. (1978). Metabolic and ultrastructural organization in *Nautilus* muscles. *J. Exp. Zool.* **205**, 51–62.

Hochachka, P. W., Mommsen, T. P., Storey, J. M., Storey, K. B., Johansen, K., and French, C. J. (1982). The relationship between arginine and proline metabolism in cephalopods. *Mar. Biol. Lett.* **3**, in press.

Irvin, J. L. (1938). Further studies on octopine. *J. Biol. Chem.* **123**, lxii.

Irvin, J. L., and Wilson, D. W. (1939). Studies on octopine. II. The precursor of octopine in autolyzing scallop muscle. *J. Biol. Chem.* **127**, 575–579.

Johansen, K., Redmond, J. R., and Bourne, G. B. (1978). Respiratory exchange and transport of oxygen in *Nautilus pompilius*. *J. Exp. Zool.* **205**, 27–36.

Leray, C., and Ladanyi, P. (1969). Histo-enzymological investigations on the muscle fibres of a cephalopode (*Octopus vulg.* Lam.). *Acta Biol. Acad. Sci. Hung.* **20**, 153–161.

Livingstone, D. R., and Bayne, B. L. (1974). Pyruvate kinase from the mantle tissue of *Mytilus edulis* L. *Comp. Biochem. Physiol. B* **48B**, 481–497.

Long, G. L. (1976). The stereospecific distribution and evolutionary significance of invertebrate lactate dehydrogenases. *Comp. Biochem. Physiol. B* **55B**, 77–83.

Long, G. L., Ellington, W. R., and Duda, T. F. (1979). Comparative enzymology and physiological role of D-lactate dehydrogenase from the foot muscle of two gastropod molluscs. *J. Exp. Zool.* **207,** 237–248.

Luisi, D. L., Baici, A., Olomucki, A., and Doublet, M. O. (1975). Temperature-determined enzymatic functions of octopine dehydrogenase. *Eur. J. Biochem.* **50,** 511–516.

Mann, T., Martin, A. W., Thiersch, J. B., Lutwak-Mann, C., Brooks, D. E., and Jones, R. (1974). D(-)Lactic acid and D(-)lactate dehydrogenase in octopus spermatozoa. *Science* **185,** 453–454.

Mann, T., Martin, A. W., Lutwak-Mann, C., and Thiersch, J. B. (1977). Glycogenolysis in octopus spermatozoa. *J. Exp. Mar. Biol. Ecol.* **27,** 155–159.

Mansour, T. (1972). Phosphofructokinase. *Curr. Top. Cell. Regul.* **5,** 1–44.

Martin, A. W., Jones, R., and Mann, T. (1976). D(-)Lactic acid formation and D(-)lactate dehydrogenase in octopus spermatozoa. *Proc. R. Soc. London, Ser. B* **193,** 235–243.

Ménagé, A., and Morel, G. (1964). Sur la présence de l'octopine dans les tissus de crown gall. *C. R. Hebd. Seances Acad. Sci.* **259,** 4795–4796.

Molla, A., Kilhoffer, M.-C., Ferraz, C., Audemard, E., Walsh, M. P., and Demaille, J. G. (1981). *Octopus* calmodulin. The trimethyllysyl residue is not required for myosin light chain kinase activation. *J. Biol. Chem.* **256,** 15–18.

Mommsen, T. P., and Hochachka, P. W. (1981). Respiratory and enzymatic properties of squid heart mitochondria. *Eur. J. Biochem.* **120,** 345–350.

Mommsen, T. P., Ballantyne, J., MacDonald, D., Gosline, J., and Hochachka, P. W. (1981). Analogues of red and white muscle in squid mantle. *Proc. Natl. Acad. Sci. U.S.A.* **78,** 3274–3278.

Monneuse-Doublet, M. O., Olomucki, A., and Buc, J. (1978). Investigations on the kinetic mechanism of octopine dehydrogenase. A regulatory behaviour. *Eur. J. Biochem.* **84,** 441–448.

Monneuse-Doublet, M. O., Lefebre, F., and Olomucki, A. (1980). Isolation and characterization of two molecular forms of octopine dehydrogenase from *Pecten maximus* L. *Eur. J. Biochem.* **108,** 261–269.

Moon, T. W., and Hulbert, W. C. (1975). The ultrastructure of the mantle musculature of the squid, *Symplectoteuthis oualaniensis*. *Comp. Biochem. Physiol. B* **52B,** 145–149.

Moore, E., and Wilson, D. W. (1937a). Nitrogenous extractives of scallop muscle. I. The isolation and study of the structure of octopine. *J. Biol. Chem.* **119,** 573–584.

Moore, E., and Wilson, D. W. (1937b). Nitrogenous extractives of scallop muscle. II. Isolation and quantitative analyses of muscles from freshly killed scallops. *J. Biol. Chem.* **119,** 585–588.

Morizawa, K. (1927). The extractive substances in *Octopus octopodia*. *Acta Sch. Med. Univ. Imp. Kioto* **9,** 285–298.

Mustafa, T., and Hochachka, P. W. (1971). Catalytic and regulatory properties of pyruvate kinase in tissues of a marine bivalve. *J. Biol. Chem.* **246,** 3196–3203.

Neely, J. R., Roverto, M. J., Whitmer, J. T., and Morgan, H. E. (1973). Effects of ischemia on function and metabolism of the isolated working rat heart. *Am. J. Physiol.* **225,** 651–659.

O'Dor, R. K., and Wells, M. J. (1978). Reproduction versus somatic growth: Hormonal control in *Octopus vulgaris*. *J. Exp. Biol.* **77,** 15–31.

O'Dor, R. K., Durward, R. D., and Wells, M. J. (1978). The protein–energy economy in cephalopods. *Proc. Can. Soc. Zool.* p. 48.

Olomucki, A., Hue, C., Lefebre, F., and Thoai, N. van. (1972). Octopine dehydrogenase. Evidence for a single chain structure. *Eur. J. Biochem.* **28,** 261–268.

Packard, A. (1969). Jet propulsion and the giant fibre response in *Loligo*. *Nature (London)* **221,** 875–877.

Pho, D. B., Olomucki, A., Hue, C. L., and Thoai, N. van. (1970). Spectrophotometric studies of binary and ternary complexes of octopine dehydrogenase. *Biochim. Biophys. Acta* **206,** 46–53.

Plaxton, W. C., and Storey, K. B. (1982). Tissue specific isozymes of alanopine dehydrogenase in the channeled whelk, *Busycotypus canaliculatum*. *Can. J. Zool.* **60,** 1568–1572.

Regnouf, F., and Thoai, N. van. (1970). Octopine and lactate dehydrogenases in mollusc muscles. *Comp. Biochem. Physiol.* **32,** 411–416.

Robertson, J. D. (1965). Studies on the chemical composition of muscle tissue: III. The mantle muscle of cephalopod molluscs. *J. Exp. Biol.* **42,** 153–175.

Robin, Y. (1964). Biological distribution of guanidines and phosphagens in marine annelida and related phyla from California, with a note on pluriphosphagens. *Comp. Biochem. Physiol.* **12,** 347–367.

Robin, Y., and Guillou, Y. (1977). An ion-exchange column chromatographic method for the isolation of octopine from *Sepia officinalis* L. muscle. *Anal. Biochem.* **83,** 45–51.

Roche, J., Thoai, N. van, Robin, Y., Garcia, I., and Hatt, J. L. (1952). Sur la nature et la répartition des guanidines monosubstitueés dans les tissus des invertébrés. Presence de dérivés métaboliques de l'arginine chez des mollusques, des crustacés et des echinodermes. *C. R. Seances Soc. Biol. Ses Fil.* **146,** 1899–1902.

Rowan, A. N., and Newsholme, E. A. (1979). Changes in the contents of adenine nucleotides and intermediates of glycolysis and the citric acid cycle on flight muscle of the locust upon flight and their relationship to the control of the cycle. *Biochem. J.* **178,** 209–216.

Sacktor, B. (1970). Regulation of intermediary metabolism with special reference to control mechanisms in insect flight muscle. *Adv. Insect Physiol.* **7,** 267–348.

Sacktor, B. (1976). Biochemical adaptations for flight in the insect. *Biochem. Soc. Symp.* No. 41, 111–131.

Sacktor, B., and Hurlbert, E. C. (1966). Regulation of metabolism in working muscle *in vivo*. II. Concentrations of adenine nucleotides, arginine phosphate, and inorganic phosphate in insect flight muscle during flight. *J. Biol. Chem.* **241,** 632–634.

Sacktor, B., and Wormser-Shavit, E. (1966). Regulation of metabolism in working muscle *in vivo*. I. Concentrations of some glycolytic, tricarboxylic acid cycle, and amino acid intermediates in insect flight muscle during flight. *J. Biol. Chem.* **241,** 624–631.

Safer, B., and Williamson, J. R. (1973). Mitochondrial–cytosolic interactions in perfused rat heart. *J. Biol. Chem.* **248,** 2570–2579.

Sakaguchi, M., Hiltz, D. F., and Dyer, W. J. (1975). Metabolism of pyruvate in postmortem adductor muscle of the sea scallop (*Placopecten magellanicus*) during iced storage. *J. Fish. Res. Board Can.* **32,** 1329–1337.

Sangster, A. W., Thomas, S. E., and Tingling, N. L. (1975). Fish attractants from marine invertebrates. Arcamine from *Arca zebra* and strombine from *Strombus gigas*. *Tetrahedron* **31,** 1135–1137.

Sato, M., Sato, Y., and Tsuchiya, Y. (1977). Studies on the extractives of molluscs. I. $\alpha$-Iminodipropionic acid isolated from the squid muscle extracts. *Nippon Suisan Gakkaishi* **43,** 1077–1079.

Scheid, M. J., and Awapara, J. (1972). Stereospecificity of some invertebrate lactic dehydrogenases. *Comp. Biochem. Physiol. B* **43B,** 619–626.

Seamon, K. B., and Moore, B. W. (1980). Octopus calmodulin. Structural comparison with bovine brain calmodulin. *J. Biol. Chem.* **255,** 11644–11647.

Smith, E. L., Austen, B. M., Blumenthal, K. M., and Nyc, J. F. (1975). Glutamate dehydrogenases. *In* "The Enzymes" (P. D. Boyer, ed.), 3rd ed., Vol. 11, pp. 293–367. Academic Press, New York.

Srere, P. A. (1974). Controls of citrate synthase activity. *Life Sci.* **15**, 1695–1710.

Storey, K. B. (1976). Purification and properties of squid mantle adenylate kinase. Role of NADH in control of the enzyme. *J. Biol. Chem.* **251**, 7810–7815.

Storey, K. B. (1977a). Purification and characterization of arginine kinase from the mantle muscle of the squid, *Symplectoteuthis oualaniensis.* Role of the phosphagen/phosphagen kinase system in a highly aerobic muscle. *Arch. Biochem. Biophys.* **179**, 518–526.

Storey, K. B. (1977b). Tissue specific isozymes of octopine dehydrogenase in the cuttlefish, *Sepia officinalis.* The roles of octopine dehydrogenase and lactate dehydrogenase in *Sepia. J. Comp. Physiol.* **115**, 159–169.

Storey, K. B. (1981). Arginine phosphate and octopine effects on glycolytic enzyme activities from *Sepia officinalis* mantle muscle. *J. Comp. Physiol.* **142**, 501–507.

Storey, K. B., and Dando, P. R. (1982). Substrate specificities of octopine dehydrogenases from marine invertebrates. *Comp. Biochem. Physiol. B.* **73B**, 521–528.

Storey, K. B., and Hochachka, P. W. (1975a). Redox regulation of muscle phosphofructokinase in a fast swimming squid. *Comp. Biochem. Physiol. B* **52B**, 159–163.

Storey, K. B., and Hochachka, P. W. (1975b). Squid muscle glyceraldehyde-3-phosphate dehydrogenase: control of the enzyme in a tissue with an active α-glycero-P cycle. *Comp. Biochem. Physiol. B* **52B**, 179–182.

Storey, K. B., and Hochachka, P. W. (1975c). Squid muscle pyruvate kinase: control properties in a tissue with an active α-GP cycle. *Comp. Biochem. Physiol. B* **52B**, 187–191.

Storey, K. B., and Hochachka, P. W. (1975b). Alpha-glycerophosphate dehydrogenase: its role in the control of the cytoplasmic arm of the alpha-glycerophosphate cycle in squid mantle. *Comp. Biochem. Physiol. B* **52B**, 169–173.

Storey, K. B., and Hochachka, P. W. (1975e). The kinetic requirements of cytoplasmic alpha-glycerophosphate (α-GP) dehydrogenase in muscle with active α-GP cycles. *Comp. Biochem. Physiol. B* **52B**, 175–178.

Storey, K. B., and Storey, J. M. (1978). Energy metabolism in the mantle muscle of the squid, *Loligo pealeii. J. Comp. Physiol.* **123**, 169–175.

Storey, K. B., and Storey, J. M. (1979a). Octopine metabolism in the cuttlefish, *Sepia officinalis:* octopine production by muscle and its role as an aerobic substrate for nonmuscular tissues. *J. Comp. Physiol.* **131**, 311–319.

Storey, K. B., and Storey, J. M. (1979b). Kinetic characterization of tissue-specific isozymes of octopine dehydrogenase from mantle muscle and brain of *Sepia officinalis. Eur. J. Biochem.* **93**, 545–552.

Storey, K. B., Baldwin, J., and Hochachka, P. W. (1975a). Squid muscle fructose diphosphatase and its role in the control of F6P-FDP cycling. *Comp. Biochem. Physiol. B* **52B**, 165–168.

Storey, K. B., Mustafa, T., and Hochachka, P. W. (1975b). Squid muscle malic enzyme. *Comp. Biochem. Physiol. B* **52B**, 183–185.

Storey, K. B., Fields, J. H. A., and Hochachka, P. W. (1978). Purification and properties of glutamate dehydrogenase from the mantle muscle of the squid, *Loligo pealeii.* Role of the enzyme in energy production from amino acids. *J. Exp. Zool.* **205**, 111–118.

Storey, K. B., Storey, J. M., Johansen, K., and Hochachka, P. W. (1979). Octopine metabolism in *Sepia officinalis:* effect of hypoxia and metabolite loads on the blood levels of octopine and related compounds. *Can. J. Zool.* **57**, 2331–2336.

Suryanarayanan, H., and Alexander, K. M. (1971). Fuel reserves in molluscan muscle. *Comp. Biochem. Physiol.* **40**, 55–60.

Suyama, M., and Kobayashi, H. (1980). Free amino acids and quaternary ammonium bases in mantle muscle of squids. *Nippon Suisan Gakkaishi* **46**, 1261–1264.

Thoai, N. van, and Robin, Y. (1959). Métabolisme des dérivés guanidylés. VIII. Biosynthèse de l'octopine et répartition de l'enzyme l'opérant chez les invertebrés. *Biochim. Biophys. Acta* **35,** 446–453.

Thoai, N. van, Huc, C., Pho, D. B., and Olomucki, A. (1969). Octopine déshydrogénase. Purification et propriétés catalytiques. *Biochim. Biophys. Acta* **191,** 46–57.

Ward, D. V., and Wainwright, S. A. (1972). Locomotory aspects of squid mantle structure. *J. Zool.* **167,** 437–449.

Williamson, J. R. (1966). Glycolytic control mechanisms. II. Kinetics of intermediate changes during the aerobic–anoxic transition in perfused rat heart. *J. Biol. Chem.* **241,** 5026–5036.

Yushok, W. D. (1971). Control mechanisms of adenine nucleotide metabolism of ascites tumor cells. *J. Biol. Chem.* **246,** 1607–1617.

Zammit, V. A. (1978). Possible relationship between energy metabolism of muscle and oxygen binding characteristics of haemocyanin of cephalopods. *J. Mar. Biol. Assoc. U.K.* **58,** 421–424.

Zammit, V. A., and Newsholme, E. A. (1976). The maximal activities of hexokinase, phosphorylase, phosphofructokinase, glycerol phosphate dehydrogenases, lactate dehydrogenase, octopine dehydrogenase, phosphoenolpyruvate carboxykinase, nucleoside diphosphatekinase, glutamate–oxloacetate transaminase and arginine kinase in relation to carbohydrate utilization in muscle from marine invertebrates. *Biochem. J.* **160,** 447–462.

# 4

# Carbohydrate Catabolism in Bivalves

## ALBERTUS DE ZWAAN

Laboratory of Chemical Animal Physiology
State University of Utrecht
3508 TB Utrecht
The Netherlands

## I. Introduction

Much of the biochemical research on bivalves before 1965 concerned a categorical survey of the presence of enzymes belonging to the main met-

THE MOLLUSCA, VOL. 1
Metabolic Biochemistry
and Molecular Biomechanics

abolic pathways of vertebrates (Goddard and Martin, 1966). The existence of classical pathways such as glycolysis, the citric acid cycle, and the respiratory chain is now well established (de Zwaan, 1977) and research has since focused on the question whether deviations or differences at the biochemical level may have evolved among different species because of differences in life-style. In this respect the bivalves were of particular interest because they are able to sustain anoxia for long periods (von Brand, 1946; Theede, 1973), and the breakdown of carbohydrate was obviously important because in the absence of oxygen, phosphorylation of ADP is still possible at the level of substrates.

The work of Hammen and Awapara in the period 1960–1970 led to early indications of biochemical deviations of the classical glycolysis in bivalves. No lactate, but succinate and alanine, appeared to be main end products of carbohydrate fermentation (de Zwaan, 1977). The observation that anaerobic glycolysis in bivalves apparently was not a copy of the Embden–Meyerhof pathway has resulted in great attention to bivalves in comparative biochemical studies.

This chapter will describe the variety and flexibility of pathways of carbohydrate catabolism in bivalves. Attention will be paid to stoichiometry, energetics, intra- and extracellular organization of pathways, transport, and excretion. The subject has been recently reviewed in detail by the author (de Zwaan, 1977; Zandee et al., 1980), and over 30 species of clams, oysters, mussels, and scallops have been mentioned and the relevant articles in the field considered. In this chapter, however, partly in order to avoid repetition, the subject is not treated from a comparative point of view but rather the sea mussel *Mytilus edulis* is selected as a model animal. There are several arguments to justify this choice and this method of approach. The first justification is the author's familiarity with this animal; second, no other bivalve has been studied so systematically and in such great detail; and third, comparative studies of several bivalves have shown that the sea mussel is a good representative for euryoxic bivalve species (de Zwaan, 1977).[1] This chapter deals only with catabolic pathways starting at the level of glucose 1-phosphate, and emphasis is placed on the findings since 1977. Glycogen metabolism and anabolic pathways in bivalves are described elsewhere. The pyruvate oxidoreductases are also discussed elsewhere (Volume 4) and for this reason are not covered in great detail in this chapter.

---

[1] The term *euryoxic* was introduced by Hammen (1976) for animals capable of tolerating a wide range of oxygen concentration including zero.

## II. Pathways of Carbohydrate Catabolism

### A. Pathways

In Fig. 1 the principle pathways of carbohydrate catabolism are presented.

### 1. Aerobic

In the presence of oxygen, glucosyl units are completely oxidized to $H_2O$ and $CO_2$ (Zaba et al., 1978; de Zwaan, 1977; Jamieson and de Rome, 1979). Dehydrogenation occurs in glycolysis and the citric acid cycle; via the coenzymes NAD or FAD and the electron-transfer chain, the hydrogen is oxidized to $H_2O$ by atmospheric oxygen; $CO_2$ is formed in the citric acid cycle and both end products are formed in the mitochondrion (Fig. 1: immersed).

### 2. Anaerobic

**a. Anoxic Endogenous Oxidation.** In nerve tissues and to a lesser extent in organs such as the heart, it is probable that in the absence of oxygen the same cooperation still exists among glycolysis, the citric acid cycle, and the electron-transfer chain, but that oxygen is replaced by a prestored electron acceptor, the yellow lipochrome pigment (Zs.-Nagy, 1977). The pigment is composed of neutral fats and phospholipids, and unsaturated fatty acids possibly serve as the endogenous electron-acceptor molecules. The organelles rich in the yellow pigment are called cytosomes. It has been postulated by Zs.-Nagy that cytosomes produce ATP during anaerobiosis by a special mechanism called anoxic endogenous oxidation (for review, see Zs.-Nagy, 1977). Cytosomes accumulate bivalent cations in the absence of oxygen. This process is inhibited by dinitrophenol and $CN^-$, suggesting the participation of cytochrome oxidase and phosphorylation in anoxic energy production. Most early studies by Zs.-Nagy were based on histological and histochemical techniques, but since 1979 biochemical research with isolated cytosomes has been started in our laboratory in collaboration with Zs.-Nagy.

Additional evidence has been obtained to show that the postulated anoxic endogenous oxidation may be of physiological importance (Holwerda et al., 1981b). Both ganglia and heart cytosomal fractions of the freshwater clam *Anodonta cygnea,* separated by means of gradient centrifugation from mitochondria, possessed succinic dehydrogenase, malic

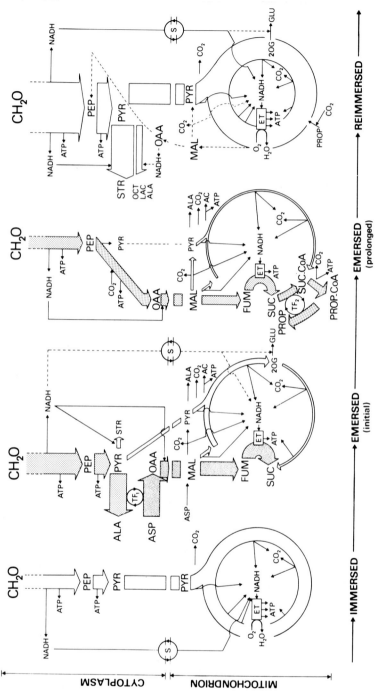

**Fig. 1.** A metabolic map to account for the degradation of carbohydrates in littoral, sessile bivalves. From the left to the right, the different pathways represent gradual transitions in the carbon flow that may occur in the course of a tidal cycle, except for stage 3 (emersed; prolonged), which only occurs with much longer periods of valve closure. The width of the bars is an indication of the relative carbon flux through that part of the pathway. Abbreviations: $CH_2O$, glycogen and/or glucose; ET, electron-transfer chain; OCT, octopine; PEP, phosphoenolpyruvate; PYR, pyruvate; S, malate shuttle; STR, strombine; $TF_1$, transfer of $NH_2$ by glutamate–oxaloacetate transferase; $TF_2$, transfer of CoA by acyl-CoA transferase.

dehydrogenase, isocitric dehydrogenase, and citrate synthase activities. Cytosomes showed an anaerobic oxidative capacity toward NADH and succinate that was of the same order as the aerobic one. However, in later experiments, in which extra precaution was taken to remove possible residual oxygen, the results could not be confirmed. Now the anaerobic oxidation capacity only appeared to be a few percent of the aerobic one (D. A. Holwerda and P. R. Veenhof, unpublished results). The anaerobic oxidation of NADH by the tissue fractions could be blocked by amytal, antimycin A, and KCN. Therefore, all known components of the terminal electron-transfer chain are likely to be involved in this anaerobic oxidation (Holwerda et al., 1981b). The anaerobic oxidative capacity, however, was only partially present in control (aerobic) animals, but increased greatly in clams kept under anaerobic conditions for 2 to 5 days. During such anoxic periods strong structural changes occur within the cytosomes (Zs.-Nagy, 1977). Membrane structures appear that seem to originate from the lipid droplets. Possibly the electron-acceptor substance is exposed by this process. For this reason it may be that anoxic endogenous oxidation is especially important to animals living in or on the bottom of small and shallow freshwater basins that show gradual seasonal changes in oxygen tension in the lower strata. It is not likely to function in intertidal bivalves, because these animals face strong and acute variations in their oxygen supply during the tidal cycle. These animals are alternately exposed to air (emersed) or covered by water (immersed).

**b. Glycogen Plus Aspartate Catabolism.** In contrast to freshwater species, marine animals have high intracellular free amino acid pools that contribute substantially to the osmotic pressure of the cell (Schofeniels and Gilles, 1972; Livingstone et al., 1979; Zurburg and de Zwaan, 1981). It appears that aspartate especially can be utilized along with carbohydrates as an immediate response to oxygen depletion (Collicutt and Hochachka, 1977; Baginski and Pierce, 1978; Ebberink et al., 1979; Ellington, 1981; Foreman and Ellington, 1981; Meinardus and Gäde, 1981). Again, in this simultaneous anaerobic carbohydrate and aspartate degradation, glycolysis, tricarboxylic acid cycle, and electron-transfer chain are involved, but now all pathways show specific alterations (Fig. 1: emersed, initial).

Glycogen carbon appears mainly as alanine, whereas the carbon skeleton of aspartate is converted to succinate by using the part of the citric acid cycle between succinate and oxaloacetate in reverse direction of its aerobic operation, and only flavoproteins are functional in the electron-transfer chain with fumarate being the terminal electron acceptor (Holwerda and de Zwaan, 1979). This situation is unique as both the mitochondrial electron acceptor (fumarate) and the electron donor (H) are formed simultaneously as catabolic products, and the process can be sustained for weeks. The involvement of aspartate, the initial precursor of fumarate, is restricted to several hours because of its relatively small pool size in comparison to glycogen. When aspartate is involved, glycolysis and the tricarboxylic acid cycle are connected indirectly by aminotransferase reactions involving the 2-oxoglutarate–glutamate couple (Fig. 1: emersed, initial). Generation of cytoplasmic malate with the maintenance of redox balance involves glutamate–pyruvate transferase, glutamate–oxaloacetate transferase, and malate dehydrogenase. Stoichiometry is obtained and there is an inverse equimolar correlation between the decrease in aspartate and the increase in alanine. However, at various times it has been observed that *in vivo* alanine accumulation exceeds aspartate utilization (Collicutt and Hochachka, 1977; Zurburg and Ebberink, 1981; Zurburg and de Zwaan, 1981). Therefore, other pathways of alanine formation have been suggested, including direct $NH_3$ fixation of pyruvate or indirect fixation via glutamate dehydrogenase (de Zwaan and van Marrewijk, 1973). The first possibility seems unlikely because attempts to detect alanine dehydrogenase activity in bivalve extracts have failed (Dando et al., 1981). Recent *in vitro* studies with inhibitors of glycolysis (monoiodoacetate), the purine nucleotide cycle (hadacidin), and amino acid transferases (aminooxyacetate) in adductor muscle of the sea mussel (de Zwaan et al., 1982a) (Fig. 2) and studies with the inhibitors monoiodoacetate, aminooxyacetate and the phosphoenolpyruvate carboxykinase inhibitor mercaptopicolinate in oyster heart (Foreman and Ellington, 1981) proved directly the importance of the aminotransferases in connecting glycolysis and the citric acid cycle at the onset of anoxia. No alanine and succinate formation was found in the presence of aminooxyacetate. In the presence of any inhibitor, no increase in glutamate was observed. Therefore, the formation of glutamate by ammonia fixation and subsequent transamination with pyruvate is also not likely to explain the discrepancy between alanine formation and aspartate utilization.

In addition to aspartate, in the adductor muscle of the sea mussel, valine, isoleucine, leucine, ornithine, and arginine are able to transaminate with 2-oxoglutarate and therefore these amino acids can also contribute to alanine formation. Apart from glutamate, no amino acid is able to transfer its amino group directly to oxaloacetate or pyruvate (de Zwaan

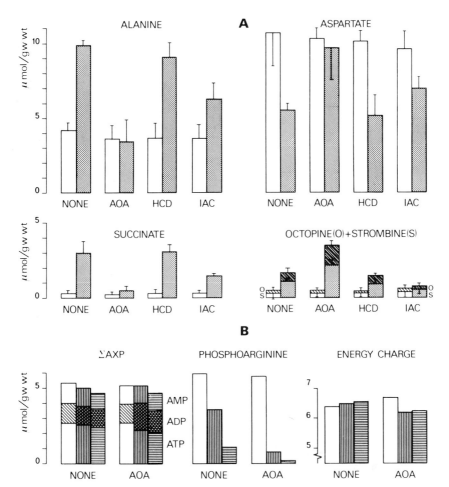

**Fig. 2.** Levels of alanine, aspartate, succinate, octopine, and strombine (**A**) and of adenylates and phosphoarginine (**B**) in excised posterior adductor muscle incubated in oxygen-free seawater with or without metabolic inhibitor. In (**B**) also the Atkinson energy charge derived from the individual adenylates is presented. Open bar, levels 30 min after the muscles were excised and immediately placed in ice-cold seawater with or without inhibitor. Code for anaerobic incubation times at 25°C is as follows: vertical lines, 1 h; dotted bar, 2 h; horizontal lines, 3 h. Code for inhibitors is as follows: none, no inhibitor added; AOA, aminooxyacetate added (5 m$M$); HCD, hadacidin added (2 m$M$); IAC, iodoacetate added (5 m$M$). Bars in (**A**) represent the average value of five groups ($t$ SD), five adductor muscles per group. Bars in (**B**) represent the average value of two groups, five adductor muscles per group. (After de Zwaan et al., 1982a.)

et al., 1983). All possible transferases can be blocked with aminooxy-acetate, and the results shown in Fig. 2A therefore prove that alanine formation is entirely formed in the glutamate–pyruvate transamination reaction, with aspartate being the main amino acid donor of 2-oxo-glutarate. The importance of aspartate as fuel is also clearly illustrated in Fig. 2B. When its utilization is blocked by aminooxyacetate, there is a much quicker drop in phosphoarginine levels in order to maintain a reasonable value of the energy charge (de Zwaan et al., 1982a). The correlation between catabolic ATP production leading to succinate accumulation and the availability of phosphoarginine pool has also been observed *in vivo* (Zurburg and Ebberink, 1981).

The importance of phosphoenolpyruvate carboxykinase and the utilization of aspartate in the anaerobic formation of succinate in the sea mussel was also investigated *in vivo* by applying the metabolic inhibitors mercaptopicolinate and aminooxyacetate. In winter, succinate formation in adductor muscle could be blocked by aminooxyacetate, whereas mercaptopicolinate displayed no effect, neither at 6 h nor at 12–18 h after the onset of aerial exposure. In summer, however, for adductor muscle a reduction in succinate formation and an increased aspartate utilization was observed after injection of mercaptopicolinate. The involvement of phosphoenolpyruvate carboxykinase in the anaerobic formation of succinate in the adductor muscle appears, therefore, to be dependent on the season (de Zwaan et al., 1983).

**c. Glycogen Catabolism.**  During prolonged anoxia a rearrangement occurs at the level of phosphoenolpyruvate in the way carbohydrates are fermented (Ebberink and de Zwaan, 1980). From this time energy metabolism is entirely carbohydrate dependent. Phosphoenolpyruvate is carboxylated by phosphoenolpyruvate carboxykinase to oxaloacetate, thus taking over the role of aspartate (Fig. 1: emersed, prolonged). However, at least in winter, this rearrangement appears to be absent in the adductor muscle (de Zwaan et al., 1983).

When carbohydrate is the sole fuel, glycolysis and the citric acid cycle are connected by malate instead of pyruvate as during aerobiosis, and malate migrates from the cytosol to the mitochondrion. Cytoplasmic redox balance in part is achieved by a 1:1 coupling of glyceraldehyde 3-phosphate dehydrogenase to malate dehydrogenase.

Malate, once within the mitochondrion, is converted to succinate by two different routes. A minor part follows the tricarboxylic acid cycle in a clockwise direction involving several oxidative steps; the remainder follows part of the cycle in the reverse direction via fumarate, which is subsequently reduced to succinate.

Anaerobic malate metabolism has been studied in detail *in vitro* with mantle mitochondria of *M. edulis* (de Zwaan et al., 1981a). Figure 3 gives a carbon and hydrogen flow scheme for the anaerobic transformation of malate. The overall reaction equation was:

$$\text{Malate } (+ \text{ } 0.06 \text{ NH}_3) \rightarrow 0.05 \text{ pyruvate } + 0.06 \text{ alanine } + 0.11 \text{ acetate}$$
$$+ 0.08 \text{ propionate } + 0.04 \text{ fumarate}$$
$$+ 0.61 \text{ succinate } + 0.61 \text{ CO}_2 \tag{1}$$

The reduction of fumarate to succinate was compensated for by reactions involving malic enzyme, malate dehydrogenase, pyruvate dehydrogenase, isocitrate dehydrogenase, and 2-oxoglutarate dehydrogenase. The relative contributions of these steps to the formation of reducing equivalents were 44, 8, 26, 9, and 13%, respectively. In the presence of arsenite, which inhibits pyruvate dehydrogenase and 2-oxoglutarate dehydrogenase, the direction of the carbon flow was more toward pyruvate and less toward succinate but without a decrease in malate turnover. The accumulation of citrate in the presence of the aconitase inhibitor monofluoroacetate was evidence for the involvement of the citrate synthase step in the anaerobic metabolism of malate. In a similar manner as for malate, the conversion of 2-oxoglutarate was followed in the presence and absence of aspartate. When adding only 2-oxoglutarate, there was an equimolar accumulation of succinate and glutamate, together accounting for about 90% of the utilized substrate, and redox balance was obtained by the 1:1

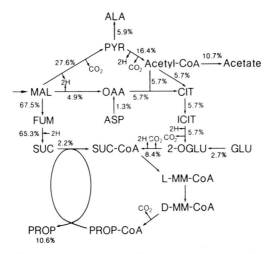

**Fig. 3.** Carbon flow scheme for the anaerobic transformation of malate by isolated mitochondria from the mantle of the sea mussel. Figures represent the substrate flow through the reaction steps after 3 h incubation, expressed as a percentage of the total malate turnover. (Modified after de Zwaan et al., 1981a.)

coupling of glutamate dehydrogenase (oxidative amination) and 2-oxoglu-
tarate dehydrogenase. When aspartate was also added, most 2-oxoglu-
tarate was aminated to glutamate by transamination with aspartate,
whereas glutamate dehydrogenase was of minor importance compared to
fumarate reductase in reoxidizing generated NADH (A. de Zwaan and
D. A. Holwerda, unpublished results). These observations show that glutamate dehydrogenase can contribute to the reoxidation of NADH during
periods in which insufficient fumarate is generated for the fumurate reductase complex, and they explain the role of small increases of glutamate
during anaerobiosis (de Zwaan, 1977). This may be particularly relevant
during the early stages of anoxia, when pyruvate is still produced in considerable amounts and may partly enter the mitochondrion. The equimolar amount of NADH generated in the cytosol for this part of pyruvate can
be reoxidized by glutamate dehydrogenase after shuttling into the mitochondrion (Fig. 1: emersed, initial).

Adductor muscle may differ considerably from mantle. In contrast to
the mantle, in adductor muscle the decrease in aspartate exceeds the increase in succinate. Only part of the anaerobically utilized aspartate is
used for succinate formation; the other part is converted via transaminations and a decarboxylation into alanine according to the overall
reaction aspartate $\rightarrow CO_2$ + alanine (de Zwaan et al., 1982a). It is suggested that this conversion may be involved in hydrogen transport
through the mitochondrial inner membrane during anaerobiosis. This hydrogen would be used for the reductive steps in the intramitochondrial
conversion of aspartate into succinate (de Zwaan et al., 1983).

**d. Formation of Volatile Fatty Acids.** During prolonged anoxia propionate and acetate are produced (Kluytmans et al., 1977, 1978; Zurburg
and Kluytmans, 1980). Hammen and Wilbur (1959) and Wijsman et al.
(1977) showed that bivalve tissues could convert [1-$^{14}$C]propio-
nate to succinate and [2,3-$^{14}$C]succinate to propionate. The route of propionate formation has been studied in detail by Schulz et al. (1982, 1983a)
for the mantle mitochondria of the sea mussel. For an optimal *in vitro* production of propionate from succinate the incubation medium required inorganic phosphate, $MgCl_2$, and malate at pH 6.8. The steps in the pathway from succinyl-CoA to propionate are shown in Fig. 3, and in
succession are catalyzed by methylmalonyl-CoA isomerase, methylmalonyl-CoA racemase, propionyl-CoA carboxylase, and acyl-CoA transferase. The final enzyme transfers CoA from propionyl-CoA to succinate.
In this manner succinate is activated without the utilization of ATP. The
formation of propionate from propionyl-CoA by a thiokinase could not be
detected.

Propionate formation could be inhibited or reduced by the use of mono-fluoracetate or arsenite, which indicates that the formation of succinyl-CoA is via the 2-oxoglutarate dehydrogenase step and explains why the addition of malate increases its formation (de Zwaan et al., 1981a; Schulz et al., 1982). Fig. 3 shows that malate is minimally transformed into succinyl-CoA, which subsequently can be converted to propionyl-CoA. Once a certain level of propionyl-CoA (and succinate) is formed, the acyl-CoA transferase reaction can take over further activation of succinate to succinyl-CoA. Malate conversion in the clockwise direction of the citric acid cycle therefore may be a prerequisite for propionate formation by serving as a primer. This fact may explain why *in vivo* a lag period is observed in the formation of propionate (Kluytmans et al., 1978). The pH may be another factor involved in the control of the lag time. The pH optimum of 6.8 for the overall conversion of succinate plus malate to propionate predicts an activation when anaerobiosis proceeds. A pH drop of 7.5 to 6.7 as a response to aerial exposure was reported by Wijsman (1975) in the extrapallial fluid of the sea mussel. The lag time *in vivo* at 13°C appears to be comparable to the time that has passed until a pH of 6.8 is reached.

**e. Postanaerobic Glycogen Degradation.** During the first hours of recovery from aerial exposure, specific anaerobic end products continue to accumulate in the adductor muscle of the sea mussel (Fig. 1: reimmersed; Fig. 5). The important feature is that the end products all originate from pyruvate, whereas succinate and propionate immediately start to drop. Strombine or *N*-(1-carboxymethyl)alanine is the most important anaerobic end product in the sea mussel, but also some octopine and lactate accumulate (de Zwaan and Zurburg, 1981; de Zwaan et al., 1982b; see also Fig. 5). In this situation there is a clear parallel with free-swimming bivalves, which accumulate octopine during recovery from swimming to exhaustion (Gäde et al., 1978; Grieshaber, 1978; de Zwaan et al., 1980). The metabolism of pyruvate in bivalves therefore is complex.

The following five different pathways of anaerobic pyruvate metabolism have been identified:

$$\text{Pyruvate} + \text{NADH} + \text{H}^+ + \text{glycine} \quad \rightarrow \text{strombine} + \text{NAD}^+ + \text{H}_2\text{O} \qquad (2)$$

$$+ \text{alanine} \quad \rightarrow \text{alanopine} \qquad (3)$$

$$+ \text{arginine} \quad \rightarrow \text{octopine} \qquad (4)$$

$$\rightarrow \text{lactate} \qquad (5)$$

$$+ \text{(aspartate)} \rightarrow \text{alanine} + \text{(malate)} \qquad (6)$$

Equations (2)–(5) are catalyzed by, respectively, strombine dehydrogenase, alanopine- or *N*-(1-carboxyethyl)alanine dehydrogenase, octopine dehydrogenase, and lactate dehydrogenase. Equation (6) occurs through the concerted action of glutamate–pyruvate transferase, glutamate–oxaloacetate transferase, and malate dehydrogenase. These different reductive conversions of pyruvate are not mutually exclusive and all enzymes involved may be present in the adductor muscle of one species (de Zwaan and Zurburg, 1981; Dando et al., 1981; Livingstone et al., 1983). In Fig. 4 it is shown that strombine is a main end product only in the adductor muscle of the sea mussel, whereas alanine accumulates also in other organs. During aerial exposure alanine accumulation exceeds strombine accumulation; the opposite is true during the first 4 h of recovery (Fig. 5). A qualitatively different conversion of pyruvate under different conditions has also been observed by Gäde (1980b) for the jumping cockle *Cardium tuberculatum*. During anoxia D-lactate accumulated, whereas octopine was the important end product during exercise. Also in *Cardium edule* D-lac-

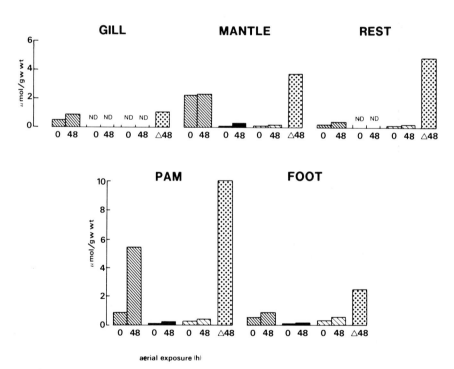

**Fig. 4.** The changes in concentration of typical end products in various organs of the sea mussel after 48 h of aerial exposure. Narrow diagonal lines, strombine (in PAM, posterior adductor muscle) or strombine/alanopine (in all other organs shown); solid bar, octopine; wider diagonal lines, lactate (D + L); dotted bar, Δalanine. (Data from Zurburg et al., 1982.)

tate is an important end product accumulating during anaerobiosis (Meinardus and Gäde, 1981). *Cardium* species therefore appear to be an exception to the rule that lactate formation is of minor importance in bivalves.

**f. Pentose Phosphate Pathway.** Another pathway of glucose degradation is the pentose phosphate pathway in which extramitochondrial $CO_2$ is formed. The importance of this pathway in comparison to the glycolytic pathway has been recently examined by radiotracer studies in the mantle of the sea mussel (Zaba and Davies, 1981). The oxidative pentose phosphate pathway was responsible for only 1–2% of glucose utilization, and therefore obviously is not of importance in gaining energy during carbohydrate catabolism. Its role in biosynthetic processes will be discussed in Volume 2, Chapter 5.

## B. Stoichiometric Equations of Glycogen Fermentation

In table IA the overall reaction equations of the main glycogen conversions just discussed are summarized, including the net yield of ATP. Exact overall equations of anaerobic invertebrate pathways and ATP yields were presented for the first time by Gnaiger (1977). Table IB explains the overall conversion of glycogen to propionate in detail by giving all series of reactions involved, taking into account cell compartmentation.

**TABLE IA**

**Stoichiometric Equations of Glycogen Fermentation in Bivalve Molluscs**[a]

| Equation | | | | | ATP yield | ATP/P | |
|---|---|---|---|---|---|---|---|
| Glc + | $6\,O_2$ + | 37 ADP → | $6\,CO_2 + 6\,H_2O$ | | + 37 ATP | | (7) |
| Glc + | $FA_u$ + | 25 ADP → | $6\,CO_2 + FA_s$ | | + 25 ATP | | (8) |
| Glc + | 2 Arg + | 3 ADP → | 2 Octopine | | + 3 ATP | 1.5 | (9) |
| Glc + | 2 PArg + | 5 ADP → | 2 Octopine | | + 5 ATP | 2.5 | (10) |
| Glc + | 2 Gly + | 3 ADP → | 2 Strombine | | + 3 ATP | 1.5 | (11) |
| Glc + | 2 Ala + | 3 ADP → | 2 Alanopine | | + 3 ATP | 1.5 | (12) |
| Glc + | 2 Asp + | 4.71 ADP → | 1.71 Succinate + 1.14 $CO_2$ + 2 Ala | | + 4.71 ATP | 2.75 | (13) |
| Glc + 0.86 $CO_2$ + | | 4.71 ADP → | 1.71 Succinate | | + 4.71 ATP | 2.75 | (14) |
| Glc | | + 6.43 ADP → | 1.71 Propionate + 0.86 $CO_2$ | | + 6.43 ATP | 3.59 | (15) |

[a] Glc, Glucosyl unit; $FA_u$, unsaturated fatty acids; $FA_s$, saturated fatty acids; P, product.

## TABLE IB

### Series of Reactions Involved in the Conversion of Glycogen to Propionate, Explaining Maintenance of Redox Balance in the Cytoplasm and Mitochondrion and the ATP Yield

Cytoplasm

$3\frac{1}{2}$ Glc + $3\frac{1}{2}$ ADP $\rightarrow$ 7 PEP + $3\frac{1}{2}$ ATP + 7 NADH

7 PEP + 7 $CO_2$ + 7 NADH + 7 ADP $\rightarrow$ 7 Mal + 7 ADP

+ _____

$3\frac{1}{2}$ Glc + 7 $CO_2$ + $10\frac{1}{2}$ ADP $\rightarrow$ 7 Mal + $10\frac{1}{2}$ ATP

Mitochondrion

1 Mal $\rightarrow$ 1 Pyr + 1 $CO_2$ + 1 NADH

1 Pry $\rightarrow$ 1 Acetyl-CoA + 1 $CO_2$ + 1 NADH

1 Mal $\rightarrow$ 1 Oxa + 1 NADH

1 Oxa + 1 Acetyl-CoA + 1 ADP $\rightarrow$ 1 Prop + 3 $CO_2$ + 2 NADH + 1 ATP

+ _____

2 Mal + 2 ADP $\rightarrow$ 1 Prop + 5 $CO_2$ + 5 NADH + 2 ATP

5 Mal $\rightarrow$ 5 Fum

5 Fum + 5 NADH + 5 ADP $\rightarrow$ 5 Suc + 5 ATP

+ _____

                              5 Suc + 5 ADP $\rightarrow$ 5 Prop + 5 $CO_2$ + 5 ATP

7 Mal + 12 ADP $\rightarrow$ 6 Prop + 10 $CO_2$ + 12 ATP

Cytoplasm + mitochondrion

Glc + 6.43 ADP $\rightarrow$ 1.71 Prop + 0.86 $CO_2$ + 6.43 ATP

## C. ATP Yield

The relative efficiencies of the glycolytic fermentations listed in Table IA are without exception extremely low compared to complete degradation of carbohydrate by oxygen. Even in case of propionate accumulation, which gives the highest ATP yield per glucosyl unit, this yield is reduced by a factor of 6. Simultaneous degradation of glycogen and aspartate to alanine and succinate, respectively, does not result in an increase of ATP per glucosyl unit compared with its conversion to succinate via carboxylation of phosphoenolpyruvate [Eqs. (13) and (14)]. In both cases the ATP yield is increased by a factor of 1.6 compared to lactate formation. The importance of aspartate utilization therefore is that the phosphoenolpyruvate carboxykinase reaction is bypassed, and therefore the extra energy-yielding reactions of the mitochondrion can be exploited without delay.

Simultaneous phosphoarginine breakdown and glycolytic octopine formation would be more advantageous [Eq. (10)] than succinate formation. However, the bivalves examined show only poor or no coupling of both processes (Gäde, 1980a). Sluggish and sessile bivalves utilize arginine phosphate during anoxia, but accumulate alanine or lactate instead of octopine (Ebberink et al., 1979; Zurburg and Ebberink, 1981; Gäde, 1980b). In scallops the utilization of phosphoarginine and the accumulation of octopine is more or less sequential, the former being the main energy source during valve snapping, the latter being the accumulating end product mainly during recovery from muscular activity (Gäde et al., 1978; Grieshaber, 1978; Livingstone et al., 1981).

Apart from the substrate-level phosphorylations of glycolysis and the citric acid cycle, the other option for ATP formation is the reduction of fumarate to succinate, the conversion of succinyl-CoA and acetyl-CoA to propionate and acetate, respectively, and the operation of the discussed anoxic endogenous oxidation. All processes are located within the mitochondrion (or cytosomes), which in case of the classical glycolysis of skeletal muscle is excluded from anaerobic ATP formation. The coupling of ATP formation to the reduction of fumarate to succinate has been proven by Holwerda and de Zwaan (1979). Anaerobic incubation of mitochondria of the sea mussel for 1 h with malate, ADP, and inorganic phosphate showed an increase of 0.44 $\mu$mol succinate compared with an increase of 0.52 $\mu$mol ATP. Because there was also an increase in AMP of 0.31 $\mu$mol, a correction on ATP consumption has been made for adenylate kinase activity by substracting an amount of ATP equimolar to AMP from the measured total ATP increase. In this way a phosphorylation ratio of 0.46 was obtained (moles ATP formed per mole fumarate reduced), indicating an *in vivo* ATP yield of 1 mol/mol fumarate reduced.

Tissue fractions of ganglia of *A. cygnea* containing cytosomes incubated in the presence of ATPase and adenylate kinase inhibitors produced ATP when oxidizing succinate. The anaerobic succinate consumption of about 20 $\mu$mol was accompanied by a net ATP production of about 2.6 $\mu$mol/g/h. These data represented a phosphorylation ratio of 0.13 (Holwerda et al., 1981b). Although this figure is not high, it is evidence for anaerobic ATP production by cytosomes. A 1:1 molar ratio of succinate oxidation to ATP synthesis is predicted on the basis of the theory of anoxic endogenous oxidation (Zs.-Nagy, 1977).

ATP synthesis along with the conversion of methylmalonyl-CoA into propionyl-CoA by purified propionyl-CoA carboxylase has been proven for the mantle of the sea mussel. A phosphorylation factor of about 0.8 was obtained (Schulz et al., 1983b).

The formation of acetate has not yet been studied in bivalves. Indications in our laboratory are obtained that the formation of acetate does not involve the acetate thiokinase reaction but depends on acyl-CoA transferase activity, which catalyzes—besides the reaction discussed in relation to propionate formation—the transfer of CoA from acetyl-CoA to succinate. For example, pieces of gill appeared to produce substantially more acetate from malate when succinate was also added.

## III. Integration of Anaerobic Carbohydrate Metabolism

### A. Organ Distribution

Three main approaches can be distinguished concerning the study of tissue-specific functions and differences in carbohydrate catabolism. First, an indication of potential can be obtained by determining patterns and activities of enzymes. On this basis Zammit and Newsholme (1976) estimated for several muscles of marine invertebrates both the maximum rate and the identity of the pathways involved in energy metabolism. They concluded that muscles of marine invertebrates form lactate and/or octopine or succinate. Accumulation of octopine would be the case in scallop snap muscle, accumulation of succinate in oyster adductor muscle. Generalizing, they distinguished two categories of bivalves: the active free-living ones that form octopine and the slow sessile ones forming succinate. However, besides lactate dehydrogenase (LDH), only octopine dehydrogenase (ODH) was involved in their study. The absence of both LDH and ODH indeed has also been confirmed for other oyster species, but instead very high strombine dehydrogenase activities could be detected (Fields, 1976; de Zwaan et al.,

1980; de Zwaan and Zurburg, 1981). Studies on the phylogenetic distribution of pyruvate oxidoreductases including a large number of bivalves have shown that their number varies from one to four (Dando, 1981; Dando et al., 1981; de Zwaan and Zurburg, 1981; de Zwaan et al., 1981b; Livingstone, 1983; Livingstone et al., 1983). The combinations and the relative activities are species dependent; for example, in the adductor muscle of the sea mussel all four dehydrogenases occur, whereas in the two *Mya* species examined only LDH could be detected. Generally speaking, bivalves possess, independent of their habitat or life-style, an impressive total pyruvate oxidoreductase activity in comparison to the phosphorylase or phosphofructokinase activity. Highest activities are found in muscular tissues (Gäde, 1980a; de Zwaan and Zurburg, 1981; see also Fig. 4). This applies also to those muscles possessing relatively high phosphoenolpyruvate carboxykinase activity and accumulating succinate during anaerobiosis. It therefore appears that muscles of bivalves adapted to long-term anaerobiosis possess two alternative pathways of anaerobic carbohydrate catabolism. One pathway is more or less specific to muscular tissue, and leads to the cytoplasmic reduction of pyruvate. This pathway supports muscular activity such as burrowing, mantle cavity cleaning, and swimming, but is also employed during recovery from exhaustive swimming of scallops or aerial exposure in sessile bivalves (see later). The other pathway leads to accumulation of succinate and propionate, and serves to supply the energy for maintenance metabolism during long-term anaerobiosis. In this respect there is no difference from other tissues and the succinate pathway therefore is nonspecific toward tissues. The recognition that the succinate- and pyruvate reductase-dependent pathways are not mutually exclusive has been overlooked in the past and explains why anaerobic pyruvate metabolism in muscles has been neglected.

Specific tissue differences in anaerobic maintenance metabolism are reflected in enzyme spectra. Indicative for the conversion of glycogen to succinate is the activity of phosphoenolpyruvate carboxykinase and the kinetic properties of pyruvate kinase. The latter has been dealt with before (de Zwaan, 1977). In the sea mussel phosphoenolpyruvate carboxykinase could be detected in the 10 different tissues tested for, but the maximal activities varied considerably (de Zwaan, 1977). Extreme low values were found for nervous tissue, gill, intestine, foot, and the anterior byssus retractor muscle. Extreme low activities have also been reported for nervous tissue of *A. cygnea,* which is in accordance with the presence of the so-called anoxic endogenous oxidation (Holwerda et al., 1981b). Collicutt (1975) reported the absence of phosphoenolpyruvate carboxykinase in oyster heart, but showed succinate accumulation concomitant with a decline in aspartate (Collicutt and Hochachka, 1977). The near absence of this enzyme therefore indicates the operation of a completely different anaerobic utilization of carbohydrates (nervous tissue of at least freshwater bivalves) or the operation of the aspartate–succinate pathway (e.g., oyster heart). In case of the gill, its low activity may reflect a low capacity of anaerobic carbohydrate utilization. This interpretation supports the assumption of Chaplin and Loxton (1976) that the gill is able to obtain oxygen during aerial exposure and will follow a predominantly aerobic fuel degradation. In contrast to other tissues, exposure to air had no influence on the distribution of carbon from glucose, bicarbonate, or glutamate. Also Booth and Mangum (1978) have shown for *Modiolus demissus* that the primary route of oxygen uptake is directly into superficial tissues, and that the blood does not have an important respiratory role. The latter has been recently confirmed for *M. edulis* by means of artificial hemolymph perfusion experiments (Famme, 1981). In *M. edulis* the activity of fumarate reductase appeared also to be an order of magnitude lower in the gill in comparison to other tissues (Holwerda and de Zwaan, 1980).

The anaerobic mitochondrial malate conversion in mantle (Fig. 3) includes the steps catalyzed by pyruvate dehydrogenase and 2-oxoglutarate dehydrogenase. However, these enzymes could not be detected in the adductor muscle of the sea mussel (Addink and Veenhof, 1975). On the other hand, activities of glutamate pyruvate transferase and glutamtate oxaloacetate transferase are relatively high in the adductor muscle in comparison to other tissues (Zurburg and Ebberink, 1981) and together with the absence of pyruvate dehydrogenase may cause the complete conversion of pyruvate into alanine. The latter is coupled to the conversion of aspartate to oxaloacetate (de Zwaan et al., 1982a). This would explain the relative importance of aspartate in providing the carbon for succinate formation. The activity of fumarate reductase in adductor muscle is of the same order of magnitude as that of other tissues (Holwerda and de Zwaan, 1980), whereas that of fumarase is much higher (Addink and Veenhof, 1975). This indicates that muscle mitochondria are especially adapted to deal with the anaerobic conversion of oxaloacetate into succinate (or with the oxidation of succinate on the return of aerobic conditions) and therefore, in contrast to mantle, may need exogenous hydrogen for the reduction of fumarase. This could be provided by NADH formed in the cytoplasm. It is proposed that transport of this hydrogen is linked to the conversion of aspartate into alanine and $CO_2$ (de Zwaan et al., 1983).

A second approach used in studying tissue-specific differences is to incubate animals anaerobically and subsequently dissect the organs for analysis of end-product levels (Collicutt and Hochachka, 1977; Zurburg and Kluytmans, 1980; Zurburg and Ebberink, 1981). Succinate appeared to accumulate in all organs of the sea mussel tested for, as well as propionate after longer periods of anaerobiosis. Differences were found in the way the organs met their energy demand during maintenance metabolism. The adductor muscle used more from its endogenous ATP and phosphoarginine pool than the foot and mantle. In the gill hardly any change in the ATP concentration was observed, whereas phosphoarginine levels were low or even undetectable in this organ (Zurburg and Ebberink, 1981). In these experiments also in the gill an increase of succinate was observed, concomitant with a decrease of aspartate and an increase of alanine. An important restriction in these studies is that transport can be involved and the sites of formation and detection therefore are not necessary identical. Transport is ruled out by the third approach, in which the organs are separated at the start of the experiment followed by individual anaerobic incubation (Collicutt and Hochachka, 1977; Zurburg, 1981; see also Fig. 2). Incubation of isolated organs of the sea mussel, including posterior adductor muscle, foot, mantle, gill, and hemocytes, indicates that these all possess the capacity to synthesize propionate (Zurburg, 1981).

## B. Transport and Excretion

Alanine release from mammalian skeletal muscle during starvation (Snell, 1980) and lactate release during activity are well established. Both metabolites are taken up by the liver for gluco(neo)genesis. In bivalves, the hepatopancreas is not analogous to the vertebrate liver in respect to storage of glycogen and regulation of blood glucose. Most bivalve organs and tissues store respectable amounts of glycogen (de Zwaan and Zandee, 1972; Zandee et al., 1980), and gluco(neo)genesis is probably a common feature of all organs.

For bivalves known to produce octopine, no increase of octopine has been detected in nonmuscular tissues or blood, whereas the sum of arginine plus phosphoarginine plus octopine within the muscle remained fairly constant (Gäde et al., 1978; Grieshaber, 1978; de Zwaan et al., 1980; Gäde, 1980a). These observations indicate that octopine does not leave the muscle, but is converted to pyruvate and arginine *in situ*. Based partly on the same arguments, strombine circulation in the sea mussel has been rejected (Zurburg et al., 1982). During anoxia the free alanine levels in the blood decline (Hanson and Dietz, 1978; Zurburg and Kluytmans, 1980), whereas steady-state concentrations and their formation during anaerobiosis differ considerably from organ to organ. Release of alanine by tissues therefore seems also unlikely. It is argued by Zurburg and Kluytmans (1980) that in addition succinate will not be transported. The concentration of succinate in the hemolymph of the sea mussel is very low compared to the tissue levels and only showed minor increases during aerial exposure. No excretion of succinate could be observed. The authors do not exclude the possibility that succinate present in the hemolymph would be taken up rapidly by some organ(s) and converted to propionate, which could keep the succinate levels low. In contrast to these results, a strong increase of succinate in the blood of *Mercenaria mercenaria* was found by Crenshaw and Neff (1969). After 2 h of aerial exposure, succinate in total tissues and the combined mantle and extrapallial fluid accumulated at equal rates. There appeared to be a good correlation between $Ca^{2+}$ and succinate accumulation in the extrapallial fluid, and it has been concluded that $CaCO_3$ from the shell has been hydrolyzed to neutralize succinic acid. $Ca^{2+}$ accumulation is also observed during valve closure of *Scrobicularia plana* (Akberali et al., 1977) and already was a well-known phenomenon for other species in older literature (de Zwaan, 1977).

Once propionate production started, its concentration increased rapidly in both tissue and hemolymph (Zurburg and Kluytmans, 1980). The concentration in the hemolymph was comparable to that of the tissues, whereas sea mussels incubated in oxygen-free seawater excreted propionate. The same applies to acetate, which is an important end product in *A. cygnea* (Gäde et al., 1975) but not in the sea mussel. The percentage distribution of propionate over different organs of the sea mussel did not change after several periods of anaerobic incubation. Together with the fact that injection of propionate into the mussel resulted in a rapid distribution over the whole body, it is obvious that propionate is transported or diffuses very easily (Zurburg, 1981).

In contrast to the *in vivo* situation, *in vitro* much more propionate was formed by the gills and the foot of *M. edulis*, whereas the succinate concentrations in these organs remained low. Apparently, in those two organs more succinate was converted into propionate (Zurburg, 1981). These results could suggest a higher *in vivo* propionate production in the gills and the foot, which may also be indicated by the observation of Kluytmans et al. (1978). These organs showed a decrease in succinate after 1 day and a concomitant increase in propionate, whereas in other organs succinate continued to increase.

## IV. Estimation of Metabolic Rates by Direct Calorimetry and Biochemical Analysis

### A. Metabolic Rates

The ability of animals to cope with wide environmental variations is connected with a great flexibility in the metabolic rate. Until recently the metabolic rate or ATP consumption rate for euryoxic invertebrates was derived from oxygen consumption rates. In bivalves a highly variable routine rate could be observed, ranging between the standard rate at rest and the active rate at maximal activity. For bivalves the reduction factor between the limits of oxygen consumption appeared to be in the order of 2 to 6 (Newell, 1979). When energy transformations by metabolic reactions rely entirely on aerobic respiration, the indirect method of metabolic rate determination is simple; for a number of vertebrates its reliability has been well established by comparing with direct calorimetry (Pamatmat, 1978). However, the lowest estimated rate based on oxygen uptake does not necessarily apply also to situations in which metabolism is partly or totally anaerobic. This restriction is recognized, and since 1976 two additional methods have been applied to estimate metabolic rates in bivalves during anaerobiosis, namely by direct calorimetry and biochemical analysis of end-product accumulation. The former method became possible because Hammen and Pamatmat have succeeded in building microcalorimeters that can be used for bivalves. A number of articles have meanwhile appeared dealing with heat production in bivalves under both aerobic and anaerobic conditions (Hammen, 1979, 1980; Pamatmat, 1979, 1980; Gnaiger, 1980a). From the reaction sequences of glycolytic fermentation (see Table IA), an ATP equivalent can be derived for each identified end product. By adding to the catabolic ATP production the net utilization of energy reserves from the phosphagen and ATP pools, the total ATP consumption rate is obtained. This indirect method has been applied to assess the metabolic rate in a number of bivalves (de Zwaan and Wijsman, 1976; Ebberink et al., 1979; de Zwaan et al., 1980; Gäde, 1980b; Meinardus and Gäde, 1981; Zurburg and Ebberink, 1981; Livingstone, 1982). The advantage of the direct method is that the metabolic rate is registered continuously, but in contrast to the biochemical method no information is given how the energy demand is metabolically met. The latter information is a prerequisite to understand possible relations between the rate of ATP utilization and the type of catabolic fermentation.

Gnaiger (1980b) has pointed out that when comparing his data of heat production for a number of euryoxic invertebrates with the caloric equivalent derived from the accumulation of end products as presented in literature, there appeared to be a great discrepancy. In a recent series of experiments in which both approaches (direct calorimetry versus biochemical end-product analysis) have been used simultaneously, it also appeared that only between 37 and 67% of the heat loss could be accounted for by the stoichiometric equations of glycogen fermentation (Shick et al., 1983). This in fact could mean that the indirect method results in an underestimate for the metabolic rate. The reason for this discrepancy is not yet understood, so this factor is neglected in the following discussion. Both approaches do support each other in a number of conclusions. During anaerobiosis the rate of ATP utilization is greatly reduced in bivalves and may be as low as 5% of the aerobic level of metabolism. This means that the limits in metabolic rates differ much more than the limits of aerobic oxygen utilization (de Zwaan and Wifsman, 1976; Pamatmat, 1979, 1980). Calorimetry has revealed that the anaerobic metabolic rate, although always below that of aerobic metabolism, is like aerobic metabolism characterized by alternating periods of resting and active rates (Pamatmat, 1980). The same conclusion has been drawn from biochemical calculations. The ATP consumption appears to be relatively high at the onset of anaerobiosis and gradually drops to lower levels. In the foot of *C. edule* the ATP consumption rate was reduced by a factor of 7.8 during 48 h of anoxia (Meinardus and Gäde, 1981). For adductor muscle of *M. edulis* it was found that energy expenditure was reduced by a factor of 5.3 in the first 12 h of shell closure and remained constant for the next 12 h (Ebberink et al., 1979). Also during anoxia there can be a remarkable contribution to the total consumed energy from the ATP pool and the breakdown of phosphoarginine. The initial glycolytic rate during catch was inversely related to the quantity of stored phosphoarginine. The latter showed strong seasonal variation (Zurburg and Ebberink, 1981). The implication of the strong reduction in energy demand during anaerobiosis is that the loss in efficiency of ATP generation per unit fuel by switching to anaerobic metabolism is compensated for or even overcompensated and explains the absence of a Pasteur effect or increased rate of fuel utilization (de Zwaan, 1977). When bivalves return to aerated water after aerial exposure the metabolic rate is temporarily increased and exceeds the average aerobic rate. This is for whole animals reflected by an increase in oxygen uptake, a phenomenon in the literature known as "oxygen debt" (de Zwaan, 1977). Of special interest are the metabolic events in the posterior adductor muscle during reimmersion. It appears that concomitant with the enhanced aerobic respiration for whole animals in this organ there is a considerable accumulation of anaerobic glycolytic end products derived from pyruvate (see later). The total increase in anaerobic end products in the adductor muscle was equivalent to about 0.05 $\mu$mol glucosyl unit/g min (data of Fig. 5B: 1–2 h). From the equation determined by Ebberink et al. (1979) for ATP utilization rates in the adductor muscle during anaerobiosis, a value of 0.6 nmol/g sec is obtained at zero time. This can be considered as a minimum aerobic rate and is equivalent to the aerobic degradation of about 0.001 $\mu$mol glucosyl unit/g min, which is about 50 times less than the calculated anaerobic rate. Although the comparison is made between two separate experiments in which different populations of mussels were used, it shows that the glycolytic rate probably will be considerably increased during the first hours of reimmersion (see Fig. 1: reimmersed).

For adductor muscle of the scallop *Placopecten magellanicus* a resting aerobic consumption rate of 0.43 $\mu$mol ATP/g min was assessed (Thompson et al., 1980). During the active swimming escape response that occurs by means of a series of valve snaps, the average energy demand until exhaustion was enhanced by a factor of about 30, whereas the glycolytic flux was increased by a factor of about 70 (de Zwaan et al., 1980). During the first hour of recovery, energy production was achieved by glycolytic rates being increased by approximately an order of magnitude (Livingstone et al., 1981). During both swimming and recovery, the anaerobic glycolytic contribution resulted in octopine accumulation.

## B. Preference for Pathways Indicated by Metabolic Rates

From the foregoing discussion it is tempting to conclude that the succinate pathway operates when there is a decreased energy demand and the glycolytic flux does not need to be increased. On the contrary, the opine pathways leading to the formation of pyruvate-derived end products (Livingstone, 1983) appear to operate when there is an increased glycolytic flux. Livingstone (1982b) has calculated the actual rates of energy production by the opine and succinate pathway in different molluscs from data in literature. For bivalves the rate of energy production ($\mu$mols ATP equivalents per gram per minute) of the former pathway was in the order of 0.1 to 1.0, compared to a range of 0.005 to 0.010 of the latter. It has been argued by Livingstone that, depending on the organism, the role of the opine pathway is to maintain resting (aerobic) rates of energy production in anoxic situations or to increase them during exercise in a manner analogous to the lactate pathway (Livingstone et al., 1981). The fact that succinate and opine formation occur on a competing basis is also illustrated by the reverse effect aminooxyacetate displayed on the formation of both compounds in the adductor muscle of the sea mussel. When aspartate utilization and succinate production were blocked by this inhibitor, opine formation was strongly enhanced (de Zwaan et al., 1983).

## V. Recovery from Periods of Extremes of Metabolic Rates

### A. Reimmersion after Aerial Exposure

The clearance of alanine, succinate, and propionate, accumulated during aerial exposure, has been studied by Widdows et al. (1979) for whole tissues excluding digestive gland of the sea mussel. A rapid decline in the succinate and propionate levels was observed and a more gradual decline in alanine concentration after reimmersion. A detailed study of the reversal of concentration changes of ATP, phosphoarginine, succinate, propionate, strombine/alanopine, octopine, malate, lactate, and free amino acids as a result of 48 h of aerial exposure has been carried out for posterior adductor muscle, foot, mantle, gills, and remaining tissue of the sea mussel (Zurburg et al., 1982).

Starting from the onset of reimmersion, for most compounds a reversal of concentration changes was found, and although the clearance rates showed variations among compounds and tissues, the most striking observation was that in the posterior adductor muscle strombine, octopine, and lactate appeared to accumulate especially during the recovery period. It also could be established that in this organ exclusively strombine accumulated, although strombine dehydrogenase also exhibits remarkable alanopine dehydrogenase activity (de Zwaan and Zurburg, 1981). Excretion of anaerobic end products during reimmersion was restricted to propionate. The patterns of accumulation of strombine, octopine, and lactate during recovery were irregular, possibly because of the relatively long period of the preceding aerial exposure. Recovery of the posterior adductor muscle was therefore reexamined after a shorter aerial exposure time (de Zwaan et al., 1982b).

In addition to the biochemical analysis, pH and $P_{O_2}$ determinations on blood taken from the adductor sinus were carried out. The results of the biochemical analysis are summarized in Fig. 5. All potential end products derived from pyruvate accumulated simultaneously in the first 4 h of recovery. The accumulation of strombine was especially remarkable and exceeded alanine accumulation; the opposite was found during aerial exposure. The latter observation is in accordance with the anaerobic *in vitro* study of the posterior adductor muscle (see Fig. 2A). During aerial exposure both $P_{O_2}$ and pH decreased in the blood. Upon reimmersion, an immediate restoration of both parameters occurred.

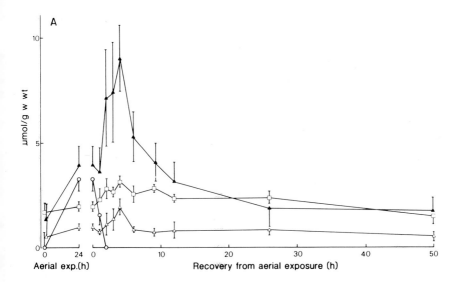

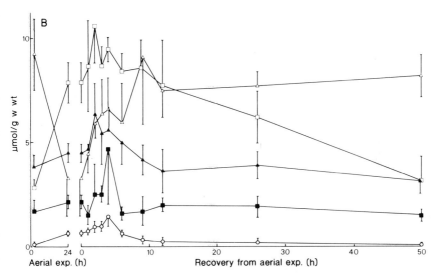

**Fig. 5** The changes in concentration of a number of metabolites and end products in the posterior adductor muscle of the sea mussel after 24 h of aerial exposure and the subsequent recovery period. (**A**) Succinate (—O—), malate (—□—), strombine (—▲—), and lactate (—△—). (**B**) Ammonia (—■—), aspartate (—△—), glutamate (—▲—), alanine (—□—), and octopine (—O—) ($n = 5$). The muscles of three animals were pooled for each sample. (Modified after de Zwaan et al., 1982b).

The operation of anaerobic pathways along with an enhanced aerobic respiration during the first 3 to 4 h of recovery is probably required because the energy demand is high as animals return to an active state. Strombine accumulation is rather low during aerial exposure, possibly because of the extremely efficient transamination reactions between aspartate and pyruvate (via the 2-oxoglutarate–glutamate couple), removing most of the formed pyruvate. This is indicated by the enhancement of the *in vitro* strombine and octopine formation when the transaminase inhibitor aminooxyacetate is added to the incubation system (Fig. 2A). During aerial exposure, however, the aspartate pool declines, and the small pool size at the onset of reimmersion is in favor of the pyruvate oxidoreductases in their competition with the enzymes involved in the amination of pyruvate. The observed inhibition of strombine dehydrogenase by succinate (Fields and Hochachka, 1981) may be another regulating factor. In this respect it is important that the clearance rate of succinate (and propionate) is extremely high. Within 1 h the formed succinate had completely disappeared (Fig. 5A). Moreover, a slightly alkaline pH due to the immediate rise in the pH after reimmersion is in favor of the operation of pyruvate kinase (de Zwaan and Holwerda, 1972) and strombine dehydrogenase (Dando, 1981). There was also a noticeable increase in malate concentration during reimmersion (Fig. 5A). This suggests that succinate was converted to malate, which in turn can be converted to pyruvate by malic enzyme as occurs during anaerobiosis (Fig. 3). Some of the carbon atoms of the pyruvate-derived end products during reimmersion may originate from succinate (and propionate). The excretion of (some of) propionate and the interconversion of anaerobic end products may explain why the so-called oxygen debt is smaller than the amount of oxygen needed for the oxidation of succinate, propionate, and alanine (de Zwaan, 1977; de Vooys and de Zwaan, 1978; Widdows et al., 1979).

## B. Recovery from Muscular Activity

Energy metabolism of muscular activity has been studied in relation to the jumping escape response of *C. tuberculatum* and the swimming escape response of scallops (Pectinidae) and file shells (Limidae), and has been recently reviewed by Gäde (1980a). It appears that the energy for jumping/swimming is provided by the dephosphorylation of phosphoarginine and the breakdown of glycogen to octopine. Hydrolysis of phosphoarginine and octopine occurs in series with a small overlap toward exhaustion.

During swimming activity a reciprocal relationship is observed between phosphoarginine and arginine, proving that the former is much more important than octopine formation in providing energy. However, the accumulation of octopine continues in the subsequent recovery period. Octopine formation during recovery is linked to the restoration of the various adenosine phosphate pools whereas rephosphorylation of arginine depends on aerobic respiration (Grieshaber, 1978; Livingstone et al., 1981). In the case of *P. magellanicus* the biochemical studies of swimming and recovery have been combined with physiological studies (Thompson et al., 1980; Livingstone et al., 1981). Adductor sinus blood $P_{O_2}$ decreased during or following swimming. The change following 5 valve snaps was small and recovery to control levels was rapid. In contrast, the decrease in $P_{O_2}$ after 10 valve snaps was marked and similar to that observed after the maximum number of valve snaps. However, the recovery time was much longer for the maximum than for 10 valve snaps, that is, between 8 and 22 h as compared to 3 h. Octopine only was found in scallops in which blood $P_{O_2}$ was low immediately after exercise. The major role of octopine production in scallops seems to be to provide energy during recovery in situations where oxygen supplies to the tissues are depleted or absent (Livingstone et al., 1981).

## VI. Regulation of the Carbon Flux and the Direction of the Flow

### A. Flux-Regulating Sites

As pointed out by Newsholme and Start (1973), reactions that are displaced from equilibrium are selected by the cell for the purpose of feedback regulation. They are described as irreversible or nonequilibrium reactions. There are two main experimental approaches for identifying nonequilibrium reactions: the comparison of equilibrium constants ($K_{eq}$) with mass action ratios (MAR) and the measurement of maximal enzyme activities. When the equilibrium constant exceeds the mass action ratio by more than 20 times, the reaction is considered to be catalyzed far from equilibrium, but at a value less than 5 the enzyme is considered to catalyze the reaction close to equilibrium (Rolleston, 1972). A nonequilibrium reaction in a metabolic pathway arises because the enzyme that catalyzes it is not sufficiently active to bring the substrates and products to equilibrium. These enzymes possess low activities (Newsholme and Start, 1973).

By one or both approaches the nonequilibrium reactions of the hepatopancreas (Cameselle, et al., 1980) and posterior adductor muscle (Ebberink and de Zwaan, 1980) of the sea mussel have been identified. It appears that in both organs hexokinase, phosphofructokinase, hexose diphosphatase, and pyruvate kinase are of the nonequilibrium type. In adductor muscle this was also stated for glycogen phosphorylase. In this organ all other glycolytic enzymes, adenylate kinase, and arginine kinase belong to the near-equilibrium type. For hepatopancreas the situation for aldolase and triose phosphate isomerase was less clear, as the values of MAR/$K_{eq}$ fall within the range 0.05–0.20 in some experiments, whereas in others aldolase seemed to be equilibrating while the isomerase appeared as nonequilibrating. The effects of 24-h aerial exposure, of cold acclimation, and of glucose, alanine, and succinate feeding were examined on the glycolytic metabolite contents. Two different patterns of variations emerged from it. One pattern fitted with an enhancement of glycolysis and applied to aerial exposure, cold acclimation, and glucose feeding. The other pattern fitted with a decreased rate of glycolysis (or an enhanced gluconeogenesis) and concerned alanine and succinate feeding (Cameselle, et al., 1980).

In the posterior adductor muscle the nonequilibrium reactions are catalyzed by enzymes possessing low maximal activities except for pyruvate kinase. In order to obtain information about how the nonequilibrium reactions may take part in the determination of the glycolytic flux during anaerobiosis, the concentrations of all intermediates and lactate were followed as a function of time of valve closure during 24 h. Calculations of the mass action ratios of the nonequilibrium reactions catalyzed by phosphofructokinase and pyruvate kinase showed that the latter decreased (moved away from equilibrium) right from the onset of valve closure, whereas that for phosphofructokinase increased (moved toward equilibrium). It was concluded that phosphofructokinase is regulating at the beginning of valve closure but loses this role as the anaerobic period progresses. The opposite would be true for pyruvate kinase. That pyruvate kinase took part in the regulation of the flux was also clear from the fact that phosphoenolpyruvate, in contrast to fructose 6-phosphate, still increased after 8 h of valve closure, whereas at this stage the carbon flow will be constant (Ebberink et al., 1979). In the adductor muscle it appears that a gradual shift between the control sites of the glycolytic flux is concomitant with a gradual shift in the route of glycogen breakdown. During the first hours of anoxia, when glycogen is converted to alanine, and aspartate to succinate, the flow of carbon will be determined by the activity of phosphofructokinase. From the moment pyruvate kinase begins to take part in the regulation of the glycolytic flux, more and more phosphoenolpyruvate will be converted via phosphoenolpyruvate carboxykinase to succinate. At this stage the activity of pyruvate kinase (plus phosphoenolpyruvate carboxykinase) is rate limiting, which leads to the accumulation of phosphoenolpyruvate (Ebberink and de Zwaan, 1980). During the period of anoxia the near-equilibrium reactions stood near equilibrium, and therefore these reactions are not major control sites of the glycolytic flux during valve closure (Ebberink and de Zwaan, 1980).

## B. Phosphofructokinase

Phosphofructokinase has been purified from adductor muscle of *Crassostrea virginica* (Storey, 1976), from the adductor muscle of *M. edulis* (Ebberink, 1982), and from the phasic and catch part of the adductor muscle of *P. magellanicus* (Ebberink et al., 1981).

By comparing the properties of the oyster enzyme (the phasic and catch muscle were homogenized together) with the others, the impression arises that it mainly reflects phasic adductor phosphofructokinase. For oyster no influence was observed for phosphoenolpyruvate, and the enzyme displayed hyperbolic saturation curves with respect to all substrates. These properties correspond with those of the enzyme from the phasic part of *P. magellanicus*. In contrast, the catch adductor of the scallop and the adductor muscle of the sea mussel—in which no clear distinction can be made between a phasic and catch part—contain enzyme species that are inhibited by phosphoenolpyruvate and show sigmoidal saturation kinetics with respect to fructose 6-phosphate. The degree of sigmoidicity increased with lowering the pH. Moreover, these enzyme species are not affected by phosphoarginine, whereas in the oyster enzyme this compound is a potent inhibitor. All enzyme species appeared to be insensitive toward inhibition by the traditional effector citrate, whereas AMP decreased the Michaelis constant for fructose 6-phosphate and counteracted ATP inhibition. In Table II the main kinetic constants of the studied enzymes are presented.

The differences in kinetic properties fit well with the metabolic events occuring within the various muscles. As explained before, in the sea mussel there is no increase of the flux during valve closure, whereas during subsequent recovery this glycolytic flux probably increases strongly. During valve closure, concentrations of ATP and phosphoarginine decrease, whereas those of ADP, inorganic phosphate, and especially AMP increase (Ebberink and de Zwaan, 1980). These changes in fact would increase the phosphofructokinase activity. How-

**TABLE II**

**Kinetic Constants of Adductor Muscle Phosphofructokinase of Three Bivalve Species[a]**

| pH | Mytilus edulis[b] posterior adductor | | | | Placopecten magellanicus[c] catch part | | | | P. magellanicus[c] phasic part | | | | Crassostrea virginica[d] phasic plus catch part | | | | |
|---|---|---|---|---|---|---|---|---|---|---|---|---|---|---|---|---|---|
| | $K_{ATP}$ | $K_{F6P}$ | $K_i^{ATP}$ | $L_o$ | $K_{ATP}$ | $K_{F6P}$ | $K_i^{ATP}$ | $L_o$ | $K_{ATP}$ | $K_{F6P}$ | $K_i^{ATP}$ | $L_o$ | $K_{ATP}$ | $K_{F6P}$ | $K_i^{ATP}$ | $K_i^{PA}$ | $L_o^f$ |
| 8.5 | — | — | — | — | 41.3 | 112.5 | 1497 | 1.32 | — | — | — | — | — | — | — | — | — |
| 8.0 | 62.7 | 71.5 | 2465 | 0.003 | 54.2 | 93.7 | 1542 | 1.43 | 53.2 | 30.9 | 3112 | 0.002 | 50 | 500 | 5000 | 1300 | — |
| 7.3 | 49.3 | 94.2 | 387 | 1.08 | 37.6 | 154.3 | 1502 | 1.76 | 47.6 | 47.2 | 576 | 0.010 | — | — | — | — | — |
| 6.9 | 33.3 | 129.6 | 254 | 2.02 | 17.2 | 206.8 | 1437 | 1.93 | 43.3 | 68.7 | 240 | 0.12 | 10 | 3500 | 1300 | 4500 | — |
| 6.7 | 15.1 | 205.6 | 120 | 3.97 | — | — | — | — | — | — | — | — | — | — | — | — | — |
| 6.5 | — | — | — | — | — | — | — | — | 40.5 | 93.2 | 187 | 0.23 | — | — | — | — | — |
| 8.0 + 2 mM PA | — | 71.5 | 2465 | 0.003 | — | 93.7 | 1542 | 1.43 | — | — | — | — | — | 3500 | — | — | — |
| 6.9 + 0.5 mM PEP | 35.0 | 225.6 | 261 | 1.98 | 19.6 | 510.3 | 1497 | 2.13 | 49.7 | 73.3 | 276 | 0.011 | — | 3500 | 1300 | 1300 | — |
| 6.9 + 0.1 mM AMP[e] | 61.0 | 82.0 | 2560 | 0.002 | 57.2 | 33.7 | 3087 | 0.002 | 54.3 | 34.4 | 3782 | 0.002 | — | 700 | ∞ | ∞ | — |

[a] $K_{F6P}$ and $K_{ATP}$, Michaelis constants (μM) for fructose 6-phosphate and for ATP, respectively; $K_i^{ATP}$, inhibition constant (μM) for ATP; $L_o$, allosteric constant (inactive/active enzyme); PA, phosphoarginine; PEP, phosphoenolpyruvate; ∞, complete reversal of inhibition.
[b] Data from Ebberink (1982).
[c] Data from Ebberink et al. (1981).
[d] Data from Storey (1976).
[e] In the case of C. virginica, 1.0 mM.
[f] Only hyperbolic saturation kinetics.

ever, the simultaneous drop of the pH (Wijsman, 1975), resulting in increased sigmoidicity and the initial drop of the concentration of fructose 1,6-bisphosphate, will counteract this stimulation (absence of a Pasteur effect). Phosphoenolpyruvate will integrate the first part of glycolysis with the second part (phosphoenolpyruvate → alanine/succinate) by feedback inhibition. During recovery from long-term valve closure there is an immediate restoration of pH and a slower restoration of the ATP and AMP levels (Zurburg et al., 1982). During the first hours of recovery, the hyperbolic kinetics at the alkaline pH, in combination with relatively low ATP and high AMP levels, will result in a strong activation of phosphofructokinase (presence of a Pasteur effect).

The free-living scallop exhibits three escape responses: swimming, jumping, and valve closure. The catch part, which is composed of smooth muscle fibers, is responsible for keeping the valves closed, but also contributes to the swimming escape response. The phasic part, which is composed of striated muscle fibers, is exclusively involved in the swimming escape response (de Zwaan et al., 1980). As described, the glycolytic flux increased by a factor of at least 50 during the swimming escape response, and almost an order of magnitude during the first hour of the recovery period. Although the hyperbolic kinetics of phosphofructokinase from the phasic part does not favor a strong activation of phosphofructokinase during valve snapping, the regulating mechanism as explained by Storey (1976) for oyster may operate in this situation. The depletion of phosphoarginine reserves and the increase of AMP will activate phosphofructokinase.

## C. Phosphoenolpyruvate Branchpoint

The control of the carbon flow through the phosphoenolpyruvate branchpoint has been dealt with in detail by the author in a previous review and will not be repeated here (de Zwaan, 1977). The work that was cited from the thesis of Zammit has been published (Zammit and Newsholme, 1976, 1978; Zammit et al., 1978). A recent review on comparative biochemistry of pyruvate kinase is available (Munday et al., 1980).

Certain mammalian pyruvate kinase species can be regulated by a reversible phosphorylation–dephosphorylation, mediated by cyclic AMP (cAMP) (Engström, 1978). It has recently been proven that this (de)phosphorylating system also operates, in certain pyruvate kinases of invertebrate sources. In 1979 Siebenaller showed this for pyruvate kinase of the adductor muscle of the sea mussel, and in 1980 it also could be proven for pyruvate kinase of the hepatopancreas of the lobster *Homarus americanus* (Trausch and Bauchau, 1980). As pointed out by Siebenaller, this mechanism represents a complement to the other regulatory mechanism of the phosphoenolpyruvate branchpoint based on the modulator effects in the reaction kinetics of the enzymes involved. In contrast to the latter, the mechanism of (de)phosphorylation does not rely on an intracellular drop in pH. The two forms in adductor muscle of the sea mussel could be electrophoretically distinguished and appeared to differ kinetically. These forms could be interconverted in response to acclimation to different salinities and also could be converted *in vitro* by a cAMP-dependent protein kinase from rabbit muscle (Siebenaller, 1979).

A possible involvement of covalent interconversion of pyruvate kinase in anaerobiosis has been further investigated (Holwerda and Veenhof, 1981; Holwerda et al., 1981a). In the course of aerial exposure, the pyruvate kinase activity decreased at low phosphoenolpyruvate concentrations, but not at a high (near-saturating) concentration. No influence of anoxia was found on the activity of phosphoenolpyruvate carboxykinase (Holwerda et al., 1981a). From an ion-exchange column pyruvate kinase activity was eluted in two separate peaks. The partition of pyruvate kinase activity between the two peaks underwent a change in response to exposure to air; the ratio of enzyme units of peak I to peak II decreased gradually from 3.7 at the onset of exposure to 1.2 after 31 h of exposure. The peaks differed kinetically: The apparent $K_m$ values for phosphoenolpyruvate of peak I and peak II were $0.13 \pm 0.05$ and $0.38 \pm 0.08$, respectively. Peak II appeared to be much more sensitive toward alanine inhibition (Holwerda et al., 1983). The same kinetic differences with respect to phosphoenolpyruvate and alanine have been found for the unphosphorylated and phosphorylated pyruvate kinase of lobster (Trausch and Bauchau, 1980). The fact that peak II was eluted from the ion-exchange column at a higher KCl concentration points to a higher molecular charge, which is in accordance with the assumption that peak II is the phosphorylated form. The results indeed strongly indicate that during aerial exposure there is a shift toward a phosphorylated form that is catalytically less active than the unphosphorylated form. The covalent phosphorylation of pyruvate kinase is probably cAMP dependent. A shift toward peak II could also be induced by incubating muscle tissue of the sea mussel with ATP and cAMP (Holwerda et al., 1983). In the same tissue, levels of cAMP and GMP start to increase during aerial exposure about 2 h after the onset of valve closure (Holwerda et al., 1981a).

## D. Fumarate Reductase

The so-called fumarate reductase complex (FR) is the key reaction in the gain of anaerobic mitochondrial ATP. In this complex fumarate replaces oxygen as terminal electron acceptor. The kinetics and mechanism of the NADH–fumarate oxidoreductase complex have been intensively studied both on intact mitochondria and on submitochondrial particles of sea mussel tissues (Holwerda and de Zwaan, 1979, 1980). On a weight basis the FR activity was the highest in the mantle and the lowest in the gill. Foot and digestive gland showed intermediate values. With NADPH as reducer no activity of FR was obtained, but when simultaneously NAD$^+$ was added, an activity equal to that of an equimolar amount of NADH was obtained. This points to the presence of a transhydrogenase and enables the NADP-dependent malic enzyme reaction (de Zwaan, 1977) to provide reducing equivalents to FR (see Fig. 3). The apparent $K_m$ values for NADH and fumarate were $4.0 \times 10^{-5}M$ and $6.3 \times 10^{-5}M$, respectively. These low values, when compared to beef heart NADH–dehydrogenase and succinate dehydrogenase, allow low levels of fumarate to be reduced by low levels of NADH. On the other hand, the apparent $K_m$ of $1.0 \times 10^{-4}M$ for succinate of mantle is also low as compared to beef heart (Singer, 1966). Obviously, when oxygen supply is sufficient the succinate–ubiquinone oxidoreductase exhibits its usual function in the citric acid cycle. The inhibition of FR by amytal, ethanol, malonate, rotenone, and succinate proved that the reaction proceeds through part of the mitochondrial respiratory chain, namely the complexes NADH–ubiquinone oxidoreductase and ubiquinone–fumarate oxidoreductase. There are no indications that the latter differs from complex II, the succinate–ubiquinone oxidoreductase. As phosphorylation site I is linked to the electron transfer of NADH to ubiquinone, it can be expected that the fumarate reductase reaction is coupled to the phosphorylation of one equivalent of ADP. Indeed it is proven (Holwerda and de Zwaan, 1979) that intact mitochondria convert malate to succinate with simultaneous production of ATP. This phosphorylation was uncoupler sensitive. The pH optimum for the formation of succinate by intact mitochondria is about 7.2. This value is intermediate between the "aerobic" and "anaerobic" pH of the extrapallial fluid, as measured *in vivo* during valve closure (Wijsman, 1975).

## References

Addink, A. D. F. and Veenhof, P. R. (1975). Regulation of mitochondrial matrix enzymes in *Mytilus edulis* L. Proc. 9th Europ. Mar. Biol. Symp. 109–119. Aberdeen Univ. Press.

Akberali, H. B., Marriott, K. R. M., and Trueman, E. R. (1977). Calcium utilization during anaerobiosis induced by osmotic shock in a bivalve molluscs. *Nature (London)* **266**, 852–853.

Baginski, R. M., and Pierce, S. K. (1978). A comparison of amino acid accumulation during high salinity adaption with anaerobic energy metabolism in the ribbed mussel, *Modiolus demissus demissus. J. Exp. Zool.* **203**, 419–428.

Booth, C. E., and Mangum, C. P. (1978). Oxygen uptake and transport in the lamellibranch *Modiolus demissus. Physiol. Zool.* **51**, 17–32.

Cameselle, J. C. Sánchez, J. L. and Carrión, A. (1980). The regulation of glycolysis in the hepatopancreas of the sea mussel *Mytilus edulis* L. *Comp. Biochem. Physiol.,* **65B** 95–102.

Chaplin, A. E., and Loxton, J. (1976). Tissue differences in the response of the mussel *Mytilus edulis* to experimentally induced anaerobiosis. *Biochem. Soc. Trans.* **4**, 437–441. pine (*meso-N*-(1 carboxyethyl)-alanine) dehydrogenase from the mussel *Mytilus edulis. Biochem. Soc. Trans.* **9**, 297–298.

Collicutt, J. M. (1975). Anaerobic metabolism in the oyster heart. M. Sc. Thesis, University of British Columbia.

Collicutt, J. M., and Hochachka, P. W. (1977). The anaerobic oyster heart: Coupling of glucose and aspartate fermentation. *J. Comp. Physiol.* **115**, 147–157.

Crenshaw, M. A., and Neff, J. M. (1969). Decalcification at the mantle-shell interface in molluscs. *Am. Zool.* **9**, 881–885.

Dando, P. R. (1981). Strombine (*N*-(carboxymethyl)-*D*-alanine)dehydrogenase and alano-

Dando, P. R., Storey, K. B., Hochachka, P. W., and Storey, J. M. (1981). Multiple cytoplasmic dehydrogenases in marine molluscs: Electrophoretic analysis of alanopine dehydrogenase, strombine dehydrogenase, octopine dehydrogenase and lactate dehydrogenase. *Mar. Biol. Lett.* **2**, 239–249.

de Vooys, C. G. N., and de Zwaan, A. (1978). The rate of oxygen consumption and ammonia excretion by *Mytilus edulis* after various periods of exposure to air. *Comp. Biochem. Physiol A* **60A**, 343–347.

de Zwaan, A. (1977). Anaerobic energy metabolism in bivalve molluscs. *Oceanogr. Mar. Biol.* **15**, 103–187.

de Zwaan, A., and Holwerda, D. A. (1972). The effect of phosphoenolpyruvate, fructose 1,6-diphosphate and pH on allosteric pyruvate kinase in muscle tissue of the bivalve *Mytilus edulis* L. *Biochim. Biophys. Acta* **276**, 430–433.

de Zwaan, A., and van Marrewijk, W. J. A. (1973). Anaerobic glucose degradation in the sea mussel *Mytilus edulis* L. *Comp. Biochem. Physiol B* **44B**, 429–439.

de Zwaan, A., and Wijsman, T. C. M. (1976). Anaerobic metabolism in bivalvia (Mollusca). *Comp. Biochem. Physiol B* **54B**, 313–324.

de Zwaan, A., and Zandee, D. I. (1972). Body distribution and seasonal changes in the glycogen content of the common sea mussel *Mytilus edulis. Comp. Biochem. Physiol. A* **43A**, 53–58.

de Zwaan, A., and Zurburg, W. (1981). The formation of strombine in the adductor muscle of the sea mussel, *Mytilus edulis. Mar. Biol. Lett.* **2**, 179–192.

de Zwaan, A., Thompson, R. J., and Livingstone, D. R. (1980). Physiological and biochemical aspects of valve snaps and valve closure responses in the giant scallop *Placopecten magellanicus*. II. Biochemistry. *J. Comp. Physiol.* **137**, 105–115.

de Zwaan, A., Holwerda, D. A., and Veenhof, P. R. (1981a). Anaerobic malate metabolism in mitochondria of the sea mussel *Mytilus edulis* L. *Mar. Biol. Lett.* **2**, 131–140.

de Zwaan, A., Leopold, M., Marteijn, E., and Livingstone, D. R. (1981b). Phylogenetic distribution of pyruvate oxidoreductases, arginine kinase and aminotransferases. *Abstr.* Third Congr. Eur. Soc. Comp. Physiol and Biochem., Noordwijkerhout. pp. 136–137. Pergamon Press.

de Zwaan, A., de Bont, A. M. T., and Verhoeven, A. (1982a). Anaerobic energy metabolism in isolated adductor muscle of the sea mussel *Mytilus edulis* L. *J. Comp. Physiol.* **149**, 137–143.

de Zwaan, A., de Bont, A. M. T., Zurburg, W., Bayne, B. L., Livingstone, D. R. (1982b). On the role of strombine formation in the energy metabolism of adductor muscle of a sessile bivalve. *J. Comp. Physiol.* (in press).

de Zwaan, A., de Bont, A. M. T., and Hemelraad, J. (1983). The role of phosphoenolpyruvate carboxykinase in the anaerobic metabolism of the sea mussel *Mytilus edulis* L. *J. Comp. Physiol.* (in press).

Ebberink, R. H. M. Control of adductor muscle phosphofructokinase activity in the sea mussel *Mytilus edulis* during anaerobiosis *Mol. Physiol.* **2**, 345–355.

Ebberink, R. H. M., and de Zwaan, A. (1980). Control of glycolysis in the posterior adductor muscle of the sea mussel *Mytilus edulis*. *J. Comp. Physiol.* **137**, 165–172.

Ebberink, R. H. M., Zurburg, W., and Zandee, D. I. (1979). The energy demand of the posterior adductor muscle of *Mytilus edulis* in catch during exposure to air. *Mar. Biol. Lett.* **1**, 23–31.

Ebberink, R. H. M., Livingstone, D. R., Thompson, R. J., and de Zwaan, A. (1981). Control of phosphofructokinase from adductor muscle of a sessile bivalve and a free living bivalve. *Abstracts Third Congr. Eur. Soc. Comp. Physiol. and Biochem. Noordwijkerhout* pp. 116–117. Pergamon Press.

Ellington, W. R. (1981). Energy metabolism during hypoxia in the isolated perfused ventricle of the whelk, *Busycon contrarium* Gonad. *J. Comp. Physiol.* **142**, 457–464.

Engström, L. (1978). The regulation of liver pyruvate kinase by phosphorylation–dephosphorylation. *Curr. Top. Cell. Regul.* **13**, 29–51.

Famme, R. (1981). Haemolymph circulation as a respiratory parameter in the mussel, *Mytilus edulis* L. *Comp. Biochem. Physiol. A* **69A**, 243–247.

Fields, J. H. A. (1976). A dehydrogenase requiring alanine and pyruvate as substrates from the oyster adductor muscle. *Fed. Proc., Fed. Am. Soc. Exp. Biol.* **37**, 1687.

Fields, J. H. A., and Hochachka, P. W. (1981). Purification and properties of alanopine dehydrogenase from the adductor muscle of the oyster, *Crassostrea gigas* (Mollusca, Bivalvia). *Eur. J. Biochem.* **114**, 615–621.

Foreman, R. A., III, and Ellington, W. R. (1981). Energy metabolism in anoxic oyster ventricles. *Fed. Proc., Fed Am. Soc. Exp. Biol.* **40**, 431.

Gäde, G. (1980a). Biological role of octopine formation in marine molluscs. *Mar. Biol. Lett.* **1**, 121–135.

Gäde, G. (1980b). The energy metabolism of the foot muscle of the jumping cockle, *Cardium tuberculatum:* sustained anoxia versus muscular activity. *J. Comp. Physiol.* **137**, 177–182.

Gäde, G., Wilps, H., Kluytmans, J. H. F. M., and de Zwaan, A. (1975). Glycogen degradation and end products of anaerobic metabolism in the fresh water bivalve *Anodonta cygnea*. *J. Comp. Physiol.* **104**, 79–85.

Gäde, G., Weeda, E., and Gabbott, P. A. (1978). Changes in the level of octopine during the escape response of the scallop, *Pecten maximum* (L.). *J. Comp. Physiol.* **124**, 121–127.

Gnaiger, E. (1977). Thermodynamic considerations of invertebrate anoxibiosis. Reprint from Applications of calorimetry in life sciences. Walter de Gruyter, Berlin, New York.

Gnaiger, E. (1980a). Das kalorische äquivalent des ATP-umsatzes im aeroben und anoxischen metabolismus. *Thermochim. Acta* **40**, 195–223.

Gnaiger, E. (1980b). Energetics of invertebrate anoxibiosis: direct calorimetry in aquatic oligochaetes. *FEBS Lett.* **112**, 239–242.

Goddard, C. K., and Martin, A. W. (1966). Carbohydrate metabolism. *In* "Physiology of Mollusca" (K. M. Wilbur and C. M. Yonge, eds.), Vol. 2, pp. 275–302. Academic Press, New York.

Grieshaber, M. (1978). Breakdown and formation of high-energy phosphates and octopine in the adductor muscle of the scallop, *Chlamys opercularis* (L), during escape swimming and recovery. *J. Comp. Physiol.* **126**, 269–276.

Hammen, C. S. (1976). Respiratory adaptations: Invertebrates. *In* "Estuarine Processes" (M. Wiley, ed.), Vol. 1, pp. 347–355. Academic Press, New York.

Hammen, C. S. (1979). Metabolic rates of marine bivalve molluscs determined by calorimetry. *Comp. Biochem. Physiol. A* **62A**, 955–959.

Hammen, C. S. (1980). Total energy metabolism of marine bivalve molluscs in anaerobic and aerobic states. *Comp. Biochem. Physiol. A* **67A**, 617–621.

Hammen, C. S., and Wilbur, K. M. (1959). Carbon dioxide fixation in marine invertebrates. *J. Biol. Chem.* **234**, 1268–1271.

Hanson, J. A., and Dietz, T. H. (1976). The role of free amino acids in cellular osmoregulation in the fresh water bivalve *Ligumia subrostrata* (Say). *Can. J. Zool.* **54**, 1927–1931.

Holwerda, D. A., and de Zwaan, A. (1979). Fumarate reductase of *Mytilus edulis* L. *Mar. Biol. Lett.* **1**, 33–40.

Holwerda, D. A., and de Zwaan, A. (1980). On the role of fumarate reductase in anaerobic carbohydrate catabolism of *Mytilus edulis* L. *Comp. Biochem. Physiol. B* **67B**, 447–453.

Holwerda, D. A., and Veenhof, P. R. (1981). Regulation of mussel pyruvate kinase during anaerobiosis. *Abstr. Third Congr. Eur. Soc. Comp. Physiol. and Biochem. Noordwijkerhout*, pp. 120–121. Pergamon Press.

Holwerda, D. A., Kruitwagen, E. C. J., and de Bont, A. M. T. (1981a). Regulation of pyruvate kinase and phosphoenolpyruvate carboxykinase activity during anaerobiosis in *Mytilus edulis* L. *Mol. Biol.* **1**, 165–171.

Holwerda, D. A., Notenboom, P., and Zs.-Nagy, I. (1981b). Biochemical characterization of cytosomes from freshwater mollusc nerve tissue. *Abstr. Third Congr. Eur. Soc. Comp. Physiol. and Biochem. Noordwijkerhout*, pp. 122–123. Pergamon Press.

Holwerda, D. A., Veenhof, P. R., van Heugten, H. A. A., and Zandee, D. I. (1983). Regulation of mussel pyruvate kinase during anaerobiosis and in temperature acclimation by covalent modification. *Mol. Physiol.* (in press).

Jamieson, D. D., and de Rome, P. (1979). Energy metabolism of the heart of the mollusc *Tapes watlingi*. *Comp. Biochem. Physiol. B* **63B**, 399–405.

Kluytmans, J. H., de Bont, A. M. T., Janus, J., and Wijsman, T. C. M. (1977). Time dependent changes and tissue specificities in the accumulation of anaerobic fermentation products in the sea mussel *Mytilus edulis* L. *Comp. Biochem. Physiol. B* **58B**, 81–87.

Kluytmans, J. H., van Graft, M., Janus, J., and Pieters, H. (1978). Production and excretion of volatile fatty acids in the sea mussel *Mytilus edulis* L. *J. Comp. Physiol.* **123**, 163–167.

Livingstone, D. R. (1983). Invertebrate and vertebrate pathways of anaerobic metabolism: evolutionary considerations. *In* "Dimensions of Palaeophysiology" (B. M. Funnell, ed.). Geol. Soc., London. In Press.

Livingstone, D. R. (1982). Energy production in the muscle tissues of different kinds of molluscs. In "Exogenous and Endogenous Influences on Metabolic and Neural Control of Activity and Energy Supply in Muscles" (A. D. F. Addink and N. Spronk, eds.), pp. 257–274. Pergamon, Oxford.

Livingstone, D. R., Widdows, J., and Fieth, P. (1979). Aspects of mitrogen metabolism of the common mussel Mytilus edulis: Adaptation to abrupt and fluctuating changes in salinity. Mar. Biol. (Berlin) **53**, 41–55.

Livingstone, D. R., de Zwaan, A., and Thompson, R. J. (1981). Aerobic metabolism, octopine production and phosphoarginine as sources of energy in the phasic and catch adductor muscles of the giant scallop Placopecten magellanicus during swimming and the subsequent recovery period: comments on the role of the octopine pathway in some marine invertebrates. Comp. Biochem. Physiol. B **70B**, 35–45.

Livingstone, D. R., de Zwaan, A., Leopold, M., and Marteijn, E. (1983). Studies on the phylogenetic distribution of pyruvate oxidoreductases. Biochem. Syst. Ecol. In press.

Meinardus, G., and Gäde, G. (1981). Anaerobic metabolism of the cockle, Cardium edule. IV. Time dependent changes of metabolites in the foot and gill tissue induced by anoxia and electrical stimulation. Comp. Biochem. Physiol. B **70B**, 271–277.

Munday, K. A., Giles, I. G., and Poat, P. C. (1980). Review of the comparative biochemistry of pyruvate kinase. Comp. Biochem. Physiol. B **67B**, 403–441.

Newell, R. C. (1979). "Biology of Intertidal Animals," Chap. 11. Mar. Ecol. Surv., Faversham, Kent, England.

Newsholme, E. A., and Start, C. (1973). "Regulation in Metabolism," pp. 1–32. Wiley, New York.

Pamatmat, M. M. (1979). Anaerobic heat production of bivalves (Polymesoda caroliniana and Modiolus demissus) in relation to temperature, body size, and duration of anoxia. Mar. Biol. (Berlin) **53**, 223–229.

Pamatmat, M. M. (1980). Facultative anaerobiosis of benthos. In "Marine Benthic Dynamics" (K. R. Tenore and B. C. Coull, eds.), Belle W. Baruch Library in Marine Science, No. 11. Univ. of South Carolina Press, Columbia.

Rolleston, F. S. (1972). A theoretical background to the use of measured concentrations of intermediates in study of the control of intermediary metabolism. Curr. Top. Cell. Regul. **5**, 47–75.

Schoffeniels, E., and Gilles, R. (1972). Ionoregulation and osmoregulation in Mollusca. In "Chemical Zoology. Vol. 7: Mollusca" (M. Florkin and B. T. Scheer, eds.), pp. 393–418. Academic Press, New York.

Schulz, T. K. F., Kluytmans, J. H., Zandee, D. I. (1982). In vitro production of propionate by mantle mitochondria of the sea mussel Mytilus edulis L.: overall mechanism. Comp. Biochem. Physiol. **73B**, 673–680.

Schulz, T. K. F., and Kluytmans, J. H. (1983a). Pathway of propionate synthesis in the sea mussel Mytilus edulis L. Comp. Biochem. Physiol. (in press).

Schulz, T. K. F., van Duin, M., and Zandee, D. I. (1983b). Propionyl-CoA carboxykinase from the sea mussel Mytilus edulis L. Some properties and its role in the anaerobic energy metabolism. Mol. Physiol. (in press).

Schick, J. M., de Zwaan, A., and de Bont, A. M. T. (1983). Anoxic metabolic rate in the mussel Mytilus edulis L. Physiol. Zool. (in press).

Siebenaller, J. F. (1979). Regulation of pyruvate kinase in Mytilus edulis by phosphorylation–dephosphorylation. Mar. Biol. Lett. **1**, 105–110.

Singer, T. P. (1966). Flavoprotein dehydrogenase of the respiratory chain. Compr. Biochem. **14**, 127–198.

Snell, K. (1980). Muscle alanine synthesis and hepatic gluconeogenesis. *Biochem. Soc. Trans.* **8**, 205–213.

Storey, K. B. (1976). Purification and properties of adductor muscle phosphofructokinase from the oyster, *Crassostrea virginica. Eur. J. Biochem.* **70**, 331–337.

Theede, H. (1973). Comparative studies on the influence of oxygen deficiency and hydrogen sulfide on marine bottom invertebrates. *Neth. J. Sea Res.* **7**, 244–252.

Thompson, R. J., Livingstone, D. R., and de Zwaan, A. (1980). Physiological and biochemical aspects of the valve snap and valve closure responses in the giant scallop *Placopecten magellanicus.* I. Physiology. *J. Comp. Physiol.* **137**, 97–104.

Trausch, G., and Bauchau, A. (1980). Some catalytic properties of the lobster hepatopancreas pyruvate kinase after phosphorylation. *Abstr. Congr. Eur. Soc. Comp. Physiol. Biochem, 2nd, Southampton,* pp. 82–83.

von Brand, T. (1946). Survey of invertebrates for anaerobiosis. *In* "Anaerobiosis in Invertebrates" (B. J. Luyet, ed.), pp. 87–94. Biodynamica, Missouri.

Widdows, J., Bayne, B. L., Livingstone, D. R., Newell, R. I. E., and Donkin, P. (1979). Physiological and biochemical responses of bivalve molluscs to exposure to air. *Comp. Biochem. Physiol. A* **62A**, 301–308.

Wijsman, T. C. M. (1975). pH fluctuations in *Mytilus edulis* L. in relation to shell movements under aerobic and anaerobic conditions. *Proc. Eur. Symp. Mar. Biol., 9th,* pp. 139–149. Aberdeen Univ. Press, Aberdeen.

Wijsman, T. C. M., de Bont, A. M. T., and Kluytmans, J. H. F. M. (1977). Anaerobic incorporation of radioactivity from 2,3-$^{14}$C-succinic acid into citric acid cycle intermediates and related compounds in the sea mussel *Mytilus edulis* L. *J. Comp. Physiol.* **114**, 167–175.

Zaba, B. N., and Davies, J. I. (1981). Carbohydrate metabolism in isolated mantle tissue of *Mytilus edulis* L. Isotopic studies on the activities of the Embden–Meyerhof and pentose phosphate pathways. *Mol. Physiol.* **1**, 97–112.

Zaba, B. N., de Bont, A. M. T., and de Zwaan, A. (1978). Preparation and properties of mitochondria from tissues of the sea mussel *Mytilus edulis* L. *Int. J. Biochem.* **9**, 191–197.

Zammit, V. A., and Newsholme, E. A. (1976). The maximum activities of hexokinase, phosphorylase, phosphofructokinase, glycerol phosphate dehydrogenase, lactate dehydrogenase, octopine dehydrogenase, phosphoenolpyruvate carboxykinase, nucleoside diphosphate kinase, glutamate–oxaloacetate transaminase and arginine kinase in relation to carbohydrate utilization in muscles from marine invertebrates. *Biochem. J.* **160**, 447–462.

Zammit, V. A., and Newsholme, E. A. (1978). Properties of pyruvate kinase and phosphoenolpyruvate carboxykinase in relation to the direction of phosphoenolpyruvate metabolism in muscles of the frog and marine invertebrates. *Biochem. J.* **174**, 979–987.

Zammit, V. A., Beis, J., and Newsholme, E. A. (1978). Maximum activities and effects of fructose biphosphate on pyruvate kinase from muscles of vertebrates and invertebrates in relation to the control of glycolysis. *Biochem. J.* **174**, 989–998.

Zandee, D. I., Holwerda, D. A., and de Zwaan, A. (1980). Energy metabolism in bivalves and cephalopods. *In* "Animals and Environmental Fitness" (R. Gilles, ed.), pp. 185–206. Pergamon, Oxford.

Zs.-Nagy, I. (1977). Cytosomes (yellow pigment granules) of molluscs as cell organelles of anoxic energy production. *Int. Rev. Cytol.* **49**, 331–377.

Zurburg, W. (1981). Environmental influences on the energy metabolism in different organs of *Mytilus edulis* L. Ph.D. Thesis, Univ. of Utrecht, Utrecht.

Zurburg, W., and de Zwaan, A. (1981). The role of amino acids in anaerobiosis and osmo-regulation in bivalves. *J. Exp. Zool.* **215**, 315–325.

Zurburg, W., and Ebberink, R. H. M. (1981). The anaerobic energy demand of *Mytilus edulis*. Organ specific differences in ATP-supplying processes and metabolic routes. *Mol. Physiol.* **1**, 153–164.

Zurburg, W., and Kluytmans, J. H. (1980). Organ specific changes in energy metabolism due to anaerobiosis in the sea mussel *Mytilus edulis* L. *Comp. Biochem. Physiol. B* **67B**, 317–322.

Zurburg, W., de Bont, A. M. T., and de Zwaan, A. (1982). Recovery from exposure to air and the occurence of strombine in different organs of the sea mussel *Mytilus edulis*. *Mol. Physiol.* **2**, 135–147.

# 5

# Carbohydrate Metabolism of Gastropods

**DAVID R. LIVINGSTONE**

Institute for Marine Environmental Research,
Plymouth PL1 3DH, Great Britain

**ALBERTUS DE ZWAAN**

Laboratory of Chemical Animal Physiology,
State University of Utrecht
3508 TB Utrecht, The Netherlands

## I. Introduction

Since the reviews of Goddard and Martin (1966) and Goudsmit (1972), considerable advances have been made in the study of the carbohydrate metabolism of the Gastropoda. Hydrolytic enzymes, glycoprotein structure, hormonal regulation, and reproduction have been particular topics of interest for this group. The studies of anaerobic metabolism and other metabolic pathways, although lagging behind those of the Bivalvia and Cephalopoda, have also advanced. In this chapter, Section II deals with the molecular forms of carbohydrate that occur; Section III follows the

THE MOLLUSCA, VOL. 1
Metabolic Biochemistry
and Molecular Biomechanics

logical progression of the finding, taking up, and utilizing of the carbohydrates; and Section IV describes the metabolic pathways by which most of these events occur.

## II. Carbohydrates

### A. Monosaccharides and Oligosaccharides

D-Glucose is the most common monosaccharide and has been detected in the blood and tissues of many gastropods (see Table IV). Other sugars also occur, in particular D-galactose, which are either stereoisomers or derivatives of monosaccharides (Table I). For example, Rao and Onnurappa (1979) identified fructose, xylose, and ribose in the digestive gland, and glucose and galactose in both the digestive gland and body fluids of the snail *Lymnaea luteola*, whereas Renwrantz et al. (1976) recorded the following concentrations of sugars in ultrafiltrates of the hemolymph of the snail *Helix pomatia*: glucose 216 $\mu$g/ml, galactose 24 $\mu$g/ml, fucose 8 $\mu$g/ml, and mannose 8 $\mu$g/ml. Sugar derivatives such as N-acetylgalactosamine and N-acetylglucosamine also occur but are usually incorporated into glycoproteins (Renwrantz et al., 1976) (see Section II,C). The percentage of sugars in the hemolymph that are free or part of macromolecules varies among species; for example, in the slug *Ariolimax columbianis*, 12% of the sugars were part of glycoproteins (Meenakshi and Scheer, 1968), compared with 90% in macromolecules in *H. pomatia* (Renwrantz et al., 1976). The oligosaccharides maltose, maltotriose, and maltotetrose were present in the hepatopancreas of *A. columbianis* (Meenakshi and Scheer, 1968), whereas the nonreducing disaccharide trehalose [$O$-$\alpha$-D-glucopyranosyl-(1 → 1)-$\alpha$-D-glucopyranoside] has been detected in several gastropod species (Fairbairn, 1958).

### B. Polysaccharides

Glycogen and galactogen are the main storage polysaccharides. Whereas glycogen occurs in most gastropod tissues and serves as a general energy source, galactogen is generally confined to the albumen gland and eggs of adult pulmonate snails and the albumen gland region of the pallial oviduct of some prosobranchs, and it serves as a specialized energy source in reproduction (see Sections III,D and E). Glycogen is a branched-chain homopolysaccharide of D-glucose in $\alpha$-1,4 linkage with branches joined through $\alpha$-1,6 linkages. The degree of

TABLE I

**Monosaccharide and Monosaccharide Derivatives Occurring Free or as Part of Macromolecules in Gastropods**[a]

Neutral sugars

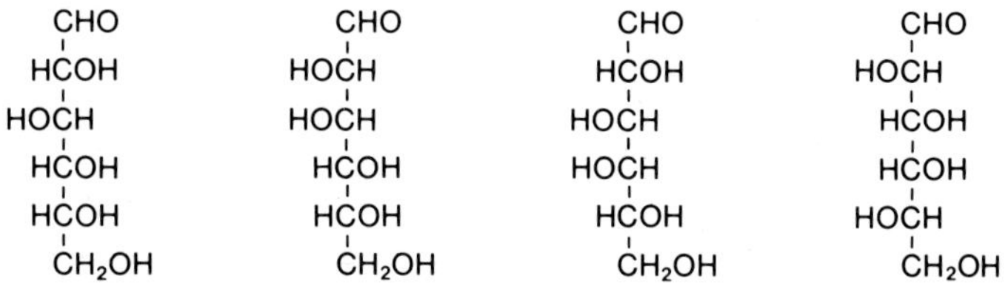

| CHO | CHO | CHO | CHO |
|---|---|---|---|
| HCOH | HOCH | HCOH | HOCH |
| HOCH | HOCH | HOCH | HCOH |
| HCOH | HCOH | HOCH | HCOH |
| HCOH | HCOH | HCOH | HOCH |
| $CH_2OH$ | $CH_2OH$ | $CH_2OH$ | $CH_2OH$ |
| D-glucose | D-mannose | D-galactose | L-galactose |

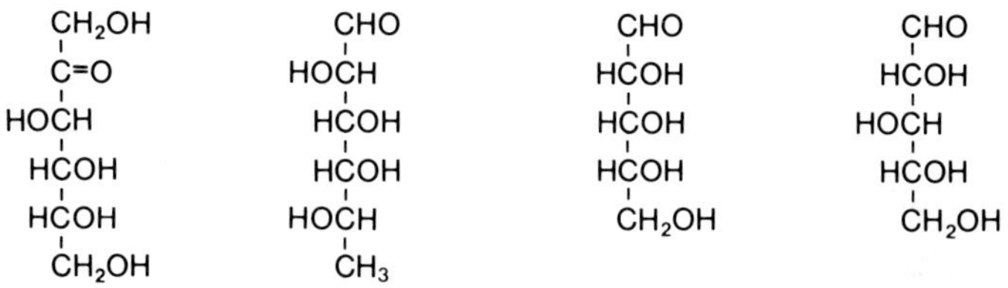

| $CH_2OH$ | CHO | CHO | CHO |
|---|---|---|---|
| C=O | HOCH | HCOH | HCOH |
| HOCH | HCOH | HCOH | HOCH |
| HCOH | HCOH | HCOH | HCOH |
| HCOH | HOCH | $CH_2OH$ | $CH_2OH$ |
| $CH_2OH$ | $CH_3$ | | |
| D-fructose | L-fucose | D-ribose | D-xylose |

Amino sugars and derivatives

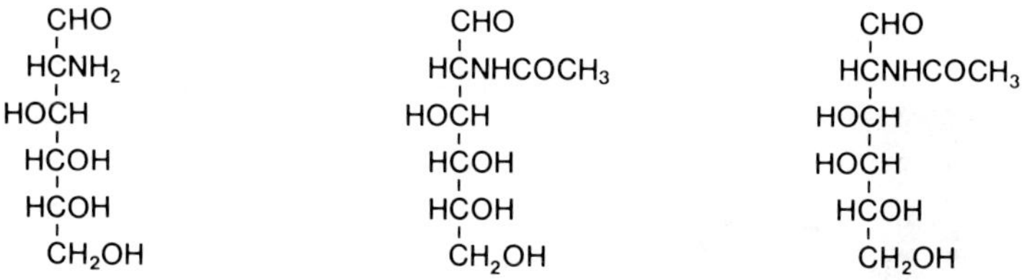

| CHO | CHO | CHO |
|---|---|---|
| $HCNH_2$ | $HCNHCOCH_3$ | $HCNHCOCH_3$ |
| HOCH | HOCH | HOCH |
| HCOH | HCOH | HOCH |
| HCOH | HCOH | HCOH |
| $CH_2OH$ | $CH_2OH$ | $CH_2OH$ |
| D-glucosamine | N-acetyl-D-glucosamine | N-acetyl-D-galactosamine |

Methylated Sugars

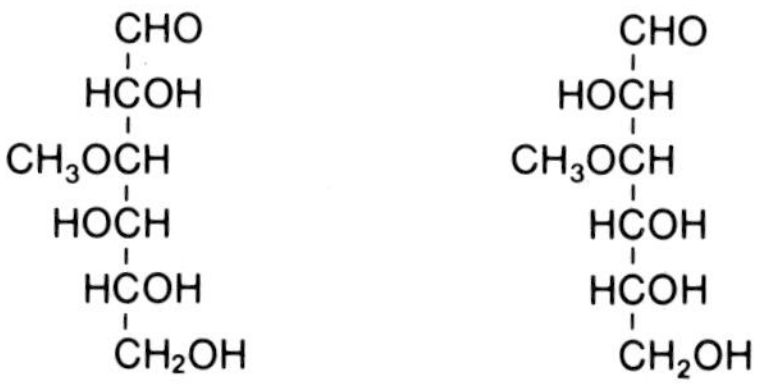

| CHO | CHO |
|---|---|
| HCOH | HOCH |
| $CH_3OCH$ | $CH_3OCH$ |
| HOCH | HCOH |
| HCOH | HCOH |
| $CH_2OH$ | $CH_2OH$ |
| 3-*O*-methylgalactose | 3-*O*-methylmannose |

[a] Sugars are represented as projection formulas but with the horizontal bonds omitted; note that the stereoisomers represented have either been identified or are the common ones.

branching and the length of the outer branches of the glycogen from the snail *Biomphalaria glabrata* are similar to those of glycogens from other animal sources including vertebrates: respectively, 9 and 40% of the total glucosyl residues (Chiang, 1977).

Galactogen is a homopolysaccharide of galactose. A number of its properties have been reviewed by Goudsmit (1972). Some galactans contain L-galactose in addition to D-galactose, for example, 36% L-galactose in the galactogen of *B. glabrata*. Chemical studies indicate that the straight chain is either $\beta$-1,3 or $\beta$-1,6 linked and that each galactopyranose unit bears a branch or side branch through the C-6 or C-3 position, respectively. Molecular weights of $4 \times 10^6$ (*H. pomatia*) and $2.2 \times 10^6$ (*Lymnaea stagnalis*) have been recorded. In more recent years the study of galactans has been greatly improved with the discovery from various invertebrate sources of antibody-like substances with anticarbohydrate specificities directed against different structures of galactans (Uhlenbruck et al., 1976a, 1978). Untreated saline extracts of tissues can be examined by the visible galactan–antigalactan precipitin reactions. This approach in combination with immunoelectrophoresis and other techniques such as precipitation with plant lectins (carbohydrate-binding proteins) suggests that previously studied galactans may have been modified by the alkali extraction procedures that were used. Uhlenbruck et al. (1977) have proposed that the native galactans in fact occur in close association with proteins as proteogalactans. They also identified in the albumen glands of the snails *Achatina fulica* and *Borus* sp. another group of galactans, termed glycoproteogalactans, which differed from the proteogalactans in being associated with glycoproteins and being heterogeneous as demonstrated with different antigalactans (Uhlenbruck et al., 1976a,b).

## C. Mucopolysaccharides and Glycoproteins

Mucopolysaccharides and glycoproteins are produced in considerable amounts by gastropods and serve a wide range of functions including providing mechanical or protective support and lubrication, and as components of egg jellies and capsules (Goudsmit, 1972). The respiratory pigments and many hydrolytic enzymes are also glycoproteins (Table II).

Mucopolysaccharides (also called acid mucopolysaccharides) are heteropolysaccharides usually composed of two types of alternating monosaccharide units at least one of which bears an acidic (carboxyl or sulfuric) group. The negatively charged molecules are soluble in water, forming highly viscous solutions. They may occur in association with specific proteins giving rise to jelly-like substances, mucoproteins or mucins. The functions of such mucins or mucous secretions include decreasing friction during locomotion, providing adhesion, decreasing dessication, and accumulating debris (Goudsmit, 1972; Grenon and Walker, 1981). The chemical composition of mucins and mucopolysaccharides varies among species. For example, the pedal mucus of the limpet *Patella vulgata* is 90% water and contains sulfated mucopolysaccharides (approximately 29% dry weight) linked to the protein moiety (33% dry weight) only by electrovalent bonds (Grenon and Walker, 1980). In contrast, the hypobranchial secretion of the whelk *Buccinum undatum* contains both glycoprotein and an acidic polysaccharide: The former contains 8% neutral sugars (glucose, galactose, fucose, and mannose) and 4.5% amino sugars (glucosamine, galactosamine); the latter is a polyglucose sulfate with an indicated cellulose-like $\beta$-1,4 structure (Hunt, 1970). A highly sulfated acid mucopolysaccharide containing galactose, fucose, and glucosamine is present in the commonduct region of the hepatopancreas of *A. columbianis* (Meenakshi and Scheer, 1968). The presence of acidic sulfate groups appears to be a feature of gastropod mucopolysaccharides, in contrast to vertebrate viscous secretions, which achieve their acidic character mainly from sialic acid residues (*N*-acyl derivatives or neuraminic acid) (Goudsmit, 1972).

Glycoproteins are conjugated proteins containing covalently bound carbohydrate groups; molecules having a very high carbohydrate content are termed proteoglycans. Of a number of molecules examined, the percentage carbohydrate composition varies from 3% for the respiratory pigment of *B. glabrata* to 62.7 mol % for the structural vitelline envelope of the eggs of the limpet *Megathura crenulata* (Table II). A variety of monosaccharide and monosaccharide derivatives are found, including unusual methylated sugars in the hemocyanins of *H. pomatia* and *L. stagnalis* (Tables I and II). As in mucopolysaccharides, sialic acid residues are generally absent (Goudsmit, 1972; Afonso et al., 1976; Hall and Wood, 1976; Colas, 1978), although they are present in the $\beta$-*N*-acetylglucosaminidase of the hepatopancreas of *P. vulgata* (Table II). The carbohydrate moiety is clearly important to function, related glycoproteins having similar amino acid compositions but different carbohydrate ones, for example, the $\beta$-fucosidases I and II of the snail *Achatina balteata* (Colas, 1978), the glycosulfatases I and II of *Charonia lampas* (Hatanaka et al., 1976a), the $\alpha$-, $\beta$-, and $\beta$-*N*-acetylglucosaminidases of *P. vulgata* (Phizackerley and Bannister, 1974), and the hemocyanins (Hall and Wood, 1976). The molecular weights of such glycoproteins can be markedly different, for example, 300,000 and

**TABLE II**

**Sugar Composition of Some Glycoproteins in Gastropods**

| Species | Source | Glycoprotein | Sugar composition[a] | Reference |
|---|---|---|---|---|
| *Patella vulgata* | Hepatopancreas | **1.** $\alpha,\beta$-N-Acetylglucosaminidase | 10% gal, 1.6% N-acetylgluNH$_2$, 0.5% man | Phizackerely and Bannister (1974) |
| | | **2.** $\beta$-N-Acetylglucosaminidase (EC 3.2.1.30) | 10.5% gal, 1.0% N-acetylgluNH$_2$, 0.6% man, 9.0% sialic acid | |
| *Lymnaea stagnalis* | Whole animal | Glycosylamidase | 8–10% Neutral sugars (ratio man:gal = 3 :2), 5.3% gluNH$_2$ | Chukhrova et al. (1974b) |
| *Littorina littorea* | Hepatopancreas | $\alpha$-L-Fucosidase (EC 3.2.1.51) | 7.4% glu, 2.2% gal, 1.8% man, 2.2% hexNH$_2$ | de Pedro et al. (1978) |
| *Achatina balteata* | Digestive juices | **1.** $\beta$-Fucosidase I | 15.5% Neutral sugars (glu, gal, man, fuc), 6.1% hexNH$_2$ | Colas (1978) |
| | | **2.** $\beta$-Fucosidase II | 7.5% Neutral sugars (glu, gal, man, fuc), 5.2% hexNH$_2$ | |
| *L. stagnalis* | Hemolymph | Hemocyanin | glu, gal, man, fuc, xyl, N-acetylgluNH$_2$, N-acetylgalNH$_2$, 3-O-methylgal | Hall et al. (1977) |
| *Helix pomatia* | Hemolymph | Hemocyanin | glu, gal, man, fuc, xyl, N-acetylgluNH$_2$, N-acetylgalNH$_2$, 3-O-methylgal, 3-O-methylman | Hall et al. (1977) |
| *Buccinum undatum*<br>*Neptunea antiqua*<br>*Colus gracilis* | Hemolymph | Hemocyanin | 1–8% glu, gal, man, fuc, N-acetylgluNH$_2$, N-acetylgalNH$_2$ | Hall and Wood (1976) |
| *Biomphalaria glabrata* | Hemolymph | Hemoglobin | 2% Neutral sugars (ratio man:gal:fuc = 2:1:1), 1% gluNH$_2$ | Afonso et al. (1976) |
| *H. pomatia* | Foot | Collagen (gelatin) | glu, gal | Stemberger (1974) |
| *Megathura crenulata* | Eggs | Vitelline envelope | 62.7% mol % Carbohydrate including gal, fuc, and galNH$_2$ | Heller and Raftery (1976) |

[a] Abbreviations: fuc, fucose; gal, galactose; glu, glucose; man, mannose; xyl, xylose; N-acetylgalNH$_2$, N-acetylgalactosamine; N-acetylgluNH$_2$, N-acetylglucosamine; galNH$_2$, galactosamine; gluNH$_2$, glucosamine; hexNH$_2$, hexosamine; 3-O-methylgal, 3-O-methylgalactosamine; 3-O-methylman, 3-O-methylmannose.

110,000, respectively, for the $\beta$-fucosidase I and II of *A. balteata* (Colas, 1978). The nature of the carbohydrate–protein linkage varies between glycoproteins; for example, in the vitelline envelope of the eggs of *M. crenulata,* the galactosamine residues are linked via *O*-glycosidic bonds to threonine residues (Heller and Raftery, 1976), whereas in the collagen from *H. pomatia,* galactose and glucosyl-galactose residues are linked to hydroxylysine and not to serine or threonine (Stemberger and Nordwig, 1974). In the vitelline envelope glycoprotein, which has a high carbohydrate composition (Table II), 61 mol % of the amino acid content is threonine and almost every threonine residue carries a saccharide moiety (Heller and Raftery, 1976). A number of other glycoproteins have been examined from egg jellies, egg capsules, and other gastropod sources, but little is known of their detailed structure (Goudsmit, 1972; Heller and Raftery, 1976). The same situation exists with regard to the actual function of the sugar residues in many of the glycoproteins, such as the hemocyanins (Hall and Wood, 1976). In some instances glycoproteins of unknown physiological function have been isolated, such as the (carbohydrate-binding) so-called agglutinins of the albumen gland of *H. pomatia* (Hammarström and Kabat, 1969; Hammarström et al., 1977).

## III. Uptake, Storage, and Utilization

### A. Chemoreception

Chemoreception (gustation, olfaction) is important in a variety of phenomena including mating, predator avoidance, prey recognition, and food selection (Kohn, 1961). Carbohydrates are a main constituent of the food of many gastropods, and in several instances sugar and sugar-like substances have been shown to be attractants in chemoreception. Glycogen and *N*-acetylglucosamine stimulated the proboscis search response of the prosobranch *Nassarius obsoletus* (Carr, 1967), and both soluble and insoluble starch fractions were potent feeding stimulants for the terrestrial slug *Ariolimax californicus* (Senseman, 1977). In the latter, the duration of the feeding bout was dependent on the starch concentration, and maltose, maltotriose, and glucose were less effective stimulants than starch. In field experiments, Woodbridge (1978) found that the periwinkle *Littorina littorea* was strongly attracted to glucose, less attracted to sucrose and galactose, and not attracted to fructose and mannitol. A different pattern of response, in terms of the time interval between stimulation and eating and the frequencies of eating movements, was observed for *L. stagnalis:* Disaccharides were better attractants than monosaccharides and the order of effectiveness was sucrose > maltose > lactose and D-fructose > D-glucose > D-galactose (Jager, 1971). The other main observations and conclusions for *L. stagnalis* were that the solubility of the sugars is probably one of the factors responsible for the relative effectiveness, high concentrations of sugars inhibited responsiveness, and starvation could increase responsiveness but in excess was inhibitory.

## B. Hydrolases

The digestive juice, salivary glands, digestive diverticula, and, where present, the crystalline style of gastropods are rich in carbohydrases. The range of enzyme activities is wide (Table III) and a few appear not to occur in higher animals, such as glycosulfatase (Roy, 1971). Several of the enzymes have been purified but the majority of activities have been studied in crude extracts in comparative surveys of many species. Assays have involved the use of both natural and synthetic substrates. The question of substrate specificity is a complicated one, and a number of the enzyme activities of Table III are certainly attributable to single enzymes with broad specificities; for example, the $\beta$-fucosidase of *A. balteata* also hydrolyzes at a lower rate $\beta$-D-galactosides and $\beta$-D-glucosides (Colas, 1978). In the few cases where natural and synthetic substrates have been used in the same study, the agreement has generally been good but with exceptions: For example, in extracts of the mud snails *Hydrobia ulvae, H. ventrosa,* and *H. neglecta,* the $\alpha$-glucosides amylose, glycogen, and 4-nitrophenyl-$\alpha$-D-glucanopyranoside were hydrolyzed equally well, but the natural $\alpha$-galactoside trisaccharide raffinose was hydrolyzed much less than the synthetic (6-bromo-2-naphthyl)-$\alpha$-D-galactopyranoside (Hylleberg, 1976); the latter difference probably arises because raffinose occurs mainly in higher plants and is unlikely to be part of the mud snails' diet (Hylleberg-Kristensen, 1972). Such differences also arise because the complete degradation of a natural substrate may require a complex of several enzymes; cellulose digestion, for example, requires a so-called $C_1$ component to first modify the crystalline cellulose ready for hydrolysis, endo- and exo-$\beta$-1,4-glucanases, and $\beta$-glucosidases (Gianfreda *et al.,* 1979a).

The carbohydrates found in highest activity are generally those hydrolyzing the reserve carbohydrates amylose, glycogen, and, in the case of marine gastropods, laminarin (Hylleberg-Kristensen, 1972; Marcuzzi and Turchetto Lafisca, 1978; Elyakova et al., 1981). A laminarin-

TABLE III

**Hydrolases Acting on Carbohydrates and Carbohydrate Derivatives in Marine and Nonmarine Gastropods**

| Enzyme | Specificity | Species[d] | |
|---|---|---|---|
| | | Marine | Nonmarine |
| Glycosulfatase (EC 3.1.6.3)[a] | Sugar–sulfate ester bonds | *Charonia lampas*[9,10] | — |
| Amylase[b] | α-1,4- And possibly α-1,6-glucan links | 39 species[c,7] | *Lymnaea luteola*[20] |
| | | 4 species[12] | |
| Cellulase (EC 3.2.1.4) | β-1,4-Glucan links in cellulose | 34 species[7] | *Amnicola limosa*[14] |
| | | 4 species[8] | *L. luteola*[20] |
| Laminarinase (EC 3.2.1.6) | β-1,3- Or β-1,4-glucan links | 40 species[7] | — |
| | | 4 species[12] | |
| | | 10 species[22] | |
| Dextranase (EC 3.2.1.11) | α-1,6-Glucan links | — | *Lymnaea stagnalis*[3] |
| Chitinase (EC 3.2.1.14) | α-1,4-Acetamido-2-deoxy-D-glucoside links in chitin | 3 species[12] | 6 species[16] |
| Polygalacturonase (pectinase) (EC 3.2.1.15) | α-1,4-Galacturonide link in pectins | 2 species[12] | 3 species[16] |
| | | | *L. luteola*[20] |
| Lysozyme (EC 3.2.1.17) | β-1,4 Links between *N*-acetylmuramic acid and 2-acetamido-2-deoxy-D-glucose residues | 3 species[11] | — |
| α-Glucosidase (EC 3.2.1.20) | α-D-Glucosides | 3 species[11] | *L. luteola*[20] |
| | | 4 species[12] | |
| | | *Littorina littorea*[18] | |
| | | *Chamelea gallina*[21] | |
| β-Glucosidase (EC 3.2.1.21) | β-D-Glucosides | 4 species[8] | *Achatina balteata*[2] |
| | | 3 species[11] | *L. luteola*[20] |
| | | 4 species[12] | |
| | | *L. littorea*[18] | |
| | | *C. gallina*[21] | |

*(Continued)*

**TABLE III** (*Continued*)

| Enzyme | Specificity | Species[a] | |
| --- | --- | --- | --- |
| | | Marine | Nonmarine |
| α-Galactosidase (EC 3.2.1.22) | α-D-Galactosides | 3 species[11]<br>4 species[12]<br>*L. littorea*[18]<br>*C. gallina*[21] | Not detected in *L. luteola*[20] |
| β-Galactosidase (EC 3.2.1.23) | β-D-Galactosides | 3 species[11]<br>4 species[12]<br>*L. littorea*[18]<br>*C. gallina*[21] | *A. balteata*[2] |
| α-Mannosidase (EC 3.2.1.24) | α-D-Mannosides | *L. littorea*[18] | *L. stagnalis*[3] |
| Trehalase (EC 3.2.1.28) | Trehalose | 4 species[12] | Not detected in *L. luteola*[20] |
| β-N-Acetylglucosaminidase (EC 3.2.1.30) | β-N-Acetylglucosaminides | *L. littorea*[18]<br>*Patella vulgata*[19]<br>*C. gallina*[20] | — |
| β-Glucuronidase (EC 3.2.1.31) | β-D-glucuronides | 3 species[11] | — |
| α,β-Acetylglucosaminidase | α-N- and β-N-Acetylglucosaminides | *P. vulgata*[19] | — |
| β-Acetylgalactosaminidase | β-N-Acetylgalactosaminides | *L. littorea*[18]<br>*C. gallina*[21] | — |
| β-Xylosidase (EC 3.2.1.37) | β-D-Xylans | 4 species[12]<br>*L. littorea*[18] | — |

| Enzyme | Substrate/linkage | Species |
|---|---|---|
| β-D-Fucosidase (EC 3.2.1.38) | β-D-Fucosides | L. littorea[18]; A. balteata[1,2] |
| Endo-1,3-β-glucanase (EC 3.2.1.39) | 1,3-β-D-Glucan links in 1,3-β-D-glucans | Littorina sp.[6]; Collisella sp.[6]; — |
| Exo-1,3-β-glucanase | 1,3-β-Glucan links in 1,3-β-D-glucans hydrolyzed to completion | —; Helix pomatia[17] |
| α-L-fucosidase (EC 3.2.1.51) | α-L-Fucosides | Turbo cornutus[13]; L. littorea[18]; C. gallina[21]; Haliotis gigantea[23]; — |
| Exo-1,4-β-glucanase (EC 3.2.1.74) | β-1,4 links from nonreducing end of chain | 4 species[8]; — |
| β-L-Aspartylglucosylaminamidohydrolase | Glyosylamide link between carbohydrate and protein where amino acid contains free α-NH$_2$ and α-COOH and a chain of four carbons | L. stagnalis[3-5] |

[a] Some sugar sulfates can also be hydrolyzed by arylsulfatase (EC 3.1.6.1), as in C. lampas (Hatanaka et al., 1976b).

[b] The exact specificity of the amylase has been examined only in a few cases [e.g., isoamylase (EC 3.2.1.68) is indicated to be present in L. littorea and Nassarius reticulatus (Hylleberg-K., 1972)].

[c] See references for species and other details; in most cases the enzymes have been extracted from the hepatopancreas or other digestive tissues.

[d] The references are given as superscript numbers: [1] Colas (1978); [2] Colas and Attias (1977); [3] Chukhrova et al. (1970); [4] Chukhrova et al. (1974a); [5] Chukhrova et al. (1974b); [6] Elyakova and Shilova (1979); [7] Elyakova et al. (1981); [8] Gianfreda et al. (1979a); [9] Hatanaka et al. (1976a); [10] Hatanaka et al. (1976b); [11] Hylleberg (1976); [12] Hylleberg-Kristensen (1972); [13] Iijima et al. (1971); [14] Kesler and Tulou (1980); [15] Koningsor and Hunsaker (1970–1971); [16] Marcuzzi and Turchetto Lafisca (1978); [17] Marshall and Grand (1975); [18] de Pedro et al. (1978); [19] Phizackerly and Bannister (1974); [20] Rao (1979); [21] Reglero and Cabezas (1976); [22] Sova et al. (1970); [23] Tanaka et al. (1968).

ase-type activity, exo-1,3-$\beta$-glucanase, is found in the terrestrial snail
*H. pomatia*, but its presence may be related to its broad specificity,
which allows it to hydrolyze a number of mixed-linkage (1,3; 1,4)-$\beta$-glu-
cans such as lichenin and cereal glucans (Marshall and Grand, 1975).
Oligosaccharides and structural carbohydrates are generally hydrolyzed
less well, although enzymes acting on cellulose derivatives and chitin
are widespread; low but significant activities are found in certain species toward structural polysaccharides such as xylan, pectin, and alginic acid (Hylleberg-Kristensen, 1972; Marcuzzi and Turchetto Lafisca, 1978; Elyakova et al., 1981). Information on the origin of enzymes
like cellulase and chitinase is limited but the evidence is that they are
not solely bacterial but are synthesized in part or in total by the tissues
(Owen, 1966). The other enzymes that occur are those that hydrolyze
molecules such as glycoproteins and sulfated polysaccharides (Table
III); the physiological significance of some of these enzymes is unknown but, for example, sulfatases may assist in the breakdown of the
sulfated polysaccharides that occur in seaweeds (Owen, 1966). The interpretation of particular patterns of enzyme activities in a species in
relation to diet and other ecological factors have been discussed by
Hylleberg-Kristensen (1972) but generally attempts to relate the two
have met with limited success. For example, the spectrum of carbohydrases of the carnivorous *Nassarius reticulatus* is very similar to that
of the detritus-feeding *H. ventrosa;* the hydrolysis of alginic acid and
alginate (components of brown algae) by the former was particularly
puzzling (Hylleberg-Kristensen, 1972). [Note that a complete cellulolytic enzyme system is present in carnivorous gastropods but the endo-$\beta$-
1,4-glucanase component is lower than in herbivores, and it has been
suggested that the "cellulase" probably represents an "evolutionary
relic" (Gianfreda et al., 1979a,b; Agnisola et al., 1981).] In contrast, in
a study of *H. ulvae, H. ventrosa,* and *H. neglecta,* it was concluded
that their coexistence could in part be explained by selection for mi-
croalgae and detritus reflecting their enzymatic potentials (Hylleberg,
1976).

A number of carbohydrases have been partially or totally purified
and they have several properties in common. Many are glycoproteins
(Table II), which can occur in variant forms that differ in their carbohydrate moieties and properties (see Section II,C). All have low pH optima, for example: exo-1,3-$\beta$-glucanase of *H. pomatia,* 4.3 (Marshall
and Grand, 1975); $\alpha$-L-fucosidase of *Turbo cornutus,* 4.0 (Iijima et al.,
1971); and $\beta$-D-fucosidase of *A. balteata,* 5.4 (Colas, 1978). Almost all
are anomerically stereospecific and hydrolyze only the $\alpha$ or $\beta$ configuration of the sugar (Iijima *et al.,* 1971; Marshall and Grand, 1975; Reg-

lero and Cabezas, 1976; Colas, 1978; de Pedro et al., 1978). The exception to this is the $\alpha,\beta$-N-acetylglucosaminidase of *P. vulgata,* which hydrolyzes both $\alpha$- and $\beta$-N-acetylglucosaminides (Phizackerley and Bannister, 1974).

## C. Absorption

The active transport of sugars in intestinal absorption has been demonstrated for the major sugars D-glucose and D-galactose for everted intestine sacs of the snail *Cryptomphalus hortensis* (Barber et al., 1975a), and for glucose for isolated intestine of the nudibranch *Aplysia californica* (Kelentey and Csáky, 1976); 3-methylglucose is also actively transported by both tissues. In contrast, in *C. hortensis,* D-fructose, L-arabinose, and D-mannitol are not actively transported, and their rate of transfer from the mucosal to the serosal side follows the kinetics of a diffusion process with no development of any accumulation gradient (Barber et al., 1975a, 1979). Surprisingly, an energy-independent transport mechanism has been reported for glucose for intestinal absorption in *H. pomatia* and the slug *Arion empiricorum* (Orive et al., 1980). Competitive inhibition among D-glucose, D-galactose, and 3-methylglucose (Barber et al., 1979), as well as inhibition of the transport of all three by phlorizine (Barber et al., 1975a), suggest that the three sugars use the same transport system in the intestine of *C. hortensis;* the common system shows the same or greater affinity for D-glucose than for D-galactose with a much lower affinity for 3-methylglucose (Barber et al., 1979). The removal of sodium ions from the solution bathing the everted intestine sacs of the *C. hortensis* strongly inhibited but did not abolish active sugar transport, indicating an important but not indispensable role for sodium (Barber et al., 1975c). Similarly, the substitution of sulfate ion for chloride ion diminished active transport in *A. californica* (Kelentey and Csáky, 1976). Active transport in *C. hortensis* is inhibited by dinitrophenol (uncoupler of oxidative phosphorylation) and sodium fluoride (inhibitor of glycolysis), and is equally effective under aerobic or anaerobic conditions (Barber et al., 1975b). The latter may be of importance during periods of limited oxygen availability brought about, for example, by changes in environmental humidity and temperature (Barber et al., 1975b).

## D. Storage

The energy metabolism of many gastropods, is carbohydrate based, for example, *L. stagnalis* (Veldhuijzen and van Beek, 1976) and the snail *Planorbis corneus* (Emerson, 1967). The carbohydrate is stored largely as glycogen and the specialized galactogen, and is tranported and available from the blood as glucose. Typical concentrations of these molecules are given in Tables IV, VA, and VB. The concentrations vary with food availability, season, and so on, and these environmental and biological interactions are described in Section III,E. The levels are subject to hormonal regulation and this is described in Section III,F.

Using all the data of Table IV, a mean blood glucose concentration of $0.74 \pm 0.56$ m$M$ ($\pm$SD; $n = 22$) can be calculated which is less than the value of $1.18 \pm 1.08$ m$M$ ($n = 16$) that can be calculated from the gastropod data of Goddard and Martin (1966); the difference probably exists because the earlier assays measured total reducing compounds and not glucose specifically. The range of concentrations recorded (Table IV) spans an order of magnitude or more, and whereas some of this variability is due to species differences and differences in animal condition, part of it is due to the wide individual variability regularly shown by the number of gastropods examined such as *B. glabrata* (Cheng and Lee, 1971; Liebsch et al., 1978) and *Australorbis glabratus* (Becker, 1972). In addition to interindividual variability, intraindividual variability in a single animal has been observed over a period of a few hours: Glucose concentration varied be-

### TABLE IV
#### Blood Glucose Levels of Some Gastropods [a]

| Species | Concentration[b] (mM) | Reference |
| --- | --- | --- |
| *Ariolimax columbianis* | 1.55 | Meenaksi and Scheer (1968) |
| *Laevicaulis alte* | 1.54 | Kulkarni (1973) |
| *L. alte* | 0.28–1.0 | Reddy et al. (1978) |
| *Lymnaea stagnalis* | 0.17 | Friedl (1968, 1971) |
| *L. stagnalis*[c] | | |
| Bemax | 0.61–1.11 | Scheerboom et al. (1978); |
| Lettuce | 0.28–0.34 | Scheerboom and Hemminga (1978) |
| *L. stagnalis*[d] | 0.02–1.97 | Scheerboom and van Elk (1978) |
| *Strophocheilus oblongus*[d] | 0.28–0.72 | Marques and Pereira (1970) |
| | 0.14 | Marques and Falkmer (1976) |
| *Biomphalaria glabrata* | 1.07–1.56 | Cheng and Lee (1971) |
| *B. glabrata* | 0.62 | Liebsch et al. (1978) |
| *Australorbis glabratus* | 0.11–0.84 | Becker (1972) |
| *Helix pomatia* | 1.20 | Renwrantz et al. (1976) |
| *Cryptozonia belangeri* | 0.27 | Rajan and Sriramulu (1978) |
| *Achatina fulica* | 0.65 | Brockelman and Sithithavorn (1980) |

[a] Concentrations are those of control animals.
[b] The original data have been converted to millimolar for comparison.
[c] *L. stagnalis* maintained on Bemax (carbohydrate-rich food) or lettuce diet.
[d] Seasonal ranges given.

**TABLE VA**

**Glycogen Levels of Different Tissues of Some Gastropods**

| Species | Tissue | Concentration[a] (mg/g dry wt) | Reference |
|---|---|---|---|
| Laevicaulis alte | Midgut gland | 178 | Kulkarni (1973) |
| Ariolimax | Hepatopancreas | 174 | Meenaksi and Scheer |
| columbianis | Foot | 110 | (1968) |
| | Albumen gland | 2.8 | |
| Biomphalaria | Hepatopancreas/ | 320 | Christie et al. (1974a) |
| glabrata | gonad | | |
| | Rest | 150 | |
| B. glabrata | Cephalopedal region | 80 | Chiang (1977) |
| | Hepatopancreas | 190 | |
| | Mantle | 130 | |
| | Ovotestis | 250 | |
| Lymnaea luteola | Hepatopancreas | 38 | Manohar and Rao |
| | Foot | 67 | (1976) |
| | Mantle | 53 | |
| Strophocheilus | Hepatopancreas | 59–45 | Marques and Falkmer |
| oblongus | Muscle | 74–61 | (1976) |
| Pila virens | Radular muscle | 320 | Suryanarayanan and |
| | Triturative stomach muscle | 218 | Alexander (1973) |
| | Pedal muscle | 205 | |
| | Opercular muscle | 211 | |

[a] Concentrations converted to dry weight using a water-content figure of 80%.

**TABLE VB**

**Galactogen Levels of the Albumen Gland of Some Pulmonate Gastropods**

| Species | Concentration[a] (mg/g dry wt) | Reference |
|---|---|---|
| Ariolimax columbianis | 715–520 | Meenakshi and Scheer (1968) |
| Biomphalaria glabrata | 450 | Christie et al. (1974a) |
| Triodopsis multilineata | 427 | Rudolph (1975) |

[a] Concentrations converted to dry weight using a water-content figure of 80%.

tween 0.06 and 0.96 m$M$ over 6 h in *A. glabratus,* for example (Becker, 1972). Marked changes in glucose concentration (two- to sixfold increase) can also result from blood-sampling procedures and other experimental manipulations (Subramanyam, 1973; Veldhuijzen, 1975b; Marques and Falkmer, 1976). The sources of these variations are possibly the different states of digestion and resorption processes, the varying activities of the musculature, and possibly "stress responses" to manipulation (Liebsch et al., 1978; Scheerboom and Doderer, 1978). In some instances, the degree of blood sugar variability has been successfully reduced by the use of special sampling equipment combined with anesthesia and muscle relaxants such as succinylcholine (Becker, 1972; Liebsch et al., 1978). Trehalose also occurs in gastropods, in quite high concentrations in some species, but no information is available on the actual concentrations in the blood: Its relationship to carbohydrate storage and mobilization is therefore unknown (Goddard and Martin, 1966).

Glycogen is distributed generally throughout the tissues, with levels being high in, for example, the hepatopancreas, foot, and mantle (Table VA). The concentrations are approximately 10 to 30% of the dry weight, which is similar to that occurring in the tissue of bivalves (de Zwaan, 1977; Livingstone, 1982). Caution must be taken in comparing values for species because different results are obtained with different extraction procedures for glycogen: The levels varied by 1.4 to 2.6 times in *L. luteola* depending on the extraction methodology (Rao et al., 1979). Glycogen also occurs in muscle tissues; in *Pila virens* the levels were highest in the triturative stomach muscles (glycogen is located uniformly between the fibers and in largest amounts on the outer and inner aspects of the muscle) and the radular muscle (glycogen is in highest concentration in the region of attachment of the radular cartilage) (Table VA) (Suryanarayanan and Alexander, 1973). Glycogen levels were very low in the albumen gland of *A. columbianis* (Table VA). However, in a number of species such as *H. pomatia,* although galactogen is the major polysaccharide of this tissue (see later), glycogen is still important in its general energy metabolism (see Section III,E).

In gastropod tissues carbohydrate occurs mainly in two types of connective tissue cells: the granular cells and the vesicular connective tissue cells (VCTCs). These have been examined in some detail in *L. stagnalis* by Sminia (1972). The granular cells are 20–30 $\mu$m wide and characterized by the presence of large numbers of granules up to 4 $\mu$m wide. The granules consist of cysteine-rich glycoproteins. The cells occur dispersed throughout the body, and although storage centers for reserve material has been suggested as a function (see Sminia, 1972), their role at present remains largely obscure. In contrast, the VCTCs have been identified as the major storage cell for glycogen with an important role in the nutrition of the tissues. They are large cells up to 60 $\mu$m wide, packed with glycogen, with the cytoplasm and organelles present as a thin rim against the cell membrane. They are not uniformly distributed about the body but are concentrated in the mantle and between the acini of the digestive gland and the ovotestis. The VCTCs have recently been isolated from the mantle tissue of *L. stagnalis* (Fig. 1) in a viable form that synthesize glycogen from glucose in a manner comparable with the intact tissue (Fig. 2). Other experiments have demonstrated that the uptake of glucose by the VCTCs is a diffusion process without the involvement of a membrane carrier (M. A. Hemminga, personal communication). The glycogen that occurs in storage cells is particulate and exists either as simple spherical particles or in large granules composed of the smaller particles: The variability in partical size is reflected in a wide molecular-weight spectrum for the glycogen (Goudsmit, 1972).

Galactogen is restricted to the albumen gland portion of the female reproductive tract of adult pulmonates and to the eggs that they produce (Goudsmit, 1972, 1973). It is synthesized solely in preparation for egg laying and is a major nutritive reserve of the embryos. In *L. stagnalis* it is metabolized during development whereas in *H. pomatia* it is metabolized just after hatching. It can be used as an emergency food source in starvation but only after the glycogen has been used up. The levels in the albumen gland are high (Table VB) but change markedly during the seasonal cycle of summer egg laying and winter hibernation that is characteristic of freshwater and terrestrial pulmonates from temperature climates (see Section III,E). In newly laid eggs galactogen can constitute 30 to 40% of the dry weight (Goddard and Martin, 1966). In addition to the species listed in Goudsmit (1972), galactogen has been identified in the albumen glands of *Catinella vermeta, Omalonyx felina,* and *Oxlyoma retusa* (Rudolph, 1974), of *Semperula maculata* (Nanaware and Varute, 1974) and of *Triodopsis multilineata* (Rudolph, 1975).

Ultrastructural observations of the albumen gland have been made in *H. pomatia* (Nieland and Goudsmit, 1969) and *T. multilineata* (Rudolph, 1975), with similar results. The gland is a compound tubular organ composed of small centrotubular cells interspersed among large ciliated secretory cells. The latter are the site of galactogen synthesis and form the tubules, which may radiate from a central duct. In the Golgi zone of the secretory cells of *H. pomatia* small vesicles containing an amorphous matrix coalesce and enlarge to form huge secretory globules, which then accumulate discrete galactogen granules 200 Å in diameter; purified galactogen consists of particles of the same size as the *in situ* granules. In *T. multilineata* galactogen first appears in the region of the central lumen and eventually fills all the tubules. The albumen gland produces the perivitelline fluid, which contains galactogen and protein: The galactogen secretion appears to be apocrine in type (Nieland and Goudsmit, 1969). The perivitelline fluid envelopes each egg as it passes down the hermaphroditic duct forming the perivitelline (albumen) layer and is also taken in by cells of the snail embryos (Goudsmit, 1972, 1973).

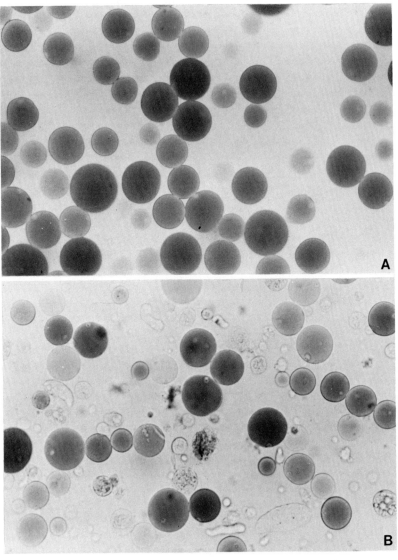

**Fig. 1.** Pictures of isolated purified (**A**) and unpurified (**B**) vesicular connective tissue cells of *Lymnaea stagnalis*. The cells are colored with a mixture of iodine and potassium iodide. The white spots occasionally visible are nuclei. ($\times$ 189) (From M. A. Hemminga, unpublished data.)

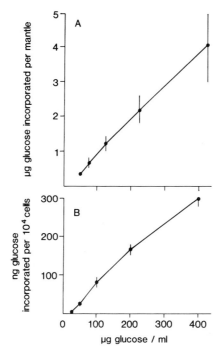

**Fig. 2.** Incorporation of D-glucose into glycogen at different medium concentrations of D-glucose in isolated mantle tissue (**A**) and isolated vesicular connective tissue cells (**B**) of *Lymnaea stagnalis*. (From M. A. Hemminga, unpublished data.)

## E. Environmental and Biological Interactions

### 1. Nutritional Factors

**a. Food Quality.** The hemolymph glucose concentration is related to the quality of food assimilated. Schwarz (1934) noted that unfed *H. pomatia* had the same blood sugar concentrations as snails fed lettuce, but that the concentrations increased on feeding lettuce and glucose or macaroni. Levels increased up to six times in *H. pomatia* fed cabbage soaked in 2% dextrose for 3.5 h and declined 2 h after feeding was stopped (Holz and von Brand, 1940). *L. stagnalis* fed on lettuce, a carbohydrate-poor diet, or starved for 15 days, maintain a rather constant glucose concentration of about 20 $\mu$g/ml: The concentration is increased in starved snails 1 day after feeding lettuce or Bemax, a carbohydrate wheat germ product, and the effect of the Bemax was about four times stronger than the lettuce and unlike the lettuce did not decline over the 7-day experimental period

(Veldhuijzen, 1975a). Scheerboom (1978) observed that when equal amounts of Bemax or lettuce were assimilated by *L. stagnalis*, a considerably higher hemolymph glucose concentration resulted in the group fed Bemax. Friedl (1971) showed that the hemolymph glucose concentration of *L. stagnalis* could be raised up to 10 times by feeding a rich starch suspension.

In contrast to these results, the blood sugar level of *B. glabrata* was not altered by feeding a carbohydrate diet for up to 12 days compared with a balanced standard diet (Stanislawski and Becker, 1979). It has been argued by the various authors that the differing results may be a consequence of the composition of the diet used before the experiments and the composition of the carbohydrate diet used subsequently. In the experiments of Stanislawski and Becker (1979), the carbohydrate diet contained only carbohydrates, salts, and vitamins, and it caused a drastic reduction of egg-laying activity, indicating that essential nutritive substances such as proteins and/or lipids are also required to sustain normal metabolic processes. Normal egg-laying activity was maintained in *B. glabrata* fed a protein or lipid diet, but blood sugar level was reduced significantly (Stanislawski and Becker, 1979). The former observation is surprising, because the production of egg masses requires the synthesis of considerable quantities of polysaccharides. In experiments on *L. stagnalis,* Veldhuijzen and van Beek (1976) observed that feeding Bemax doubled total polysaccharides in 15 days: The greatest increases were in the digestive gland/ovotestis, the mantle, and the muscle fraction (head-foot and penis), each plateauing at different times, and least in the female organs, prostate gland, and the rest fraction. However, egg production was also doubled compared to lettuce-fed snails, indicating a clear stimulation of polysaccharide synthesis and turnover in the female organs.

Direct addition of glucose to the hemolymph by injection leads only to a temporal increase in concentration. In *Patella* sp. (Barry and Munday, 1959), in *Strophocheilus oblongus* (Marques and Pereira, 1970; Marques and Falkmer, 1976), and in *L. stagnalis* (Veldhuijzen, 1975b), injected glucose was rapidly removed from the hemolymph and concentrations returned to preinjection levels within a few hours. The rapid removal is probably mainly a result of carbohydrate-dependent processes in the female accessory sex glands, and to some extent, to glycogen synthesis as a reserve material. Meenakshi and Scheer (1968) found that [$^{14}$C]glucose injected into *A. columbianis* was incorporated into the female organs and the glycogen of the digestive gland and foot. Similarly, 60–83% of the label of [$^{14}$C]glucose injected into *L. stagnalis* was incorporated into the female accessory sex organs (Veldhuijzen and Dogterom, 1975).

**b. Food Quantity.** The relationship between hemolymph glucose levels in *L. stagnalis* and the assimilation of different quantities of lettuce has been studied by Scheerboom (1978). In snails fed small lettuce rations the glucose concentrations equal those of standard snails. However, when assimilation was between 20 and 30 mg (dry weight), the glucose levels increased with increasing amount of food assimilated. The consumption of food is itself under the control of the hemolymph glucose concentration and is inhibited by concentrations above 120 $\mu$g/ml. This control mechanism is probably related to the degree of saturation of glycogen storage in muscles and the VCTCs (Scheerboom and Doderer, 1978). Food consumption is particularly high during the first hours after a change of diet from lettuce to Bemax and subsequently declines to a stable plateau. This decrease in consumption appears to be paralleled by a gradual increase in the hemolymph glucose concentration, the levels in one experiment increasing from 57 to 761 $\mu$g/ml and subsequently declining to about 200 $\mu$g/ml (Scheerboom et al., 1978). In this case, the decrease in consumption was not due to the rise in blood sugar levels because in other experiments the pattern of Bemax consumption was not influenced by a continuous infusion of glucose solution producing a high blood sugar level (Scheerboom and Hemminga, 1978). It is therefore assumed that in *L. stagnalis* at least two types of postingestion stimuli are involved in the regulation of food intake: (a) short-term regulation by mechanoreceptors in the digestive gland and (b) long-term regulation via the hemolymph glucose concentration (Scheerboom and Doderer, 1978).

## 2. Seasonal Changes, Hibernation, Estivation, and Starvation

The carbohydrate metabolism of many gastropods is seasonally variable. The changes are usually in response to changes in environmental temperature and food availability and are often linked to a seasonal reproductive cycle. The pattern of events may involve periods of inactivity: For example, pulmonate snails from temperate regions hibernate during winter and those from the tropics or deserts estivate during dry periods. Carbohydrate levels are generally highest in the summer and autumn and lowest in the winter (Marques and Pereira, 1970; Chatterjee and Ghose, 1973; McLachlan and Lombard, 1980). The cycles of storage and utilization of glycogen and galactogen are well documented (Goddard and Martin, 1966; Meenakshi and Scheer, 1969; Goudsmit, 1972; Nanaware and Varute, 1974; Rudolph, 1975) and can be illustrated by the events occurring in *H. pomatia* (Goudsmit, 1973, 1975). In autumn glycogen is synthesized and stored in several tissues, including the albumen gland, and subsequently is slowly catabolized during winter hibernation. In spring the snails begin refeeding and accumulate galactogen specifically in the albumen gland. In summer the galactogen is transferred to the eggs and no more galactogen is then synthesized until the next egg-laying season. The accumulation of carbohydrate generally parallels increasing food availability and food intake, and may be reflected in higher blood suger levels, as in *L. stagnalis* (Scheerboom and van Elk, 1978), *S. oblongus* (Marques and Pereira, 1970), and *Thais lamellosa* (Lambert and Dehnel, 1974). To an extent galactogen formation may be independent of nutritional conditions (Meenaksi and Scheer, 1969), possibly implying a synthesis from stored glycogen, but generally starvation will rapidly bring egg production to a halt (Veldhuijzen, 1975a).

Estivation is characterized by a drop in oxygen consumption and Krebs cycle oxidations, and a consumption of tissue and hemolymph body reserves, particularly carbohydrate (Singh and Nayeemunnisa, 1976; Heeg, 1977; Krupanidhi et al., 1978; Swami and Reddy, 1978; Horne, 1979). In a comparison of three species, Horne (1979) found that the least resistant organism, the slug *Limax flavus,* had the highest oxygen consumption and consumed relatively more protein and less carbohydrate than the freshwater prosobranch *Marisa cornaurietis* and the land snail *Bulimulus dealbatus.* The carbohydrate metabolism of estivating gastropods may be partially (Swami and Reddy, 1978) or totally anaerobic: The view that estivating *P. virens* consumes no oxygen and is totally anaerobic has been contested by experiments with *Pila globosa* in which injected [U-$^{14}$C]glucose was converted to $^{14}CO_2$ (Reddy and Ramamurthi, 1973); however, in this case a totally anaerobic metabolism is still possible as $CO_2$ can be produced via the succinate pathway (see Chapter 4, this volume). Starvation also results in a reduced metabolic rate, but the rates are higher than in estivating animals; for example, *Bulinus africanus* (Heeg, 1977). Similarly, carbohydrate consumption is increased in starved gastropods (Emerson, 1967; Christie et al., 1974a; Stanislawski and Becker, 1979), although in some cases levels have been maintained; for example, *A. fulica* (Brockelman and Sithithavorn, 1980). Hemolymph glucose concentrations were maintained for 15 days in starved *L. stagnalis,* probably at the expense of stored glycogen (Veldhuijzen, 1975a); a greatly reduced incorporation of injected glucose into stored carbohydrates was also observed (Veldhuijzen and Dogterom, 1975).

In addition to seasonal and other changes, circadian fluctuations in carbohydrates have been observed. In the slug *Laevicaulis alte,* the total carbohydrate of several tissues was highest during the inactive light phase and lowest during the active dark phase: The changes were greatest in the foot, suggesting a correlation with locomotor activity (Kumar et al.,

1981). An inverse relationship was seen in *L. alte* between hepatopancreatic glycogen and blood glucose levels (Reddy et al., 1978). Galactogen synthesis and ovipository activity in *L. stagnalis* are dependent on daylight hours (see Section IV,A,1).

### 3. Parasitism

Many gastropods serve as intermediate hosts for the larval forms of digenetic trematode parasites. The presence of the parasites can lead to giant growth of the host and decreased ovipository activity (Meuleman, 1972; Sluiters et al., 1975; van Elk and Joosse, 1981; Mohamed and Ishak, 1981), as well as alterations in carbohydrate metabolism. Possible causes of these effects include interference with the host's hormonal system, drainage of its nutrients and metabolic intermediates, physical blocking of its circulatory processes, and the release of toxic substances by the parasites (Wijsman, 1979; Narayanan and Venkateswararao, 1980; Mohamed and Ishak, 1981).

Decreased tissue carbohydrate levels and/or decreased blood glucose concentrations as a result of infection have been observed in a number of cases such as *Biomphalaria alexandrina* (Mohamed and Ishak, 1981) and *B. glabrata* (Cheng and Lee, 1971; Christie et al., 1974a; Stanislawski and Becker, 1979) infected with *Schistosoma mansoni, L. littorea* infected with *Cryptocotyle lingua* or *Himasthla leptosoma* (Robson and Williams, 1971; Thomas, 1974) and with *Renicola rosovita* (Robson and Williams, 1971), *Indoplanorbis exustus* infected with Cercariae indicae (Vaidya, 1979), and *N. obsoletus* infected with *Stephanostomum tenue* (Fried and Blumenthal, 1967). The lowered carbohydrate levels of the female sex organs of infected *B. alexandrina* are probably partly responsible for the decreased ovipository activity (Mohamed and Ishak, 1981). The depletion of carbohydrate reserves appears to be due partly to consumption by the larval parasites (McDaniel and Dixon, 1967; Christie et al., 1974b; Hoskin and Cheng, 1974; Reader, 1974; Mohamed and Ishak, 1981) and partly to metabolic rearrangements as a consequence of infection (see later). In *L. luteola* infected with *Xiphidio cercaria, Amphistome cercaria,* or *Furcocerus cercaria,* a relationship was seen between the level of carbohydrate depletion and the type and intensity of infection (Krishna and Simha, 1977). In contrast, in other cases, carbohydrate and sugar levels have either not changed with infection or in fact increased—for example, in *L. luteola* infected with *Prosothoginimus* sp. (Manohar and Rao, 1976). The reasons for these discrepancies are not clear, but possibilities include different individual tissue responses, the differential dependence of parasite species on carbohydrate or protein (Manohar and Rao, 1976), the effects of the different stages (larval forms) of parasitic infection, the host levels of carbohydrate [cercariae only develop if adequate glucose is present (Cheng and Lee, 1971; Manohar and Rao, 1976)], biochemical adaptation by the host (Brockelman and Sithithavorn, 1980), and the methodologies of carbohydrate extraction (Rao et al., 1979).

At the molecular level, *B. alexandrina* and *Bulinus truncatus* infected with *S. mansoni* show a lowered capacity for Krebs cycle oxidations (particularly succinate), lower cytochrome oxidase activity, a lowered capacity for gluconeogenesis, an increased capacity for lactate production, and depleted glycogen reserves (Ishak et al., 1975). The shift from an aerobic to an anaerobic metabolism presumably accounts for part of the glycogen depletion in this organism because of the low energetic efficiency of the latter process (see Section IV,B,2 and 5). Similarly, it must be partly responsible for the decreased gluconeogenesis. The reason for the shift is unknown but it may result from direct inhibition or stimulation of host enzymes or processes by substances from the parasite, or may simply be a compensatory response for the reduced aerobic energy production. The response to infection for different gastropods is not consistent, however. For example, *L. luteola* infected with *Prosthogonimus* sp. showed an enhanced oxidation of succinate (Narayanan and Venkateswararao, 1980). *B. glabrata* infected with *S. mansoni* showed an increase in oxygen consumption and total heat production (measured calorimetrically), but no shift to anaerobic metabolism (Becker and Lamprecht, 1977; Becker, 1980a,b). *L. stagnalis* infected with *Trichobilharzia ocellata* showed no accumulation of anaerobic end products (T. C. M. Wijsman, personal communication). Other observations on the effects of infection are few but include an elevation of glycolytic enzyme activities in the host tissue (Marshall et al., 1974; Teng et al., 1979) and a possible secretion of parasite phosphoglucoisomerase (EC 5.3.1.9) into the host (Wium-Anderson and Simonsen, 1974).

Other parasitic infections studied include the effect of the nematode *Angiostrongylus cantonensis* on *A. fulica:* Infection resulted in a temporary lowering of hemolymph sugars (Brockelman and Sithithavorn, 1980).

## F. Hormonal Regulation

Hemolymph glucose concentrations vary greatly within individual species and may be influenced by food quality and quantity, handling, and parasitic infection (Section III,D and E), and indirectly by temperature (Veldhuijzen, 1975b) and photoperiod (Bohlken et al., 1978) via their effect on reproductive activity. In *L. stagnalis,* for example, concentrations in animals starved or fed lettuce or Bemax stabilized at 20, 50, and 200

$\mu$g/ml, respectively (Scheerboom and Hemminga, 1978), and values for field animals varied between 4 and 355 $\mu$g/ml (Scheerboom and van Elk, 1978). Despite this variation, several lines of evidence indicate the existence of mechanisms for regulating blood glucose levels and for counteracting hypo- and hyperglycemia. In *L. stagnalis*, although fixed maximum blood glucose concentrations are not apparent, a minimum level is maintained in starved animals (Veldhuijzen, 1975a). Insulin-like activity and homologs of pancreatic $\beta$ cells in the digestive tract have been identified in several gastropods, namely *Helix aspersa* (Goddard et al., 1964), *H. pomatia* (Ammon et al., 1967), *B. undatum* (Boquist et al., 1971; Davidson et al., 1971), *A. fulica* (Gomih and Grillo, 1976), and *L. stagnalis* (M. A. Hemminga, personal communication). Mammalian insulin injected into several gastropods resulted in a decrease in blood glucose and an increase in glycogen deposition in the tissues [e.g., *L. alte* (Kulkarni, 1973)], although in some cases a subsequently injected glucose load was required to demonstrate any or a marked effect [e.g., *S. oblongus* (Marques and Falkmer, 1976) and *Cryptozonia belangeri* (Rajan and Sriramulu, 1978)]. Generally the responses to insulin are moderate compared to vertebrates but this may be due to the nonspecific nature of the insulin used and may also reflect the lower metabolic rates of gastropods. However, dose–response relationships have been demonstrated (Kulkarni, 1973; Marques and Falkmer, 1976; Rajan and Sriramulu, 1978). Injection of alloxan, which in vertebrates has a specific destructive effect on the pancreatic $\beta$ cells, also raised blood glucose levels (Kulkarni, 1973; Gomih and Grillo, 1976).

The interactions between hemolymph glucose concentration and other hormonal systems have been studied in *L. stagnalis*. Feeding experiments indicate that over the long term, blood glucose concentration represents a balance between dietary intake and removal by glycogen synthesis, body growth, and female reproduction. It has been suggested that these utilizing processes will themselves be regulated by blood sugar levels through the release of dorsal body hormone (involved in the control of female reproduction) and light green cell hormone (involved in the control of body growth) from endocrine centers in the cerebral ganglia (Scheerboom and Hemminga, 1978). In experiments by Veldhuijzen (1976) and Veldhuijzen and Cuperus (1976), injection of [$^{14}$C]glucose resulted *in vitro* and *in vivo* in a 60–80% incorporation of radiolabel into the albumen gland and some incorporation into the mantle. Starvation greatly reduced incorporation into the albumen gland but had little effect on the mantle. Removal and reimplantation of the dorsal bodies simulated the effects of starvation and refeeding, respectively. It was concluded that dorsal body hormone stimulates galactogen synthesis in the albumen gland, and its release is probably reduced in starved snails; furthermore, as glycogen synthesis in the mantle was not affected by the hormone, this synthesis is probably influenced directly by blood sugar levels. Carbohydrate metabolism is also affected by the removal of the growth hormone-producing light green cells, which resulted in an increase in polysaccharide deposition and an occasional increase in blood sugar levels (Dogterom, 1980). However, it was concluded that these effects were the result of growth stoppage rather than any direct hormonal influence. Other hormones are produced by the lateral lobes (paired structures on the sides of the cerebral ganglia). The cauterization of these lobes in *L. stagnalis* (Geraerts, 1976) and *B. truncatus* (Mohamed and Geraerts, 1980) resulted in a number of effects, including a decrease in female reproductive activity, giant growth, a decrease in the polysaccharide content of blood and some tissues, a decrease in the rate of oxidation of the Krebs cycle intermediates, and production of the end products of anaerobic metabolism, namely, succinate, D-lactate, and alanine. The latter observations are significant because they suggest that the balance of ATP production by aerobic and anaerobic mechanisms is under hormonal control. This conclusion is doubtful, however, because the original observations of Geraerts (1976) could not be confirmed by Wijsman (personal communication), and it now seems likely that they were in fact the results of experimental stress.

The hormonal control of galactogen synthesis in *H. pomatia* has been studied by Goudsmit (1975), and Goudsmit and Feldman (1974). The albumen gland and brain were maintained in organ culture and the incubation of a brain from a reproductively active snail with the albumen gland explant from a hibernating snail resulted in an activation of galactogen synthesis. The neurosecretion stimulating galactogen synthesis has been partially purified and is a protease-labile substance with a molecular weight of about 3000–6000 (Goudsmit and Ram, 1979).

## IV. Metabolism

## A. Anabolism

### 1. Glycogenesis and Galactogenesis

Glycogen is present in a number of different tissues and its synthesis from glucose has been demonstrated in several of them by the incorporation of radiolabel from [$^{14}$C]glucose into the glycogen fraction, namely the hepatopancreas, foot, and albumen gland of *A. columbianis* (Meenakshi and Scheer, 1968), the hepatopancreas of the abalone *Haliotus rufescens* (Bennett and Nakada, 1968), the albumen gland of *H.*

*pomatia* (Goudsmit and Feldman, 1974; Goudsmit, 1975), and the mantle of *L. stagnalis* (Veldhuijzen, 1976; see also Fig. 2). Glycogen synthesis has also been observed in the sperm cell of *H. aspersa* (Personne and Anderson, 1970). Maltose, maltotriose, and maltotetrose are present in the hepatopancreas of *A. columbianis,* and radiolabel experiments indicate that they are intermediates in glycogen biosynthesis rather than degradation products (Meenakshi and Scheer, 1968). The same has been found for several other species from the incorporation of radioactivity from [$^{14}$C]glucose into maltose and glucose-containing oligosaccharides (Bryant et al., 1964). Most if not all of the intermediates and enzymes of glycogen synthesis (Fig. 3, reactions 1–4) have been found in some species of young and adult gastropods (Goudsmit, 1972); more recently, the following enzyme activities have been detected: hexokinase in the pedal retractor and radular retractor muscles of several gastropods (Zammit and Newsholme, 1976a); hexokinase and phosphoglucomutase in the hepatopancreas of the winkle *Littorina saxatilis rudis* (Marshall et al., 1974) and the developing oocytes of *Aplysia depilans* and *Pisania maculosa* (Bolognari et al., 1979); phosphoglucomutase in muscle of *Concholepas concholepa* (Hernandez et al., 1970); UDP-D-glucose pyrophosphorylase in the albumen glands of *H. pomatia* and *L. stagnalis* (Fantin and Gervaso, 1971; Goudsmit and Friedman, 1976); and glycogen synthetase in the hepatopancreas and cephalopedal region of *B. glabrata* (Chiang, 1977).

The UDP-D-glucose pyrophosphorylase of *H. pomatia* is inactive with ATP, CTP, and TTP, but forms GDPglucose from GTP (the significance of the latter reaction in albumen gland metabolism is unknown): It is also inactive with galactose 1-phosphate and UTP, ATP, or GTP

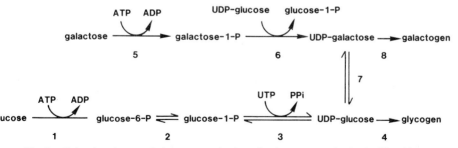

**Fig. 3.** Related pathways of glycogen synthesis and galactogen synthesis. **1**, -Hexokinase (EC 2.7.1.1); **2**, phosphoglucomutase (EC 2.7.5.1); **3**, UDP-D-glucose pyrophosphorylase (EC 2.7.7.9); **4**, glycogen synthetase EC 2.4.1.11); **5**, galactokinase (EC 2.7.1.6); **6**, galactose 1-phosphate uridyltransferase (EC 2.7.7.12); **7**, UDP-D-galactose 4-epimerase (EC 5.1.3.2); **8**, galactogen synthetase.

(Goudsmit and Friedman, 1976). The glycogen synthetases of *B. gla-brata* have similar properties, both showing cooperative activation by glucose 6-phosphate with Hill coefficients of about 2; in the presence of 5 m$M$ glucose 6-phosphate they show Michaelis–Menten kinetics with respect to UDPglucose with apparent $K_m$ values for UDPglucose of 1.74 m$M$ for the hepatopancreas enzyme and 1.36 m$M$ for the cephalo-pedal enzyme (Chiang, 1977). These properties indicate that at least part of the glycogen synthetase activity is normally controlled by intra-cellular concentrations of glucose 6-phosphate.

Galactogen synthesis takes place largely in the albumen gland. The predominant dietary carbohydrate of land snails is cellulose, and there-fore glucose, and glucose serves as the major carbon source for galac-tose and galactogenesis (Goudsmit and Friedman, 1976). The incorpora-tion of radioactivity from [$^{14}$C]glucose into galactose and galactogen has been demonstrated for *A. columbianis* (label from maltose also incor-porated) (Meenakshi and Scheer, 1968), *H. pomatia* (Goudsmit and Feldman, 1974; Goudsmit, 1975), *Bulimnaea megasoma* (Goudsmit, 1976), and *L. stagnalis* (Veldhuijzen and Cuperus, 1976). In *B. mega-soma*, the transfer of labeled galactogen from adults to eggs was also observed (Goudsmit, 1976). As in the case of glycogenesis, most inter-mediates and enzymes of galactogenesis (Fig. 3, reactions 5–8) have been detected in gastropod tissues (Goudsmit, 1972); again, more re-cently galactokinase and galactose 1-phosphate uridyltransferase have been detected in the albumen gland of *H. pomatia* (Goudsmit and Friedman, 1976), and UDP-D-galactose 4-epimerase in the albumen gland of *H. pomatia* (Fantin and Gervaso, 1971; van Elk and Joosse, 1981). In *H. pomatia*, galactose 1-phosphate uridyltransferase is spe-cific for UDPglucose, showing no activity with GDP-, ADP-, or TDP-D-glucose, whereas the UDP-D-galactose 4-epimerase will also convert TDP-D-glucose to TDP-D-galactose but is inactive with ADP- and GDP-D-glucose (Goudsmit and Friedman, 1976). The UDPgalactose 4-epi-merase of the albumen gland of *L. stagnalis* has several properties in common with the enzyme from bacterial and vertebrate sources: an ab-solute requirement for NAD (activity without NAD is 2% of the maxi-mum velocity), which also protects the enzyme against inactivation by freezing or heat treatment; inhibition by NADH; and an optimal pH of 8.75 (its apparent $K_m$ values for UDPgalactose and NAD are of the order of 300 to 400 $\mu M$ and 100 $\mu M$, respectively, and it is inhibited by UDPglucose) (van Elk and Joosse, 1981). The inhibition by NADH, which occurs at physiological concentrations, is possibly of regulatory significance. Very little is known about the galactogen synthetase sys-tem, except that galactogen is required as an acceptor molecule and the enzymes must exhibit some specificity with respect to the primer as activity is much reduced when galactogen from another species is used (Goudsmit, 1972). Goudsmit (1972) has suggested that at least three enzymes will be involved in the synthetase system: one to form the $\beta$-D-1,3 linkages, one to form the $\beta$-D-1,6 linkages, and one to incorporate L-galactose into certain galactogens (see Section II,B).

The utilization of glucose for both glycogenesis and galactogenesis means that where these pathways occur in the same tissue, as in the albumen gland, intracellular regulatory mechanisms must exist to control the carbon flow between them. Aspects of this have been studied in intact *L. stagnalis* (van Elk and Joosse, 1981), and in explants in organ culture of the albumen gland of *H. pomatia* (Goudsmit, 1975). In the latter work, the flexibility of the pathways was well illustrated by the incubation of an albumen gland from a winter-hibernating snail with the circumpharyngeal ganglia from a reproductively active snail: This resulted in a marked redirection of carbon flux from glucose into galactogen and away from glycogen. An important enzyme in the pathways is UDP-D-galactose epimerase as it results in the net conversion of glucose into galactose (Fig. 3). The two sugars can also be interconverted via galactose 1-phosphate uridyltransferase but this does not represent a net conversion. The reversibility of the system is demonstrated by the incorporation of radiolabel from [$^{14}$C]galactose into glycogen (Bennett and Nakada, 1968). The activity of the epimerase in *L. stagnalis* is high in the albumen gland and low in other tissues such as the foot and mantle. van Elk and Joosse (1981) demonstrated a positive correlation between ovipository activity, and hence galactogen synthesis, and the specific activity of the epimerase: Thus in fed snails held under conditions of increased daylight hours, the size of the albumen gland remains constant, but more eggs are produced and epimerase activity is higher; no such correlation exists in starved or parasitized snails in which egg production is reduced or absent. It is very unlikely, however, that the epimerase is rate limiting in galactogen synthesis because of the near-equilibrium nature of the reaction, and this control is most likely to reside in the galactogen synthetase system.

## 2. Synthesis of Mucopolysaccharides and other Sugar Derivatives

Very little is known about the synthesis of these compounds. The activated precursors for the syntheses are almost certainly nucleoside diphosphate sugars, and several have been identified in gastropod tissues: UDP-acetylglucosamine, UDPacetylgalactosamine, GDP-D-mannose, and

GDP-L-galactose (the latter two can be interconverted) (Goudsmit, 1972). The incorporation of radiolabel from [$^{14}$C]glucose and [$^{14}$C]maltose into the galactose, fucose, and glucosamine residues of an acid mucopolysaccharide has been observed in A. *columbianis;* [$^{14}$C]galactose also radiolabeled the macromolecule (Meenakshi and Scheer, 1968).

## 3. Gluconeogenesis

The synthesis of glucose from noncarbohydrate precursors has been demonstrated for whole tissues of the snails B. *alexandrina* and B. *truncatus* for the substrates pyruvate, lactate, α-ketoglutarate, L-glutamate, L-aspartate, and L-alanine (Ishak et al., 1975; Sharaf et al., 1975). Glucose 6-phosphate was synthesized from lactate, pyruvate, fructose 1,6-bisphosphate, succinate, α-ketoglutarate, L-ornithine, L-arginine, L-alanine, and L-glutamate by tissue extracts of C. *concholepa* (Morán and González, 1967). However, although the enzymes of glycolysis have been detected (see Section IV,B,2), there are few reports of the other enzymes needed to catalyze the "irreversible" reactions and effect gluconeogenesis. Glucose 6-phosphatase (EC 3.1.3.9) is present in the hepatopancreas of L. *saxatilis rudis* (Marshall et al., 1974) and H. *pomatia* (Kasprzyk et al., 1978), and malic enzyme (EC 1.1.1.40) in the foot of B. *glabrata* (Bacila, 1970). Phosphoenolpyruvate (PEP) carboxykinase is present in muscular tissue (Zammit and Newsholme, 1976a; Ellington, 1981), but the only gluconeogenic tissue it has been recorded in is the hepatopancreas of L. *saxatilis rudis* (McManus and James, 1975c). There appear to be no data on fructose 1,6-bisphosphatase (EC 3.1.3.11) or pyruvate carboxylase (EC 6.4.1.1). The most likely route of oxaloacetate formation from pyruvate is via the pyruvate carboxylase reaction because malic enzyme usually operates in the direction of pyruvate formation. Oxaloacetate will also be available from the free amino acid pool by transamination; transaminases are widespread in gastropod tissues (Sollock et al., 1979). The conversion of oxaloacetate to phosphoenolpyruvate via the phosphoenolpyruvate carboxykinase reaction may be complicated by the involvement of this enzyme in anaerobic metabolism (see Section IV,B,5).

## B. Catabolism

### 1. Glycogenolysis and Galactogenolysis

Phosphorylase (EC 2.4.2.1) is widely distributed in gastropod tissues (Bennett and Nakada, 1968; Hernandez et al., 1970; Zammit and Newsholme, 1976a; Chiang, 1977; Bolognari et al., 1979) but little is known of its properties. The enzyme of the hepatopancreas of *B. glabrata* exists in an inactive form that can be activated by an endogenous phosphorylase kinase; both the hepatopancreas and cephalopedal phosphorylases are stimulated by AMP (Chiang, 1977). Less is known of the enzyme or enzymes required to break down galactogen. The fact that snail embryos are capable of galactogenolysis was demonstrated by incubating homogenates of embryos of *B. megasoma* with [$^{14}$C]galactogen resulting in the production of radiolabeled dialyzable metabolites (these were not characterized further) (Goudsmit, 1976). The same homogenates contained two acid $\beta$-galactosidases with pH optima of 3 and 5, respectively, which cleaved synthetic *o*-nitrophenyl-$\beta$-D-galactopyranoside but not galactogen. Goudsmit (1976) concluded that these enzymes were probably specific for a $\beta$-1,4 linkage, whereas the true galactosidase would have to cleave $\beta$-1,3-6 bonds. In contrast, Barnett (1971) found a $\beta$-galactosidase of pH 5 optimum in the liver and intestine of *H. aspersa*, and one of pH 2 optimum in the albumen gland and other tissues, and the latter enzyme could hydrolyze galactogen.

## 2. Glycolysis

A complete classical glycolytic pathway converting glucose 6-phosphate to pyruvate is almost certainly present in all gastropod tissues. Its presence has been demonstrated or indicated by the measurement of enzyme activities (Bennett and Nakada, 1968; Bacila, 1970; Goudsmit, 1972; Marshall et al., 1974; Zammit and Newsholme, 1976a; Zammit et al., 1978; Avelar et al., 1978), glycolytic intermediates (Goudsmit, 1972; Beis and Newsholme, 1975), lactic acid, or other anaerobic end-product formation (see Section IV,B,5) and radiolabeling experiments with [$^{14}$C]glucose (Bryant et al., 1964).

Fructose-bisphosphate aldolase (EC 4.1.2.13) has been purified from the foot muscle of *H. pomatia* and been shown to be a class I aldolase, that is, one that functions catalytically via a Schiff base involving an active-site lysyl residue (this type is normally present in eukaryotic cells) (Buczylko et al., 1980a). It is a tetramer of 40,000-MW subunits, and although it differs slightly in amino acid composition and secondary structure from mammalian aldolases, there appear to be no major structural differences (Buczylko et al., 1980b). Its apparent $K_m$ values for fructose 1,6-bisphosphate and fructose 1-phosphate at pH 7.5 are 0.3 $\mu M$ and 1 m$M$, respectively, and the fructose bisphosphate:fructose 1-phosphate activity ratio is 20; in this and other properties it is therefore intermediate between the mammalian skeletal isoenzyme (tailored for a glycolytic role) and the mammalian liver isoenzyme (tailored for a gluconeogenic role) (Buczylko et al., 1980a). Pyruvate kinase (EC 2.7.1.40) has been studied and these properties are discussed in Section IV,B,5 in relation to anaerobic metabolism.

### 3. Krebs Cycle and Other Mitochondrial Reactions

A considerable body of information exists on the presence of the Krebs cycle in gastropod tissues (Goddard and Martin, 1966; Coles, 1969; Bacila, 1970; Bolognari et al., 1979). McManus and James (1975a) detected all the enzymes of the cycle except $NAD^+$-isocitrate dehydrogenase (EC 1.1.1.41) in the hepatopancreas of *L. saxatilis rudis*. Alp et al. (1976) measured the activities of $NAD^+$-isocitrate dehydrogenase, $NADP^+$-isocitrate dehydrogenase (EC 1.1.1.42), and citrate synthase (EC 4.1.3.7) in the radular retractor muscle of several gastropods. The activities in these tissues were high, indicating that such muscles, which tend to be mechanically active over long periods of time, probably operate aerobically. Radiolabel from [2-$^{14}$C]acetate and [1,4$^{14}$C]succinate was incorporated into citrate, malate, and fumarate in whole homogenates of several gastropod species (Bryant et al., 1964). Pyruvate, succinate, and glutamate with malate stimulated oxygen uptake by tissue homogenates of *B. alexandrina* and *B. truncatus* (Ishak et al., 1975). The cycle has not been studied in any detail in any one organism, however, and in some instances anomalies have been observed; for example, glucose 6-phosphate was not metabolized to Krebs cycle intermediates in hepatopancreas homogenates of *Melanerita melanotragus* and *Austrocochlea obtusa* (Bryant, 1965). Only in the case of the $NAD^+$-isocitrate dehydrogenase of the radular muscle of the whelk *B. undatum* have the regulatory properties of the enzyme been examined (Zammit and Newsholme, 1976b). The enzyme shows sigmoidal kinetics with respect to isocitrate, and the cooperativity is increased by $Ca^{2+}$ and decreased by ADP. The $Ca^{2+}$ inhibition is pronounced at pH 7.1 and absent at pH 6.5. The authors argue for a central role for $Ca^{2+}$ in controlling $NAD^+$-isocitrate dehydrogenase activity during contraction, and for a central role for ADP when the muscle is at rest.

The mitochondria of the hepatopancreas of *H. aspersa* contain an oxidase that converts D-mannitol to D-mannose in the absence of pyridine nucleotides (Vorhaben et al., 1980). The rate of conversion was high relative to other substrates, suggesting that plant alditols may be important dietary carbohydrates for herbivorous gastropods.

### 4. Pentose Phosphate Pathway

Glucose 6-phosphate dehydrogenase (EC 1.1.1.49) has been detected in tissues of *L. stagnalis* (Horstmann, 1960), *H. rufescens* (Bennett and Nakada, 1968), *H. pomatia* (Hunger and Horstmann, 1968), *B. glabrata* (Bacila, 1970), *C. concholepa* (Hernandez et al., 1970), *N. obsoletus* (Fried and Levin, 1973; Schilansky et al., 1977), *L. saxatilis rudis* (Marshall et al., 1974), *L. littorea* (Schilansky et al., 1977), and *A. depilans* and *P. maculosa* (Bolognari et al., 1979). Other enzymes of the shunt have also been found, including 6-phosphogluconate dehydrogenase (EC 1.1.1.43) and a *trans*-ketolase-*trans*-aldolase activity (Horstmann, 1960; Bennett and Nakada, 1968; Coles, 1969). Isoenzymes of glucose 6-phosphate dehydrogenase are present in the hepatopancreas of several species (Coles, 1969). The operation of the pathway *in vivo* has been demonstrated using positionally labeled glucose (Bennett and Nakada, 1968). Nothing is known concerning the regulation of carbon flux in the pathway.

### 5. Anaerobic Metabolism

**a Occurrence of Anaerobiosis.** Anaerobic metabolism is required in gastropod tissues during periods of oxygen limitation (intertidal aerial exposure, waterlogging), escape responses (freshwater snails in danger can expel air from the lung cavity and sink to the bottom of ponds, which may be anoxic; vigorous mechanical activity may be required to escape predators), estivation (Section III,E,2), and possibly parasitism (Section III,E,3). Anaerobic functions are generally thought to be less well developed than in bivalves (von Brand, 1946), and unlike bivalves direct calorimetry indicates metabolism is fully aerobic in the presence of plentiful oxygen (Pamatmat, 1978; Hammen, 1980). The ability to survive experimental anoxia without damage is variable among species. Freshwater Lymnaeidae and Physidae survived only 6 h, whereas Planorbidae and operculate snails survived between 24 and 64 h (von Brand et al., 1950). The terrestrial pulmonate *Strophocheilus oblongus musculus* survived 48 h (Haeser and De Jorge, 1971), and the mud snails *N. obsoletus* and *Nassarius trivittatus* had a mortality percentage after 10 days of 100 and 80%, respectively (Kushins and Mangum, 1971). Longer periods of 1 and 6 months, respectively, have been recorded for *Cepaea nemoralis* (van der Horst, 1974) and estivating *P. virens* (Meenakshi, 1964). A number of marine gastropods, particularly intertidal species, show compensatory respiratory responses to hypoxia and are capable of aerial respiration (McMahon and Russell-Hunter, 1978). In experiments on several *Littorina* species, for example, no anaerobic end products were formed during aerial exposure (Kooijman et al., 1982).

**b. Multiple End Products.** A number of anaerobic end products have been detected in gastropods, namely, lactate, octopine, alanine, succinate, acetate, propionate, and butyrate. Anaerobic glycolysis coupled to lipid synthesis has been proposed for some species (Section IV,B,5,e), and indirect evidence points to the importance of strombine and alanopine as anaerobic end products (Section IV,B,5,d). The particular end products formed appear to be related to animal life-style and, in the case of alanopine and strombine, to phylogenetic position.

Classical studies by von Brand and co-workers examined, in freshwater snails, carbohydrate consumption and lactate accumulation (von Brand et al., 1950), and the production and excretion of volatile fatty acids (Mehlman and von Brand, 1951). Lactate was an excreted end product in all 17 species studied, but it accumulated only in the tissues of Lymnaeidae and Physidae. Lactate was the main end product of *L. stagnalis* and *Lymnaea natalensis*. The volatile fatty acids acetic and propionic acid were also produced and partly excreted, and were the main end products of some species. Meenakshi (1958) found that lactate was also a major anaerobic end product in the snail *P. virens*.

During the period 1959–1966, work by Hammen and co-workers (see de Zwaan, 1977) and Awapara and co-workers identified the special position of succinate in molluscan intermediary metabolism and its origin via $CO_2$ fixation. In the hepatopancreas of the snail *Otala lactea,* radiolabel from $NaH^{14}CO_3$ and [U-$^{14}$C]glucose was incorporated into glutamate and alanine and from $NaH^{14}CO_3$ into aspartate (Awapara et al., 1963). A ready incorporation of $CO_2$ into amino acids and Krebs cycle intermediates and an active glycolysis was therefore evident. Other studies of *O. lactea* and the marine gastropods *Littorina irrorata, Thais haemostoma,* and *Siphonaria lineolata* examined the metabolism of $^{14}CO_2$, [2-$^{14}$C]pyruvate, L-[U-$^{14}$C]aspartate, and L-[U-$^{14}$C]glutamate, and the activities of $CO_2$-fixating enzymes and transaminases (Awapara and Campbell, 1964; Simpson and Awapara, 1964, 1966). Except for *T. haemostoma,* phosphoenolpyruvate carboxykinase activity was considerably higher than in rat liver, whereas malic enzyme and propionyl-CoA carboxylase (EC 6.4.1.3) could not be detected. The authors concluded the high PEP-carboxykinase activity could account for the high rate of $CO_2$ fixation leading to alanine, glutamate, and aspartate formation, and that phosphoenolpyruvate would be the major source of four-carbon dicarboxylic acids terminating in succinate; succinate would replace lactate as the dominant anaerobic end product in certain species.

Different results were obtained for the terrestrial gastropods *Pomatias elegans, C. nemoralis, Cepaea hortensis,* and *H. pomatia* (Bryant et al., 1964). In digestive glands, the incorporation of radiolabel from [2-$^{14}$C]acetate was restricted to citrate, malate, and glutamate, and in *C. nemoralis* a major portion occurred in lactate. In this species there was also the greatest incorporation of radiolabel from [U-$^{14}$C]glucose into lactate. A distinction between nonmarine and marine gastropods was therefore becoming evident, in the former lactate rather than alanine being the major terminal end product of the glycolytic pathway. McManus and James (1975b) studied anaerobic metabolism in the digestive gland of *L. saxatilis rudis,* following the *in vitro* conversion of [U-$^{14}$C]glucose into amino acids and carboxylic acids, and obtained an intermediate result. The major portion of radioactivity was incorporated into alanine, succinate, and lactate in an approximate 2:1:1 ratio. On the basis of this result and enzyme studies (Marshall et al., 1974; McManus and James, 1975c), they proposed that pyruvate was formed by two routes: directly from the glycolytic pathway by the action of pyruvate kinase on phosphoenolpyruvate and by a circuitous route involving PEP carboxykinase, malate dehydrogenase (EC 1.1.1.37), and malic enzyme. In fact, 17% of the radiolabel was incorporated into two unknown compounds, and it is possible that one of these was alanopine as indicated by the recently discovered alanopine dehydrogenase activity (see Table IX).

The work of McManus and co-workers closed a period of about 15 years in which the research had been concerned primarily with following the distribution of radiolabel from specific precursors. Subsequent studies returned to the approach of von Brand and quantified changes in the levels of substrates and end products. J. H. F. M. Kluytmans and P. R. Veenhof (unpublished data) studied changes in metabolite levels during experimental anaerobiosis in seven species of marine, freshwater, and terrestrial gastropods (Table VI). The most striking difference was between marine and nonmarine species. Lactate was by far the major end product of the terrestrial *C. nemoralis* and *H. pomatia,* and was significant in the freshwater *L. stagnalis,* which also excreted it. In contrast, succinate and alanine were dominant in the marine species; the accumulation of alanine was species dependent. The volatile acids propionic and acetic were additional end products in all aquatic species and were partly excreted; acetate was particularly significant in *L. stagnalis* (in contrast, volatile fatty acids were not formed in the freshwater snail *M. cornaurietis:* Table VII). Unlike volatile fatty acids, succinate formation was not related to environment but was similar in all species.

The relative unimportance of alanine formation in the freshwater and terrestrial species may be related to the limited availability of amino-group donors, the concentrations of free amino acids being low in such animals (Wieser and Schuster, 1975). In the whelk *Busycon contrarium,* for example, alanine formation is inversely related to aspartate utilization (see Section IV,B,5,c). Alanine formation in marine gastropods has also been studied by Wieser (1980). In the intertidal species *L. littorea* and *Monodonta lineata,* whereas alanine levels increased following exposure to both air and nitrogen, succinate increased only after exposure to nitro-

## TABLE VI

### Accumulation and Excretion of Anaerobic End Products in Several Gastropods after 24-Hour Experimental Anaerobiosis at 12°C[a]

| Species | Environment[b] | Net accumulation (total tissues) (μmol/g dry wt) | | | | | Excretion (medium) (μmol/g dry wt) | | | |
| --- | --- | --- | --- | --- | --- | --- | --- | --- | --- | --- |
| | | Succinate | Lactate | Alanine | Acetate | Propionate | Succinate | Lactate | Acetate | Propionate |
| Buccinum undatum | M | 16.9 | 5.3 | 19.9 | 3.6 | 11.6 | — | — | 9.8 | 28.2 |
| Crepidula fornicata | M | 37.3 | 4.5 | 6.9 | 2.3 | — | — | — | 2.0 | — |
| Littorina littorea | M | 27.2 | 6.6 | 27.0 | 32.5 | 1.9 | 4.7 | 1.4 | 14.6 | — |
| Nucella lapillus | M | 28.9 | 0.5 | 18.9 | —[c] | — | — | — | 6.3 | 0.9 |
| Lymnaea stagnalis | F | 21.6 | 20.6 | 2.3 | 51.6 | 6.4 | 42.8 | 26.0 | 58.0 | — |
| Cepaea nemoralis | T | 15.0 | 104.3 | 9.8 | — | — | — | — | — | — |
| Helix pomatia | T | 23.2 | 122.3 | 8.0 | 5.4 | — | — | — | — | — |

[a] Data of J. H. F. M. Kluytmans and P. R. Veenhof (unpublished data).
[b] M, Marine; F, freshwater; T, terrestrial.
[c] No change.

gen. Wieser concluded that alanine formation was concerned primarily with the maintenance of osmotic balance, as might be required during desiccation conditions of aerial exposure, and not, like succinate, with the maintenance of redox balance and energy production during anoxia. The picture was complicated, however, by the study of the subtidal *N. reticulatus,* which accumulated only alanine and no succinate during anoxia. In this case, a role for alanine in anoxia energy production was concluded. The results with *N. reticulatus* are surprising, however, because anoxic succinate production has been demonstrated in other related members of the Neogastropoda (Ellington, 1981; see also Table VI). Octopine is also an important anaerobic end product in this group of animals (Koormann and Grieshaber, 1980; Ellington, 1981).

Two species that have been investigated in depth during recent years are *L. stagnalis* and *H. pomatia.* Aspects studied include the nature and stoichiometry of the anaerobic pathways, tissue and seasonal differences, changes in blood metabolites, and the excretion of end products. *L. stagnalis* has a reported tolerance to anoxia of several days (von Brand, 1946), but in studies by T. C. M. Wijsman (personal communication) the conditions became lethal after about 40 h, and during this time a considerable stress response was observed including the production of large amounts of mucus and a maximal effort to facilitate gas exchange. Other changes included a decrease in heart rate from 30 to 10 beats/min, hyperglycemia during the first 6 h—hemolymph glucose increased from 40 to 674 $\mu$g/ml (T. C. M. Wijsman, personal communication)—and a decrease in ammonia excretion rate to 20% of that of the control animals, indicating a reduced amino acid (protein) catabolism. Carbohydrate consumption increased, and after 1 day the levels had decreased by about 50%. The ratio of anaerobic glycogen consumption to aerobic was 2.25 after 24 h at 20°C, which compares reasonably well with other reported values of 1.2 after 6 h at 30°C (von Brand et al., 1950) and 1.8 after 15 h at 20°C (de Zwaan et al., 1976). The anaerobic rates of glycogen consumption for the three studies were, respectively, 1.90, 1.05, and 1.61 $\mu$mol glucosyl units/h. The changes in anaerobic end products in the studies of T. C. M. Wijsman (personal communication) and de Zwaan et al. (1976) are given in Table VII and essentially agree with the work of J. H. F. M. Kluytmans and P. R. Veenhof (unpublished data; see Table VI). Succinate, D-lactate, and acetate were the major end products, and in addition to being excreted, also increased markedly in the hemolymph. Although in both cases the data are incomplete, it is possible using all the data to obtain an approximation of the stoichiometry of anaerobic glycogen catabolism (Table VIII). The identified end products account for about 45% of the degraded glycogen in the study of de Zwaan et al. (1976) (no data on blood levels)

## TABLE VII

### Consumption of Glycogen and Accumulation in the Blood and Tissues, and Excretion of Anaerobic End Products in Two Species of Freshwater Snails during Oxygen Deprivation

| Species and experimental conditions | Fraction | Glycogen (mg/g wet wt) | Anaerobic end product (μmol/g wet wt) | | | | | | |
|---|---|---|---|---|---|---|---|---|---|
| | | | Succinate | D-Lactate | Acetate | Propionate | Aspartate | Alanine | Glycine |
| *Lymnaea stagnalis* | | | | | | | | | |
| *In vivo* | | | | | | | | | |
| Control[a] | Tissues[e] | 15.6 | 0.24 | 0.17 | 4.95 | 0.29 | — | 0.26 | — |
| Control | Foot | —[f] | 0.14 | 0.21 | — | — | 0.13 | 0.13 | 0.07 |
| Anaerobic (15 h) | Tissues | 10.7 | 2.49 | 1.97 | 8.94 | — | — | 0.38 | — |
| Anaerobic (15 h) | Foot | — | 2.05 | 3.86 | — | 0.90 | — | 0.40 | 0.09 |
| Anaerobic (15 h) | Excretion | — | 4.71 | 2.79 | 4.34 | — | 0.27 | n.d. | — |
| Control[b] | Tissues | 20.5 | 0.50 | 0.25 | — | — | — | — | — |
| Control | Blood | — | n.d.[g] | n.d. | — | Trace | — | — | — |
| Anaerobic (15 h) | Tissues | — | 3.26 | 4.02 | — | — | — | — | — |
| Anaerobic (15 h) | Blood | — | 11.93 | 8.75 | — | — | — | — | — |
| Anaerobic (24 h) | Tissues | 11.4 | 4.70 | 4.02 | — | — | — | — | — |
| Anaerobic (24 h) | Blood | — | 15.20 | 8.40 | — | — | — | — | — |

214

*Marisa cornaurietis*

| | | | | | | | | | |
|---|---|---|---|---|---|---|---|---|---|
| In vivo | | | | | | | | | |
| Control[c] | Muscle[h] | 2.1 | 0.07 | n.d. | 2.60 | 0.50 | 0.31 | 0.32 | 0.18 |
| Control | Nonmuscle | 3.1 | 0.39 | 0.31 | 0.92 | 0.02 | 0.19 | 0.30 | 0.27 |
| Anaerobic (17 h) | Muscle | 1.4 | 4.02 | 1.40 | 2.20 | n.d. | 0.35 | 0.54 | 0.26 |
| Anaerobic (17 h) | Nonmuscle | 1.6 | 2.06 | 1.60 | 0.06 | 0.05 | 0.24 | 0.50 | 0.30 |
| In vitro | | | | | | | | | |
| Control[d] | Muscle | −[i] | 0.29 | 0.79 | n.d. | n.d. | 0.24 | 0.23 | 0.17 |
| Anaerobic (3 h) | Muscle | − | 3.25 | 1.96 | n.d. | n.d. | 0.14 | 0.84 | 0.17 |
| Anaerobic (3 h) | Muscle | + | 2.96 | 2.98 | n.d. | n.d. | 0.21 | 0.34 | 0.16 |

[a] After de Zwaan et al. (1976). Snails were starved 24 h before the anaerobic incubation at 20°C.
[b] After T. C. M. Wijsman (personal communication). Snails were starved 24 h before the anaerobic incubation at 20°C.
[c] After A. de Zwaan and J. H. E. M. Kluytmans (unpublished data). Snails were starved 72 h before the anaerobic incubation at 20°C.
[d] After A. de Zwaan and A. M. T. de Bont (unpublished data). Experimental conditions as described in de Zwaan et al. (1982b).
[e] Whole tissues except foot.
[f] Not determined.
[g] Not detectable.
[h] Includes head, foot, and pedal retractor muscle.
[i] Absence (−) or presence (+) of aminooxyacetate (5 mM).

TABLE VIII

Balance of Glycogen Consumption and End Product Accumulation in *Lymnaea stagnalis* after 15 H in Oxygen-Free Water at 20°C

| | End product[c] | | | | |
|---|---|---|---|---|---|
| | Succinate | Lactate | Alanine | Volatile fatty acids | Glycosyl units consumed[c] |
| Tissue[a] | 0.15 | 0.12 | 0.04 | 0.31 | |
| Excretion | 0.31 | 0.23 | — | 0.26 | |
| Total | 0.46 | 0.35 | 0.04 | 0.60 | 1.61 |
| Percentage of carbohydrate accounted for | 14.3 | 10.8 | 1.2 | 18.6 | |
| Tissue[b] | 0.18 | 0.25 | 0.05 | — | |
| Blood[d] | 0.32 | 0.23 | n.d.[f] | — | |
| Excretion | —[e] | 0.31 | — | — | |
| Total | 0.50 | 0.79 | 0.05 | — | 1.90 |
| Percentage of carbohydrate accounted for | 13.2 | 20.8 | 1.3 | — | |

[a] After de Zwaan et al. (1976).

[b] After Wijsman (personal communication).

[c] All data are reduced to $\mu$moles accumulated or consumed per gram fresh tissue per hour of anaerobiosis and represent the difference between the control and anaerobic group.

[d] The hemolymph volume of adult *L. stagnalis* is 40% of the total wet weight (T. C. M. Wijsman, personal communication).

[e] Not determined or not present.

[f] Not detectable.

and 35% in the study of T. C. M. Wijsman (personal communication) (no data on excreted succinate or volatile fatty acids). Combining the two results, a figure of about 60% is obtained, which means that a substantial fraction of the consumed glycogen is unaccounted for. A significant portion of this fraction may be attributable to two as yet unidentified volatile fatty acids (see de Zwaan et al., 1976). An important observation of T. C. M. Wijsman (personal communication) was that during anaerobiosis hemolymph concentrations of D-lactate and succinate were two to three times higher than tissue levels. The authors therefore argue that in *L. stagnalis* anaerobic end products must be actively transported from certain tissues into the hemolymph. Similar results have been obtained for *H. pomatia*, and utilization of hemolymph lactate and succinate by the liver has been suggested (Wieser, 1981).

The relationship between lactate and succinate production has been studied in *H. pomatia* and a seasonal dependence has been observed in the responses of the foot to experimental anoxia (Wieser, 1978), though not in the liver or the hemolymph (Wieser, 1981). In their natural conditions the snails are thought to be aerobic in spring and summer but anoxic in winter when in waterlogged soil. During experimental anoxia D-lactate and succinate appeared immediately in the hemolymph, reaching concentrations of 60 m$M$ or more. The rates of appearance in the period 10 to 30 h after the onset of anoxia were similar. Propionate formation in the blood and liver was negligible, and acetate was present in the liver, but the levels were unrelated to the duration of anoxia. The patterns of accumulation in the foot were different in spring and winter animals. In the former there were small rises in succinate and alanine, and D-lactate accumulation followed a complicated course, reaching a maximum of 40 $\mu$mol/g wet weight after 14 h, then declining and finally accumulating to a second maximum of 60 $\mu$mol/g. In contrast, in winter animals D-lactate production was initially rapid (maximum reached between 6 and 11 h), but the levels then returned to control values, and succinate production was more marked, particularly during the latter stages of anoxia. The author concluded that winter snails were better able to cope with anoxia than spring snails, and that a two-phase response was involved, lactate production providing the energy in the first phase and succinate production (or lipid synthesis) in the second phase. Simultaneously with the measurement of foot metabolites, an estimate of "*in vivo*" D-lactate dehydrogenase (EC 1.1.1.28) activity was obtained by excising the foot and measuring the changes in D-lactate in the excised tissue after 5 min at room temperature. The results showed that for *in vivo* D-lactate concentrations (i.e., concentration on excision) of less than about 4 m$M$, more D-lactate was subsequently formed during the 5 min, but that for concentrations above 4 m$M$ the D-lactate levels decreased; thus in the latter the net reaction was operating in the direction of D-lactate oxidation. It was concluded that this mechanism would eventually contribute to (result in) the decline in D-lactate observed in the foot after *in vivo* levels reached 30–40 m$M$. Given this mechanism, the question arises as to why *in vivo* D-lactate levels ever exceed 4 m$M$. One answer may be that implied by the author, which is that other intracellular factors would ensure that the reaction operated in the direction of D-lactate formation up to *in vivo* concentrations of 40 m$M$. Another is that the excised experiments do not represent an *in vivo* situation. A third possibility is that D-lactate is oxidized above 4 m$M$ but that D-lactate is transported to the foot from other tissues via the hemolymph. The supportive evidence for this speculation is very limited, but it includes the high hemolymph levels of D-lactate (Wieser, 1981) and the observation that glycogen levels declined very little in the foot of the pulmonate *S. oblongus musculus* after 25 h anoxia (Haeser and De Jorge, 1971).

**c. Two Routes of Succinate Formation.** It is clear from the preceding section that although a certain amount is known about the anaerobic end products formed and the pathways involved, the state of knowledge is considerably less than for marine bivalves (see Chapter 4; this volume). However, similarities are apparent between marine gastropods and bivalves in that succinate may originate from aspartate or glycogen, and carbon flow in the latter route is channeled through PEP-carboxykinase. The brief details of the two routes of succinate formation are as follows (see Chapter 4, this volume for full description): Initially glycogen is converted to pyruvate by glycolysis and pyruvate is transaminated to alanine, with aspartate serving as the source of amino groups: the deamination of aspartate produces oxaloacetate, which is reduced to malate in the cytosol. At a later stage carbon flow is diverted at the level of the PEP-carboxykinase reaction to produce oxaloacetate, which is similarly reduced to malate. Malate passes into the mitochondria and is converted to succinate. The carbon skeleton for succinate therefore originates from both aspartate and glycogen depending on the stage of anoxia. The aspartate–succinate route has been demonstrated in the isolated perfused ventricle of *B. contrarium* during contractile activity under hypoxia (Ellington, 1981). Aspartate decreased and alanine increased in a 1:1 stoichiometric relationship. A correlation was also seen between alanine accumulation (aspartate utilization) and succinate accumulation, but no stoichiometry was observed; that is, aspartate utilization greatly exceeded succinate formation. This discrepancy was also observed in studies of the isolated adductor muscle of the mussel *Mytilus edulis* and is a result of carbon from aspartate flowing to pyruvate via a route involving malic enzyme (de Zwaan et al., 1982a). The carbon for alanine formation in the *in vitro* situation therefore originates from both glycogen and aspartate. In the *in vivo* situation the discrepancy is less marked. In the case of the ventricle of *B. contrarium*, PEP-carboxykinase activity is very high (Ellington, 1981), indicating the glycogen–succinate route will also be important *in vivo*.

The aspartate–succinate route is likely to be less important or absent in nonmarine gastropods. Aspartate levels are much lower—for example, 12 and 0.2 $\mu$mol/g wet weight in *B. contrarium* (Ellington, 1981) and *M. cornaurietis* (Table VII), respectively—and lactate replaces alanine as the major glycolytic end product. This view is supported by experiments on the snail *M. cornaurietis* (Table VII). If the adductor muscle of *M. edulis* is anoxically incubated with the transaminase inhibitor aminooxyacetate, aspartate utilization and alanine and succinate formation are all inhibited. (de Zwaan et al., 1982a). In contrast, in muscle of *M. cornaurietis*, succinate formation is little affected and only about 0.1 $\mu$mol of the 3.25 $\mu$mol/g formed was derived from aspartate, the rest coming from glycogen (Table VII). The accumulation of alanine, although small, exceeded the utilization of aspartate. *In vivo* no aspartate was utilized but alanine was formed (A. de Zwaan and A. M. T. de Bont, unpublished data). A possible source of nitrogen for this alanine formation is ammonia via the action of alanine dehydrogenase (EC 1.4.1.1), which has been detected electrophoretically in gastropod tissue (Coles, 1969). However, caution is required in the use of the latter data, as similarly placed bands in the electropherograms were observed for glutamate and lactate dehydrogenases.

**d. Amino Acid-Dependent Oxidoreductases.** In addition to lactate dehydrogenase (LDH) and octopine dehydrogenase (ODH) (EC 1.5.1a), other pyruvate oxidoreductases, alanopine dehydrogenase (ADH) and strombine dehydrogenase (SDH), have recently been discovered in invertebrate tissues. The amino acid-dependent oxidoreductases catalyze the reductive condensation of pyruvate with an amino acid to give an imino acid derivative, a so-called opine compound (see Livingstone, 1982). The enzymes are individual proteins and can occur in the same tissue (Dando et al., 1981). ODH utilizes arginine to form octopine; ADH has a high affinity for alanine, forming the product alanopine; and SDH uses alanine and glycine equally well, giving rise to alanopine and strombine, respectively. [The difference in substrate specificity for the latter two enzymes has been taken as a functional definition (Dando, 1981; Dando et al., 1981).]

ADH has been purified from the foot muscle of *L. littorea* (Plaxton and Storey, 1982a). This enzyme has a molecular weight of about 42,000 and is a monomer. The preferred substrates are alanine and pyruvate, the activity with glycine is less than 37% of that with alanine, and in the reverse direction meso-alanopine is oxidized but D-strombine is not. A number of kinetic parameters have been determined including the following. The absolute $K_m$ values for pyruvate and L-alanine were 0.17 and 14.9 m$M$ at pH 6.5 rising to 0.26 and 23.8 m$M$ at pH 7.5, respectively. The apparent $K_m$ values for meso-alanopine were 6.5 m$M$ at pH 6.5 and 50 m$M$ at pH 8.5 whereas those for NADH (9 $\mu$m) and NAD$^+$ (0.18 mM) were pH-independent. Product-inhibition by meso-alanopine was observed and succinate also inhibited the foreward reaction (formation of alanopine). ADH has also been examined from several tissues of the whelk *Busycotypus canaliculatum* and the apparent $K_m$ values for L-alanine were 8.84, 10.64 and 13.12 m$M$ for the hepatopancreas, ventricle, and gill, respectively (Plaxton and Storey, 1982b). The activities with glycine also differed between the tissues.

The phylogenetic distribution of the oxidoreductases has been a topic of interest and the data for gastropods are summarized in Table IX. The problem of high endogenous amino acid levels interfering with the assays has been overcome by removing the amino acids by gel filtration (Gäde, 1980; de Zwaan et al., 1982b; Livingstone, et al., 1982), testing also for the product of the reaction (Regnouf and Thoai, 1970), or using electrophoresis (Dando et al., 1981). Only in the study of Zammit and Newsholme (1976a) were crude supernatants used, and with the exception of these data there is good qualitative agreement among the results. The distribution of the enzymes at the level of orders indicate that it is at least partly phylogenetically determined. LDH is most generally distributed and is the only oxidoreductase of the Anaspidea and Nudibranchia (Opisobranchia) and the Basommatophora and Stylommatophora (Pulmonata). The Archaegastropoda contain both LDH and ODH but they appear to be mutually exclusive: ADH and SDH are absent and therefore only one oxidoreductase is present in each species. The Mesogastropoda are the only group containing all four enzymes with LDH and ODH again being mutually exclusive: ADH and SDH are also mutually exclusive and generally therefore two oxidoreductases are present per species. The Neogastropoda contain more than one oxidoreductase but are characterized by high activities of ODH. In the case of the family Strombidae (order Mesogastropoda), immunological enzyme inhibition studies have shown that the ODH of *Strombus luhuanus* is only antigenically related to ODHs from other species in the Strombidae (Baldwin, 1982).

The overall picture of oxidoreductase distribution is that, with the exception of LDH, it is restricted. LDH and ODH are mutually exclusive, implying a similarity of function. A part of this is probably related to muscular activity. In the case of marine gastropods, ODH is the selected enzyme when the mechanical activity is vigorous, such as escape movements of *B. undatum* (Koormann and Grieshaber, 1980) and burrowing activities of *Turritella communis* (this was the only Mesogastropod containing ODH; see Table IX). This selection may be related to a requirement for high phosphoarginine levels (i.e., arginine) for these activities. The functions of ADH and SDH are less well understood because of their recent discovery (Livingstone, 1982; see also Chapter 4, this volume) (see also Section IV,B,5,h), but it is clear that their prerequisite for high intracellular concentrations of glycine and alanine (apparent $K_m$ values of the order 50–100 m$M$) would prevent their use in nonmarine gastropods. The evolutionary picture may have been that originally in primitive molluscs there was one or several broadly specific amino acid- (and keto acid-) dependent oxidoreductases, and subsequently selection and specialization have taken place, all the opine dehydrogenases being lost with the migration to fresh water and land. The specialization has ultimately resulted in isoenzymes of ODH in cephalopods (see elsewhere)—these were absent in *B. undatum* (Gäde, 1980)—, isoenzymes of ADH in gastropods (Plaxton and Storey, 1982b) and isoenzymes of LDH in gastropods (Section IV,B,5,g) and others.

**e. Unusual Electron Acceptors.** Van der Horst (1974) and Oudejans and van der Horst (1974) have proposed in *C. nemoralis* an integration of glycolysis and fatty acid biosynthesis during anoxia, the latter serving as an electron-acceptor mechanism for glycolytically generated NADH (via NADPH). The incorporation of radiolabel from [$^{14}$C]glucose, [$^{14}$C]pyruvate, [$^{14}$C]alanine, [$^{14}$C]lactate, and [$^{14}$C]acetate into lipid components during aerobiosis and anoxia was studied. There was some incorporation from all precursors, but with the exception of pyruvate and acetate it was much reduced under anaerobic conditions. Incorporation from [1-$^{14}$C]acetate into unsaturated fatty acids was very low after 30 min anoxia but increased afterwards, whereas the specific activity of the saturated fatty acid fraction exceeded by far the aerobic incorporation (van der Horst, 1974). Qualitative differences were also observed. In saturated fatty acids, radioactivity was principally associated with palmitic acid and stearic acids during normoxia compared with a dominant labeling of the former and a reduced labeling of the latter during anoxia: In unsaturated fatty acids, radioactivity in oleic acid was dominant during normoxia but reduced in anoxia concomitant with an increased labeling of palmitoleic acid (Oudejans and van der Horst, 1974). The distribution of radiolabel from pyruvate and acetate was similar and from this it was concluded that part of the pyruvate would be converted to acetyl-CoA, implying an apparently oxygen independent decarboxylation step, and that this would be the key reaction coupling anaerobic glycolysis with fatty acid biosynthesis.

There is a need for an evaluation of this work, particularly in view of the fact that authors (Storey, 1977; Wieser, 1978; Wieser and Wright, 1978) have cited it as an important anaerobic route. Indeed, Storey (1977) wrongly assumed that lactate is not produced in *C. nemoralis* (see Table VI) and obtained kinetics for the foot LDH of *H. aspersa* compatible with a system in which glucose (and lactate derived from glucose) is predominantly channeled into lipid synthesis (see Section IV,B,5,g). No data are offered in Oudejans and van der Horst (1974) on how much precursor is actually converted to lipids, but incorporation as a percentage of the administered total doses is given. These were low, varying between 0.25 and 4% for lactate and acetate, respectively, with a value of 0.5% for glucose. In comparable experiments with molluscs, the total conversion of glucose into end products is usually of the order of over 80% (see, e.g., Simpson

**TABLE IX**

**Activities of Pyruvate Oxidoreductases in the Gastropoda[a]**

| Species classification | Environment[b] | Activity[c] | | | | Reference |
|---|---|---|---|---|---|---|
| | | LDH | ODH | SDH | ADH | |
| **Prosobranchia** | | | | | | |
| **Archaeogastropoda** | | | | | | |
| Calliostoma zizyphinum | M | 0.0 | 0.0 | 0.0 | 0.3 | de Zwaan et al. (1982b) |
| C. zizyphinum | | 0.0 | 0.0 | —[e] | — | Regnouf and Thoai (1970) |
| Gibbula umbilicalis (F)[d] | M | 3.6 | 289 | — | — | Barrett and Körting (1981) |
| (R) | | 3.1 | 515 | — | — | |
| Haliotis tuberculata | M | 9.6 | 0.0 | — | — | Regnouf and Thoai (1970) |
| Monodonta lineata (F) | M | 8.3 | 268 | — | — | Barrett and Körting (1981) |
| (R) | | 7.5 | 675 | — | — | |
| M. lineata (R) | | 0.4 | 0.0 | — | — | Zammit and Newsholme (1976a) |
| M. lineata | | 0.1 | 25.7 | 0.0 | — | de Zwaan et al. (1982b) |
| M. turbinata (R) | M | 0.0 | 0.0 | — | 0.0 | Zammit and Newsholme (1976a) |
| Patella aspersa (F) | M | 89 | 12 | — | — | Barrett and Körting (1981) |
| (R) | | 41 | 19 | — | — | |
| P. aspersa (F) | | +++[f] | 0.0 | 0.0 | 0.0 | Dando et al. (1981) |
| Patella vulgata | M | 1.6 | 0.0 | 0.0 | 0.0 | de Zwaan et al. (1982b) |
| P. vulgata (F) | | +++ | 0.0 | 0.0 | 0.0 | Dando et al. (1981) |
| P. vulgata (F) | | 30 | 4.8 | — | — | Barrett and Körting (1981) |
| (R) | | 36 | 5.9 | — | — | |
| P. vulgata (R) | | 1.6 | 0.0 | — | — | Regnouf and Thoai (1970) |
| P. vulgata (RR) | | 1.4 | 17.7 | — | — | Zammit and Newsholme (1976a) |
| Patella depressa (F) | M | +++ | 0.0 | 0.0 | 0.0 | Dando et al. (1981) |
| Trochocodea crassa | M | 1.6 | 27.2 | — | — | Regnouf and Thoai (1970) |
| **Mesogastropoda** | | | | | | |
| Aporrhais pes-pelicani | M | 300 | 0.0 | — | — | Regnouf and Thoai (1970) |
| Cupulus hungaricus | M | 0.24 | 0.0 | — | — | Regnouf and Thoai (1970) |

222

| | | | | | | Reference |
|---|---|---|---|---|---|---|
| Crepidula fornicata | M | 9.7 | 0.0 | 3.9 | 4.1 | de Zwaan et al. (1982b) |
| C. fornicata (F) | M | Trace | 0.0 | +++ | 0.0 | Dando et al. (1981) |
| Strombus gigas | M | — | — | Present | — | Sangster et al. (1975) |
| Littorina littorea | M | 16.0 | 0.0 | — | — | Regnouf and Thoai (1970) |
| L. littorea (F) | | 289 | 0.0 | — | — | Barrett and Körting (1981) |
| (R) | | 169 | 0.0 | — | — | Zammit and Newsholme (1976a) |
| L. littorea (R) | | 4.8 | 0.0 | 2.8 | 13.3 | de Zwaan et al. (1982b) |
| L. littorea (R) | | 7.3 | 0.0 | 0.0 | 2.8 | Dando et al. (1981) |
| L. littorea (F) | | 14.2 | 0.0 | 0.0 | ++ | de Zwaan et al. (1982b) |
| Littorina littoralis obtusata | M | 3.1 | 0.0 | 1.4 | 15.7 | Barrett and Körting (1981) |
| L. littoralis (F) | | 70 | 0.0 | — | — | Barrett and Körting (1981) |
| (R) | | 32 | 0.0 | — | — | |
| Littorina saxatilis rudis (F) | M | 71 | 0.0 | — | — | |
| (R) | | 36 | 0.0 | — | — | |
| Trivia monocha | M | 1.0 | 0.0 | 2.2 | 12.0 | Livingstone et al. (1982) |
| Turritella communis | M | 0.0 | 4.7 | 0.6 | 1.9 | de Zwaan et al. (1982b) |
| Neogastropoda | | | | | | |
| Buccinum undatum (F) | M | 44 | 195 | — | — | Barrett and Körting (1981) |
| (R) | | 27 | 216 | — | — | |
| B. undatum | | 0.8 | 12.0 | 0.0 | 9.3 | Regnouf and Thoai (1970) |
| B. undatum | | 3.1 | 86.8 | — | — | de Zwaan et al. (1982b) |
| B. undatum (RR) | | 23.3 | 22.4 | — | — | Zammit and Newsholme (1976a) |
| B. undatum (RR) | | 24.0 | 229.0 | — | — | Gäde (1980) |
| (F) | | 4.4 | 159.0 | — | — | |
| Busycon contrarium | M | 28 | 282.9 | — | 96.6 | Ellington (1981) |
| Nassarius reticulum | M | 4.2 | 24.0 | — | — | Regnouf and Thoai (1970) |
| Neptuna antiqua | M | 4.0 | 72.0 | — | — | Regnouf and Thoai (1970) |
| Nucella lapillus | M | 0.4 | 29.4 | 1.2 | 6.7 | de Zwaan et al. (1982b) |
| N. lapillus (F) | | 10 | 248 | — | — | Barrett and Körting (1981) |
| (R) | | 19 | 633 | — | — | |

(Continued)

**TABLE IX** (*Continued*)

| Species classification | Environment[b] | Activity[c] LDH | ODH | SDH | ADH | Reference |
|---|---|---|---|---|---|---|
| *Purpura lapillus* | M | 1.0 | 168.0 | — | — | Regnouf and Thoai (1970) |
| *Ocenebra erinacea* | M | 0.8 | 8.7 | 0.2 | 0.7 | Livingstone et al. (1982) |
| Opistobranchia | | | | | | |
| Anaspidea | | | | | | |
| *Aplysia punctata* | M | 72.0 | 0.0 | — | — | Regnouf and Thoai (1970) |
| Cephalaspidea | | | | | | |
| *Philine aperta* | M | 0.16 | 0.0 | — | — | Regnouf and Thoai (1970) |
| *Scapander lignarius* (F) | M | +++ | 0.0 | 0.0 | 0.0 | Dando et al. (1981) |
| Nudibranchia | | | | | | |
| *Aeolidia papillosa* | M | 3.5 | 0.0 | 0.0 | 0.0 | de Zwaan et al. (1982b) |
| *Archidoris pseudoargus* (F) | M | 196 | 0.0 | — | — | Barrett and Körting (1981) |
| *Goniodoris nodosa* | M | 0.5 | 0.0 | 0.0 | 0.0 | Livingstone et al. (1982) |
| *Jorunna tomentosa* | M | 0.5 | 0.0 | 0.0 | 0.0 | Livingstone et al. (1982) |
| Pulmonata | | | | | | |
| Basommatophora | | | | | | |
| *Lymnaea stagnalis* | FR | 31.6 | 0.0 | 0.0 | 0.0 | Livingstone et al. (1982) |
| *Lymnaea ovata* | FR | 1.5 | 0.0 | 0.0 | 0.0 | Livingstone et al. (1982) |

224

| Species | Habitat[b] | LDH | ODH | SDH | ADH | Reference |
|---|---|---|---|---|---|---|
| *Planorbis corneus* | FR | 6.4 | 0.0 | 0.0 | 0.0 | Livingstone et al. (1982) |
| *P. corneus* (F) | | 5 | 0.0 | — | — | Barrett and Körting (1981) |
| (R) | | 3 | 0.0 | — | — | |
| *Marisa cornuariëtis* | FR | 7.4 | 0.0 | 0.0 | 0.0 | Livingstone et al. (1982) |
| Stylommatophora | | | | | | |
| *Agriolimax reticulatus* (F) | T | 24 | 0.0 | — | — | Barrett and Körting (1981) |
| *Arion rufus* | T | 0.9 | 0.0 | 0.0 | 0.0 | Livingstone et al. (1982) |
| *Cepea nemoralis* | T | 15.4 | 0.0 | 0.0 | 0.0 | Livingstone et al. (1982) |
| *C. nemoralis* (F) | | 81 | 0.0 | — | — | Barrett and Körting (1981) |
| (R) | | 85 | 0.0 | — | — | |
| *Succinea elegans* | T | 7.9 | 0.0 | 0.0 | 0.0 | Livingstone et al. (1982) |

[a] ADH, Alanopine dehydrogenase; LDH, lactate dehydrogenase; SDH, strombine dehydrogenase; ODH, octopine dehydrogenase. The enzyme species that either dominate or are exclusively present have been enclosed in boxes. For the discrimination between SDH and ADH, the functional definition presented by Dando et al. (1981) has been applied (see text).

[b] FR, Freshwater; M, marine; T, terrestrial.

[c] Activities are given in μmol/min/g wet wt. at 25°C (Zammit and Newsholme, 1976a; Gäde, 1980; de Zwaan et al., 1982b); in μmol/min/g wet wt. at 30°C (Regnouf and Thoai, 1970); and in nmol/min/mg protein at 30°C (Barrett and Körting, 1981).

[d] (F), Foot; (R), foot retractor muscle; (RR), radular retractor muscle.

[e] Not determined.

[f] +++, Very strong; ++, strong (Dando et al., 1981).

and Awapara, 1966). Assuming, therefore, a high total conversion, the bulk of glucose-carbon (probably over 80%) must have flowed not to lipids but to other end products. Given the high production of lactate and the significant contribution of succinate in organisms such as *C. nemoralis* and *H. pomatia* (Table VI), a relative unimportance for lipid synthesis in anaerobic energy production must be concluded. Furthermore, because of the observed high accumulation of lactate, it is difficult to conclude that lactate dehydrogenase would significantly act as a lactate oxidase during anoxia, as has been proposed (Storey, 1977; Wieser, 1978). Further arguments against a general importance for this coupled system are given by a consideration of the stoichiometries of *de novo* fatty acid synthesis [Eq. (1)] and fatty acid desaturation [Eq. (2)].

8 Acetyl-CoA + 14 NADPH + 14 $H^+$ + 7 ATP + $H_2O$ →
$$\text{palmitic acid} + 8 \text{ CoA} + 14 \text{ NADP}^+ + 7 \text{ ADP} + 7 \text{ P}_i \quad (1)$$

$$\text{Palmitoyl-CoA} + \text{NADPH} + H^+ + O_2 \rightarrow \text{palmitoleyl-CoA} + \text{NADP}^+ + 2 H_2O \quad (2)$$

Equation (1) requires considerable amounts of ATP and Eq. (2) requires molecular oxygen, both conditions being unlikely to be met during anoxia. The observed formation of unsaturated monoenoic acids in *C. nemoralis* during anoxia (van der Horst, 1974) must either have proceeded by different reactions or involved some residual supply of oxygen.

A rather unique system for replacing oxygen as the terminal electron acceptor during anoxia has been described by Zs.-Nagy (1977) in the form of a lipochrome pigment located in granules called cytosomes (for a full description of the system see Chapter 4, this volume). Redox potential experiments identified the electron-accepting capability of the pigment in isolated ganglia slices of *L. stagnalis* (Zs.-Nagy, 1971a). Zs.-Nagy proposed that the cytosomes produce ATP during anoxia by a mechanism termed "anoxic endogenous oxidation" and that strong pigmentation is likely to be a feature of molluscs with high anoxia tolerances. Heavy pigmentation was found in the central nervous systems of *Murex trunculus* and *Aplisia limacia,* which had anoxia-survival times of 100–200 and 10–12 h, respectively (Zs.-Nagy, 1971b).

**f. Pyruvate Kinase.**   This is a key enzyme in anaerobic metabolism occurring at the PEP branchpoint (see Chapter 4, this volume). Along with PEP-carboxykinase, it functions to direct glycolytic carbon flow to pyruvate (and hence to oxidation, lactate or opine formation) or to oxaloacetate (and succinate formation). Data on the enzyme in gastropods are limited. The pyruvate kinase of the digestive gland of *L. saxatilis rudis* possesses simple hyperbolic substrate kinetics (apparent $K_m^{PEP} = 0.44$ m$M$), is strongly activated by fructose 1,6-bisphosphate, and inhibited by alanine (noncompetitive) and ATP: The pH optimum is 7.2 and activity drops markedly below this pH (McManus and James, 1975c). The production of protons and alanine during anaerobiosis will inhibit the enzyme and channel the flux toward oxaloacetate and the observed succinate production (McManus and James, 1975b). Similar results were obtained for the pyruvate kinase of the foot of *L. stagnalis* (A. de Zwaan and M. Kammüller, unpublished data). The apparent $K_m$ values for PEP at neutral pH were 0.37 m$M$ in the absence of modulators, and 0.02 and 1.83 m$M$ in the presence of 0.01 m$M$ fructose 1,6-bisphosphate and 2 m$M$ alanine, respectively. Saturation curves were hyperbolic (Hill coefficients between 0.92 and 0.96), lowering the pH from 7.5 to 6.5 decreased $V_{max}$ and increased apparent $K_m$, and alanine and protons both acted as mixed noncompetitive inhibitors. Hyperbolic and sigmoidal kinetics with respect to PEP were obtained at pH 7 and pH 6, respectively, for the pyruvate kinase of the foot of *H. pomatia* (Wieser and Lackner, 1977). Fructose 1,6-bisphosphate activated at pH 6 and restored hyperbolic kinetics. Inhibition by phosphoarginine was observed but this result has been questioned (see de Zwaan and Ebberink, 1978).

Hoffmann (1976) studied the effects of assay temperature on the apparent $K_m$ for PEP of a number of invertebrates. The values for the foot enzymes of *L. littorea, B. undatum,* and *H. pomatia* fell within the narrow range of 0.05 to 0.10 m$M$ and changed little between 5 and 40°C. In contrast, the foot enzyme of *Haliotis fulgens* showed a U-shaped curve with apparent $K_m$ values between 0.15 and 0.25 m$M$. A correlation was observed between the absolute $K_m$ values and the recorded capacities of the tissues for aerobic glycolysis. Wieser and Wright (1979) also observed that the apparent $K_m$ value for the foot enzyme of *H. pomatia* did not vary between 10 and 25°C (unlike those of D-lactate dehydrogenase and arginine kinase, which increased threefold). In this study, the activity of pyruvate kinase was measured at 20°C over a year and found to be positively correlated with the environmental temperature at the time of collection. Hoffmann and Rädeke (1978) examined the temperature stability of the pyruvate kinases of a number of invertebrates, including *B. undatum* and *H. pomatia,* in terms of changes in activity at saturating and half-saturating substrate concentrations. The gastropods were regarded as facultative anaerobic invertebrates, and from these and other results it was concluded that regulatory pyruvate kinases displayed greater *in vitro* lability than nonregulatory ones.

**g. Lactate Dehydrogenase.** The lactate dehydrogenase of gastropods, as for all molluscs, has been found to be exclusively D-stereospecific (Long and Kaplan, 1968; Michejda *et al.,* 1969; Scheid and Awapara,

1972; Long, 1976), although a low rate of L-lactate utilization has been observed in *C. nemoralis* (Oudejans and van der Horst, 1974). Isoenzymes of LDH have been detected in *H. aspersa* (Storey, 1977; Long et al., 1979), *H. pomatia* (Wieser and Wright, 1978), *Pila ovata* (Coles, 1969), and *C. nemoralis* (Gill, 1978).

The majority of studies have concerned the *Helix* sp. Storey (1977) examined D-LDH from eight major tissues of *H. aspersa* and found them all to be tetramers of 140,000 MW (subunits were 34,000 MW). Electrophoresis revealed the same complement of two bands of activity in all tissues. The apparent $K_m$ values of the foot enzyme for pyruvate, lactate, NAD$^+$, and NADH were, respectively, 0.42, 3.2, 0.02, and 0.015 m$M$, which are in good agreement with the data of Michejda et al. (1969), Wieser and Wright (1978), and Long et al. (1979). The most significant observation was pyruvate inhibition of the enzyme from all tissues ($K_i$ of 10 m$M$) and no inhibition by lactate up to 100 m$M$. These characteristics closely resemble those of the heart-type LDH of vertebrate tissue and led Storey to conclude that the enzyme of *H. aspersa* would function under anoxia both to divert pyruvate to lipid synthesis and to actively convert formed lactate to pyruvate. The same enzyme has been examined by Long et al. (1979) in a comparison with the foot LDH of the abalone *Haliotis cracherodii*. The purified enzymes from both sources had physical and catalytic properties similar to one another and to other known D-LDH enzymes. The molecular weights were approximately 80,000, corresponding to a dimeric subunit organization. The main differences between the enzymes of *H. aspersa* and *H. cracherodii* were, respectively, the presence of two and one electrophoretic bands of activity, and the absence and presence of pyruvate inhibition. The results of Long et al. (1979) for *H. aspersa* are therefore in conflict with those of Storey (1977) with respect to molecular weight and pyruvate inhibition. Michejda et al. (1969) did not observe pyruvate inhibition in crude extracts of *H. pomatia*. Long et al. (1979) argue the validity of their molecular-weight determinations and reject the proposed function for LDH of Storey (1977). Rather, they relate the kinetic differences between their two species to different locomotor requirements. They conclude that the foot LDH of *H. aspersa* would be suitable for generating high rates of energy production (anaerobic glycolysis) for intense foraging activity, whereas that of *H. cracherodii* would be suitable for the (aerobic) metabolism of a slow, continually contracting muscle involved in securing attachment of the animal to the rocky substratum. Wieser and Wright (1978) studied the kinetics of the D-LDH of the foot of *H. pomatia,* and concluded from the pH dependence and other factors that it would operate in the direction of lactate formation under the slightly acidic conditions to be expected during anaerobiosis. They also speculated that at higher *in vivo* lactate concentrations the reverse reaction would become significant (see Section IV,B,5,b).

Seasonal variations in LDH activity and isoenzymes have been observed. In the foot of *H. pomatia,* the activity of winter animals was twice that of summer animals, and of the five isoenzymes, the two most cathodically migrating ones were absent in winter animals; the latter were suggested to be functional in locomotion and therefore not required by hibernating animals (Wieser and Wright, 1978). Five isoenzymes were identified in the digestive gland of *C. nemoralis* that varied during the year and according to whether the snails were feeding or estivating, and that showed different kinetics with respect to pyruvate inhibition (Gill, 1978). The changes were considered to be nongenetic and were interpreted in relation to changing requirements for glycogen synthesis from lactate.

The existence of an $NADP^+$-dependent LDH has been claimed for *B. glabrata* (Bacila, 1970) and *B. africanus* (Coles, 1969).

**h. Anaerobic Energy Metabolism.** The major sources of energy for metabolism without oxygen are phosphoarginine (and to a small extent the ATP pool) and glycogen. For example, the energy in the columnellar muscle of *B. undatum* during escape movements is provided 36% by phosphoarginine and 64% by the octopine pathway (calculated from Koormann and Grieshaber, 1980), and in the ventricle of *B. contrarium* during hypoxia, 37% by phosphoarginine, 50% by the succinate pathway, and 13% by the octopine pathway (calculated from Ellington, 1981).

An important consideration in the catabolism of glycogen is the anaerobic pathway employed. Three types of pathways—the lactate, opine (octopine, alanopine, and strombine), and succinate pathways—are present in gastropod and other molluscan tissues; they differ in their energetic efficiencies (amount of ATP produced per glucosyl unit) and/or the rate at which the energy is produced (Livingstone, 1982). The lactate and opine pathways are essentially linear pathways of low energetic efficiency (3 ATP per glucosyl unit) giving rise to medium or high rates of energy production, and tend to be employed in situations where energy demand is relatively high, such as vigorous mechanical activity, recovery from aerial exposure (sessile bivalves), and swimming (free-swimming bivalves). In contrast, the succinate pathway is a branched pathway of relatively high energetic efficiency (6 to 7 ATP per glucosyl unit), giving rise to low rates of energy production, and is used during survival of anoxia. In freshwater gastropods, for example, species resistant to anoxia produce mainly volatile fatty acids, whereas less resistant species produce mainly lactate (Mehlman and von Brand, 1951). Examples of rates of energy production of the pathways in gastropods are (in $\mu$mol ATP equivalents/g wet weight/min): succinate pathway, 0.13; lactate pathway, 0.46; and octopine pathway, 0.49 (Livingstone, 1982). The high rates of energy production are obtained by an acceleration of glycolytic rate: For example, *in vivo* anaerobic pyruvate flux in the foot of *H. pomatia* is about 20 times higher than the aerobic pyruvate flux (Wieser and Wright, 1978). Contributory factors to the high rates of energy production are that the resulting organic acid end products may be discharged into the hemolymph and there neutralized by the mobilization of $CaCO_3$ (Wieser, 1981; see also Sminia et al., 1977), and subsequently possibly excreted (Wijsman, 1979) or metabolized by other tissues (see Section IV,B,5,b).

## Acknowledgments

The authors gratefully acknowledge M. A. Hemminga, J. H. F. M. Kluytmans, P. R. Veenhof, A. M. T. de Bont, T. C. M. Wijsman, and M. Kammüller for the use of unpublished data.

## References

Afonso, A. M. A., Arrieta, M. R., and Neves, A. G. A. (1976). Characterization of the hemoglobin of *Biomphalaria glabrata* as a glycoprotein. *Biochim. Biophys. Acta* **439**, 77–81.

Agnisola, C., Savadore, S., and Scardi, V. (1981). On the occurrence of cellulolytic activity in the digestive gland of some marine carnivorous molluscs. *Comp. Biochem. Physiol. B* **70B**, 521–525.

Alp, P. R., Newsholme, E. A., and Zammit, V. A. (1976). Activities of citrate synthase and $NAD^+$-linked and $NADP^+$-linked isocitrate dehydrogenase in muscle from vertebrates and invertebrates. *Biochem. J.* **154**, 689–700.

Ammon, J., Melani, F., and Gröschel-Stewart, U. (1967). Nachweis von immunologisch hemmbarer Insulinaktivität bei Schnecken (*Helix pomatia* L.). In "Die Pathogenese des *Diabetes mellitus*. Die endokrine Regulation des Fettstoffwechsels" (E. Klein, ed.), pp. 96–98. Springer-Verlag, Berlin and New York.

Avelar, P. M. F., Giacometti, D., and Bacila, M. (1978). Comparative levels of muscle glycolytic enzymes in mammals, fish, echinoderm and molluscs. *Comp. Biochem. Physiol. B* **60B**, 143–148.

Awapara, J., and Campbell, J. W. (1964). Utilization of $C^{14}O_2$ for the formation of some amino acids in three invertebrates. *Comp. Biochem. Physiol.* **11**, 231–235.

Awapara, J., Campbell, J. W., and Peck, E. (1963). Formation of amino acids from $C^{14}O_2$ and D-glucose-4-$C^{14}$ in invertebrates. *Fed. Proc., Fed. Am. Soc. Exp. Biol.* **22**, 553.

Baldwin, J. (1982). An immunochemical study of structural and evolutionary relationships among molluscan octopine dehydrogenases. *Pacific. Sci.* (in press).

Bacila, M. (1970). Anaplerotic mechanisms and metabolic regulation in *Biomphalaria glabrata*. *An. Acad. Bras. Cienc.* **42**, 161–169.

Barber, A., Jordana, R., and Ponz, F. (1975a). Active transport by the intestine of snail (*Cryptomphalus hortensis* Müller). *Rev. Esp. Fisiol.* **31**, 119–124.

Barber, A., Jordana, R., and Ponz, F. (1975b). Effect of anaerobiosis, dinitrophenol and fluoride on the active intestinal transport of galactose in snail. *Rev. Esp. Fisiol.* **31**, 125–130.

Barber, A., Jordan, R., and Ponz, F. (1975c). Sodium dependence of intestinal active transport of sugars in snail (*Cryptomphalus hortensis* Müller). *Rev. Esp. Fisiol.* **31**, 271–276.

Barber, A., Jordana, R., and Ponz, F. (1979). Competitive kinetics of sugar active transport in snail intestine. *Rev. Esp. Fisiol.* **35**, 243–248.

Barnett, J. E. G. (1971). An acid $\beta$-galactosidase from the albumen gland of *Helix aspersa*. *Comp. Biochem. Physiol. B* **40B**, 585–592.

Barrett, J., and Körting, W. (1981). Octopine dehydrogenase in gastropods from different environments. *Experientia* **37**, 958–959.

Barry, R. J. C., and Munday, K. A. (1959). Carbohydrate levels in *Patella*. *J. Mar. Biol. Assoc. U.K.* **38**, 81–95.

Becker, W. (1972). The glucose content in haemolymph of *Australorbis glabratus*. *Comp. Biochem. Physiol. A* **43A**, 809–814.

Becker, W. (1980a). Metabolic interrelationships of parasitic trematodes and molluscs, especially *Schistosoma mansoni* in *Biomphalaria glabrata*. *Z. Parasitenkd.* **63**, 101–111.

Becker, W. (1980b). Microcalorimetric studies in *Biomphalaria glabrata:* The influence of *Schistosoma mansoni* on basal metabolism. *J. Comp. Physiol. B* **135**, 101–105.

Becker, W., and Lamprecht, J. (1977). Microcalorimetric investigation of the host–parasite relationship between *Biomphalaria glabrata* and *Schistosoma mansoni*. *Z. Parasitenkd.* **53**, 297–305.

Beis, I., and Newsholme, E. A. (1975). The contents of adenine nucleotides, phosphagens and some glycolytic intermediates in resting muscles from vertebrates and invertebrates. *Biochem. J.* **152**, 23–32.

Bennett, R., Jr., and Nakada, H. I. (1968). Comparative carbohydrate metabolism of marine molluscs—I. The intermediary metabolism of *Mytilus californianus* and *Haliotus rufescens*. *Comp. Biochem. Physiol.* **24**, 787–797.

Bohlken, S., Anastácio, S., van Loenhout, H., and Popelier, C. (1978). The influence of daylight on body growth and female reproductive activity in the pond snail (*Lymnea stagnalis*). *Gen. Comp. Endocrinol.* **34**, 109.

Bolognari, A., Carmignani, M. P. A., Zaccone, G., and Minniti, F. (1979). Cytochemical detection of some enzymes of the carbohydrate metabolism in the yolk of molluscan oocytes. *Cell. Mol. Biol.* **24**, 265–266.

Boquist, L., Falkmer, S., and Mehrota, B. K. (1971). Ultrastructural search for homologues of pancreatic $\beta$-cells in the intestinal mucosa of the mollusc *Buccinum undatum*. *Gen. Comp. Endocrinol.* **17**, 236–239.

Brockelman, C. R., and Sithithavorn, P. (1980). Carbohydrate reserves and hemolymph sugars of the African giant snail, *Achatina fulica* in relation to parasitic infection and starvation. *Z. Parasitenkd.* **62**, 285–291.

Bryant, C. (1965). The metabolism of the digestive glands of two species of marine gastropod (*Melanerita melanotragus* and *Austrocochlea obtusa*). *Comp. Biochem. Physiol.* **14**, 223–230.

Bryant, C., Hines, W. J. W., and Smith, M. J. H. (1964). Intermediary metabolism in some terrestrial molluscs (*Pomatia, Helix* and *Cepaea*). *Comp. Biochem. Physiol.* **11**, 147–153.

Buczylko, J., Hargrave, P. A., and Kochman, M. (1980a). Fructose-biphosphate aldolase from *Helix pomatia*—I. Purification and catalytic properties. *Comp. Biochem. Physiol. B* **67B,** 225–232.

Buczylko, J., Hargrave, P. A., and Kochman, M. (1980b). Fructose-biphosphate aldolase from *Helix pomatia*—II. Chemical and physical properties. *Comp. Biochem. Physiol. B* **67B,** 233–238.

Carr, W. E. S. (1967). Chemoreception in the mud snail, *Nassarius obsoletus*. II. Identification of stimulatory substances. *Biol. Bull. (Woods Hole, Mass.)* **133,** 106–127.

Chatterjee, B., and Ghose, K. C. (1973). Seasonal variation in stored glycogen and lipid in the digestive gland and genital organs of two freshwater prosobranchs. *Proc. Malacol. Soc. London* **40,** 407–412.

Cheng, T. C., and Lee, F. O. (1971). Glucose levels in the mollusc *Biomphalaria glabrata* infected with *Schistosoma mansoni*. *J. Invertebr. Pathol.* **18,** 395–399.

Chiang, P. K. (1977). Glycogen metabolism in the snail *Biomphalaria glabrata*. *Comp. Biochem. Physiol. B* **58B,** 9–12.

Christie, J. D., Foster, W. B., and Stauber, L. A. (1974a). The effect of parasitism and starvation on carbohydrate reserves of *Biomphalaria glabrata*. *J. Invertebr. Pathol.* **23,** 55–62.

Christie, J. D., Foster, W. B., and Stauber, L. A. (1974b). $^{14}$C uptake by *Schistosoma mansoni* from *Biomphalaria glabrata* exposed to $^{14}$C-glucose. *J. Invertebr. Pathol.* **23,** 297–302.

Chukhrova, A. I., Kaverzneva, E. D., and Tyutrina, G. V. (1970). Isolation of enzymes splitting the carbohydrate–peptide bond of the amide type from the snail *Limnaea stagnalis*. *Biochemistry (Engl. Transl.)* **35,** 78–82.

Chukhrova, A. I., Kiseleva, V. V., and Kaverzneva, E. D. (1974a). The substrate-specificity of *Limnaea stagnalis* enzymes which split a carbohydrate–protein bond of the amide type. *Biochemistry (Engl. Transl.)* **39,** 147–151.

Chukhrova, A. I., Kiseleva, V. V., and Kaverzneva, E. D. (1974b). Characteristics of *Limnaea stagnalis* enzymes which split carbohydrate–protein bonds of amide type. *Biochemistry (Engl. Transl.)* **39,** 428–432.

Colas, B. (1978). Some physicochemical and structural properties of two $\beta$-fucosidases from *Achatina balteata*. *Biochim. Biophys. Acta* **527,** 150–158.

Colas, B., and Attias, J. (1977). Purification de deux $\beta$-D-glycosidases du suc digestif d'*Achatina balteata*. *Biochemie* **59,** 577–585.

Coles, G. C. (1969). Isoenzymes of snail livers—II. Dehydrogenases. *Comp. Biochem. Physiol.* **31,** 1–14.

Dando, P. R. (1981). Strombine [*N*-(carboxymethyl)-D-alanine] dehydrogenase and alanopine [*meso*-N-(1-carboxyethyl)-alanine] dehydrogenase from the mussel *Mytilus edulis* L. *Biochem. Soc. Trans.* **9,** 297–298.

Dando, P. R., Storey, K. B., Hochachka, P. W., and Storey, J. M. (1981). Multiple dehydrogenases in marine molluscs: electrophoretic analysis of alanopine dehydrogenase, strombine dehydrogenase, octopine dehydrogenase and lactate dehydrogenase. *Mar. Biol. Lett.* **2,** 249–257.

Davidson, J. K., Falkmer, S., Mehrotra, B. K., and Wilson, S. (1971). Insulin assays and light microscopical studies of digestive organs in promostomian and deuterostomian species and in coelenterates. *Gen. Comp. Endocrinol.* **17,** 388–401.

de Pedro, M. A., Reglero, A., and Cabezas, J. A. (1978). Purification and some properties of $\alpha$-L-fucosidase from *Littorina littorea* L. *Comp. Biochem. Physiol. B* **60B,** 379–382.

de Zwaan, A. (1977). Anaerobic energy metabolism in bivalve molluscs. *Oceanogr. Mar. Biol.* **15,** 103–187.

de Zwaan, A., Ebberink, H. M. (1978). Apparent inhibition of pyruvate kinase by arginine phosphate. *FEBS Lett.* **89**, 301–303.

de Zwaan, A., Mohamed, A. M., and Geraerts, W. P. M. (1976). Glycogen degradation and the accumulation of compounds during anaerobiosis in the fresh water snail *Lymnaea stagnalis. Neth. J. Zool.* **26**, 549–557.

de Zwaan, A., de Bont, A. M. T., and Verhoeven, A. (1982a). Anaerobic energy metabolism in isolated adductor muscle of the sea mussel *Mytilus edulis* L. *J. Comp. Physiol.* (in press).

de Zwaan, A., Leopold, M., Marteijn, E., and Livingstone, D. R. (1982b). Phylogenetic distribution of pyruvate oxidoreductases, arginine kinase and aminotransferases. "Exogenous and endogenous influences on metabolic and neural control" (A. D. F. Addink and N. Spronk, eds.), Vol. 2. Abstr. Congr. Eur. Soc. Comp. Physiol. Biochem. 3rd. pp 136–137. Noordwijkerhout, The Netherlands. Pergamon Press, Oxford.

Dogterom, A. A. (1980). The effect of the growth hormone of the freshwater snail *Lymnaea stagnalis* on biochemical composition and nitrogenous wastes. *Comp. Biochem. Physiol. B* **65B**, 163–167.

Ellington, W. R. (1981). Energy metabolism during hypoxia in the isolated, perfused ventricle of the whelk, *Busycon contrarium* Conrad. *J. Comp. Physiol. B* **142**, 457–464.

Elyakova, L. A., and Shilova, T. G. (1979). Characterization of the type of action of β-1,3-glucanases from marine invertebrates. *Comp. Biochem. Physiol. B* **64B**, 245–248.

Elyakova, L. A., Shevchenko, N. M., and Avaeva, S. M. (1981). A comparative study of carbohydrase activities in marine invertebrates. *Comp. Biochem. Physiol. B* **69B**, 905–908.

Emerson, D. N. (1967). Carbohydrate orientated metabolism of *Planorbis corneus* (Mollusca, Planorbidae) during starvation. *Comp. Biochem. Physiol.* **22**, 571–579.

Fairbairn, D. (1958). Trehalose and glucose in helminths and other invertebrates. *Can. J. Zool.* **36**, 787–795.

Fantin, A. M. B., and Gervaso, M. V. (1971). A histoenzymatic investigation of galactogen synthesis in the albumen gland of *Helix pomatia* and *Lymnaea stagnalis. Histochemie* **28**, 88–94.

Fried, B., and Blumenthal, A. B. (1967). The influence of larval *Stephanostomum tenue* (Trematoda) infection on the carbohydrate content of the marine snail, *Nassarius obsoletus. Proc. Pa. Acad. Sci.* **41**, 42–45.

Fried, G. H., and Levin, N. L. (1973). Enzymatic activity in hepatopancreas of *Nassarius obsoletus. Comp. Biochem. Physiol. B* **45B**, 153–157.

Friedl, F. E. (1968). Basal hemolymph glucose and dietary hyperglycemia in the snail *Lymnaea stagnalis. Am. Zool.* **8**, 763–764.

Friedl, F. E. (1971). Hemolymph glucose in the freshwater pulmonate snail *Lymnaea stagnalis:* Basal values and an effect of ingested carbohydrate. *Comp. Biochem. Physiol. A* **39A**, 605–610.

Gäde, G. (1980). A comparative study of octopine dehydrogenase isoenzymes in gastropod, bivalve and cephalopod molluscs. *Comp. Biochem. Physiol. B* **67B**, 575–582.

Geraerts, W. P. M. (1976). Control of growth by the neurosecretory hormone of the light green cells in the freshwater snail *Lymnaea stagnalis. Gen. Comp. Endocrinol.* **29**, 61–71.

Gianfreda, L., Imperato, A., Palescandolo, R., and Scardi, V. (1979a). Distribution of β-1,4-glucanase and β-glucosidase activities among marine molluscs with different feeding habits. *Comp. Biochem. Physiol. B* **63B**, 345–348.

Gianfreda, L., Tosti, E., and Scardi, V. (1979b). A comparison of $\alpha$- and $\beta$-glucanase activities in fourteen species of marine molluscs. *Biochem. Syst. Ecol.* **7,** 57–59.

Gill, P. D. (1978). Non-genetic variation in isoenzymes of lactate dehydrogenase of *Cepaea nemoralis. Comp. Biochem. Physiol. B* **59B,** 271–276.

Goddard, C. K., and Martin, A. W. (1966). Carbohydrate metabolism. *In* "Physiology of Mollusca" (K. M. Wilbur and C. M. Yonge, eds.), Vol. 2, pp. 275–308. Academic Press, New York.

Goddard, C. K., Nichol, P. I., and Williams, J. F. (1964). The effect of albumen gland homogenate on the blood sugar of *Helix aspersa. Comp. Biochem. Physiol.* **11,** 351–366.

Gomih, Y. K., and Grillo, T. A. I. (1976). Insulin-like activity of the extract of the digestive gland and the pylorus of the Giant African Snail *Achatina fulica*. A preliminary report. *Evol. Pancreatic Islets, Proc. Symp., Leningrad, 1975* pp. 153–162.

Goudsmit, E. M. (1972). Carbohydrates and carbohydrate metabolism in Mollusca. *In* "Chemical Zoology" (M. Florkin and B. T. Scheer, eds.), Vol. 7, pp. 219–243. Academic Press, New York.

Goudsmit, E. M. (1973). The role of galactogen in pulmonate snails. *Malacol. Rev.* **6,** 58–59.

Goudsmit, E. M. (1975). Neurosecretory stimulation of galactogen synthesis within the *Helix pomatia* gland during organ culture. *J. Exp. Zool.* **191,** 193–198.

Goudsmit, E. M. (1976). Galactogen catabolism by embryos of the freshwater snails, *Bulimnaea megasoma* and *Lymnaea stagnalis. Comp. Biochem. Physiol. B* **53B,** 439–442.

Goudsmit, E. M., and Feldman, S. L. (1974). Organ culture of *Helix pomatia* albumen gland in a defined medium. *Malacol. Rev.* **7,** 53.

Goudsmit, E. M., and Friedman, T. B. (1976). Enzymatic synthesis and interconversion of UDP-D-glucose and UDP-D-galactose in the albumen gland of the snail, *Helix pomatia. Comp. Biochem. Physiol. B* **54B,** 135–139.

Goudsmit, E. M., and Ram, J. L. (1979). Preliminary purification of a neurosecretion that stimulates galactogen synthesis in the albumen gland of the land snail, *Helix pomatia. J. Gen. Physiol.* **74,** 7a–8a.

Grenon, J.-F., and Walker, G. (1980). Biochemical and rheological properties of the pedal mucus of the limpet, *Patella vulgata. Comp. Biochem. Physiol. B* **66B,** 451–458.

Grenon, J.-F., and Walker, G. (1981). The tenacity of the limpet, *Patella vulgata* L.: an experimental approach. *J. Exp. Mar. Biol. Ecol.* **54,** 277–308.

Haeser, P. E., and De Jorge, F. B. (1971). Anoxic anoxia in *Strophocheilus* (Pulmonata Mollusca). *Comp. Biochem. Physiol. B* **38B,** 753–757.

Hall, R. L., and Wood, E. J. (1976). The carbohydrate content of gastropod haemocyanins. *Biochem. Soc. Trans.* **4,** 307–309.

Hall, R. L., Wood, E. J., Kamberling, J. P., Gerwing, G. J., and Vliegenthart, F. G. (1977). 3-0-Methyl sugars as constituents of glycoproteins. Identification of 3-0-methylgalactose and 3-0-methylmannose in pulmonate gastropod haemocyanins. *Biochem. J.* **165,** 173–176.

Hammarström, S., and Kabat, E. A. (1969). Purification and characterization of a bloodgroup A reactive hemagglutinin from the snail *Helix pomatia* and a study of its combining site. *Biochemistry* **8,** 2696–2705.

Hammarström, S., Murphy, L. A., Goldstein, I. J., and Etzler, M. E. (1977). Carbohydrate binding specificity of four *N*-acetyl-D-galactosamine-"specific" lectins: *Helix pomatia* A hemagglutinin, soy bean agglutinin, lima bean lectin and *Dolichos biflorus* lectin. *Biochemistry* **16,** 2750–2755.

Hammen, C. S. (1980). Total energy metabolism of marine bivalve molluscs in anaerobic and aerobic states. *Comp. Biochem. Physiol. A* **67A,** 617–621.

Hatanaka, H., Ogawa, Y., and Egami, F. (1976a). Two glycosulfatases from the liver of a marine gastropod, *Charonia lampas*. *J. Biochem.* (*Tokyo*) **79**, 27–34.

Hatanaka, H., Ogawa, Y., and Egami, F. (1976b). Arylsulphatase and glycosulphatase of *Charonia lampas*. Substrate specificity towards sugar sulphate derivatives. *Biochem. J.* **159**, 445–448.

Heeg, J. (1977). Oxygen consumption and the use of metabolic reserves during starvation and aestivation in *Bulinus* (*Physopsis*) *africanus* (Pulmonata: Planorbidae). *Malacologia* **16**, 549–560.

Heller, E., and Raftery, M. A. (1976). The vitelline envelope of eggs from the Giant Keyhole Limpet *Megathura crenulata*. I: Chemical composition and structural studies. *Biochemistry* **15**, 1194–1198.

Hernandez, F., Sanchez, R., and Pavesi, L. (1970). Study of some properties of glycogen phosphorylase, phosphoglucomutase and glucose-6-phosphate dehydrogenase from muscle of *Concholepas concholepa*. *Arch. Biol. Med. Exp.* **1**, No. 75. (Abstr.)

Hoffmann, K.-H. (1976). Catalytic efficiency and structural properties of invertebrate muscle pyruvate kinases: Correlation with body temperature and oxygen consumption rates. *J. Comp. Physiol. B* **110**, 185–195.

Hoffmann, K.-H., and Rädeke, U. (1978). Stability of invertebrate muscle pyruvate kinases: correlation with enzyme regulatory properties. *Comp. Biochem. Physiol. B* **61B**, 321–325.

Holz, F., and von Brand, T. (1940). Quantitative studies upon some blood constituents of *Helix pomatia*. *Biol. Bull.* (*Woods Hole, Mass.*) **79**, 423–431.

Horne, F. R. (1979). Comparative aspects of estivating metabolism in the gastropod, *Marisa*. *Comp. Biochem. Physiol. A* **64A**, 309–311.

Horstmann, H. J. (1960). Untersuchungen zum Stoffwechsel der Lungenschecken, I Glykolyse bei den Embryonen von *Lymnaea stagnalis* L. *Hoppe-Seyler's Z. Physiol. Chem.* **319**, 110–119.

Hoskin, G. P., and Cheng, T. C. (1974). *Himasthla quissetensis:* Uptake and utilization of glucose by rediae as determined by autoradiography and respirometry. *Exp. Parasitol.* **35**, 61–67.

Hunger, V. J., and Horstmann, H. J. (1968). Sauerstoffverbrauch und Aktivität einiger Enzyme des Kohlenhydrat-Stoffwechsels während der Embryonalentwicklung der Weinbergschnecke (*Helix pomatia* L.) *Z. Biol.* (*Munich*) **116**, 90–104.

Hunt, S. (1970). "Polysaccharide-Protein Complexes in Invertebrates." Academic Press, New York.

Hylleberg, J. (1976). Resource partitioning on basis of hydrolytic enzymes in deposit-feeding mud snails (Hydrobiidae). *Oecologia* **23**, 115–125.

Hylleberg-Kristensen, J. (1972). Carbohydrates of some marine invertebrates with notes on their food and on the natural occurrence of the carbohydrates studied. *Mar. Biol.* (*Berlin*) **14**, 130–142.

Iijima, Y., Muramatsu, T., and Egami, F. (1971). Purification of α-L-fucosidase from the liver of a marine gastropod, *Turbo cornutus*, and its action on blood group substances. *Arch. Biochem. Biophys.* **145**, 50–54.

Ishak, M. M., Mohamed, A. M., and Sharaf, A. A. (1975). Carbohydrate metabolism in uninfected and trematode-infected snails *Biomphalaria alexandrina* and *Bulinus truncatus*. *Comp. Biochem. Physiol. B* **51B**, 499–505.

Jager, J. C. (1971). A quantitative study of a chemoresponse to sugars in *Lymnaea stagnalis* (L.). *Neth. J. Zool.* **21**, 1–59.

Kasprzyk, A., Mackowiak, D., and Obuchowicz, L. (1978). The preliminary study on the activity of glucose-6-phosphatase (E.C. 3.1.3.9.) in the hepatopancreas of *Helix poma-tia. Bull. Soc. Amis Sci. Lett. Poznan, Ser. D* No. 17, 115–121.

Kelentey, B., and Csáky, T. (1976). Transport of $^{14}$C-labelled glucose and 3-m-glucose in isolated intenstine of *Aplysia californica. Acta Physiol. Acad. Sci. Hung.* **48**, 218.

Kesler, D. H., and Tulou, C. A. G. (1980). Cellulase activity in the freshwater gastropod *Amnicola limosa, Nautilus* **94**, 135–137.

Kohn, A. J. (1961). Chemoreception in gastropod molluscs. *Am. Zool.* **1**, 291–308.

Koningsor, R. L., Jr., and Hunsaker, D., II (1970–1971). Cellulase from the crop of *Aplysia vaccaria* Winkler, 1955. *Veliger* **13**, 285–289.

Kooijman, D., van Zoonen, H., Zurburg, W., and Kluytmans, J. (1982). On the aerobic and anaerobic energy metabolism of *Littorina* species in relation to the pattern of intertidal zonation. "Exogenous and endogenous influences on metabolic and neural control" (A. D. F. Addink and N. Spronk, eds.), Vol. 2. Abstr. *Congr. Eur. Soc. Comp. Physiol. Biochem. 3rd,* pp. 134–135. Noordwijkerhout, The Netherlands. Pergamon Press, Oxford.

Koormann, R., and Grieshaber, M. (1980). Investigations on the energy metabolism and on octopine formation of the common whelk, *Buccinum undatum* L., during escape and recovery. *Comp. Biochem. Physiol. B* **65B**, 543–547.

Krishna, G. V. R., and Simha, S. S. (1977). Effects of parasitism on the carbohydrate reserves of fresh water snail, *Lymnaea luteola. Comp. Physiol. Ecol.* **2**, 242–244.

Krupanidhi, S., Reddy, V. V., and Naidu, B. P. (1978). Organic composition of tissues of the snail *Cryptozona ligulata* (Ferussac) with special reference to aestivation. *Indian J. Exp. Biol.* **16**, 611–612.

Kulkarni, A. B. (1973). A study on the carbohydrate metabolism in the land slug, *Laevicaulis alte. Broteria, Ser. Trimest.: Cienc. Nat.* **42**, 111–120.

Kumar, T. P., Ramamurthi, R., and Babu, K. S. (1981). Circadiae fluctuations in total protein and carbohydrate content in the slug *Laevicaulis alte* (Ferrussac, 1821). *Biol. Bull. (Woods Hole, Mass.)* **160**, 114–122.

Kushins, L. J., and Mangum, C. P. (1971). Responses to low oxygen conditions in two species of the mud snail, *Nassarius. Comp. Biochem. Physiol. A* **39A**, 421–435.

Lambert, P., and Dehnel, P. A. (1974). Seasonal variations in biochemical composition during the reproductive cycle of the intertidal gastropod *Thais lamellosa* Gmelin (Gastropoda, Prosobranchia). *Can. J. Zool.* **52**, 306–318.

Liebsch, M., Becker, W., and Gagelmann, G. (1978). An improvement of blood sampling technique for *Biomphalaria glabrata* using anaesthesia and long-term relaxation and the role of this method in studies of the regulation of hemolymph glucose. *Comp. Biochem. Physiol. A* **59A**, 169–174.

Livingstone, D. R. (1982). Energy production in the muscle tissues of different kinds of molluscs. "Exogenous and endogenous influences on metabolic and neural control" (A. D. F. Addink and N. Spronk, eds.), Vol. 1. Invited lectures, *Proc. Congr. Eur. Soc. Comp. Physiol. Biochem., 3rd,* pp. 257–274. Noordwijkerhout, The Netherlands. Pergamon Press, Oxford.

Livingstone, D. R., de Zwaan, A., Leopold, M., and Marteijn, E. (1982). Studies on the phylogenetic distribution of pyruvate oxidoreductases (lactate dehydrogenase, opine dehydrogenases). In preparation.

Long, G. L. (1976). The stereospecific distribution and evolutionary significance of invertebrate lactate dehydrogenase. *Comp. Biochem. Physiol. B* **55B**, 77–83.

Long, G. L., and Kaplan, N. O. (1968). D-Lactate specific pyridine nucleotide lactate dehydrogenases in animals. *Science* **162**, 685–686.

Long, G. L., Ellington, W. R., and Duda, T. F. (1979). Comparative enzymology and physiological role of D-lactate dehydrogenase from the foot muscle of two gastropod molluscs. *J. Exp. Zool.* **207,** 237–248.

McDaniel, J. S., and Dixon, K. E. (1967). Utilization of exogenous glucose by the rediae of *Parorchis acanthus* (Digenea: Philophthalmidae) and *Cryptocotyle lingua* (Digenea: Heterophyidae). *Biol. Bull. (Woods Hole, Mass.)* **133,** 591–599.

McLachlan, A., and Lombard, H. W. (1980). Seasonal variations in energy and biochemical components of an edible gastropod, *Turbo sarmaticus* (Turbinidae). *Aquaculture* **19,** 117–125.

McMahon, R. F., and Russell-Hunter, W. D. (1978). Respiratory responses to low oxygen stress in marine littoral and sublittoral snails. *Physiol. Zool.* **51,** 408–424.

McManus, D. P., and James, B. L. (1975a). Tricarboxylic acid cycle enzymes in the digestive gland of *Littorina saxatilis rudis* (Maton) and in the daughter sporocysts of *Microphallus similis* (Jäg). *Comp. Biochem. Physiol. B* **50B,** 491–495.

McManus, D. P., and James, B. L. (1975b). Anaerobic glucose metabolism in the digestive gland of *Littorina saxatilis rudis* (Maton) and in the daughter sporocysts of *Microphallus similis* (Jäg) (Digenea: Microphallidae). *Comp. Biochem. Physiol. B* **51B,** 293–297.

McManus, D. P., and James, B. L. (1975c). Pyruvate kinases and carbon dioxide fixating enzymes in the digestive gland of *Littorina saxatilis rudis* (Maton) and in the daughter sporocysts of *Microphallus similis* (Jäg) (Digenea: Microphallidae). *Comp. Biochem. Physiol. B* **51B,** 299–306.

Manohar, L., and Rao, P. V. (1976). Physiological response to parasitism. I. Changes in carbohydrate reserves of the molluscan host. *Southeast Asian J. Trop. Med. Pubic Health* **7,** 395–404.

Marcuzzi, G., and Turchetto Lafisca, M. (1978). Contribute to the knowledge of polysaccharases in soil animals. *Rev. Ecol. Biol. Sol* **15,** 135–145.

Marques, M., and Falkmer, S. (1976). Effects of mammalian insulin on blood glucose level, glucose tolerance, and glycogen content of musculature and hepatopancreas in a gastropod mollusk *Strophocheilus oblongus, Gen. Comp. Endocrinol.* **29,** 522–530.

Marques, M., and Pereira, S. (1970). Seasonal variations in blood glucose and glycogen levels of some tissues of *Strophocheilus oblongus* (Mollusca, Gastropoda). *Rev. Bras. Biol.* **30,** 43–48.

Marshall, I., McManus, D. P., and James, B. L. (1974). Glycolysis in the digestive gland of healthy and parasitized *Littorina saxatilis rudis* (Maton) and in the daughter sporocysts of *Microphallus similis* (Jäg) (Digenea: Microphallidae). *Comp. Biochem. Physiol.* **49,** 291–299.

Marshall, J. J., and Grand, R. J. A., (1975). Comparative studies on β-glucan hydrolases. Isolation and characterization of an Exo (1 → 3)-β-glucanase from the snail, *Helix pomatia. Arch. Biochem. Biophys.* **167,** 165–175.

Meenakshi, V. R. (1958). Anaerobiosis in the South Indian apple snail *Pila virens* during estivation. *J. Zool. Soc. India* **9,** 62–71.

Meenakshi, V. R. (1964). Aestivation in the Indian apple snail *Pila.* I. Adaptation in natural and experimental conditions. *Comp. Biochem. Physiol.* **11,** 379–386.

Meenakshi, V. R., and Scheer, B. T. (1968). Studies on the carbohydrates of the slug *Ariolimax columbianis* with special reference to their distribution in the reproductive system. *Comp. Biochem. Physiol.* **26,** 1091–1097.

Meenakshi, V. R., and Scheer, B. T. (1969). Regulation of galactogen synthesis in the slug *Ariolimax columbianis. Comp. Biochem. Physiol.* **29,** 841–845.

Mehlman, B., and von Brand, T. (1951). Further studies on the anaerobic metabolism of some fresh water snails. *Biol. Bull. (Woods Hole, Mass.)* **100,** 199–205.

Meuleman, E. A. (1972). Host–parasite interrelationships between the freshwater pulmonate *Biomphalaria pfeifferi* and the trematode *Schistosoma mansoni*. *Neth. J. Zool.* **22**, 355–427.

Michejda, J. W., Wala, R., Zerbe, T., and Tilgner, H. (1969). D(-)Lactate dehydrogenase in foot and heart muscle of a snail, *Helix pomatia* L. *Bull. Soc. Amis Sci. Lett. Poznan, Ser. D* No. 9, 181–191.

Mohamed, A. M., and Geraerts, W. P. M. (1980). Role of the lateral lobes in the control of certain physiological and biochemical processes in *Bulinus truncatus* (Gastropoda, Pulmonata). *Hydrobiologia* **75**, 267–271.

Mohamed, A. M., and Ishak, M. M. (1981). Growth rate and changes in tissue carbohydrates during schistosome infection of the snail *Biomphalaria alexandrina*. *Hydrobiologia* **76**, 17–21.

Morán, A., and González, R. (1967). Exploration of the carbohydrate metabolism in *Concholepas concholepa*. II.—Glyconeogenesis and utilization of some glucidic metabolites and amino acids. *Arch. Biol. Med. Exp.* **4**, No. 107, p. 219. (Abstr.)

Nanaware, S. G., and Varute, A. T. (1974). Histochemical studies on galactogen in the albumen gland of a land pulmonate, *Semperula maculata* in the seasonal breeding-aestivation cycle. *Folia Histochem. Cytochem.* **12**, 21–28.

Narayanan, R., and Venkateswararao, P. (1980). Effect of xiphidiocercarial infection on oxidation of glycolytic and Krebs cycle intermediates in *Lymnaea luteola* (Mollusca). *J. Invertebr. Pathol.* **36**, 21–24.

Nieland, M. L., and Goudsmit, E. M. (1969). Ultrastructural observations on galactogen formation and secretion in the albumin gland of *Helix pomatia*. *J. Invest. Dermatol.* **52**, 392. (Abstr.)

Orive, E., Berjon, A., and Otero, M. P. F. (1980). Metabolism of nutrients during intestinal absorption in *Helix pomatia* and *Arion empiricorum* (Gastropoda: Pulmonata). *Comp. Biochem. Physiol. B* **66B**, 155–158.

Oudejans, R. C. H. M., and van der Horst, D. J. (1974). Aerobic–anaerobic biosynthesis of fatty acids and other lipids from glycolytic intermediates in the pulmonate land snail *Cepaea nemoralis* (L.). *Comp. Biochem. Physiol. B* **47B**, 139–147.

Owen, G. (1966). Digestion. *In* "Physiology of Mollusca" (K. M. Wilbur and C. M. Yonge, eds.), Vol. 2, pp. 53–96. Academic Press, New York.

Pamatmat, M. M. (1978). Oxygen uptake and heat production in a metabolic conformer (*Littorina irrorata*) and a metabolic regulator (*Uca pugnax*). *Mar. Biol. (Berlin)* **48**, 317–325.

Personne, P., and Anderson, W. (1970). Localisation mitochondriale d'enzymes liées au metabolisme du glycogene dans le spermatozoide de l'escargot. *J. Cell Biol.* **44**, 20–28.

Phizackerley, P. J. R., and Bannister, J. V. (1974). The characterization of N-acetylglucosaminidases from the limpet *Patella vulgata* (L.). *Biochim. Biophys. Acta* **362**, 129–135.

Plaxton, W. C. and Storey, K. B. (1982a). Alanopine dehydrogenase: Purification and characterization of the enzyme from *Littorina littorea* foot muscle. *J. Comp. Physiol. B.* (in press).

Plaxton, W. C. and Storey, K. B. (1982b). Tissue specific isozymes of alanopine dehydrogenase in the channeled whelk *Busycotypus conaliculatum*. *Can. J. Zool.* **60**, 1568–1572.

Rajan, R. K., and Sriramulu, V. (1978). Effects of insulin on blood glucose level, glucose tolerance and glycogen content of the foot and hepatopancreas in *Cryptozonia belangeri* (Deshayes) (Mollusca: Gastropoda). *Curr. Sci.* **47**, 248–249.

Rao, M. B. (1979). Preliminary studies on the natural diet and carbohydrates in the digestive gland of the tropical aquatic pulmonate snail *Lymnaea luteola* Lamarck. *Malacologia* **18**, 421–422.

Rao, P. V., and Onnurappa, J. (1979). Qualitative analysis of sugars and their effect on tissue respiration in selected usual and unusual fresh water gastropod snail hosts of larval trematodes. *Indian J. Exp. Biol.* **17**, 294–297.

Rao, P. V., Chowdary, V. D., and Babu, G. R. (1979). Effect of xiphidiocercarial infection on ethanol precipitable tissue carbohydrate levels of the snail host *Lymnaea luteola*. *Indian J. Exp. Biol.* **17**, 1230–1232.

Reader, T. A. J. (1974). Autoradiographic studies on the uptake of [$^{14}$C]-glucose by *Bithynia tentaculata* (Mollusca: Gastropoda) and its larval digeneans. *J. Helminthol.* **48**, 235–240.

Reddy, D. C. S., Jayaram, V., Sowijanya, K., and Naidu, B. P. (1978). Dial variations in levels of blood-glucose and hepatopancreatic glycogen in slug *Laevicaulis alte*. *Experientia* **34**, 606.

Reddy, S. R. R., and Ramamurthi, R. (1973). Oxidation of C$^{14}$-glucose by the aestivating snail *Pila globosa* (Swainson). *Veliger* **15**, 355–356.

Reglero, A., and Cabezas, J. A. (1976). Glycosidases of molluscs. Purification and properties of α-L-fucosidase from *Chamelea gallina* L. *Eur. J. Biochem.* **66**, 379–387.

Regnouf, F., and Thoai, N. V. (1970). Octopine and lactate dehydrogenases in mollusc muscles. *Comp. Biochem. Physiol.* **32**, 411–416.

Renwrantz, L., Glöckner, W., Mitterer, K.E., and Uhlenbruck, G. (1976). Eine quantitative Bestimmung der Hämolymph-Kohlenhydrate von *Helix pomatia*. *J. Comp. Physiol.* **105**, 185–188.

Robson, E. M., and Williams, I. C. (1971). Relationships of some species of digenea with the marine prosobranch *Littorina littorea* (L.) III. The effect of larval digenea on the glycogen content of the digestive gland and foot of *L. littorea*. *J. Helminthol.* **45**, 381–401.

Roy, A. B. (1971). The hydrolysis of sulfate esters. *In* "The Enzymes" (P. D. Boyer, ed.), 3rd ed., Vol. 5, pp. 1–19. Academic Press, New York.

Rudolph, P. H. (1974). Histochemical identification of galactogen in the Succineidae. *Malocol. Rev.* **7**, 52. (Abstr.)

Rudolph, P. H. (1975). Accumulation of galactogen in the albumen gland of *Triodopsis multilineata* (Pulmonata: Stylommatophora) after dormancy. *Malacol. Rev.* **8**, 57–63.

Sangster, A. W., Thomas, S. E., and Tingling, N. L. (1975). Fish attractants from marine invertebrates. Arcamine from *Arca zebra* and strombine from *Strombus gigas*. *Tetrahedron* **31**, 1135–1137.

Scheerboom, J. E. M. (1978). The influence of food quantity and food quality on assimilation, body growth and egg production in the pond snail *Lymnaea stagnalis* (L.) with particular reference to the haemolymph-glucose concentration. *Proc. K. Ned. Akad. Wet., Ser. C* **81**, 184–197.

Scheerboom, J. E. M., and Doderer, A. (1978). The effects of artificially raised haemolymph-glucose concentrations on feeding, locomotory activity, growth and egg production of the pond snail *Lymnaea stagnalis* (L.). *Proc. K. Akad. Wet., Ser. C* **81**, 377–386.

Scheerboom, J. E. M., and Hemminga, M. A. (1978). Regulation of hemolymph-glucose concentration in the pond snail (*Lymnaea stagnalis*). *Gen. Comp. Endocrinol.* **34**, No. 133, p. 112. (Abstr.)

Scheerboom, J. E. M., and van Elk, R. (1978). Field observations on the seasonal variations in the natural diet and the haemolymph-glucose concentration of the pond snail *Lymnaea stagnalis* (L.). *Proc. K. Akad. Wet., Ser. C* **81**, 365–376.

Scheerboom, J. E. M., Hemminga, M. A., and Doderer, A. (1978). The effects of a change of diet on consumption and assimilation and on the haemolymph-glucose concentration of the pond snail *Lymnaea stagnalis* (L.). *Proc. K. Akad. Wet. Ser. C* **81,** 335–346.

Scheid, M. J., and Awapara, J. (1972). Stereospecificity of some invertebrate lactate dehydrogenases. *Comp. Biochem. Physiol. B* **43B,** 619–626.

Schilansky, M. M., Levin, N. L., and Fried, G. H. (1977). Metabolic implications of glucose-6-phosphate dehydrogenase and lactic dehydrogenase in two marine gastropods. *Comp. Biochem. Physiol. B* **56B,** 1–4.

Schwarz, K. (1934). Uber den Blutzucker der Weinbergschnecke. *Biochem. Z.* **275,** 262–269.

Senseman, D. M. (1977). Starch: a potent feeding stimulant for the terrestrial slug *Ariolimax californicus*. *J. Chem. Ecol.* **3,** 707–715.

Sharaf, A. A., Mohamed, A. M., Elghar, M. R. A., and Mousa, A. H. (1975). Control of snail hosts of bilharziasis in Egypt. 2. Effect of triphenyltin hydroxide (Du-Ter) on carbohydrate metabolism of the snails *Biomphalaria alexandrina* and *Bulinas tuncatus*. *Egypt. J. Bilharziasis* **2,** 37–47.

Simpson, J. W., and Awapara, J. (1964). Phosphoenolpyruvate carboxykinase in invertebrates. *Comp. Biochem. Physiol.* **12,** 457–464.

Simpson, J. W., and Awapara, J. (1966). The pathway of glucose degradation in some invertebrates. *Comp. Biochem. Physiol.* **18,** 537–548.

Singh, I., and Nayeemunnisa (1976). Neurochemical correlates of aestivation: Changes in the levels of glycogen and glucose in the cerebral, pleuropedal and visceral ganglia of the Indian apple snail, *Pila globosa. Proc.—Indian Acad. Sci., Sect. B* **84B,** 56–59.

Sluiters, J. F., Brussaard-Wüst, C. C. M., and Meuleman, E. A. (1975). Effects of exposure of *Lymnaea stagnalis* to different numbers of miracidia of *Trichobilharzia ocellata*. *Proc. Eur. Multicoll. Parasitol. Trogir., 2nd.,* p. 40.

Sminia, T. (1972). Structure and function of blood and connective tissue cells of the fresh water pulmonate *Lymnaea stagnalis* studied by electron microscopy and enzyme histochemistry. *Z. Zellforsch. Mikrosk. Anat.* **130,** 497–526.

Sminia, T., de With, N. D., Bos, J. L., van Nieuwmegen, M. E., Witter, M. P., and Wondergem, J. (1977). Structure and function of the calcium cells of the freshwater pulmonate snail *Lymnaea stagnalis*. *Neth. J. Zool.* **27,** 195–208.

Sollock, R. L., Vornhaben, J. E., and Campbell, J. W. (1979). Transaminase reactions and glutamate dehydrogenase in gastropod hepatopancreas. *J. Comp. Physiol. B* **129,** 129–135.

Sova, V. V., Elyakova, L. A., and Vaskovsky, V. E. (1970). The distribution of laminarinases in marine invertebrates. *Comp. Biochem. Physiol.* **32,** 459–464.

Stanislawski, E., and Becker, W. (1979). Influences of semi-synthetic diets, starvation and infection with *Schistosoma mansoni* (Trematoda) on the metabolism of *Biomphalaria glabrata* (Gastropoda). *Comp. Biochem. Physiol. A* **63A,** 527–533.

Stemberger, A., and Nordwig, A. (1974). Invertebrate collagens: qualitative and quantitative studies on their carbohydrate moieties. *Hoppe-Seyler's Z. Physiol. Chem.* **355,** 721–724.

Storey, K. B. (1977). Lactate dehydrogenase in tissue extracts of the land snail, *Helix aspersa:* unique adaptation of LDH subunits in a facultative anaerobe. *Comp. Biochem. Physiol. B* **56B,** 181–187.

Subramanyam, O. V. (1973). Neuroendocrine control of metabolism of a pulmonate snail, *Ariophanta* sp. *Experientia* **29,** 1150–1151.

Suryanarayanan, H., and Alexander, K. M. (1973). Biochemical studies on red muscles of the gastropod *Pila virens*, with a note on its histochemistry. *Comp. Biochem. Physiol.* A **44A**, 1157–1162.

Swami, K. S., and Reddy, Y. S. (1978). Adaptive changes to survival during aestivation in the gastropod snail, *Pila globosa*. *J. Sci. Ind. Res.* **37**, 144–157.

Tanaka, K., Nakano, T., Noguchi, S., and Pigman, W. (1968). Purification of α-L-fucosidase of abalone livers. *Arch. Biochem. Biophys.* **126**, 624–633.

Teng, Y.-S., Palmieri, J. R., and Sullivan, J. T. (1979). Changes in snail carbohydrate and nucleic acid metabolism due to trematode infection. *Southeast Asian J. Trop. Med. Public Health* **10**, 151–152.

Thomas, J. S. (1974). The effect of rediae of *Cryptocotyle lingua* (Creplin, 1825) and *Himasthla leptosoma* (Creplin, 1825) on the glycogen and free sugar levels of the digestive gland and gonad of *Littorina littorea* (Linnaeus, 1758). *Veliger* **17**, 207–210.

Uhlenbruck, G., Steinhausen, G., and Baldo, B. A. (1976a). Galactans and anti-galactans from invertebrates. *Z. Naturforsch. C: Biosci.* **31C**, 205–206.

Uhlenbruck, G., Steinhausen, G., and Kareem, H. A. (1976b). Different glycosubstances and galactans in the albumin gland and eggs of *Achatina fulica*. *Z. Immunitaetsforsch.* **152**, 220–230.

Uhlenbruck, G., Steinhausen, G., and Palatnik, M. (1977). Similarity of glycoproteo-galactans in the albumin glands from *Achatina* and *Borus* snails. *Comp. Biochem. Physiol.* B **57B**, 335–339.

Uhlenbruck, G., Steinhausen, G., Geserick, G., and Prokop, O. (1978). Further comparative studies of glycosubstances and proteins from different snail albumin glands. *Comp. Biochem. Physiol.* B **59B**, 285–288.

Vaidya, D. P. (1979). The pathological effects of trematodes on the biochemical components of hepatopancreas and blood glucose of freshwater snail *Indoplanorbis exustus*. *Riv. Parassitol.* **40**, 267–272.

van der Horst, D. J. (1974). *In vivo* biosynthesis of fatty acids in the pulmonate land snail *Cepaea nemoralis* (L.). under anoxic conditions. *Comp. Biochem. Physiol.* B **47B**, 181–187.

van Elk, R., and Joosse, J. (1981). The UDP-galactose 4-epimerase of the albumen gland of *Lymnaea stagnalis* and the effects of photoperiod, starvation and trematode infection on its activity. *Comp. Biochem. Physiol.* B **70B**, 45–52.

Veldhuijzen, J. P. (1975a). Effects of different kinds of food, starvation and restart of feeding on the haemolymph-glucose of the pond snail *Lymnaea stagnalis*. *Neth. J. Zool.* **25**, 89–102.

Veldhuijzen, J. P. (1975b). Glucose-tolerance in the pond snail *Lymnaea stagnalis* as affected by temperature and starvation. *Neth. J. Zool.* **25**, 206–218.

Veldhuijzen, J. P. (1976). The influence of the dorsal body hormone and of starvation on the synthesis of galactogen and glycogen in the pond snail *Lymnaea stagnalis*. *Gen. Comp. Endocrinol.* **29**, 290.

Veldhuijzen, J. P., and Cuperus, R. (1976). Effect of starvation, low temperature and the dorsal body hormone on the *in vitro* synthesis of galactogen and glycogen in the albumen gland and mantle of the pond snail *Lymnaea stagnalis*. *Neth. J. Zool.* **26**, 119–135.

Veldhuijzen, J. P., and Dogterom, G. E. (1975). Incorporation of ¹⁴C-glucose in the pond snail *Lymnaea stagnalis* as affected by starvation. *Neth. J. Zool.* **25**, 247–260.

Veldhuijzen, J. P., and van Beek, G. (1976). The influence of starvation and of increased carbohydrate intake on the polysaccharide content of various body parts of the pond snail *Lymnaea stagnalis*. *Neth. J. Zool.* **26**, 106–118.

von Brand, T. (1946). "Anaerobiosis in Invertebrates." Biodynamica, Missouri.
von Brand, T., Baernstein, H. D., and Mehlman, B. (1950). Studies on the anaerobic metabolism and the aerobic carbohydrate consumption of some fresh water snails. *Biol. Bull.* (*Woods Hole, Mass.*) **98**, 266–276.
Vorhaben, J. E., Scott, J. F., and Campbell, J. W. (1980). D-Mannitol oxidation in the land snail, *Helix aspersa*. *J. Biol. Chem.* **255**, 1950–1955.
Wieser, W. (1978). The initial stage of anaerobic metabolism in the snail, *Helix pomatia* L. .*FEBS Lett.* **95**, 375–378.
Wieser, W. (1980). Metabolic end products in three species of marine gastropods. *J. Mar. Biol. Assoc. U.K.* **60**, 175–180.
Wieser, W. (1981). Responses of *Helix pomatia* to anoxia: Change of solute activity and other properties of the haemolymph. *J. Comp. Physiol. B* **141**, 503–509.
Wieser, W., and Lackner, R. (1977). Inhibition of the pyruvate kinase of *Helix pomatia* L. by phospho-L-arginine. *FEBS Lett.* **80**, 299–302.
Wieser, W., and Schuster, M. (1975). The relationship between water content, activity and free amino acids in *Helix pomatia*. *J. Comp. Physiol. B* **98**, 169–181.
Wieser, W., and Wright, E. (1978). D-Lactate formation, D-LDH activity and glycolytic potential of *Helix pomatia* L. *J. Comp. Physiol. B* **126**, 249–255.
Wieser, W., and Wright, E. (1979). The effects of season and temperature on D-lactate dehydrogenase, pyruvate kinase and arginine kinase in the foot of *Helix pomatia* L. *Hoppe-Seyler's Z. Physiol. Chem.* **360**, 533–542.
Wijsman, T. C. M. (1979). Anaerobic metabolism in the freshwater pulmonate *Lymnaea stagnalis*. *In* "Proceedings of the International Symposium on the Physiology of Euryoxic Animals" (D. Holwerda, ed.), pp. 60–62. Univ. of Utrecht. Utrecht.
Wium-Anderson, G., and Simonsen, V. (1974). Phosphoglucoseisomerases in the sporocyst of the trematode *Schistosoma mansoni* and its intermediate host *Biomphalaria alexandrina* (Pulmonata). *Parasitology* **68**, 189–192.
Woodbridge, R. G., III (1978). The common periwinkle, *Littorina littorea*, Linne, attracted by sugars. *Experientia* **34**, 1445.
Zammit, V. A., and Newsholme, E. A. (1976a). The maximum activities of hexokinase, phosphorylase, phosphofructokinase, glycerol phosphate dehydrogenases, lactate dehydrogenase, octopine dehydrogenase, phosphoenolpyruvate carboxykinase, nucleoside diphosphatekinase, glutamate–oxaloacetate transaminase and arginine kinase in relation to carbohydrate utilisation in muscles from invertebrates. *Biochem. J.* **160**, 447–462.
Zammit, V. A., and Newsholme, E. A. (1976b). Effects of calcium ions and adenosine diphosphate on the activities of NAD+-linked isocitrate dehydrogenase from the radular muscles of the whelk and flight muscles of insects. *Biochem. J.* **154**, 667–687.
Zammit, V. A., Beis, I., and Newsholme, E. A. (1978). Maximum activities and effects of fructose biphosphate on pyruvate kinase from muscles of vertebrates and invertebrates in relation to the control of glycolysis. *Biochem. J.* **174**, 989–998.
Zs.-Nagy, I. (1971a). Pigmentation and energy dependent $Sr^{2+}$-accumulation of molluscan neurons under anaerobic conditions. *Ann. Biol. Tihany* **38**, 117–129.
Zs.-Nagy, I. (1971b). The lipochrome pigment of molluscan neurons as a specific electron acceptor. *Comp. Biochem. Physiol. A* **40A**, 595–602.
Zs.-Nagy, I. (1977). Cytosomes (yellow pigment granules) of molluscs as cell organelles of anoxic energy production. *Int. Rev. Cytol.* **49**, 331–377.

# 6

# Amino Acid Metabolism in Molluscs[1]

STEPHEN H. BISHOP
LEHMAN L. ELLIS
JAMES M. BURCHAM

Department of Zoology
Iowa State University
Ames, Iowa



[1] This paper was written during the tenure of a National Science Foundation Grant-in-Aid (PCM-80-22606) and represents contribution number 156 from the Tallahassee, Sopchoppy, and Gulf Coast Marine Biological Association.



THE MOLLUSCA, VOL. 1
Metabolic Biochemistry
and Molecular Biomechanics


ISBN 0-12-751401-5

# I. Introduction

This chapter will focus on amino acid metabolism in molluscan species. Aspects related to the neurobiology, energy budget, reproduction, digestion, feeding, and salt or water balance are discussed in other chapters. This discussion will review the literature and development of concepts since the earlier reviews of this subject (Campbell and Bishop, 1970; Florkin and Bricteux-Grégoire, 1972; Schoffeniels and Gilles, 1972; Campbell, 1973).

As heterotrophs, the various molluscan species occupy a complex variety of niches ranging from herbivores to detritivores, carnivores, and parasites. Apparently most species have a requirement for the 10 dietarily essential amino acids common to most metazoans (arginine, histidine, lysine, threonine, phenylalanine, tryptophan, methionine, valine, leucine, and isoleucine), but some species may also require serine, glycine, cysteine, and proline. Most species can catabolize these amino acids, but as a group they have limited capacities for interconversion and biosynthesis. We will try to relate amino acid metabolism to the overall metabolism of the organisms so as to present a reasonably integrated picture.

# II. Ammonia Formation and Fixation

The catabolism of amino acids involves two interrelated processes: the removal with subsequent transfer or elimination of the amino group and the metabolism of the carbon skeleton. For aquatic molluscs most of the nitrogen in the amino group of the amino acids is eventually excreted as ammonia, whereas in terrestrial or aquatic species exposed to xeric environments, substantial amounts of nitrogen are excreted as urea, or as uric acid and other purine compounds. The process of nitrogen excretion will be treated as a separate section.

Amino-group removal from amino acids can be direct or indirect. The direct schemes are through action of amino acid oxidases or other oxidase systems, specific dehydrogenases such as glutamate dehydrogenase (GDH), or deaminating dehydrases such as serine dehydrase. The indirect schemes are through "transdeaminase" pathways. The transdeaminase pathways involve channeling of the amino nitrogen through a series of transaminases to a single intermediate that is then acted on by a deaminating enzyme (Lowenstein, 1972). Two major transdeaminase schemes are known to be active in mammalian tissues. A glutamate dehydrogenase-linked transdeaminase system seems to be the predominant ammonia-forming system in mammalian liver (Krebs et al., 1978), and a purine nucleotide cycle transdeaminase scheme seems to be the primary ammonia-forming system in mammalian skeletal muscle (Meyer and Terjung, 1979). There is good evidence for operation of both direct and indirect deaminating steps in molluscan tissues.

## A. L-Amino Acid Oxidase (EC 1.4.3.2)

L-Amino acid oxidase has been found in a number of molluscan species (see Campbell and Bishop, 1970). The best characterized is the L-amino acid oxidase (EC 1.4.3.2) from the tissues of *Mytilus edulis* (Blaschko and Hope, 1956; Hope and Horncastle, 1967; Hope et al., 1967) and *Modiolus demissus* (Burcham et al., 1980). Some aspects of the earlier studies on this enzyme have been reviewed (Campbell and Bishop, 1970). The products of the reaction are peroxide, ammonia, and the $\alpha$-ketoacid analog corresponding to the L-amino acid substrate. In mussels, this deaminating oxidase is particulate in that it sediments at low speed (5000 $g$ for 15 min) and cannot be solubilized by solvent or detergent extraction (Burcham et al., 1980; Blaschko and Hope, 1956). This enzyme shows greatest activity with the L-amino acids that have fairly long nonacidic R groups and lack methyl, carboxyl, or hydroxyl substitutions on the $\beta$-carbon. In this respect, it is quite similar to the enzyme in chicken liver microsomes (Struck and Sizer, 1960) and in mycelia from *Neurospora crassa* (Bender and Krebs, 1950). Another soluble L-amino acid oxidase that may be a decarboxylating oxygenase [EC 1.13.12(?)] in combination with an amidase has been reported in tissues of an aquatic pulmonate snail (*Lymnaea stagnalis*) (Olomucki et al., 1960). However, experiments with catalase-free preparations and with azide added to preparations containing catalase indicated that perioxide was a reaction product. Peroxide can cause an oxidative decarboxylation of $\alpha$-ketoacids to the shortened organic acid and may account for the products identified by Olomucki et al. (1960). From these data and the recent studies by Freidl (1979), it appears that the L-amino acid oxidase in *L. stagnalis* is a soluble form of the deaminating oxidase rather than the decarboxylating oxygenase in combination with an amidase. Some caution must be exercised here, because Flashner and Massey (1974) and Nakazawa et al. (1972) have shown that the decarboxylating oxygenase from *Pseudomonas putida* can behave as a deaminating oxygenase under certain circumstances. In any case, it appears that the L-amino acid oxidase in the tissues of molluscs is the deaminating oxidase.

Similar nonparticulate L-amino acid oxidases have been found in *Popenaias buckleyi* but not in *Anodonta couperiana* by Falany and Freidl (1981), and in *Cardium tuberculatum* by Glahn et al. (1955). The partially purified activity from *C. tuberculatum* is optimally active at about pH 8 and is activated by $Mg^{2+}$ (Roche et al., 1959). Blaschko and Hope (1956) report L-amino acid oxidase activity in *Cyprina islandica* but not *Mya arenaria, Pecten maximus,* or *Helix pomatia.*

The physiological role of L-amino acid oxidase is uncertain. When present, it is apparently found in most tissues (Burcham et al., 1980). However, the variations with pH optima, solubility, and metal ion activation make assessment of the distribution among species and the relative activity in the tissues difficult. More interestingly, this oxidase is most active on those amino acids that do not accumulate in the tissues of marine or freshwater bivalves subjected to hyperosmotic stress. In this regard, the enzyme may have a major role in the catabolism of these amino acids and in the regulation of the species of amino acids that accumulate in the intracellular free amino acid pool.

## B. D-Amino Acid Oxidase (EC 1.4.3.3)

Small amounts of D-amino acids are found in the tissues and some peptides of many animal species (see Corrigan, 1969). D'Aniello and Giuditta (1977, 1978) report high levels (3–14 μmol/g wet weight) of D-aspartate in the axoplasm and ganglia of some cephalopods (*Loligo vulgaris, Sepia officinalis,* and *Octopus vulgaris*) and no D-aspartate in the hepatopancreas of two gastropods (*Aplysia lemacina* and *Haliotis lamellosa*) or a bivalve (*Penna nobilis*). The levels of D-aspartate were comparable to the levels of L-aspartate in the cephalopod tissues. A search for D-alanine indicated an absence in the tissues of the molluscs examined and modest levels in crustacean hepatopancreas.

Blaschko and Hawkins (1952) reported a D-amino acid oxidase activity with broad specificity in the tissues of two cephalopods, *S. officinalis* and

$O.$ *vulgaris,* and of the snail, *H. pomatia.* The activity was not found in any of the bivalve molluscs tested (see Campbell and Bishop, 1970). Blaschko and Himms (1955) found that the hepatopancreas from cuttlefish *(Sepia)* had two D-amino acid oxidases, one with broad specificity and the other active only with D-aspartate and D-glutamate.

The broadly specific oxidase has been partially purified from octopus $(O.$ *vulgaris)* by Rava et al. (1981). The enzyme is most active with D-alanine, shows no activity with glycine, is most active between pH 9 and 10, has a molecular weight of 80,000–90,000 by gel filtration, and is strongly inhibited by benzoate.

Oxidases specific for D-glutamate and D-aspartate have been found in tissues of three cephalopods—*Loligo forbesi, O. vulgaris,* and *Eledone cirrhosa*—and one has been purified from hepatopancreas of $O.$ *vulgaris* (Rocca and Ghiretti, 1968; D'Aniello and Rocca, 1972). However, recent studies by D'Aniello et al. (1975) indicate that the highest activities are found in octopod cephalopods. The prosthetic group is a tightly bound flavin adenine dinucleotide (FAD) molecule with one FAD per 50,000 MW; the $K_m$ values for D-aspartate and D-glutamate are 5.19 and 6.24 m$M$, respectively, and the native enzyme has a molecular weight of 100,000. This oxidase differs from the mammalian kidney oxidase (Dixon and Kenworthy, 1967) in that it is not inhibited by benzoate and is inactivated by removal of the FAD. The octopus enzyme is inhibited by barbiturates in much the same manner as the NADPH oxidases in mammalian tissue (Casola et al., 1964). The physiological role of the D-amino acid oxidases in cephalopods is uncertain but may be related to feeding habit (D'Aniello et al., 1975).

## C. Glutamate Dehydrogenase (EC 1.4.1.2)

The existence and physiological importance of glutamate dehydrogenase as an ammonia-forming or -fixing mechanism in the tissues of molluscan species was in doubt in 1970 (Campbell and Bishop, 1970; Campbell et al., 1972). Since then GDH activity has been demonstrated in tissues of a number of species (see Table I). In all cases, the enzyme is apparently localized in the mitochondria, is activated by ADP, and is inhibited by GTP. L-Leucine has also been shown to be an activator (Freidl, 1979; P. M. Reiss, unpublished data; Storey et al., 1978). The enzymes from tissues of *H. pomatia, M. demissus,* and *Loligo pealeii* are partially active with NADPH as a substrate in place of NADH. When assayed with NADPH, the pH optimum is somewhat lower, the activity is 2–20% of that with NADH, and there is no activation with ADP. When assayed in the ammonia-forming direction, the enzyme in gastropod and bivalve tis-

**TABLE I**

**Glutamate Dehydrogenase Activity in Tissues of Some Molluscan Species**

| Species | Activity (glutamate formation) ($\mu$mol/g wet wt./h) | Experimental conditions and comment | Reference |
|---|---|---|---|
| Gastropoda | | | Sollock et al. (1979) |
| *Helix aspersa* | | | |
| Hepatopancreas | 12.7 | pH 8; 0.5 m$M$ ADP; NADH at low $\alpha$-ketoglutarate concentration (0.4 m$M$); high concentration of other substrates using a radiometric assay at 30°C | |
| Bivalvia | | | |
| *Mytilus edulis* | | | |
| Hepatopancreas | 40 | pH 8; NADH, 3 m$M$ | Addink and |
| Abductor muscle | 49 | ADP, and high | Veenhof |
| Mantle | 27 | concentrations of other substrates; spectrophotometric assay at 25°C | (1975) |
| *Modiolus demissus* | | | |
| Gill | 11 | pH 8.5; NADH, 1 m$M$ | Reiss et al. |
| Heart | 23 | ADP, and high | (1977) |
| Hepatopancreas | 13.5 | concentrations of | |
| Mantle | 7.6 | other substrates using a spectrophotometric assay at 25°C | |
| Cephalopoda | | | |
| *Loligo pealeii* | | | |
| Hepatopancreas | 738 | pH 8; NADH, 5 m$M$ | Storey et al. |
| Brain | 222 | ADP, and high | (1978) |
| Systemic heart | 3126 | concentrations of | |
| Branchial heart | 1198 | other substrates | |
| Testes | 19 | using a | |
| Gill | 357 | spectrophotometric | |
| Tentacle muscle | 630 | assay; temperature | |
| Mantle muscle | 752 | not given | |
| Beak muscle | 2004 | | |
| Digestive pouch | 3318 | | |

sues operates at about 20% of the rate in the glutamate-forming direction; with the squid mantle muscle enzyme, the maximal rate in the ammonia-forming direction would appear to be about 50% of the rate in the opposite direction. Storey et al. (1978) estimate the molecular weight (gel filtration) of the squid mantle muscle enzyme to be about 300,000. All data presented thus far are somewhat preliminary but are consistent with the notion that the properties of the GDH in these molluscan tissues are similar to those of GDHs from mammalian liver (Fisher, 1973; Smith et al., 1975).

Low GDH levels similar to those in the tissues of mussels (see de Zwaan, 1977) and land snails (Table II) have been found in the tissues of the prosobranch snails, *Pila globosa* by Reddy and Swami (1975) and *Littorina saxatilus* by McManus and James (1975a), the aquatic pulmonate snails *L. stagnalis* and *Biomphalaria glabrata* by Sollock et al. (1979) and Freidl (1979), respectively, and the American oyster, *Crassostrea virginica* by Wickes and Morgan (1976). High GDH levels similar to those found in the mantle muscle of *L. pealeii* (Table II) have been found in the muscles of other cephalopods *Loligo opalescens, Illex illecebrosus, Ommostrephes* sp., *Berryteuthis magister,* and *Symplectoteuthis oualaniensis* by Mommsen et al. (1981).

Some differences appear to be extant when comparing GDH activities in bivalve and gastropod tissues with those in cephalopod tissues. The activity levels in cephalopod tissues tend to be one to several orders of magnitude higher than those in bivalve and gastropod tissues; on the other hand, the $K_a$ for ADP activation is low (90–170 $\mu M$) in mussel tissues (Reiss et al., 1977) and high (750 $\mu M$) in squid mantle muscle (Storey et al., 1978). In mussel and gastropod tissues, the maximal activation of GDH activity by ADP is 2- and 10-fold, whereas ADP activation of most squid tissue enzymes is between 50- and 100-fold (ref. in Table I). If these differences in the $K_a$ for ADP activation and the ADP rate enhancement are real, then one would predict that the GDHs in mussel and gastropod tissues are in an activated state most of the time and that the activity of GDH in the squid tissues would vary considerably as a function of ADP and GTP levels. With regard to the cephalopods, Mommsen et al. (1981) find that the levels of GDH and other oxidative enzymes in mantle muscle fibers correlate directly with the number of mitochondria and oxidative capacity of the various fibers. Because bivalves and gastropods have very slow metabolic rates compared to cephalopods, the low GDH levels in bivalve and gastropod tissues may reflect their overall low metabolic activity and not argue against a primary role for GDH in glutamate metabolism in molluscs from these two classes.

**TABLE II**

**Alanine and Aspartate Aminotransferase Levels in Tissues of Some Molluscan Species**

| Species | Tissue | AIAT[a] | AAT[a] | Reference |
|---|---|---|---|---|
| Gastropods | | | | |
| Otala lactea | Hepatopancreas | 250 | 696 | Sollock et al. (1979) |
| Helix aspersa | Hepatopancreas | – | 593 | Sollock et al. (1979) |
| Helix pomatia | Hepatopancreas | + | – | Zandee et al. (1958) |
| | Foot muscle | + | – | Zandee et al. (1958) |
| Strophocheilus oblongus | Hepatopancreas | 183 | 757 | Sollock et al. (1979); Christie and Michelson (1975) |
| Biomphalaria glabrata | Hepatopancreas | 372 | 508 | Sollock et al. (1979); Christie and Michelson (1975) |
| Lymnaea stagnalis | Whole body | – | + | Freidl (1979) |
| Pila globosa | Foot-active animals | – | + | Swami and Reddy (1978) |
| | Mantle-active animals | + | + | Swami and Reddy (1978) |
| | Hepatopancreas-active animals | + | + | Swami and Reddy (1978) |
| | Nerve-active animals | + | + | Swami and Reddy (1978) |
| Viviparus viviparus | Hepatopancreas | + | + | Sollock et al. (1979) |
| Bivalves | | | | |
| Mytilus edulis | Hepatopancreas | – | 637–684 | Read (1962); Addink and Veenhof (1975) |
| | Gill | – | 472–613 | Read (1962); Addink and Veenhof (1975) |
| | Byssus retractor muscle | – | 401–519 | Read (1962); Addink and Veenhof (1975) |
| | Mantle | – | 377–590 | Read (1962); Addink and Veenhof (1975) |
| | Bojanus organ | – | 707–1179 | Read (1962); Addink and Veenhof (1975) |
| | Blood | – | 0 | Read (1962); Addink and Veenhof (1975) |
| Modiolus demissus | Whole body | 174–218 | 84–111 | Read (1963); Hammen (1968) |
| | Heart | 120 | 126 | Greenwalt (1981) |
| Modiolus modiolus | Whole body | + | + | Read (1962, 1963) |
| Brachidontes recurvus | Whole body | – | + | Read (1963) |
| Mya arenaria | Gill | + | + | DuPaul and Webb (1974) |
| | Hemolymph | + | + | Rodrick (1979) |

| Species | Tissue | | | Reference |
|---|---|---|---|---|
| *Isognomon alatus* | Whole body | – | + | Read (1963) |
| *Crassostrea rhizophorae* | Whole body | + | + | Read (1963) |
| *Crassostrea virginica* | Mantle | 30 | 100 | Awapara and Campbell (1964) |
| | Whole body | 8 | 121 | Awapara and Campbell (1964); Chambers et al. (1975) |
| | Adductor muscle | – | + | Wickes and Morgan (1976) |
| | Gill | | + | Wickes and Morgan (1976) |
| *Lima scabra* | Whole body | – | + | Wickes and Morgan (1976) |
| *Pinna carnea* | Whole body | – | + | Read (1963) |
| *Phacoides pectinatus* | Whole body | – | + | Read (1963) |
| *Rangia cuneata* | Mantle | 145 | 82 | Awapara and Campbell (1964) |
| *Solemya velum* | Whole body | 784–813 | 372–388 | Hammen (1968) |
| *Donax variabilis* | Whole body | 192–209 | 381–480 | Hammen (1968) |
| *Tagelus plebius* | Whole body | 127–144 | 141–144 | Hammen (1968) |
| *Mercenaria mercenaria* | Whole body | 81–91 | 43–56 | Hammen (1968) |
| | Gill | + | + | DuPaul and Webb (1974) |
| *Teredo navalis* | Gill | 86–125 | 54–69 | DuPaul and Webb (1974) |
| *Anodonta couperiana* | Hepatopancreas | 60–80 | + | Falany and Freidl (1981) |
| *Popenaias buckleyi* | Hepatopancreas | + | + | Falany and Freidl (1981) |
| *Spisula solidissima* | Gill | + | + | DuPaul and Webb (1974) |
| Cephalopods | | | | |
| *Loligo pealeii* | Axoplasm | – | 917 | Roberts et al. (1958) |
| | Axon sheath | – | 602 | Roberts et al. (1958) |
| *Loligo opalescens* | Inner mantle muscle | 228 | 1140 | Mommsen et al. (1981) |
| | Middle mantle muscle | 66 | 516 | Mommsen et al. (1981) |
| | Outer mantle muscle | 186 | 876 | Mommsen et al. (1981) |
| | Tentacle | 36 | 984 | Ballantyne et al. (1981) |
| | Fin | 40 | 2760 | Ballantyne et al. (1981) |
| | Heart | 1050 | 6540 | Ballantyne et al. (1981) |
| | Brain | 480 | 1380 | Ballantyne et al. (1981) |
| | Gill | 138 | 570 | Ballantyne et al. (1981) |
| | Skin | 18 | 186 | Ballantyne et al. (1981) |

*(Continued)*

**TABLE II (Continued)**

| Species | Tissue | AlAT[a] | AAT[a] | Reference |
|---|---|---|---|---|
| Ommastrephes sp. | Inner mantle muscle | 156 | 4955 | Mommsen et al. (1981) |
| | Middle mantle muscle | 108 | 1752 | Mommsen et al. (1981) |
| | Outer mantle muscle | 66 | 288 | Mommsen et al. (1981) |
| Berrytouthis magister | Inner mantle muscle | 258 | 1128 | Mommsen et al. (1981) |
| | Middle mantle muscle | 162 | 744 | Mommsen et al. (1981) |
| | Outer mantle muscle | 288 | 1168 | Mommsen et al. (1981) |
| Symplectoteuthis oualaniensis | Inner mantle muscle | 228 | 8760 | Mommsen et al. (1981) |
| | Middle mantle muscle | 126 | 7620 | Mommsen et al. (1981) |
| | Outer mantle muscle | 108 | 1812 | Mommsen et al. (1981) |
| | Whole mantle muscle | 96 | 1200 | Hochachka et al. (1975) |
| Illex illecebrosus | Inner mantle muscle | 312 | 5160 | Hochachka et al. (1975) |
| | Middle mantle muscle | 300 | 2730 | Hochachka et al. (1975) |
| | Outer mantle muscle | 258 | 7500 | Hochachka et al. (1975) |
| Nautilis pompilius | Spadix muscle | — | 270 | Hochachka et al. (1978) |
| | Funnel muscle | — | 528 | Hochachka et al. (1978) |
| | Retractor muscle | — | 678 | Hochachka et al. (1978) |

[a] The numbers represent units as $\mu$moles of product formed per gram wet tissue per hour under the reported assay conditions (see references). The (+) and (−) refer to presence or not measured, respectively, in cases where calculation of actual units was unclear. AAT, Aspartate aminotransferase; AlAT, alanine aminotransferase.

## D. Alanine Dehydrogenase (EC 1.4.1.1)

Although there is no report of the actual measurement of this activity in molluscan tissues, the existence of such an enzyme in some bivalves (*M. edulis* and *Rangia cuneata*) has been postulated by several authors (Livingstone et al., 1979; Henry et al., 1980; de Zwaan and van Marrewijk, 1973; Zurburg and de Zwaan, 1981) to account for alanine biosynthesis and breakdown in tissues with low GDH activities and with high rates of alanine turnover and accumulation. However, in studies using transaminase inhibitors in *M. demissus*, [14]C tracer and amino acid accumulation experiments with isolated heart and gill tissues support the idea that alanine synthesis and catabolism is transaminase linked and not direct through an alanine dehydrogenase system (Greenwalt and Bishop, 1980; Bishop et al., 1981).

GDH can potentially behave as an alanine dehydrogenase. GDHs from livers of some mammalian species under appropriate conditions will use alanine or pyruvate as alternate substrates at 2–5% of the rate with glutamate or α-ketoglutarate, respectively (Smith et al., 1975). Rates with alanine and pyruvate with the GDH from *M. demissus* gill tissue were below detectable levels of 0.5% the rate with glutamate and α-ketoglutamate (Bishop, unpublished). Additionally we found no evidence for a separate alanine dehydrogenase activity corresponding to protist or bacterial systems in the tissues of *M. demissus*.

Until there is a clear demonstration of a true alanine dehydrogenase in molluscan tissues, we must assume that none exists. Alanine metabolism will be discussed in context with the metabolism of the other amino acids (see later).

## E. Glutamine Synthetase (EC 6.3.1.2) and Glutaminase (EC 3.5.1.2)

Glutamine has a major role in the biosynthesis of amino sugars, purines, pyrimidines; in protein synthesis; in substrate transfer between tissues for gluconeogenesis; and possibly in buffering of acid production in urine. Among molluscs, glutamine synthetase activity was first reported in the tissues of some land snails (Campbell et al., 1972). More recently, Campbell and Vorhaben (1979) have shown that glutamine synthetase is localized in the cytosol of the hepatopancreas of *Helix aspersa* rather than the mitochondia. Glutaminase activity has been reported in the tissues of some cephalopods (Potts, 1967; Campbell and Bishop, 1970); however, levels in land snails were not detectable (Campbell et al., 1972). The levels of glutamine in the tissues and blood of pulmonate snails vary considerably with degree of starvation or feeding and with estivation (Campbell and Bishop, 1970; Chetty et al., 1979). The level of glutamine synthetase in hepatopancreas tissue of the slug *Limax flavus* did not change with starvation (Horne, 1977b), even though overall urea production increased. During estivation–starvation of the prosobranch *P. globosa*, Reddy and Swami (1975) report a doubling in the glutamine synthetase level in the hepatopancreas. Studies with $^{14}$C-labeled precursor molecules (glutamate, TCA-cycle intermediates, and glucose) in a number of molluscan species indicate a weak glutamine biosynthetic capacity in most ammonotelic species (bivalves) and a reasonably rapid glutamine synthesis or turnover in species that excrete reasonable amounts of purine (terrestrial pulmonate snails and some prosobranch gastropods). These studies will be developed in a subsequent section.

## F. L-Serine Dehydrase (EC 4.2.1.13)

L-Serine dehydrase catalyzes the formation of pyruvate and ammonia from serine. The activity could serve as the primary deaminating step at the end of a series of transaminase steps that couple the reaction to the triose metabolic pathways (Gilles, 1969; Bishop, 1976). Low levels of this activity have been found in the hepatopancreas of *Murex trunculus* by Bargoni and Sisini (1962) and in the adductor muscle, gill, mantle, heart, and hepatopancreas of both *M. demissus* and *M. edulis* (J. M. Burcham and S. H. Bishop, unpublished data). The activity from *M. demissus* gill tissue was enhanced by added pyridoxal phosphate and was optimal at pH 8.5; the $K_m$ for L-serine was 14–15 m$M$, and the tissue activity levels varied between 0.4 and 10 $\mu$mol of pyruvate formed/g wet tissue/h at 25 m$M$ L-serine. This gill enzyme was poorly inhibited by L-cycloserine (20% inhibition at 5 m$M$) and more strongly inhibited by aminooxyacetic acid ($I_{50} = 0.16$ m$M$).

Bargoni and Sisini (1962) have reported more activity with D-serine than L-serine with the extracts of the *Murex* hepatopancreas, indicating the presence of a D-serine dehydrase in this tissue. Both L- and D-serine dehydrases have been found in the gut tissue of the earthworm *Lumbricus terrestris* (Fujimoto and Adams, 1965). In preliminary studies with the gill enzyme from *M. demissus*, no activity was seen with either D-serine or L-threonine as a substrate (S. H. Bishop, unpublished data).

Overall, the low tissue activities and high relative $K_m$ for serine argue against a significant physiological role for the L-serine dehydrase as part of a major transdeaminase system; however, it may have a role in L-serine catabolism.

## G. Purine Nucleotide Cycle Transdeaminase Scheme

The three enzymes of the purine nucleotide cycle—adenylosuccinate synthetase [EC 6.3.4.4 IMP: L-aspartate lyase (GDP-forming)], adenylosuccinate lyase (EC 4.3.2.2), and AMP deaminase (EC 3.5.4.6 AMP-aminohydrolase)—have been detected in the hepatopancreas of *H. aspersa* by Campbell and Vorhaben (1979). The levels for the synthetase and the lyase were quite low (40–106 nmol/g tissue/h and 4–7 $\mu$mol/g tissue/h, respectively), whereas the AMP deaminase levels approximated the levels in other invertebrate tissues and nonskeletal muscle tissues of vertebrates (Gibbs and Bishop, 1977). AMP deaminase activity has also been found in gill and adductor muscle tissue of *M. demissus* and *C. virginica* by Gibbs et al. (1976), and in the hepatopancreas of *H. pomatia* by Umiastowski (1964). When tested, ADP and ATP activate the AMP deaminase activities to varying degrees and GTP is inhibitory (Gibbs et al., 1976; Campbell and Vorhaben, 1979). Overall, the activities seem similar to those found in nonskeletal muscle tissue of mammals. In rat tissues, the cycle appears to operate as a major ammonia-forming scheme in the fast-twitch fibers of skeletal muscle and as a mechanism for adenine nucleotide biosynthesis and turnover in the other tissues (Krebs et al., 1978; Meyer and Terjung, 1979; Lowenstein and Goodman, 1978). Although the extremely low synthetase activity would argue against a major role for the purine nucleotide cycle as a major transdeaminase pathway in molluscs, it has been investigated in only one tissue of one molluscan species. In preliminary radiotracer experiments with gill tissue from *M. demissus,* oxidation of L-aspartate was inhibited 80% by the transaminase inhibitor aminooxyacetic acid (Bishop et al., 1981). If the purine nucleotide cycle was a major route for aspartate catabolism, then one would have predicted little or no inhibition because no transaminase step would have been involved.

## III. Transaminations

### A. Alanine Aminotransferase (EC 2.6.1.2) and Aspartate Aminotransferase (EC 2.6.1.1)

Both transaminase activities have been detected in all tissues of all molluscan species investigated (see Table II). The method of assay and conditions of assay (pH, buffer, temperature, substrate concentration, etc.) varied with the investigation and were not at all times optimal. Therefore, the data in Table II should be viewed as semiquantitative. Partially purified preparations of the enzymes from mussels were stimulated by added pyridoxal phosphate (Read, 1963), and studies by Awapara and Campbell (1964) indicate that the activities in the tissues of *C. virginica, R. cuneata,* and *Otala lactea* are reversible. None of these activities has been purified or studied in any great detail.

The optimal pH for the aspartate aminotransferase (AAT) activity seems to be between 7.7 and 8.5 (Read, 1963; Roberts et al., 1958). The $K_m$ values for $\alpha$-ketoglutarate and aspartate with the AAT from squid axoplasm were less than 0.2 and 0.5 m$M$, respectively. The $K_m$ values are lower than the $K_m$ values obtained with the rabbit brain AAT (Roberts et al., 1958). L-Cycloserine and aminooxyacetic acid are quite reactive with carbonyl compounds and have been used in metabolic experiments as relatively specific inhibitors of the pyridoxal phosphate-containing transaminases (Meijer et al., 1975, 1978; Wong et al., 1973). Both L-cycloserine and aminooxyacetic acid inhibit the AAT and alanine aminotransferase (AIAT) activities from the gill and heart tissues of *M. demissus* (Greenwalt and Bishop, 1980; Bishop et al., 1981). The $I_{50}$ values (concentration of inhibitor causing a 50% inhibition of the activity) for the two inhibitors were similar to the $I_{50}$ values for the AAT and AIAT activities from mammalian tissues.

The AAT levels tend to be quite a bit higher than the AIAT levels in the hepatopancreas of some gastropods (Sollock et al., 1979; Swami and Reddy, 1978) and in cephalopod muscle tissues (Mommsen et al., 1981). With the exception of the American oyster, levels of the two enzymes in bivalves vary with the particular tissue but seem to be reasonably similar in amount of activity. Hammen (1968) finds that the levels of activity (units/g) in the tissues vary with the size of the animal, indicating that animals of uniform size should be used in making comparisons of the levels of tissue activity under various experimental regimens. Zammit and Newsholme (1976) reported variable levels of AAT in muscle tissue from a number of bivalves.

Sollock et al. (1979) report a partitioning of the AAT activity into two cellular fractions, the cytosol and mitochondria, in the hepatopancreas from four gastropod species (*O. lactea, H. aspersa, Strophocheilus oblongus,* and *B. glabrata*). Cytosolic and mitochondrial AAT activities have also been reported for osyter (Chambers et al., 1975) and littorine tissues (McManus and James, 1975a). Isoenzymes for AAT activities have been found in the tissues of several molluscan species (Johnson and Utter, 1973; Johnson et al., 1972; Levinton and Koehn, 1976). Both the cytosolic and mitochondrial AAT loci appear to be polymorphic, and six different alleles have been discerned in the tissues of *M. edulis.* Isoenzymes for AIAT may occur in some species (Sollock et al., 1979) but the pattern is not well documented. There are no studies on the differences or similarities in kinetic behavior or other properties of these isoenzyme forms of AAT or AIAT in molluscan tissues.

DuPaul and Webb (1974) found little change in AAT or AIAT levels in the gills of bivalves (*Mya, Mercenaria, Spisula*) transferred from $20\%_{00}$ seawater to $30\%_{00}$ seawater for periods up to 6 days. On the other hand, AIAT levels in the gills and muscles of oysters collected from areas of low and high salinity tended to be highest in the tissues of oysters collected at the higher salinities (Wickes and Morgan, 1976). These contrasting results need reinvestigation; they could be a result of differences in species and/or other factors related to season, food, size of the animals, and assay procedure. For instance, with pulmonate land snails (*P. globosa*), the tissue levels of AIAT and AAT were greater in active animals than in starving estivating animals (Swami and Reddy, 1978).

Read (1963) examined the thermostability of AAT partially purified from whole-body homogenates of a number of bivalve species. The bivalve species (see Read refs. in Table II) living in the more exposed environments where temperatures reach the extremes tended to have the more thermostable AAT activities. Among the Mytilidae, *M. demissus* had the most thermostable activity and *Modiolus modiolus* had the most thermolabile activity, and *M. edulis* had an intermediate thermostability. Activities from the other species had a thermostability equal to or less than *M. modiolus*.

## B. Ornithine Aminotransferase (EC 2.6.1.13)

Ornithine $\Delta$-aminotransferase (OAT) catalyzes the reaction $\alpha$-ketoglutarate + L-ornithine $\rightleftharpoons$ L-glutamate semialdehyde + glutamate. L-Glutamate semialdehyde cyclizes spontaneously to $\Delta^1$-pyrroline-5-carboxylate, which can be measured colorimetrically. Low levels of the enzyme have been detected in the tissues of several molluscs using either the colorimetric method or a radiometric procedure to estimate glutamate formation from $\alpha$-ketoglutarate (Table III).

OAT activities in the hepatopancreas of the snails, *H. aspersa, O. lactea,* and *S. oblongus,* are localized in the mitochondria (Campbell et al., 1972). This mitochondrial localization has also been found for OAT in gill and heart tissue of *M. demissus* (J. M. Burcham and S. H. Bishop, unpublished data). In mammalian liver tissue, most of the OAT is mitochondrial (Peraino and Pitot, 1963), whereas in some insect tissues, the activity may be localized in the cytosol (Campbell et al., 1972). Preliminary studies (J. M. Burcham and S. H. Bishop, unpublished data) with OAT partially purified from gill tissues of *M. demissus*

TABLE III

Levels of Ornithine Aminotransferase (OAT) in Some Molluscan Tissues

| Species | Tissue | OAT units[a] | Reference |
|---------|--------|--------------|-----------|
| Gastropods | | | |
| Otala lactea | Hepatopancreas | 7.6–27 | Campbell and Speeg (1968a) |
| Helix aspersa | Hepatopancreas | 42 | Campbell et al. (1972) |
| Strophocheilus oblongus | Hepatopancreas | 3.5–67 | Sollock et al. (1979) |
| Viviparus viviparus | Hepatopancreas | 0.27 | Sollock et al. (1979) |
| Biomphalaria glabrata | Hepatopancreas | 3.0 | Sollock et al. (1979) |
| Bivalves | | | |
| Modiolus demissus | Heart | 6 | Greenwalt and Bishop (1980) |
| M. demissus | Gill | 3 | Bishop et al. (1981) |
| Popenaias buckleyi | Hepatopancreas | 2.4 | Falany and Freidl (1981) |
| Anodonta couperiana | Hepatopancreas | 1.9 | Falany and Freidl (1981) |

[a] Units are $\mu$moles of product formed per hour per gram wet tissue under reported assay conditions.

indicate optimal activity at pH 7.5 in both Tris and imidazole buffer, and $K_m$ values of 4.8 and 2 m$M$ for L-ornithine and $\alpha$-ketoglutarate, respectively. The gill and heart OAT from *M. demissus* are strongly inhibited by both L-cycloserine and aminooxyacetic acid (Greenwalt and Bishop, 1980; Bishop et al., 1981). Addition of pyridoxal phosphate to the assay mixture stimulated the activities in the snail hepatopancreas (Campbell and Speeg, 1968a). Pyridoxal phosphate must be included in the buffers to prevent loss of activity during purification of the OAT activity in the ribbed mussel gill tissue (J. M. Burcham and S. H. Bishop, unpublished data).

Although no extensive characterization of the OAT in molluscan tissues has been published, the mitochondrial localization, $K_m$ values for substrates, pH optimum, pyridoxal phosphate requirement, and inhibition by L-cycloserine and aminooxyacetic acid indicate a close similarity to the activities in mammalian tissues.

## C. Tyrosine Aminotransferase (EC 2.6.1.5)

Low levels of this activity (TAT) have been found in the hepatopancreases of several molluscan species (Table IV). The number of species investigated is quite small but those investigated contain some TAT activity in their tissues. Read (1962), using a qualitative assay procedure with whole-body homogenates of *M. edulis*, reported questionable levels of TAT activity. There are no studies on the purification or characterization of TAT from molluscan tissues.

**TABLE IV**

Tyrosine Aminotransferase Activities in Tissues of Molluscan Species

| Species | Tissue | Units | Reference |
|---------|--------|-------|-----------|
| Gastropods | | | |
| *Helix pomatia* | Hepatopancreas | 2 | Michalek-Moricca (1965) |
| *Otala lactea* | Hepatopancreas | 27 | Sollock et al. (1979) |
| *Strophocheilus oblongus* | Hepatopancreas | 0.9 | Sollock et al. (1979) |
| *Viviparus viviparus* | Hepatopancreas | 6.5 | Sollock et al. (1979) |
| *Biomphalaria glabrata* | Hepatopancreas | 1.2 | Sollock et al. (1979) |
| *Pila globosa* | Hepatopancreas | + | Swami and Reddy (1978) |
| *P. globosa* | Foot | + | Swami and Reddy (1978) |
| *P. globosa* | Mantle | + | Swami and Reddy (1978) |
| *P. globosa* | Nerve tissue | + | Swami and Reddy (1978) |
| Bivalves | | | |
| *Popenaias buckleyi* | Hepatopancreas | 4.7 | Falany and Freidl (1981) |
| *Anodonta couperiana* | Hepatopancreas | 1.8 | Falany and Freidl (1981) |
| *Mytilus edulis* | Whole tissue | ± | Read (1962) |

## D. Other Aminotransferases

Three investigators have surveyed molluscan tissues for transaminase activities transferring the amino group from a variety of amino acids to $\alpha$-ketoglutarate (Read, 1962; Sollock et al., 1979; Falany and Freidl, 1981). All species tested have the aforementioned AAT, AIAT, OAT, and TAT activities. The study by Read (1962) using whole-body homogenates of *M. edulis* indicated a questionable activity with all of the other amino acids except lysine and arginine. Sollock et al. (1979), using cytosol preparations of the hepatopancreas of *O. lactea* and *Viviparus viviparus*, found some activity with all the amino acids tested. Falany and Freidl (1981), using bivalve hepatopancreas homogenates (species in Table II) found activity with the branched-chain amino acids. In other studies, Bishop et al. (1981) found a leucine aminotransferase and a glycine : pyruvate aminotransferase in the hearts and gills of the ribbed muscle. Studies on the aminotransferases related to glycine and serine are discussed in Section IV,G.

## IV. Biosynthesis and Catabolism

Amino acid biosynthetic and catabolic capacities in the tissues of molluscan species are poorly understood. Most of the studies (cited later) involve a combination of enzyme and tracer experiments using

[$^{14}$C]glucose, $^{14}CO_2$, [$^{14}$C]acetate, [$^{14}$C]succinate and other organic acids, or the individual amino acids. Differences in the amount of radioactivity used, purity of the $^{14}$C tracer, microbiological contamination, and sensitivity of the analytical procedures used to study incorporation, make final judgments on the metabolic capacities uncertain in some cases. For instance, studies with tracer injected or fed to whole animals can be somewhat equivocal because of the chance for microbiological contamination, whereas studies with isolated tissues may reduce this problem but tend to eliminate possible tissue specialization. In any case, tracer studies using [$^{14}$C]glucose with pulmonate snails, mussels, clams, oysters, and prosobranch snails, indicate that all these molluscs have a strong capacity for rapid biosynthesis of glutamate, alanine, and aspartate, and a sometimes weaker or nonexistent capacity for asparagine, glutamine, serine, glycine, and proline biosynthesis in most species (see Campbell and Bishop, 1970; Bryant et al., 1964; Bryant, 1965; Simpson and Awapara, 1965; Campbell and Speeg, 1968a; McManus and James, 1975a,b; de Zwaan et al., 1975; Kerkut et al., 1969; de Zwaan, 1977; Baginski and Pierce, 1978; Ahmad and Chaplin, 1979; Trytek and Allen, 1980; Stokes and Awapara, 1968; Chen and Awapara, 1969a,b; Hammen, 1966; Hammen and Wilbur, 1959; Awapara and Campbell, 1964; Allen and Kilgore, 1975; Orive et al., 1980; Wijsman et al., 1977; McManus and James, 1975a,b,c).

In these studies, aspartate is most strongly labeled with [$^{14}$C]succinate and $^{14}CO_2$ tracers, whereas alanine is most heavily labeled with [$^{14}$C]glucose and [$^{14}$C]pyruvate tracers. Glutamate is generally the most weakly labeled of the three in most of these tracer studies. These results indicate a considerable capacity for $CO_2$ fixation into the dicarboxylic acids of the TCA cycle and a tendency for most molluscs to accumulate modest amounts of alanine under anaerobic conditions.

After injection of $^{14}CO_2$ into snails (*O. lactea* and *H. aspersa*), the specific radioactivity of free glutamate and free aspartate increased and decreased, respectively, with time of incubation (Campbell and Speeg, 1968a). A similar time dependence with regard to the labeling pattern was found by Allen and Kilgore (1975) for *Haliotis* and by Trytek and Allen (1980) for *Bankia* in a series of [$^{14}$C]glucose-tracer studies. This constant turnover of intermediates with time of incubation probably accounts for some of the differences in relative labeling patterns among the various investigators but does not alter the basic conclusion concerning biosynthetic capacities. These data are then in agreement with the ubiquitous distribution of GDH, the alanine and aspartate aminotransferases, and a full complement of enzymes in the glycolytic and TCA-cycle pathways.

The capacity for biosynthesis of the other "dietarily nonessential amino acids" (glycine, serine, proline, glutamine, asparagine, ornithine, cysteine, and arginine) may be nonexistent in some species. In order to give a reasonably well-integrated view, we will discuss the various amino acids in groups and indicate probable pathways for biosynthesis and/or catabolism.

## A. Alanine and Iminodipropionic Acid

It seems clear from the $^{14}$C-tracer studies that any gluconeogenic intermediate compound can eventually give rise to pyruvate for conversion to alanine. The absolute levels of alanine and other amino acids in the tissues of molluscs vary with the season, food availability, anaerobiosis, and salinity of the bathing medium (Schoffeniels and Gilles, 1972; Bayne, 1976; Kluytmans et al., 1977, 1980; Zandee et al., 1980; de Zwaan, 1977; see also Chapter 4, this volume and Volume 2, Chapter 5). The $^{14}$C tracer experiments with tissues of *M. demissus* indicated that transaminase inhibitors (aminooxyacetic acid and L-cycloserine) block alanine biosynthesis, inhibit catabolism about 80–90%, and inhibit alanine accumulation in the tissues under stress (Greenwalt and Bishop, 1980; Bishop et al., 1981; Greenwalt, 1981). Until the presence of an alanine dehydrogenase is firmly established, the data are in agreement with the idea that all alanine is synthesized through transamination of pyruvate. In some species, a small amount of alanine may be catabolized by the amino acid oxidase, but catabolism of alanine in most species involves transamination to pyruvate. The considerable controversy and disagreement on the metabolic origins of the pyruvate and the amino group used for alanine biosynthesis is dealt with elsewhere (see Section VI on amino acid accumulation; see also Chapter 3, this volume and Volume 2, Chapter 2).

An imino amino acid identified as *meso*-$\alpha,\alpha$-iminodipropionic acid was isolated from squid muscle extracts by Sato et al. (1977). The compound has the same stereochemical configuration about the $\alpha$-carbon as naturally occurring octopine. The compound has been detected in some bivalves (Collicutt and Hochachka, 1977; Sato et al., 1978; Fields et al., 1980; Bishop et al., 1980) and gastropod tissues (Ellington, 1981). A dehydrogenase has been purified from oyster (*Crassostrea gigas*) adductor muscle (Fields and Hochachka, 1981) and ribbed mussel (*M. demissus*) gill tissue (Bishop et al., 1980) that catalyzes the reversible reduction of the Schiff base formed between L-alanine and pyruvate to yield $\alpha,\alpha$-iminodipropionic acid. The enzyme is specific for pyruvate and L-alanine but shows some activity with glycine and glyoxylate.

L-Ornithine and L-arginine would not serve as substrates with the purified enzyme, indicating that it is distinct from octopine dehydrogenase. Iminodipropionoacetic acid has been found in scallop aductor muscle (Sato et al., 1978).

## B. Glutamine

The $^{14}$C-tracer studies indicate a capacity for glutamine biosynthesis in the tissues of all prosobranch and pulmonate snail species investigated (Campbell and Speeg, 1968a; Campbell and Bishop, 1970; Huggins et al., 1967; McManus and James, 1975a; Campbell et al., 1972; Orive et al., 1980). Huggins et al. (1967) report $^{14}$C incorporation into glutamine from [$^{14}$C]glutamate in ganglia and adductor muscle of *M. edulis*. However, in tracer studies with *M. edulis* and other bivalve species, most investigators fail to report glutamine labeling from $^{14}$C-labeled TCA-cycle intermediates, aspartate, glutamate, bicarbonate, or glucose (Wijsman et al., 1977; de Zwaan et al., 1975; Ahmad and Chaplin, 1979; Hammen and Wilbur, 1959; Baginski and Pierce, 1978; Collicutt and Hochachka, 1977; Trytek and Allen, 1980; Simpson and Awapara, 1965; Awapara et al., 1963; Greenwalt, 1981). Although glutamine synthetase has been found in the tissues of several gastropod species (see earlier), there has been no report of the activity in tissues of molluscan species from the other classes.

## C. Asparagine

Asparagine biosynthesis may not occur in molluscan tissues. Except for a tracer study with [4-$^{14}$C]succinate by Bryant (1965) with homogenates of periwinkle digestive glands, there is no mention of asparagine labeling in any of the other tracer studies referenced earlier. However, in more recent studies McManus and James (1975b), using [$^{14}$C]glucose with hepatopancreas tissue from *Littorina saxatilus rudis,* report labeling of aspartate and two unknown amino acids but do not mention labeling of asparagine. There is no mention of asparagine labeling in [4-$^{14}$C]aspartate tracer studies with tissues from oysters (Collicutt and Hochachka, 1977) or from ribbed mussels (Baginski and Pierce, 1978), or in [$^{14}$C]succinate tracer studies with sea mussel tissues (Wijsman et al., 1977).

## D. Arginine and Urea Biosynthesis

*De novo* arginine biosynthesis has been demonstrated in only a few invertebrate animals (Campbell, 1973). The only known arginine biosynthetic pathway is the urea or ornithine cycle (Scheme 1). The first four

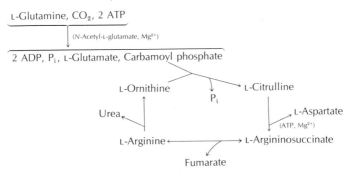

Scheme 1. Urea cycle in pulmonates.

steps of the pathway involve arginine biosynthesis, with the last step, arginase, hydrolyzing arginine to form urea and ornithine. The $^{14}CO_2$-tracer technique has been a convenient method for evaluating the probable operation of a complete set of urea enzymes in a tissue when technical or other difficulties preclude actual assay of all of the individual enzymatic activities. The method involves use of tissue incubations or injected animals followed by measurement of $^{14}C$ transfer from $H^{14}CO_3^-$ to the carbon of urea or the guanidino-carbon of free or protein-bound arginine.

Tracer studies with the $^{14}CO_2$ method in pulmonates indicate that most have the capacity for *de novo* urea biosynthesis (Campbell and Speeg, 1968a; Speeg and Campbell, 1969; Horne and Barnes, 1970; Tramell and Campbell, 1972; Horne, 1973b, 1977b; Schmale and Becker, 1977; Meyer and Becker, 1980). Confirming tracer studies using L-[$^{14}C$]ornithine and [$^{14}C$]ureidocitrulline with tissues and whole animals indicated transfer of $^{14}C$ to arginine by the tissues of *O. lactea*, *H. aspersa*, and *S. oblongus*. Some caution must be exercised in the interpretation of these results because the transfer of $^{14}C$ from L-[$^{14}C$]ornithine to L-arginine can proceed through an exchange reaction involving an amidinotransferase (see Scheme 2). The incorporation of $^{14}C$ from the ureido group of citrulline into arginine and urea is therefore supporting evidence for the operation

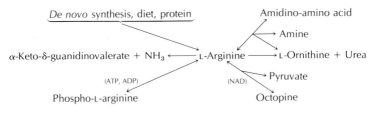

Scheme 2. Arginine catabolism and anabolism.

of only part of the pathway. Radiocarbon incorporation into urea from $^{14}CO_2$ was not found in the three prosobranch species investigated by Horne and Barnes (1970). Tracer studies using $H^{14}CO_3^-$ with some bivalve species (oysters, mussels, and clams) by a number of other investigators (see references given earlier on the general tracer studies) have been inconclusive because no special efforts were made to measure the radioactivity in arginine or urea. Recent studies in this laboratory (Ellis et al., 1981) indicated that no $H^{14}CO_3^-$ incorporation into arginine occurred in gill tissue (*M. demissus*) incubations; whereas there was significant incorporation into aspartate, glutamate, and alanine.

A complete complement of urea cycle enzymes has been found in the tissues of *O. lactea, H. aspersa,* and *S. oblongus* (Campbell and Speeg, 1968b; Campbell et al., 1972; Tramell and Campbell, 1970a, 1972), and in the tissues of *Bulimulus dealbatus* and *L. flavus* by Horne (1973b, 1977a,b). Schmale and Becker (1977) were able to demonstrate all of the activities except carbamoyl phosphate synthetase (CPS) in the tissues of *B. glabrata.* Results similar to those of Schmale and Becker (1977) were obtained by Andrews and Reid (1972) for some bivalve species (*Mytilus californianus, Anodonta kennerlgi, Saxidomus giganteus,* and *Composomyax subdiaphana*), by Freidl (1974) and Freidl and Bayne (1966) for *L. stagnalis,* and by Horne and Boonkoon (1970) for some other gastropod species. As with mammalian and amphibian tissues, the CPS and ornithine transcarbamylase activities are mitochondrial and the other activities, argininosuccinic acid synthetase (ASAS), argininosuccinic acid lyase, and arginase, are cytosolic (Speeg and Campbell, 1968b; Campbell et al., 1972). Both CPS and ASAS are present in low levels in the tissues, and both present assay problems. The most convincing assay methods involve use of $H^{14}CO_3^-$ in the CPS assay and [$^{14}C$]citrulline in the ASAS assay. The colorimetric procedures for these activities can be compromised by the presence of unknown chromogenic materials in the tissue extracts of molluscs and other invertebrates. Therefore, confirmation of the levels of these enzymes in bivalve tissues is required. The actual tissue levels and rates of biosynthesis of urea and arginine in some of the pulmonates change markedly with change in environment, water, and food availability (see Section VII).

The CPS in the mitochondria of snail hepatopancreas and the tissues of other invertebrates with known ability to synthesize arginine and urea via the urea cycle (Tramell and Campbell, 1970a, 1972) differs from the CPS-I in mammalian or frog liver mitochondria or the CPS-II in the cytosol of most vertebrate tissues (Ratner, 1973). CPS-I utilizes ammonia and requires *N*-acetylglutamate as an activator whereas CPS-II utilizes the amide nitrogen from glutamine and has no *N*-acetylglutamate requirement. CPS from the mitochondria of these invertebrates has been termed CPS-III in that it requires $N$-acetylglutamate as an activator and utilizes the amide nitrogen of glutamine rather than ammonia. The enzyme is inhibited by glutamine antagonists such as azaserine and $o$-carbamyl-L-serine. The $K_m$ for glutamine is about 2.5 m$M$, asparagine does not substitute for glutamine as a substrate, and the apparent $K_a$ for $N$-acetylglutamate is 0.25 m$M$.

## E. Arginine and Urea Catabolism

Arginine serves a central role in the metabolic economy of most molluscan species (Scheme 2). The guanidino group can be used in a series of transamidinase reactions for the biosynthesis of a number of putative phosphogens (see Chapter 3, this volume). The most common phosphogen in molluscan tissues is phosphoarginine and most of the non-protein-bound arginine in molluscan tissues exists as phosphoarginine. During anaerobic metabolism, some of the phosphoarginine is used for ATP biosynthesis and the released arginine can accumulate as octopine by reductive condensation with pyruvate. Octopine accumulation is most pronounced in the "fast" muscle tissues of molluscan species (cephalopods and some bivalves) that have significant amounts of this high glycolytic capacity tissue (Gäde, 1980).

Arginase (EC 3.5.3.1) catalyzes the hydrolysis of L-arginine to ornithine and urea and is widely distributed in the tissues of molluscan species (Baldwin, 1935; Gaston and Campbell, 1966; Hanlon, 1975; Baret et al., 1965, 1972; Hammen et al., 1962; Horne and Boonkoon, 1970; Andrews and Reid, 1972; Swami and Reddy, 1978; Bayne and Freidl, 1968). In general, different investigators report different tissue levels for the same species. For instance, with the hepatopancreas from $M. edulis,$ Andrews and Reid (1972) reported moderately high levels whereas Baret et al. (1965) and Hanlon (1975) found none. With $M. demissus$ tissues, Hammen et al. (1962), Bishop et al. (1981), and Hanlon (1975) report tissue levels ranging from 2.4 to 30 units/g, whereas the level of arginase was at or below the detectable limit in the assays by Gaston and Campbell (1966). With the pulmonates, the activity levels in the tissues vary greatly among individual snails of the same species. In a series of feeding–starvation–estivation experiments at different temperatures, arginase levels in the hepatopancreas of $O. lactea$ were highest in starved or estivating animals at 23°C and lowest in feeding animals and animals held at 3 or 32°C (Gaston and Campbell, 1966). With $L. flavus,$ the arginase levels were more variable between individual slugs than with either feeding or fasting slugs (Horne, 1977a,b). On the other hand, with $B. dealbatus,$ activity levels in the hepatopancreas increased almost linearly with the days of estivation (200 to 5000 units/g in 15 days) (Horne, 1973b). In general, the tissue levels in pulmonate snail tissues are very much higher (100–50,000 units/g wet tissue) than the levels in bivalves and prosobranch gastropods (below detectable levels to 60 units/g wet tissue). Activity levels in cephalopod tissues are quite variable (Needham, 1935; Gaston and Campbell, 1966).

Arginase from the hepatopancreas of *O. lactea* has been purified by Campbell (1966) and the properties compared to the arginase from rat liver. Both rat and snail arginase activities migrate as double bands during electrophoresis. The migration of the snail activities indicated that they had a considerably lower isoelectric point than the rat liver enzymes and that both species had isozymes of arginase. The mixed isoarginases from *O. lactea* will hydrolyze D-arginine and L-canavanine at about 10% the rate with L-arginine; no activity was seen with L-homoarginine or L-$\alpha$-amino-$\gamma$-guanidinobutyrate. Baret et al. (1972) have purified the individual isoarginases from the hepatopancreas of *H. pomatia* and *H. aspersa* and compared their kinetic properties. With the isoarginases from *H. pomatia* and *H. aspersa*, L-homoarginine was a reactive substrate with isozyme I but not isozyme II, and the individual isoarginases had slightly different $K_m$ values for L-arginine of about 12 m$M$ and 8–9 m$M$ with the isoarginase I and isoarginase II, respectively (Baret et al., 1972). Baret et al. (1972) find that the mixed-isozyme preparations show substrate inhibition at high substrate concentrations ($>100$ m$M$), whereas the individual isozymes do not show substrate inhibition. Campbell (1966) found substrate inhibition above 150–170 m$M$ with the *O. lactea* mixed-isozyme preparation.

The free energy of activation was 8788–9138 cal/mol for the *O. lactea* (mixed isozymes) activity (Campbell, 1966). Slightly different activation energies between 7700 and 8100 cal/mol were obtained for the individual isoarginases from *H. pomatia* and *H. aspersa* (Baret et al., 1972). Optimal activity was seen at pH 9.5–10 for all these activities and all were strongly activated by $Mn^{2+}$ or $Co^{2+}$. Campbell (1966) finds a change in relative electrophoretic mobility after heating in the presence of $Mn^{2+}$ and postulates conversion of a proarginase to an active arginase with the binding of the metal ion ($K_m = 5.6 \times 10^{-5} M$). Baret et al. (1972) find some differences in the degree and pH dependency of the metal ion activation process between the isoarginases of *H. pomatia* and *H. aspersa*.

L-Ornithine and L-lysine were strong competitive inhibitors of the *O. lactea* arginase (mixed isozymes) with respective $K_i$ values of 0.72 and 0.6 m$M$ (Campbell, 1966). Reddy and Baby (1976) reported similar types of inhibition for the arginase activity in *Ariophanta legulata* and found that valine, isoleucine, and leucine can be strong arginase inhibitors. In other studies, high concentrations of thiol-reactive reagents were inhibitory and thiol compounds (glutathione, cysteine, etc.) tend to activate or enhance the activity (Campbell, 1966; Reddy and Baby, 1976).

Reddy and Campbell (1968, 1970), in estimating the molecular weights of the arginase from the hepatopancreas of *O. lactea* and *H. aspersa,* find values of 244,000 and 232,000, respectively. These molecular weights are twice that of the tetrameric rat liver enzyme, eight times that of the monomeric earthworm gut arginase, and approximately the same as the octomeric activity found in the land planarian, chicken, and lizard liver, *Neurospora,* and the silkworm fat body. Pulmonates appear to have the octomeric arginase.

The conversion of arginine to α-keto-γ-guanidovalerate (KGV) by the L-amino acid oxidase has been discussed already. Oxidative decarboxylation of KGV yields γ-guanidinobutyrate. Baret et al. (1965) have found an amidinase activity in the hepatopancreas of a number of molluscan species that will hydrolyze γ-guanidinobutyrate to γ-aminobutyrate and urea. The γ-guanidinobutyrate amidinohydrolase and the arginase activities from the hepatopancreas of *H. pomatia* can be separated by chromatography on DEAE-cellulose and by heat lability. Additionally the arginase is cytosolic (Campbell, 1966; Gaston and Campbell, 1966; Porembska et al., 1968), whereas the γ-guanidinobutyrate amidinohydrolase is mitochondrial (Porembska et al., 1968). Except for the studies of Baret et al. (1965), there are no confirming observations on the presence of γ-guanidinobutyrate amidinohydrolase in species from other molluscan classes. Tracer studies using L-[U-¹⁴C]arginine with the hepatopancreas from *H. aspersa* and *O. lactea* (Campbell and Speeg, 1968a) and from *M. demissus* (Greenwalt, 1981) indicated little or no synthesis of γ-aminobutyrate from L-arginine. The functional significance of γ-aminobutyrate synthesis from L-arginine in molluscan tissues is therefore questionable. Tracer studies indicating conversion to ornithine, proline, and glutamate are discussed in Section IV,F.

Urease activity has been reported in the tissues of a number of invertebrate species (Campbell and Bishop, 1970; Campbell et al., 1972; Hanlon, 1975; Loest, 1979b). A continuing problem is the assurance that the activity is a gene product of the molluscan species and not of a microbial symbiont. Fetal mammals and germ-free chicks and rats have no urease activity in their gut tissue, whereas conventionally fed mammals and birds have considerable urease activity associated with the gut, presumably with the flora and microorganisms (possibly intracellular) inhabiting the brush border (Delluva et al., 1968; Rahmen and Decker, 1966). The molluscan tissue origin of the urease in the tissues of *O. lactea* and *H. aspersa* was established with reasonable certainty by Speeg and Campbell

(1968a). Many urease assays involving the measurement of ammonia formation from urea added to cell-free tissue extracts can give ambiguous results unless the activity is present at high levels. An improved assay method using [$^{14}$C]urea has been developed by Campbell's group (McDonald et al., 1972).

The highest urease levels (20–200 units/g unit tissue) are found in the tissues of some pulmonate snails (Campbell et al., 1972; Loest, 1979b), whereas lower levels (0–10 units/g) are reported in tissues from some other molluscan species including some pulmonates (Campbell and Bishop, 1970; Andrews and Reid, 1972; Campbell and Boyan, 1976; Horne and Boonkoon, 1970; Hanlon, 1975; Swami and Reddy, 1978). Those with significant amounts of urease—*O. lactea, H. aspersa,* and *H. pomatia* (Campbell et al., 1972; Tramell and Campbell, 1972; Speeg and Campbell, 1969)—have little or no urea in their excreta. In those with small amounts or no measurable urease activity—*B. glabrata* (Schmale and Becker, 1977), *B. dealbatus* (Horne, 1973a,b); *Achitina fulcata* and *Entosphenus tridentatus* (Horne and Boonkoon, 1970); *L. flavus* (Horne, 1977a,b); and *L. stagnalis* (Freidl, 1974, 1979)—urea can account for a major fraction of the nitrogen in the excreta. The relative amount of urea nitrogen in the excreta varies with diet, estivation, starvation, water availability, and degree of parasitic infestation (see Section VII).

The urease in the tissues of *O. lactea* and the other snails is cytosolic (Campbell et al., 1972). The enzyme purified from *O. lactea* hepatopancreas had a molecular weight of about 260,000, contained nickel as a coenzyme, and had a $K_m$ for urea of 0.11 m$M$ at the optimal pH (8.5) (McDonald et al., 1980). The amount of nickel–snail enzyme molecule and the subunit structure of the enzyme is uncertain. The jackbean urease is also a nickel metalloprotein (Dixon et al., 1975). The $K_m$ for urea is similar to the $K_m$ reported for the lugworm urease (Cooley et al., 1976), but is much lower than the $K_m$ values for bacterial or plant ureases (McDonald et al., 1980). This snail urease will hydrolyze hydroxyurea and is strongly inhibited by acetohydroxamic acid and similar hydroxamates. Acetohydroxamate forms a complex in the active sites of the enzyme ($K_i$ of about $10^{-6}$ $M$). The reaction of acetohydroxamate is fairly specific for urease and has been used to block urea catabolism *in vivo* in a variety of studies on ammonia production from urea, and to evaluate urea synthesis and arginine turnover in snails (Speeg and Campbell, 1969; Campbell et al., 1972; Tramell and Campbell, 1972; Campbell and Boyan, 1976).

The urease from bean seeds is extremely sensitive to sulfhydryl-reactive reagents and heavy metal ions. Although the urease activity from *O. lactea* was not inhibited by relatively high concentrations of thiol-reactive organic reagents, silver and mercurous ions inhibited strongly at very low concentrations (5–100 $\mu M$). These results may reflect some interference with the action of the nickel ions or possible complexing by specific basic residues (i.e., histidine) rather than a specific reaction with cysteine residues (McDonald and Campbell, 1980).

*H. pomatia, H. aspersa,* and *O. lactea,* with high urease activity and little or no urea in their excreta, produce variable amounts of gaseous ammonia, which may be derived from urea (Speeg and Campbell, 1968a). The urea formed by action of the urea cycle or from arginine released from proteins or phosphoarginine is hydrolyzed by the relatively high urease levels in the tissues. The rate of urea decomposition could be reduced considerably by injection of the urease inhibitor acetohydroxamate (AHA) (Speeg and Campbell, 1969). However, the snails tended to concentrate the injected AHA in the kidney so that it had only a temporary effect on urease activity in the whole animals. With isolated tissue incubations, AHA blocked urease activity and urea accumulation from citrulline, and arginine could be measured easily. In the whole animals, the addition of AHA reduced ammonia production to a considerable extent, which resulted in an inhibition of the shell-regeneration process and a reduction in the $^{14}CO_2$ exchange in the $CaCO_3$ of the shell (Speeg and Campbell, 1968a; Campbell and Speeg, 1969; Campbell and Boyan, 1976). From these and other data Campbell and Speeg (1969) have proposed that ammonia production is linked to the calcification process in calcifying tissues by acting as a buffer to conjugate protons produced during the formation of $CaCO_3$.

## F. Ornithine, Proline, and Glutamate

As noted already, most molluscs can produce ornithine from arginine by action of arginase. Tracer studies with [U-$^{14}$C]arginine and [U-$^{14}$C]ornithine with snail (*H. aspersa, O. lactea,* and *Bulimulus alternata*) hepatopancreas tissue by Campbell and Speeg (1968a) indicated rapid transfer of label to ornithine, proline, glutamate, glutamine, aspartate, and alanine. Given the presence of ornithine aminotransferase (Table III) in these tissues and the discovery of pyrroline-5-carboxylate (P-5-C) reductase in a slug (Greenberg, 1962), an early version of the pathway for the metabolism of these amino acids was postulated (Campbell and Bishop, 1970) (Scheme 3).

In a series of enzyme and $^{14}$C-tracer studies using gill tissue from the ribbed mussel, *M. demissus,* by Bishop et al. (1980), radiolabel from [U-$^{14}$C]arginine and [U-$^{14}$C]ornithine was transferred to ornithine, proline, glutamate, aspartate, and alanine. Addition of aminooxyacetic acid blocked transfer of label from arginine and ornithine to proline, gluta-

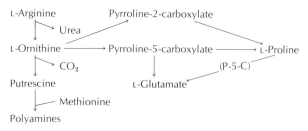

**Scheme 3.** Ornithine and proline metabolism.

mate, aspartate, and alanine, and caused an accumulation of label in ornithine. Addition of arsenite ($\alpha$-ketoacid dehydrogenase antagonist) blocked
transfer of label from arginine and ornithine to aspartate and alanine,
and caused an accumulation of label in proline, glutamate, and ornithine. Radiocarbon from [U-$^{14}$C]pyrroline-5-carboxylate and [U-$^{14}$C]pyr-
roline-2-carboxylate (P-2-C) was transferred to $CO_2$, glutamate, and proline but not ornithine or arginine. Because P-2-C is the product of L-amino
acid oxidase action on L-ornithine, this indicates that the ornithine transaminase or P-5-C pathway is the major route for ornithine catabolism and
that P-2-C, if formed *in vivo,* can be metabolized. With [U-$^{14}$C]proline, radiolabel was transferred to glutamate, aspartate, and alanine but not to
ornithine or arginine. Using [U-$^{14}$C]glutamate and [$^{14}$C]glucose, radiolabel
was incorporated into alanine and aspartate but not into proline, ornithine, or arginine. Enzyme assays indicated the presence of ornithine
aminotransferase, pyrroline-5-carboxylate reductase, an L-amino acid oxidase that would convert ornithine to pyrroline 2-carboxylate, a mitochondrial proline oxidase, and an indication of a weak pyrroline-5-carbox-
ylate dehydrogenase activity (Bishop et al., 1981).

Subsequent studies in this laboratory (J. M. Burcham and S. H.
Bishop, unpublished data), using [$^{14}$C]proline with gill tissues from other
bivalves, indicated a rapid catabolism in *C. virginica* and in *M. demissus*
but none in *M. edulis* or *Mercenaria mercenaria.* Zaba et al. (1978) found
no proline oxidase in the mitochondria from *M. edulis* hepatopancreas tissue, whereas Ballantyne et al. (1981) reported proline oxidase activity in
cuttlefish mantle muscle mitochondria.

This lack or low level of proline and ornithine biosynthesis from glutamate by the tissues of *M. demissus* has also been noted by Baginski (Baginski, 1978; Baginski and Pierce, 1978). In the [$^{14}$C]bicarbonate-tracer
experiments, Campbell and Speeg (1968a) report weak incorporation of
$^{14}CO_2$ into ornithine isolated from blood, kidney, lung, and reproductive
tract and into proline isolated from the blood of snails (*O. lactea*). With
isolated hepatopancreas tissue of *O. lactea, H. aspersa,* and *B. alternata*

(*B. dealbatus*), no [14]C from [[14]C]bicarbonate was detected in either proline or ornithine (Campbell and Speeg, 1968a). With *S. oblongus* no [14]C was incorporated into ornithine from [[14]C]glutamate, [[14]C]bicarbonate, or [[14]C]glucose with whole-body injections or hepatopancreas tissue incubations (Tramell and Campbell, 1972). With the studies on the other species, incorporation of [14]C label from [14]$CO_2$ or [[14]C]glucose, or glutamate into proline and ornithine is absent or not mentioned (see earlier references).

The general conclusion from the two or three studies that have investigated ornithine and proline metabolism specifically would be that most molluscan species cannot synthesize much if any proline or ornithine from glutamate or glucose, but that most can probably produce both from arginine. The missing enzyme in the pathway from glutamate to proline or ornithine seems to be the ATP-dependent dehydrogenase that converts glutamate to glutamic semialdehyde (pyrroline-5-carboxylate). The capacity for proline catabolism may vary with the species in that some may lack the proline oxidase system.

Glutamate decarboxylase (EC 4.1.1.15) has not been detected in nerve or other tissues of molluscan species. Although γ-aminobutyric acid (GABA) has been shown to be a neurotransmitter in crustaceans and vertebrates (Gershenfeld, 1973; Osborne, 1976b), conclusive evidence that it has a similar function in nervous systems of molluscan tissue is lacking. For instance, GABA levels in ganglia from *H. pomatia*, *H. aspersa*, *Buccinum undatum*, and the octopus, *E. cirrhosa*, were between 0.7 and 1.3 m$M$, which is 10 to 20 times lower than the levels in crustacean ganglia (Osborne, 1971; Osborne et al., 1971). Tracer studies using [[14]C]glutamate with ganglionic tissues from *H. pomatia* and *M. edulis* (Huggins et al., 1967; Osborne et al., 1971), and other tissues from *M. demissus* (Greenwalt, 1981) and *M. edulis* (de Zwaan et al., 1975), indicated an absence of GABA biosynthesis from glutamate. In other studies using [[14]C]amino acids, [14]$CO_2$, or [[14]C]glucose with tissues from *H. aspersa*, *O. lactea*, *H. pomatia*, *M. edulis M. demissus*, *C. virginica*, and *C. gigas*, glutamate is labeled strongly, but there is no mention of GABA labeling (Greenwalt, 1981; Kerkut et al., 1969; Campbell and Bishop, 1970; Campbell and Speeg, 1968a; Awapara and Campbell, 1964; Collicutt and Hochachka, 1977; de Zwaan, 1977; Bishop et al., 1981).

Ornithine decarboxylase (EC 4.1.1.17) activity is apparently present in the tissues of many molluscs. The product, putrescine, is used in polyamine (spermine and spermidine) biosynthesis and has been found in the tissues of *H. pomatia* and some other molluscs at low levels (Gould and Cottrell, 1974). The polyamines are thought to play a regulatory role in gene expression during growth and differentiation (Tabor and Tabor, 1976). Preliminary evidence for putrescine, spermidine, and spermine biosynthesis has been obtained by Greenwalt (1981) in a series of [$^{14}$C]orni-thine and [$^{14}$C]arginine-tracer studies with tissues from *M. demissus*.

## G. Serine, Glycine, and Threonine

Although the mechanisms for serine and glycine biosynthesis and ca-tabolism have not been studied in any detail in molluscan species, $^{14}$C-tracer studies indicate slow turnover in the tissues of most gastropod and bivalve species. Both compounds are precursors for purine biosynthesis in pulmonates (discussed in Section IV,L). The metabolic patterns seen in the gastropods and bivalves are discussed here. Radiolabeling of threo-nine has not been observed in any of the studies to be described.

Allen and Kilgore (1975) report very modest $^{14}$C incorporation into gly-cine and serine 71–96 h after injection of [U-$^{14}$C]glucose into *Haliotis ru-fescens*. No radiolabeling of glycine or serine from [U-$^{14}$C]glucose was ev-ident in short-term incubations (2 h) of hepatopancreas tissue from *L. saxatilus* (McManus and James, 1975b). With the pulmonates, however, there was reasonably strong incorporation of radiolabel from [U-$^{14}$C]glu-cose into both glycine and serine with relatively short-term incubations of hepatopancreas tissue from *O. lactea* (Campbell and Bishop, 1970). Both glycine and serine were labeled after [$^{14}$C]bicarbonate injection into *O. lactea;* when the hepatopancreas tissues of *O. lactea* and *H. aspersa* were incubated with [$^{14}$C]bicarbonate (Campbell and Speeg, 1968a). Tramell and Campbell (1972) report strong labeling of free glycine from [$^{14}$C]bicarbonate by the hepatopancreas of *S. oblongus*. Although these authors do not mention incorporation into free serine, they do report radiolabeling of both protein-bound serine and glycine.

With bivalves, Hammen and Wilbur (1959) find labeling of serine from [$^{14}$C]bicarbonate with oyster (*C. virginica*) mantle tissue but do not men-tion labeling of glycine. Following prolonged incubation (90 h) of ship-worm (*Bankia setacea*) mantle tissue with [$^{14}$C]glucose, incorporation of $^{14}$C was greatest in serine, which was 10 times the amount in glycine and about twice that in alanine and aspartic acid (Trytek and Allen, 1980). Short-term (2–4 h) incubations of gill tissue from *M. demissus* with [U-$^{14}$C]glucose indicated only a small amount of $^{14}$C in glycine and serine compared to alanine, aspartate, and glutamate, whereas no $^{14}$C from [U-$^{14}$C]alanine was detected in glycine or serine (Greenwalt, 1981). de Zwaan and van Marrewijk (1973), using mussels (*M. edulis*) injected with [U-$^{14}$C]glucose, found no radioactivity in free tissue glycine or serine with long-term incubation (48 h). With mussels (*M. edulis*) held for 6 h in an atmosphere containing $^{14}$CO$_2$, Ahmad and Chaplin (1979) observed strong radiolabeling of alanine, glutamate, aspartate, and succinate but no incorporation into glycine or serine.

These studies indicate a capacity for serine biosynthesis in some gastropod and bivalve species but do not indicate the pathway. In vertebrates and microorganisms, serine can be synthesized from the trioses or from glycine. The first pathway involves transamination of hydroxypyruvate by alanine to form serine and pyruvate, or transmination of phosphohydroxypyruvate by glutamate, which is followed by phosphatase action on the resulting phosphoserine. The second pathway involves hydroxymethylation of glycine in the serine hydroxymethyltransferase reaction.

In the transaminase survey by Sollock et al. (1979) and by Falany and Freidl (1981) with pulmonate snail hepatopancreas and bivalve hepatopancreas, serine transamination was assayed using $\alpha$-ketoglutarate rather than pyruvate (Rowsell et al., 1979), and little or no activity was found. There are no other studies on the enzymology of serine biosynthesis from the trioses. With regard to serine catabolism, serine dehydrase activity has been found in the few molluscs investigated (see Section II,F). Serine conversion to pyruvate by this enzyme has been confirmed by *in vivo* studies using [U-$^{14}$C]serine with gill tissue pieces from *M. demissus* (Ellis et al., 1981).

Another route for both serine and glycine biosynthesis or catabolism is through the serine–glycine exchange reaction by serine hydroxymethyltransferase (SHMT). In vertebrates, this enzyme is of primary importance in the generation of C-1 fragments for various biosynthetic reactions (purine biosynthesis, etc.). Reasonably high levels of SHMT activity have been found in tissues of *Pectin caurinus* by Whiteley (1960) and in the gills of *M. demissus* by Ellis et al. (1981). The enzyme is readily reversible and can provide a mechanism for serine biosynthesis if the methylene-THFA is available. Molluscs have modest levels of the formyl-THFA synthetase and other C-1 THFA enzymes (Whiteley, 1960). Tracer studies using [$^{14}$C]glycine with gill tissue from *M. demissus* indicate rapid labeling of serine with lesser amounts of $^{14}$C in aspartate, alanine, and glutamate (Ellis et al., 1981). With clams (*R. cuneata*) held anaerobically, all of the radioactivity from injected [2-$^{14}$C]glycine remained in a single spot with the $R_f$ of glycine on a one-dimensional paper chromatogram (Philley, 1978). Because this spot was large and had a leading edge, it could be a mixture of glycine and serine. In any event, the lack of $^{14}$CO$_2$ production from [2-$^{14}$C]glycine in short-term incubations is in agreement with similar studies with gill tissue from *M. demissus* by Ellis et al. (1981).

Schirch and Gross (1968) found that the serine hydroxymethyltransferase has threonine aldolase activity and can cleave threonine to glycine and acetaldehyde. This aldolase pathway may be the primary catabolic route for threonine in mammalian liver (Bird and Nunn, 1979). Using SHMT partially purified from gill tissue of *M. demissus,* L. L. Ellis (unpublished data) finds a low level (1–5%) of threonine aldolase activity associated with the enzyme. Although the overall metabolism of threonine in molluscan tissue has not been evaluated, it could provide an alternate pathway for the synthesis of a small amount of glycine. There is no report of a threonine dehydrase activity in molluscan tissue and we have found none in the tissues of *M. demissus.*

With regard to glycine metabolism, glycine can be interconverted with serine (see earlier) for subsequent metabolism, or metabolized directly through the glycine synthase–oxidase pathway or by transamination to and from glyoxylate. Evidence summarized by de Zwaan (1977) indicates an absence of a glyoxylate cycle (dicarboxylic acid cycle) in the tissues of molluscan species. All species assayed have low levels of glycine or glyoxylate transaminase activities (see Section III on transaminases). It would appear that the capacity for glycine turnover by transamination may be limited by glyoxylate metabolism and availability. In recent studies on $^{14}CO_2$ production from [U-$^{14}$C]glyoxylate and [U-$^{14}$C]glycine with gill tissue pieces (*M. demissus*) indicated that the rate of glyoxylate catabolism was 10–40% of the rate of glycine catabolism (Ellis et al., 1981). Ellis et al. (1981) suggest that the route for glyoxylate catabolism may be through transamination to glycine.

The glycine synthease–oxidase multienzyme complex (Yoshida and Kikuchi, 1972) is of major importance to both glycine and serine metabolism in vertebrate liver. A glycine synthase–oxidase system has been detected in the mitochondria of the gills of *M. demissus* (Ellis et al., 1981). Exchange experiments using mitochondria isolated from gill tissue and gill tissue with [$^{14}$C]bicarbonate indicate an enhanced $^{14}$C incorporation into glycine with added serine. This enrichment is not inhibited by added glyoxylate but is blocked by arsenite and aminooxyacetic acid. Complementary studies with [U-$^{14}$C]glycine, [1-$^{14}$C]glycine, and [2-$^{14}$C]glycine indicated that 95% of the $^{14}CO_2$ formed from [U-$^{14}$C]glycine arose from the C-1 position, and that most of the C-2 of glycine could be trapped as the formyl derivative of dimedon. Addition of glyoxylate did not inhibit $^{14}CO_2$ production from [1-$^{14}$C]glycine in isolated gill mitochondria. Because the mitochondria showed the same pattern as results with tissue slices, the glycine synthase–oxidase system appears to be the major glycine catabolic pathway in *M. demissus.* This activity is apparently under metabolic control in that it is activated in tissues held at low salinities and inhibited in tissues held at high salinities. The existence of the glycine synthase–oxidase system has not been investigated in other molluscan species.

The general pathways for serine and glycine biosynthesis, catabolism, and interconversion are described in Scheme 4. The presence of several

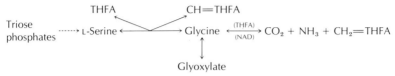

Scheme 4. Serine and glycine metabolism.

biosynthetic and catabolic options may explain some of the labeling patterns with the $^{14}C$-tracer studies. For instance, prolonged incubation with [$^{14}C$]glucose tends to enhance labeling of serine with a lesser amount in glycine, which would suggest a slow turnover of both amino acids. Presumably the $^{14}C$ from glucose would be channeled through the trioses to serine and then glycine. With [$^{14}C$]bicarbonate, $^{14}C$ could exchange directly into glycine through the glycine oxidase and, with long-term incubation, result in the observed strong incorporation into glycine and a lesser incorporation into serine through the SHMT step. The lack of $^{14}C$ incorporation into glycine from [$^{14}C$]alanine, aspartate, and glutamate and the glyoxylate dilution experiments with *M. demissus* tissues argues against the operation of a glyoxylate cycle and against glyoxylate as a major source of carbon for glycine biosynthesis. As mentioned before, these conclusions are made from very few data on a limited number of species but may provide some insight into the apparently confused biosynthetic patterns for glycine and serine in molluscs.

## H. Sulfur Amino Acids: Taurine and Methylated Amines

The few studies on the complex interconversions of the sulfur amino acids in molluscs have focused on taurine and isethionic acid (ISA) biosynthesis or the methyltransferases. The high concentrations of taurine and putative metabolites of taurine in the cells of many marine molluscs probably play a major role in regulating cellular osmotic pressure (Allen and Garrett, 1971a; Lange, 1972; Amende and Pierce, 1978; Norton, 1979). The studies with the $S$-adenosylmethionine-linked methyltransferases have emphasized the production and possible inactivation of putative neurotransmitters rather than methionine metabolism.

With regard to methionine, the $O$-methyltransferases and $N$-methyltransferases using $S$-adenosylmethionine have been found in several gastropod species, and label from [*methyl*-$^{14}$C]methionine is incorporated into several of the catacholamine metabolites (Cardot, 1979). Although formation of CH$_3$-THFA and subsequent transfer to homocysteine to form methionine has not been investigated, Ericson (1960a,b) has detected a betaine-homocysteine methyltransferase from the hepatopancreas of *Anodonta cygnea* that uses glycine-betaine as the preferred substrate. Shieh (1968), using [*methyl*-$^{14}$C]methionine, found no incorporation of label into choline of phosphatidylcholine of scallop (*Placopecten magellanicus*) phospholipid. Marine molluscs can accumulate a variety of methylated amines, particularly quaternary amines (for earlier literature, see Florkin, 1966; Florkin and Bricteux-Grégoire, 1972; Campbell and Bishop, 1970). These include $N$-methylpyridine and $N$-methylpicolinic acid, and some quaternary amines such as glycine-betaine in most marine clams, oysters, mussels, snails, and cephalopods; $\gamma$-butyrobetaine in *Conus*; tetraamine in the snail *Neptuna* and squid axoplasm; $\beta$-alanine-betaine in scallops and oysters (Abe and Kaneda, 1975a; Konosu and Hayashi, 1975); and trimethylamine oxide in cephalopods and some bivalves including scallops (Norris and Benoit, 1945). Sarcosine ($N$-methylglycine) has been found in gastropods, bivalves, and cephalopods (Campbell and Speeg, 1968a; Livingstone et al., 1979; Hanley, 1975). Other betaines such as taurobetaine and ulvaline (Abe and Kaneda, 1975b) are produced by marine algae and may accumulate in the tissues of some species after dietary intake. There is no conclusive evidence that these molluscs synthesize these betaines or choline, and there is no report of choline oxidase activity in molluscs. In general, it appears that molluscan species may have a limited capacity for methylation through the methyltransferases for neurotransmitter metabolism (discussed more fully in another chapter). Additionally, some species, possibly only freshwater species, may be able to salvage methyl groups from betaines and choline as suggested by Florkin and Bricteux-Grégoire (1972).

Allen and Awapara (1960) studied the fate of [$^{35}$S]methionine, [$^{35}$S]cysteine, and [$^{35}$S]taurine after injection into the bivalved molluscs *M. edulis* and *R. cuneata*. Radioactivity from [$^{35}$S]cysteine was found in cysteic acid, taurine, and cysteine sulfinic acid with *R. cuneata,* and in cysteic acid and taurine with *M. edulis.* Allen and Garrett (1972) report labeling of cysteic acid and taurine with [$^{35}$S]methionine in tissues of *Mya arenaria* and *M. edulis.* The chromatograms in both of these studies contained some unidentified radioactive compounds. Sulfate labeling from [$^{35}$S]cysteine was small compared to the labeling of taurine.

Incubation of squid axons and optic ganglia with [$^{14}$C]cysteine resulted in modest labeling of bicarbonate, cysteine sulfinic acid, cysteic acid, hypotaurine, and taurine, but no labeling of isethionic acid; most of the label was incorporated into other acid-stable metabolites (organic acids, amino acids, etc.) (Hoskin et al., 1975; Hoskin and Brande, 1973). Allen and Awapara (1960) do not report $^{35}$S incorporation into ISA in the bivalve tissues from the precursors they used. In more recent studies with squid axons, Hoskin and Kordik (1977) report incorporation of $^{35}$S from $H_2{}^{35}$S and [$^{35}$S]cysteine into taurine and ISA, which would seem to contradict the $^{14}$C-tracer studies. Hoskin and Kordik (1977) suggest that direct fixation of sulfide by rodanese followed by transsulfuration through an as yet unknown process may be involved in ISA biosyntheses. In any event, they conclude that taurine and L-cysteine are not precursor molecules in the biosynthesis of ISA in squids or other molluscs.

The apparent contradiction between the $^{35}$S-and-$^{14}$C-tracer studies by Hoskin and associates may result from the chemical reactivity of some of the sulfur compounds used. These "ambiguities" in the metabolism and chemistry of the sulfur amino acids have been summarized by Cavallini et al. (1979). The possibility for the formation of sulfite–sulfide adducts with the same chromatographic mobility as the products of interest would be a particular problem with $H_2{}^{35}$S-tracer experiments. With [$^{35}$S]cysteine, the sulfhydryl group can migrate, causing similar labeling or mislabeling patterns. Problems not considered by investigators thus far are the reactivity of cysteine sulfinic acid with the aspartate aminotransferase for eventual pyruvate production, and the apparent absence (in molluscs) of glutamate decarboxylase, which may be identical to the cysteine sulfinic acid decarboxylase (converts cysteine sulfinic acid to hypotaurine) (Wu, 1976; Cavallini et al., 1976). Although the pathway involving taurine production from the cysteine released during CoA turnover (Cavallini et al., 1976) has not been studied in molluscan tissues, cysteamine has been found in the tissues of *M. edulis,* and the oxidase converting cysteamine to taurine has been partially purified (Yoneda, 1967, 1968). In mammals, this cysteamine pathway appears to be a minor route of taurine biosynthesis (Awapara, 1976).

From these meager data the overall pattern indicates a modest capacity for taurine biosynthesis from L-cysteine and for transfer of sulfur from methionine to cysteine through cystathionine. Taurine synthesis from L-cysteine in the tissues of *R. cuneata* and the squid probably proceeds through the cysteine sulfinic acid–hypotaurine pathway, whereas in *M. areneria* and *M. edulis* it may proceed through decarboxylation of cysteic acid. Overall, the rate of taurine biosynthesis in the marine molluscs seems to be too slow to account for the high levels in the tissues. Much of the taurine in these marine species is probably obtained in the diet or by action of the gut flora on dietary foodstuffs. The similar low rates of taurine biosynthesis in rat heart and brain tissues has led to a similar explanation for the taurine levels in mammalian tissues (Huxtable, 1976, 1978 ;Chubb and Huxtable, 1978).

The catabolism of taurine in marine molluscs is very slow in isolated tissues but seems to proceed at a modest rate in whole animals (Greenwalt, 1981; Baginski and Pierce, 1978). The catabolic scheme found in bacteria involves a transamination of taurine to sulfoacetaldehyde followed by cleavage to sulfate and acetaldehyde (Kondo et al., 1971). In some bacteria, including the gut bacteria from mammals, the sulfoacetaldehyde can be reduced to ISA (Fellman et al., 1978, 1980). This pathway has been proposed (Jacobsen and Smith, 1968) as the major pathway for ISA biosynthesis in vertebrate tissues. However, recent studies with germ-free rats and mice (Fellman et al., 1978, 1980) using [2-$^3$H]taurine indicate a lack of conversion of taurine to ISA and an absence of the transaminase converting taurine to sulfoacetaldehyde. Fellman et al. (1980) find that taurine is aminoacylated and that one of these compounds cochromatographs with ISA in some chromatographic systems. Because there is some ISA and $^3H_2O$ produced after injection of [2-$^3$H]taurine into non-germ-free rats and mice, Fellman et al. (1978, 1980) conclude that ISA production in mammals probably results from ISA recovery after action of the gut bacteria on the taurine secreted with the bile to the gut. Awapara (1976) concludes in his review that, if present at all in mammals, ISA formation from taurine by isolated tissues can only account for a small fraction of the actual ISA accumulated.

From the studies with squids by Hoskin and with bivalves by Awapara (summarized earlier), it appears that molluscan tissues may also lack the ability to catabolize taurine and to synthesize ISA from taurine. Although the catabolic route for taurine is not established, it may involve decomposition by the microbial flora associated with most molluscs, and in some species, some taurine may be converted to ISA by this flora, then reabsorbed and stored in particular tissues. This model means that the gut and its microflora could serve as both the source and the site of the catabolism of taurine and ISA. Metabolism would then be regulated by the transport systems in the tissues and the associated flora. Experiments to resolve this dilemma are required if we are to determine the origin and fate of this important metabolite.

Glutathione (γ-L-glutamyl-L-cysteinylglycine) serves several important functions in animal cells. These include reaction with a variety of halogenated, nitro, or sulfated aromatic compounds in the glutathione S-thiotransferase (ligandin) reaction to ultimately give rise to the detoxified mercapturic acid compounds (Kaplowitz, 1980), in the γ-glutamyltranspeptidase reaction to form γ-glutamyl amino acids which may be important to the membrane transport of some amino acids (Meister, 1973), as an intercellular reducing agent to maintain the thiol groups of cysteine and the ferrous ion state of hemoglobin (McIntyre and Curthoys, 1980); and as a free radical "scavenger" associated with some peroxidase reactions (Flohé and Günzler, 1976). The role of glutathione in the ligandin reactions and as a reducing agent in the tissue of molluscan species has not been investigated. Low levels of γ-glutamyltranspeptidase activity have been found in gill and mantle tissue of oysters, mussels, clams, and scallops plus the visceral mass and foot muscle of *Patela vulgata* (limpet) (Glynn and Johnson, 1981). Although a possible role of the γ-glutamyl cycle in amino acid transport in molluscs has not been addressed directly, Weinrich (1979) reports γ-glutamylhistamine to be a major product of histamine metabolism in the ganglia of *Aplysia californica* in a series of tracer experiments. Double labeling of the γ-glutamylhistamine by both [$^{14}$C]glutamate and [$^{14}$C]histamine or [$^3$H]histamine indicate an active γ-glutamyltranspeptidase and a capacity for glutathione biosynthesis in *A. californica*.

## I. Histidine

Histamine production from histidine by histidine decarboxylase has been found in a number of molluscs. Reite (1972) has reviewed the earlier literature on histamine metabolism and role of histamine as a putative neurotransmitter. Histidine decarboxylase activity and histamine production in *Aplysia* seems to be restricted to a limited population of nerve cells (Weinreich et al., 1975; Weinreich and Yu, 1977; Weinreich and Rubin, 1981). Therefore, the absence of decarboxylase in the nonneural and some neural tissues of some species does not mean that the histamine production is absent or unimportant. Mettrick and Telford (1965) detected the activity in some species but not others, and Hartman et al. (1960) found considerable activity in the posterior salivary gland of *Octopus apollyan*.

Histamine catabolism has been studied in only two molluscan species. In a series of [$^{14}$C]histamine- and [$^3$H]histamine-tracer studies with ganglia from *A. californica*, Weinreich (1979) has demonstrated histamine conversion to γ-glutamylhistamine, imidazole acetic acid, methylimidazole acetic acid, *N*-acetylhistamine, and methylhistamine as major metabolites. Earlier studies by Huggins and Woodruff (1968) with ganglia from *H. aspersa* indicated conversion of histamine to imidazole acetic acid and two unknown compounds. One of these unknown compounds had histamine-like activity in a bioassay but was not methylhistamine or acetylhistamine. Weinreich (1979) suggests that one of these unknown compounds (substance C) may be γ-glutamylhistamine considering its chromatographic mobility, and that there may be differences in histidine catabolism in different molluscan species.

Histidine is a substrate for the L-amino acid oxidase that is found in some bivalved molluscs (see Section II,A). The product of this reaction, imidazole pyruvic acid and compounds resulting from the oxidation of imidazole pyruvic acid, such as imidazole acetic acid, hydroxymethylimidazole, and formylimidazole were found in the tissues of *M. edulis* by Thoai et al. (1954). In mammals, histidine is converted to urocanic acid through the histidase reaction (Kolenbrander and Berg, 1967). Murexine (urocanylcholine) has been found in hypobranchial gland extracts of some Muricids (prosobranchs); none has been detected in tissue extracts of a number of other molluscan species (Florkin and Bricteux-Grégoire, 1972; Campbell and Bishop, 1970). Synthesis of murexine from histidine has not been demonstrated in molluscs and there is no report of histidase activity in molluscan tissues.

## J. Aromatic Amino Acid Decarboxylations, Hydroxylases, Amine Oxidases, and Metabolism of Aromatic Amino Acids in Neural Tissues

The decarboxylases and amine oxidases for histidine and histamine plus tyrosine, tryptophan, and the hydroxylated aromatic amino acids and amines have been found in molluscan tissues and have been associated with the production and inactivation of neurotransmitters or venom components. This aspect of neurotransmitter production, release, and function has been reviewed recently by Cardot (1979) and Osborne (1976b), and is dealt with in other chapters in this monograph.

Using $^{14}$C-tracer and microfluorometric techniques in combination with specific inhibitors such as $\alpha$-methyl-$p$-tyrosine, Cardot and others with *H. pomatia* (Cardot, 1974, 1979, 1980) have shown conversion of phenylalanine to tyrosine and DOPA, and conversion of tyrosine and DOPA to dopamine, norepinephrine, octopamine, dihydroxylphenylacetic acid, and methoxydopamine in the ganglionic tissue. These studies indicate the presence of tyrosine hydroxylase (tyrosinase), dopamine $\beta$-hydroxylase, phenylalanine hydroxylase, DOPA-decarboxylase (aromatic amino acid decarboxylase), monoamine oxidase, catachol-$o$-methyltransferase, and catechol-$N$-methyltransferase in molluscan ganglia. Some of these activities have been found in the tissues of *B. glabrata* by Guchhait et al. (1980) and Mermel et al. (1981). Serotonin synthesis from hydroxytryptophan and tryptophan was first shown in ganglia of *Busycon canaliculata* by Mirolli (1968), and serotonin has been shown to be a substrate for the monoamine oxidase (Goldman and Schwartz, 1977; Mermel et al., 1981) in other molluscan species.

## K. Aromatic Amino Acid Metabolism in Nonneural Tissues

Melanin formation and protein tanning processes with tyrosinase and phenoloxidases have been studied with regard to pigmentation in shells, melanophores and melanocytes, ink production, shell formation, bivalve hinge ligament assembly, and byssus thread production. These special aspects of protein processing will be developed in other chapters in this series.

A polyphenoloxidase activity from gills of M. edulis was first described by Blaschko and Milton (1960), Blaschko and Levine (1960), and then by Aiello (1965) as a "hydroxyindole oxidase." It causes hydroxylation and possibly subsequent polymerization of a number of aromatic amines and amino acids (specificities summarized in Campbell and Bishop, 1970). Kampa and Peisach (1980), using a homogeneous preparation of this enzyme, report a native molecular weight of 220,000 with seven to nine subunits plus one heme iron and one nonheme copper ion per subunit. The substrates are the phenolic or indolic compounds and molecular oxygen; the products seem to be the hydroxylated or polymerized phenolic or indolic compounds and peroxide. As with other phenoloxidase-like enzymes, the activity is strongly inhibited by copper-chelating agents and carbon monoxide, and weakly inhibited by cyanide and azide. Preliminary studies on activation by peroxide and ascorbate need further investigation. Although Kampa and Peisach (1980) feel that the enzyme is probably a monooxygenase, properties similar to the tryptophan 2,3-dioxygenase, and the weak epr spectrum for the heme iron (possible ferric state) lead to the suggestion that it may resemble cytochrome $a_3$ in some respects. In any event, the enzyme behaves as a phenoloxidase with tyrosinase, hydrotryptophan hydroxylase, and DOPA-oxidase activities.

Prota et al. (1981) have examined the tyrosinase activity in the ink glands of three cephalopods (O. vulgaris, S. officinalis, and L. vulgaris). The whole ink in the gland sacs contains active soluble enzyme that is not bound to the melanin or other particles in the ink. The enzyme activities from the three species had different molecular weights (205,000–125,000). Dihydroxyphenylalanine oxidase and tyrosinase activity seemed to be associated with the same protein and both substrates ultimately formed melanin and dopachrome. Hydroxyindole-containing compounds were not tested as substrates. The enzyme is apparently secreted into the ink sac by the gland and forms melanin from substrates secreted into the sac.

With no evidence to the contrary, the activity seems to be similar to that found in *M. edulis* gills.

Another phenoloxidase activity from mantle tissue and the periostracum of *M. demissus* has been partially characterized by Waite and Wilbur (1976). The enzyme is apparently produced by the mantle and periostracal cells as a proenzyme that can be converted to two active forms of differing molecular weight by chymotrypsin treatment. Some of the enzyme in the periostracum is apparently cross-linked in the protein matrix and is partially active without chymotrypsin treatment. The active mantle enzyme, however, is associated with a large particle, requires chymotrypsin treatment for activation, and is solubilized as an active enzyme in 0.5% sodium dodecyl sulfate. Waite and Wilbur (1976) surmise that after secretion and activation by an endogenous protease, the oxidase plays a primary role in cross-linking and tanning the proteins of the periostracum. This production of phenoloxidases in proenzyme form followed by protease activation has been found for some insect cuticular and frog skin phenoloxidases.

The chymotrypsin-activated periostracal enzyme has a curious substrate specificity. It is active with dihydroxycatechol and alkylcatechol substrates but shows no activity with tyrosine, tyramine, or norepinephrine. It was not tested with the hydroxyindole-type substrates. This activity is blocked by low-level cyanide ($K_i = 10\ \mu M$) or diethyldithiocarbamate ($K_i = 15\ \mu M$), and addition of excess cupric ion restored activity to the diethyldithiocarbamate-inhibited enzyme. Although this enzyme appears to have some similarities to the *M. edulis* gill activity and the cephalopod ink gland enzymes, the lack of activity with tyrosine or tyramine and inhibition by low levels of cyanide set it apart from these activities. Chymotrypsin is not the native endogenous protease activator. Because all specificity measurements were performed using the chymotrypsin-activated enzyme, possibly chymotrypsin treatment altered the normal substrate specificity. Investigations as to the nature of this protease and the substrate specificity of the naturally activated oxidase in the periostracum would be of considerable interest.

Other aspects of tryptophan and tyrosine catabolism by pulmonate snails have been discussed previously (Campbell and Bishop, 1970; Florkin and Bricteux-Grégoire, 1972).

## L. Purine Metabolism

Uric acid can comprise a major fraction of the nitrogen in the excreta of some pulmonates and gastropods with a terrestrial or semiterrestrial existence (Needham, 1935). The correlation between this "uricotelism" and the terrestrial or semiterrestrial existence has weakened somewhat in recent years, but has prompted an investigation of purine biosynthesis in pulmonates (nitrogen excretion is discussed in Section VII).

A series of $^{14}C$-tracer studies indicate that the biosynthetic route for purines in pulmonates is the same pathway as that found in other animals and microorganisms. The most complete studies are those by Lee and Campbell (1965) with *O. lactea* and by Gorzkowski (1969) with *H. pomatia*, who used specifically labeled $^{14}C$-precursor molecules, then chemically degraded the isolated uric acid to determine the specific incorporation into the individual carbon atoms of the uric acid. Radiocarbon from formate, glycine, and bicarbonate caused $^{14}C$ enrichment into C-2 and C-8, into C-4 and C-5, and into C-6, respectively. Using [3-$^{14}C$]serine injected into snails (*O. lactea*), Lee and Campbell (1965) found specific labeling of C-2 and C-8, the same carbons labeled by formate; whereas [2-$^{14}C$]-4-amino-5-imidazole carboxamide enhanced labeling of C-8. Azaserine and diazooxynorleucine, inhibitors of L-glutamine utilization in amidotransferase reactions, blocked uric acid synthesis in *O. lactea* (Lee and Campbell, 1965; Speeg and Campbell, 1968a). By using low levels of azaserine, which inhibit the glutamine-dependent formylglycine-amide ribotide amidotransferase more strongly than the phosphoribosyl pyrophosphate glutamine amidotransferase, Lee and Campbell (1965) were able to detect an accumulation of $^{14}C$ from [$^{14}C$]formate in the intermediate compound, formylglycineamide ribotide. These studies and other tracer experiments using [$^{14}C$]glycine by Jezewska et al. (1964) with *H. pomatia*, by Horne and Beck (1979) with *L. flavus*, and by Speeg and Campbell (1968b) with *O. lactea*, and using [$^{14}C$]formate with *H. aspersa* by Clark and Rudolph (1979), indicate a *de novo* biosynthetic capacity for all of the purine bases (guanine, adenine, hypoxanthine, uric acid, and xanthine) in these pulmonate species. There are no similar studies with molluscan species from the other classes or subclasses.

Conway et al. (1969) found accumulation of uric acid in developing embryos of a pulmonate pond snail, *Lymnaea palustris*, and strong $^{14}C$ labeling of uric acid and xanthine after injection of [8-$^{14}C$]hypoxanthine and [2-$^{14}C$]uridine. Because no $^{14}C$ from [$^{14}C$]urea was incorporated into uric acid, they propose that the incorporation of label into the C-2 of uric acid from [2-$^{14}C$]uridine was by an as yet undefined pathway and not by degradation of [2-$^{14}C$]uracil to $^{14}CO_2$ followed by the incorporation of the $^{14}CO_2$ into the uric acid. Studies by Freidl (1979), summarized earlier, indicate an absence of urease in lymneids, thereby accounting for the lack of $^{14}C$ transfer from urea into uric acid. Studies summarized later (under pyrimidine metabolism) indicate a capacity for pyrimidine catabolism in most molluscs. Most likely the $^{14}CO_2$ formed from the catabolism of [2-$^{14}$C]uracil was incorporated into uric acid through the *de novo* biosynthetic route.

The only enzymes in the *de novo* pathway that have received any attention are the phosphoribosyl pyrophosphate amidotransferase, which has been found in *L. flavus* (Horne and Beck, 1979), and enzymes of the purine nucleotide cycle (adenylosuccinate synthetase, adenylosuccinate lyase, and AMP deaminase), which have been found in the hepatopancreas of *H. aspersa* (Campbell and Vorhaben, 1979).

The older literature on the enzymes involved in purine catabolism will not be reviewed here (see Florkin, 1966; Campbell and Bishop, 1970). A 5'-nucleotidase activity has been found in the octopus and some bivalve species (Ishida et al., 1969; Umemori, 1967; Umemori-Aikawa, 1971; Arch and Newsholme, 1978). The inosine-guanosine phosphorylase and adenosine phosphorylase activities, and the hypoxanthine-guanine phosphoribosyltransferase have been partially purified from hepatopancreas tissue of *H. pomatia* by Barankiewicz and Jezewska (1973, 1976).

Adenosine deaminase (ADA) has been found in the tissues of all species investigated (Arch and Newsholme, 1978). Campbell and Boyan (1976) find reasonably high levels of ADA in the tissues of those gastropod species with low urease activity and have suggested that this enzyme may play a role in the production of ammonia for buffering of hydrogen ions in the calcification process in some molluscs. These observations on ADA and urease activity have been extended by Loest (1979b). ADAs from the tissues of various bivalves seem to have very similar molecular weights (130,000–140,000) but show variations in relative stability and in the pH for optimal activity (pH 4–7) (Harbison and Fisher, 1973a,b, 1974; Aikawa, 1966; Aikawa et al., 1977; Umemori-Aikawa and Aikawa, 1974). On the other hand, the ADA in pulmonate tissues seemed to be most active between pH 6 and 9 (Loest, 1979b). The relative activities in the tissues of the pulmonates ranged from less than 10 to over 1600 units/g wet tissue when assayed between pH 8.5 and 9 (Loest, 1979b), whereas the activities in the bivalves ranged between 300 and 600 units/g wet tissue when assayed between pH 5 and 6. The ADAs from bivalve tissues seem to be equally active with adenosine and deoxyadenosine. In a survey with an antibody to beef intestinal mucosal ADA, Lee et al. (1973) report reasonably strong inhibition of the ADA activity from the tissues of other mammals and little or no inhibition of the activity in the tissues of a toad, chicken, fish, or scallop.

The studies just summarized indicate that molluscs can convert purine nucleotides to purine nucleosides, and the purine nucleosides to individual purine bases, and can also salvage the purine bases as purine nucleotides. Conversion of adenine to hypoxanthine by adenine deaminase has not been found in most tissues, and the presence of guanase in some pulmonate species has been questioned (Clark and Rudolph, 1979). Early reports of the presence of these two enzymes in molluscan tissues have been reviewed (Florkin, 1966; Campbell and Bishop, 1970). The enzyme converting hypoxanthine and xanthine to uric acid (xanthine oxidase or xanthine dehydrogenase) has been found in the tissues of most prosobranchs, pulmonates, and bivalves (Florkin, 1966; Florkin and Bricteux-Grégoire, 1972; Campbell and Bishop, 1970; Swami and Reddy, 1978). From the early studies on the uricolytic pathway (see Scheme 5), one can conclude that the bivalves and some prosobranchs tend to have the

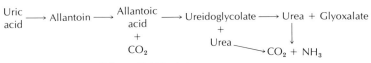

Scheme 5. Uricolytic sequence.

enzymes in the uricolytic sequence whereas the pulmonates and prosobranchs, which accumulate uric acid, are deficient at the uricase step (Campbell and Bishop, 1970). There are precious little data to support this general theme. The experiments by Duerr (1967, 1968) on the relative amount of urea, ammonia, and uric acid excreted by some prosobranchs would seem to suggest a wide variation in uric acid biosynthetic and uricolytic capacity among the various prosobranch species. Among bivalves, Andrews and Reid (1972) have found uricase, allantoinase, allantoicase, and urease in the hepatopancreas tissue from *A. kennerlyi, Mytilus californianus, S. giganteus,* and *Compsomyax subdiaphana.* We could find no reports on the ureidoglycolase activity but it is presumed to be present in those molluscs that can degrade uric acid to urea or to $CO_2$ and ammonia.

## M. Pyrimidine Metabolism

The pyrimidine biosynthetic pathway scheme has not been investigated in any detail in any molluscan species. In tracer studies with snails (*H. pomatia*), $^{14}C$ from labeled bicarbonate, aspartate, and orotate was incorporated into the CMP and UMP from RNA of the hepatopancreas (Porembska et al., 1966). This study indicated a general ability for *de novo* biosynthesis of pyrimidine nucleotides in at least one molluscan species.

With regard to the enzymes of the pathway, mitochondrial carbamoyl-phosphate synthesis by carbamoyl phosphate synthetase III (CPS-III) has been shown in the tissues of several snails (see Section IV, D). However, the enzyme generally associated with carbamoyl phosphate formation for pyrimidine nucleotide biosynthesis is the cytosolic, glutamine-dependent, and acetylglutamate-independent carbamoyl phosphate synthetase II (CPS-II) (Jones, 1980). In mammalian tissues (Mori and Tatibani, 1975; Jones, 1980). CPS-II activity copurifies with aspartate carbamoyltransferase and dihydroorotase as a single trienzyme protein that is coded by a single gene, the *CAD* gene. In *Ascaris, Schistosoma mansoni,* and *Drosophila,* the *CAD* gene activities are apparently also associated with a single trienzyme (Aoki et al., 1975; Aoki and Oya, 1979; Rawls, 1979; Jarry, 1976; Brothers et al., 1978). Among molluscs, aspartate carbamoyltransferase has been detected in the gonadal tissue of some bivalves (*M. edulis, Venus verrucosa,* and *P. maximus*) by Bergeron and Alayse-Ganet (1981), and in the hepatopancreas tissue of two snails, *S. oblongus* and *H. pomatia,* by Tramell and Campbell (1970a) and Porembska et al. (1966). Although these investigators did not assay for the other two activities associated with the *CAD* gene, they should be present if the gene structuring in molluscs is similar to other animals.

Catabolism of uracil and thymidine probably follows the pathway used by other animals (see Scheme 6). The end products ($\beta$-alanine and

$$\text{Uracil} \longrightarrow \text{Dihydrouracil} \longrightarrow \text{Carbamoyl-}\beta\text{-alanine} \longrightarrow CO_2 + NH_3 + \beta\text{-Alanine}$$

Scheme 6. Pyrimidine catabolism (uracil).

$\beta$-aminoisobutyric acid) have been detected in the tissues of many molluscan species (Campbell and Bishop, 1970). Tracer studies with [2-$^{14}$C]uracil using the hepatopancreas tissue from gastropods *Thais hemostoma, B. canaliculata, Siphonaria pectinata, H. aspersa,* and *O. lactea* indicated radiolabeling of both carbamoyl-$\beta$-alanine and bicarbonate (S. H. Bishop, unpublished data). J. W. Campbell and K. Allen (unpublished data) found an increase in tissue $\beta$-alanine content following injection of clams (*R. cuneata*) with uracil, dihydrouracil, and carbamoyl-$\beta$-alanine.

## V. Amino Acid Transport

Although the transport of amino acids across cellular membranes is not generally thought to involve biochemical transformations, amino acid transport undoubtedly plays a significant role in the regulation of intracellular levels of amino acids. Unlike mammalian transport systems, which are categorized into various sodium-dependent and -independent systems (Christensen, 1969), molluscan transport systems have not been sufficiently characterized to assign specific classes. Because many molluscs are aquatic, the potential capacity for direct uptake and loss of amino acids through exposed surfaces adds a dimension to the consideration of amino acid transport and is the most frequently studied aspect (Stephens, 1972; Stewart, 1979).

*M. arenaria* gill fragments have been shown to remove neutral amino acids (alanine, leucine, methionine, phenylalanine, serine, and glycine), acidic amino acids (aspartate), basic amino acids (lysine, histidine), and imino acids (proline) from incubation media (Stewart and Bamford, 1975, 1976; Stewart, 1978a,b). Inhibition studies indicate that there is a specific transport site for each of the previously mentioned classes of amino acids (Stewart, 1978a); however, there appears to be considerable overlap in the specificity of the sites, particularly with respect to the basic and neutral sites. Furthermore, only methionine and leucine were shown to have a common transporter using the criteria of Schriver and Wilson (1964) and Christensen (1975), even though other amino acids were examined for identify of transport sites. Isolated gill preparations of *Cerastroderma edule* have also been shown to have separate neutral and basic amino acid transporters (Bamford and McCrea, 1975). Neutral amino acid transporters have also been detected in isolated gill preparations of *M. californianus, M. edulis,* and *M. demissus* (Swinehart et al., 1980; Crowe et al., 1977; Bamford and Campbell, 1976; Wright et al., 1975).

Inhibition studies with NaCN and 2,4-DNP indicate that alanine transport in gills from *M. arenaria* and *C. edule* is energy dependent (Stewart and Bamford, 1975; Bamford and McCrea, 1975). Sodium-dependent transport in gill tissue has been indicated for cycloleucine in *M. edulis* (Wright and Stephens, 1977), for alanine in *M. arenaria* (Stewart and Bamford, 1975), and for glycine in *M. edulis* and *M. modiolus.* Cycloleucine was shown also to have a substantial sodium-independent component in *M. edulis* (Wright and Stephens, 1977).

*In vivo* techniques have shown that intact animals can absorb amino acids directly from the medium. *R. cuneata, M. edulis,* and *Hydrobia neglecta* have been shown to absorb glycine from the medium (Anderson and Bedford, 1973; Péquignat, 1973; Jorgensen, 1980). Absorption of alanine and glycine is energy and sodium dependent in *Bankia gouldi* (Stewart and Dean, 1980). Hatchlings of *Octopus joubini* accumulate $\alpha$-aminoisobutyric acid, L-valine, and D-valine from seawater (Castille and Lawrence, 1980). Rice et al. (1980) have shown that larval and juvenile oysters, *Ostrea edulis,* absorb glycine and alanine from seawater. It has also been shown that the principal site of amino acid absorption in intact animals is through the gills (Anderson and Bedford, 1973; Pequignat, 1973; Stewart, 1978a).

Using whole animals (*M. demissus*), all the protein amino acids and taurine can be absorbed (tyrosine, tryptophan, cysteine, and leucine were not tested) from seawater; uptake of glycine, threonine, and glutamine was most rapid whereas arginine was the slowest (Crowe et al., 1977). Recently, flow-through techniques have been developed by Wright and Stephens (1977) for the mantle chamber and gill compartment of *M. demissus*, that allow a detailed analysis of amino acid transport in intact animals. They have shown that the $K_T$ (*in vivo*) is one to two orders of magnitude lower than those measured *in vitro* with gill piece incubations. The apparently higher affinity of transport systems for amino acids *in vivo* is thought to be due to the reduction of unstirred boundary layers *in vivo* (Wright and Stephens, 1978; Wright et al., 1980).

Dipeptides have also been shown to be absorbed by molluscs. Glycylphenylalanine absorption in *M. edulis* (Stewart, 1981), alanyl-alanine absorption in *B. gouldi* (Stewart and Dean, 1980) and in *M. arenaria* (Stewart and Bamford, 1975), and valylleucine absorption in *M. arenaria* (Stewart and Bamford, 1975) have been noted. However, it is not clear whether or not these peptides are subjected to hydrolysis before transport. Peptidases are known to be associated with the surfaces of the gills (Pequignat, 1973; Stewart, 1978a; Stewart et al., 1979), and the dipeptides examined apparently compete for the same transport site as the constituent amino acids (Stewart and Bamford, 1975; Stewart, 1981).

Amino acid transport has also been examined in molluscan nervous tissue. Results in these tissues seem to indicate a heterogeneity of transport systems for particular amino acids that is probably related to their putative neurotransmitter activity. In *A. californica*, specific neurons ($R_3-R_{14}$) take up glycine by a carrier-mediated process that is sodium dependent, calcium dependent, and sensitive to mercury and 2,4-DNP inhibition (Price et al., 1978, 1979; McAdoo et al., 1978), but there also appears to be a sodium-independent carrier in these cells. The sodium-independent carrier appears to be the only glycine transport system in other neurons (McAdoo et al., 1978). Carrier-mediated transport of other amino acids in *Aplysia* neurons has been reported by McAdoo et al. (1978) for leucine, alanine, proline, and serine; by Zeman et al. (1975) for γ-aminobutyrate; and by Schwartz et al. (1975) for choline. In *H. pomatia*, glutamate is transported by a combination of carrier-mediated processes that are both sodium dependent and independent (Osborne et al., 1978). Glutamate transport is partially inhibited by 2,4-DNP, and the partial inhibition is probably a function of the transport and tissue heterogeneity. Leucine is apparently transported by a sodium-dependent carrier (Kostenko et al., 1979). Carrier-mediated transort of other amines has also been reported by Osborne et al. (1978) for dopamine and 5-hydroxytryptamine, and by Osborne (1976a) for choline. In *Loligo* giant axons, glycine transport is apparently via at least two transporters, both of which are sodium independent. One transporter is inhibited by $CN^-$ and ouabain and is apparently specific for glycine; the other appears to be a general neutral amino acid carrier that has highest affinity for short aliphatic or polar R groups and essentially no affinity for basic or acidic amino acids, imino acids, or peptides (Caldwell and Lea, 1978). On the other hand, glutamate transport is sodium dependent and inhibited by $CN^-$ and ouabain (Baker and Potashner, 1973; Caldwell and Lea, 1975). Alanine, leucine, serine, proline, phenylalanine, tyrosine, aspartate, arginine, and cysteine have also been shown to be absorbed (Caldwell and Lea, 1978; Hoskin and Brande, 1973).

Environmental factors are known to affect rates of amino acid transport in molluscs. Increases in temperature increase the $K_T$ and maximal influx rates with a $Q_{10}$ of 5 in *M. californianus* (Wright et al., 1980). The high $Q_{10}$ in this system is thought to relate to ciliary activity reducing unstirred boundary layers. Temperature dependence on glutamate transport has been demonstrated in squid giant axon (*L. forbesi*) (Caldwell and Lea, 1975). Free amino acid efflux has been shown to be decreased in *Noetia ponderosa* blood cells upon exposure to 4°C (Amende and Pierce, 1980). In mussel (*M. demissus*) hearts, Pierce and Greenberg (1972, 1973) have shown that amino acid efflux is increased upon exposure to hyposmotic media, and that the increased efflux is dependent on $Ca^{2+}$ and $Mg^{2+}$ concentration and on ATP (Pierce and Greenberg, 1973, 1976; Watts and Pierce, 1978). Studies with intact *M. californianus* have shown that uptake of glycine is dependent on $Mg^{2+}$ concentration, and the efflux of primary amines is affected by $Ca^{2+}$ and $Mg^{2+}$ (Swinehart et al., 1980). Efflux of free amino acids from whole *R. cuneata* and from *N. ponderosa* blood cells has been shown to be dependent on $Ca^{2+}$ and $Mg^{2+}$ (Otto and Pierce, 1981a,b; Amende and Pierce, 1980).

## VI. Free Amino Acid Levels in Tissues and Hemolymph

The concentration of free amino acids in the tissues and extracellular fluid compartments of molluscs varies with the diet, season, temperature, reproductive and developmental stage, and environmental stresses related to desiccation, anaerobiosis, osmotic pressure, pollution, and parasitism. Most of these changes do not deal with the pathways of amino acid metabolism per se and are discussed in other chapters. Table V presents a list of some of the species investigated.

With marine molluscs, the intracellular concentrations of free amino acids are high (50–400 m$M$) in the tissues and low (0.2–5 m$M$) in the hemolymph. Most are osmoconformers in that the osmotic pressure of the hemolymph and the tissues are essentially equivalent to the osmotic pressure of the bathing media. The tissues of the stenoholine marine species,

TABLE V

Amino Acid Composition of Blood and Tissue Fluids: List of Species with Individual Analyses[a]

| Species | Reference |
|---|---|
| Amphineurons | |
|     *Acanthochitonia discrepans* | Hoyaux et al. (1976); Gilles (1972) |
| Gastropods | |
|   Prosobranchs | |
|     *Ampullaria glauca* | See Florkin and Bricteux-Grégoire (1972) |
|     *Aplysia californica* | Lombardini et al. (1979) |
|     *Buccinum undatum* | See Florkin and Bricteux-Grégoire (1972) |
|     *Busycon perversum* | Simpson et al. (1959) |
|     *Fascioloria distans* | Simpson et al. (1959) |
|     *Haliotus tuberculata* | See Florkin and Bricteux-Grégoire (1972) |
|     *Littorina littorea* | Hoyaux et al. (1976); Watts (1971) |
|     *Nassarius obsoletus* | Kasschau (1975a,b) |
|     *Oliva sayana* | Simpson et al. (1959) |
|     *Patella vulgata* | Hoyaux et al. (1976) |
|     *Polinices duplicata* | Simpson et al. (1959) |
|     *Purpura lapillus* | Hoyaux et al. (1976) |
|     *Pyrazus ebeninus* | Ivanovici et al. (1981) |
|     *Tegula funebralis* | Peterson and Duerr (1969) |
|     *Thais hemostoma* | Simpson et al. (1959) |
|   Pulmonates | |
|     *Biomphalaria glabrata* | Stanislowski et al. (1979) |
|     *Bulimnea megasoma* | Gilbertson and Schmid (1975) |
|     *Helisoma trivolvis* | Gilbertson and Schmid (1975) |
|     *Helix aspersa* | Campbell and Speeg (1968a) |
|     *Helix pomatia* | Weiser and Schuster (1975) |
|     *Lymnaea catascopium* | Gilbertson and Schmid (1975) |
|     *Lymnaea stagnicola* | Gilbertson and Schmid (1975); and see Florkin and Bricteux-Grégoire (1972) |
|     *Otala lactea* | Campbell and Speeg (1968a) |
|     *Siphonaria lineolata* | Simpson et al. (1959) |
|     *Siphonaria zelandica* | Bedford (1969) |
|     *Stagnicola exilis* | Gilbertson and Schmid (1975) |
| Bivalves | |
|     *Anadara trapezia* | Ivanovici et al. (1981) |
|     *Anodonta cygnea* | Potts (1954, 1958) |
|     *Arca unbonata* | Simpson et al. (1959) |
|     *Cardium edule* | Shumway et al. (1977) |
|     *Chlamys opercularis* | Shumway et al. (1977) |
|     *Corbicula manilensis* | Gainey (1978a,b) |
|     *Crassostrea gigas* | Riley (1976, 1980); Shumway et al. (1977) |
|     *Crassostrea virginica* | Simpson et al. (1959); Lynch and Wood (1966) |
|     *Glycymeris glycymeris* | Hoyaux et al. (1976) |
|     *Gryphea angulata* | Bricteux-Grégoire et al. (1964c) |
|     *Hydridella menziese* | Bedford (1973) |
|     *Hydrobia ulvae* | Negus (1968) |

TABLE V (*Continued*)

| Species | Reference |
|---|---|
| *Ligumia subrostrata* | Hanson and Dietz (1976) |
| *Lithophaga bisulcata* | Simpson et al. (1959) |
| *Macoma ponderosa* | Roesijadi and Anderson (1979) |
| *Melanopsis trifasciata* | Bedford (1971a,b,c) |
| *Mercenaria mercenaria* | Shumway et al. (1977); Jefferies (1972) |
| *Modiolus demissus* | Simpson et al. (1959); Baginski and Pierce (1975, 1977); Shumway and Youngson (1979); Greenwalt and Bishop (1980) |
| *Modiolus modiolus* | Pierce (1971); Shumway et al. (1977) |
| *Mya arenaria* | DuPaul and Webb (1970); Virkar and Webb (1970); Shumway et al. (1977) |
| *Mytilus edulis* | Lange (1963); Livingstone et al. (1979); Hoyaux et al. (1976); Shumway et al. (1977); Bricteux-Grégoire et al. (1964a); see others in text |
| *Mytilus galloprovincialis* | Sansone et al. (1978) |
| *Noetia ponderosa* | Amende and Pierce (1978) |
| *Ostrea edulis* | Bricteux-Grégoire et al. (1964b) |
| *Polymesoda caroliniana* | Gainey (1978a,b) |
| *Rangia cuneata* | Allen (1961); Anderson (1975); Fyhn (1976); Otto and Pierce (1981a,b); Henry and Mangum (1981a,b) |
| *Saccostrea commercialis* | Ivanovici et al. (1981) |
| *Scrobicularia plana* | Hoyaux et al. (1976); Shumway et al. (1977) |
| *Tapes japonicum* | Sato et al. (1979) |
| Cephalopods | |
| *Dosidicus gigas* | Deffner (1961a,b); see Florkin (1966) |
| *Illex argentinus* | Suyama and Kobayashi (1980) |
| *Loligo opalescens* | Ballantyne et al. (1981); Suyama and Kobayashi (1980) |
| *Loligo pealeii* | Storey and Storey (1978); Deffner (1961a,b) |
| *Loliguncula brevis* | Simpson et al. (1959) |
| *Nototodarus sloani* | Suyama and Kobayashi (1980) |
| *Octopus ochellatus* | See Florkin and Bricteux-Grégoire (1972) |
| *Octopus vulgaris* | See Florkin and Bricteux-Grégoire (1972) |
| *Ommastrephes baritrami* | Suyama and Kobayashi (1980) |
| *Ommastrephes sloani* | See Florkin and Bricteux-Grégoire (1972) |
| *Sepia esculenta* | Suyama and Kobayashi (1980) |
| *Sepia officinalis* | Robertson (1965) |
| *Sepia pharaonis* | Suyama and Kobayashi (1980) |
| *Todarodes pacificus* | Suyama and Kobayashi (1980) |

[a] References are not complete but will lead the reader to the bulk of the literature.

including the cephalopods and many of the bivalve and gastropod species, show irreversible cellular swelling, leak salts and amino acids, and die when placed in diluted seawater. On the other hand, some euryhaline marine species (bivalves and a few gastropod species) survive in diluted seawater as osmoconformers. When placed in diluted seawater these animals show transient cellular swelling with a controlled efflux of salt and amino acids from the cells, followed by increased catabolism of these amino acids and a return of the cells to near-normal size. Taurine, alanine, and glycine plus proline, aspartate, glutamate, and the quaternary amines in some cases comprise the bulk of the organic solutes in the cell. With osmotic adjustment, alanine, glycine, and proline levels rise or fall most rapidly, whereas taurine and the quaternary amines accumulate slowly and are lost slowly. These intracellular amino acids and amines play an important role in the regulation of cell volume and intracellular osmotic pressure during osmotic stress, and may have a salt-sparing effect in some tissues (Allen and Garret, 1971b; Lange, 1972; Schoffeniels and Gilles, 1972; Gilles, 1974, 1979; Treherne, 1980).

In the freshwater bivalves (Potts, 1958; Simpson et al., 1959; Bedford, 1973; Hanson and Dietz, 1976; Gainey, 1978a,b) and gastropods (Simpson et al., 1959; Gilbertson and Schmid, 1975; Reddy and Swami, 1978; Stanislowski et al., 1979), the total free amino acid concentrations are low (4–10 mM) in the tissues and even lower (0.3–1.5 mM) in the hemolymph. Although all of the amino acids are present in small amounts, alanine, glycine, glutamate, asparatate, serine, and threonine are in greatest abundance. Little or no taurine or quaternary amines have been detected. The concentrations of solutes, including ions, in the blood and tissues are much greater than in the environment and are regulated fairly closely. These animals are able to concentrate these solutes in the tissues and produce a dilute urine. When these animals are subjected to brackish water or to desiccation, the osmotic pressure of the blood increases and the levels of amino acids in the tissues rise in much the same manner as the marine molluscs (Bedford, 1973; Hanson and Dietz, 1976; Gainey, 1978a,b).

The bivalves *R. cuneata* and *Polymesoda caroliniana* inhabit brackish water estuaries rather than purely marine or freshwater environments. At salinities between 10 and 20‰, the free amino acid levels in the tissues contribute 25–30% of the intracellular solute (Allen, 1961; Fyhn, 1976; Gainey, 1978a,b; Henry et al., 1981; Otto and Pierce, 1981a,b). In both species alanine, glycine, proline, and glutamate are in greatest abundance in the tissues. With transfer to higher salinities the animals behave as osmoconformers and the concentrations of these amino acids in the tissues increase. With transfer to fresh water, the animals behave as osmotic and ionic regulators by maintaining high levels of ions in the tissues and blood, and the amino acid levels in the tissues fall to near the levels found in freshwater bivalves.

Terrestrial molluscs such as *H. pomatia* and *O. lactea* show changes in free amino acid levels in the tissues and hemolymph during desiccation (Campbell and Speeg, 1968a; Weiser and Schuster, 1975). Amino acid levels were low in the hemolymph (0.37–2.0 m$M$) and higher in the tissues (12–60 m$M$). The tissues have small amounts of taurine similar to the levels reported for freshwater and brackish water bivalves.

Adjustment of these free amino acid levels in marine, euryhaline, and freshwater species is a complicated process. Taurine and the quaternary amines seem fairly inert metabolically, are probably of dietary or symbiont origin (Section IV), and are retained in the cells at high salinities by membrane-related processes (Amende and Pierce, 1980; Pierce and Amende, 1981; Pierce and Greenberg, 1972, 1973; see also Section V). On the other hand, the amino acids glycine, alanine, proline, aspartate, and glutamate show rapid metabolic turnover (Section IV). It appears that at high salinities, these amino acids are accumulated in the tissues by membrane-related processes (Pierce and Greenberg, 1972, 1973; Shumway et al., 1977; Livingstone et al., 1979; Shumway and Youngson, 1979; Strange and Crowe, 1979a,b; Amende and Pierce, 1980; Crowe, 1981; Otto and Pierce, 1981a,b; Pierce and Amende, 1981) in combination with a decreased catabolism and possibly an increased biosynthesis (Gilles, 1969, 1970, 1974, 1979; Baginski and Pierce, 1975, 1977, 1978; Bedford, 1971c; Bishop, 1976; Henry et al., 1981; Bishop et al., 1981; Greenwalt, 1981; Zurburg and de Zwaan, 1981). As discussed in Section IV, aspartate, glutamate, glycine, alanine, and serine arise from the metabolism of a variety of gluconeogenic compounds including glycogen, whereas proline is probably derived from arginine and ornithine but not from glutamate. All can be derived from the amino acids released during peptide and protein turnover and may accumulate if catabolism is slowed.

It appears that the marine as well as the brackish and freshwater species have the abilities to increase the free amino acid levels in the tissues in response to hyperosmotic stress. The penetration of freshwater species into the marine environment, and marine species into the freshwater environment, seems limited more by their abilities to regulate ion concentrations than to mobilize amino acids for cellular osmoregulation (Gainey and Greenberg, 1977).

## VII. Nitrogen Loss and Excretion

Physiological aspects of nitrogenous excretory product elimination related to kidney and other excretory organ functions are discussed in another chapter. This section deals with the metabolic aspects.

## A. Bivalves and Cephalopods

Ammonia is generally considered to be the major nitrogenous excretory product of bivalves and cephalopods (Delauny, 1931, 1934; Baldwin and Needham, 1934; Potts, 1967; Campbell, 1973; Mangum et al., 1978; Weiser, 1980). However, a variety of studies indicate great variation both in the amount of nitrogen lost as ammonia and in the amounts of ammonia relative to urea, amino acids, and other nitrogenous compounds (including small amounts of uric acid) in the excreta. Most of the literature on cephalopods is reviewed in the article by Campbell and Bishop (1970). Using a variety of clams, mussels, and oysters held in aquaria, Hammen (1968) found that although ammonia was the major nitrogenous excretory product, the amount of amino acid excreted was substantial in some species and seemed proportional to the relative surface area : mass ratio as well as the transaminase levels in the tissues. In a separate study with one of these species (*M. demissus*), Bartberger and Pierce (1976) found little or no free amino acid loss and substantially more ammonia excretion per gram of animal than that reported by Lum and Hammen (1964) or Hammen (1968). Bartberger and Pierce (1976) suggest that both laboratory holding conditions and factors used in calculating nonshell tissue weight may account for some of the differences in their measurements. In this regard, oysters (*C. virginica*) (Hammen et al., 1966), steamers (*M. arenaria*) (Allen and Garrett, 1971b), and mussels (*M. edulis*) (Bayne, 1976) can excrete 4–28% of the nitrogen as urea, whereas most other bivalves excrete less than this amount or no urea at all.

In early studies, Spitzer (1937) demonstrated profound differences in nitrogen loss between summer and winter marine bivalves. Srna and Baggaley (1976) report that the rate of ammonia excretion by oysters (*C. virginica*) and clams (*M. mercenaria*) was greater than that reported by Hammen and increased with increased weight of the animal and with abrupt shift of animals (acclimated to 20°C) to higher or lower temperatures. They also established that the ammonia released by the animals was produced by the bivalves rather than by any microbial associates. In another study, ammonia production by the surf razor clam, *Siliqua patula,* was found to vary with time of the year and presumably with food availability (Lewin et al., 1979). Lewin et al. (1979) found no marked effect of temperature on ammonia production with razor clams collected in the field and that, on a per-gram of weight basis, smaller animals produced more ammonia than larger animals. The ammonia produced by these clams was probably essential to the nitrogen economy of the algae living in the surf and surf–sand environment. A similar relationship in nitrogen economies between oysters and food organisms has been postulated for the oyster reef community (Saijo and Mitamura, 1971).

Bayne and Scullard (1977a,b) have evaluated some of the environmental factors affecting rates of nitrogen loss from *M. edulis* and some other Mytilidae in order to rationalize some of the apparent inconsistencies described earlier. With summer animals (*M. edulis*), up to 30% of the nitrogenous waste is lost as amino acid nitrogen and the remainder as ammonia. In winter and spring, the proportion of excreted amino acid nitrogen fell to zero and 3.4%, respectively, and the remainder was excreted as ammonia. In overall terms, ammonia excretion was maximal in the late spring and early summer when the animals were feeding, and minimal in the winter during gametogenesis and spawning. On a per-gram weight basis, winter animals weighing less than 0.5 g excreted very little nitrogen compared to large winter animals, which is in agreement with the report by Srna and Baggaley (1976). However, when these winter animals were starved, the ammonia excretion rate increased 5- to 10-fold for the small animals (0.2–0.5 g) and less than 50% or not at all for animals weighing more than 2 g. Starved winter animals had a uniformly lower rate of oxygen consumption compared to fed winter animals. In contrast, ammonia excretion rates and $O_2$-consumption rates of summer animals were lower in starved animals when compared to fed animals. Following refeeding of starved summer animals, there was a transient increase in both oxygen consumption and ammonia excretion (Bayne and Scullard, 1977b). In experiments with spring animals adapted at various temperatures, the metabolic rate (oxygen consumption) showed some degree of compensation, whereas the ammonia excretion rate did not acclimate and maintained a $Q_{10}$ of about 2.

These results indicate shifts in physiological capacity with change in temperature, season, and reproductive cycle that affect the nitrogen economy and the metabolic rate in somewhat disparate fashions. The results reinforce Hammen's (1968) suggestion on using animals of uniform size and physiological experience for experimental purposes. Bayne and Scullard (1977a) calculate that from 0.5 to 2% per day of the total body nitrogen is lost to turnover and must be replaced to maintain the balance. The lack of temperature adaptation with regard to ammonia production and the marked increase in rate of ammonia loss with starved small winter animals means that in the winter, small animals may undergo severe nitrogen stress and experience survival problems. In other experiments by Bayne and Scullard (1977a,b) with *Mytilus galloprovincialis* and *M. californianus*, the amount of nitrogen lost as amino acids relative to ammonia varied with the season and location of collection, the time held in the laboratory, and the feeding regimen.

Another factor affecting the rate of loss of nitrogenous compounds is the salinity of the bathing medium. Studies by Bartberger and Pierce (1976) with *M. demissus*, by Henry and Mangum (1981b) with *R. cuneata*, by Livingstone et al. (1979) with *M. edulis*, and by Emerson (1969) with a variety of other species indicate a transient increase in the rate of ammonia loss a few hours after transfer of animals from high-salinity media to low-salinity media. This transient increase more or less accounts for the overall decline in the total nitrogen loss from the intracellular free amino acid pool. If these animals are held at the lowered salinity for longer periods, both the metabolic and ammonia excretion rates are maintained 1.5–2 times greater than the rates in animals held at higher salinities.

With air exposure or periods of hypoxia–anoxia, most bivalves become somewhat anaerobic and show a small Pasteur effect when returned to aerated water (Widdows et al., 1979). The ammonia produced during these hypoxic bouts is released as a transient pulse after return to aerated water and seems to be proportional to the duration of air exposure or hypoxia rather than to the degree of the Pasteur effect (de Vooys and de Zwaan, 1978; Widdows et al., 1979). In this regard, most of these animals seem to exhibit a form of metabolic shutdown during anaerobiosis (Pamatmat, 1978, 1979) and show very little Pasteur effect (de Zwaan and Wijsman, 1976). During this "shutdown," ammonia production appears to continue and to be somewhat independent of the overall metabolic rate.

In general, the data support the notion that bivalves are ammonotelic and that a considerable portion of the ammonia production may not be linked to the overall energy budget of the animal. In any case, the variety of ammonia-forming enzyme systems through both direct and indirect pathways (Section II) would seem to argue for a constant fraction of the total ammonia production to be a consequence of the direct deaminating systems (influenced markedly by food availability) and a variable portion that may be linked to the metabolic rate through the indirect transdeaminating systems.

## B. Prosobranch Gastropods

As in bivalves, ammonia appears to be the major nitrogenous component in the excreta of most aquatic prosobranch gastropods. However, some intertidal species and those with a semiterrestrial existence may accumulate relatively large amounts of uric acid in their kidneys and other tissues (Needham, 1935; Potts, 1967; Campbell and Bishop, 1970).

All marine prosobranchs examined excrete ammonia and very little or no free amino acid nitrogen. In a survey of seven marine prosobranchs, Duerr (1968) found no urea and from 0.3 to 6 $\mu$mol ammonia produced/g wet tissue day. The metabolic rates of these snails varied between 12 and 58 ml of $O_2$ consumed/h/g total weight, which is a considerably narrower range than the wide range in ammonia production. Weiser (1980) and Mangum et al. (1978) report similar ammonia excretion rates in some other species. Feeding snails, *Nassarius reticulatus* and *B. undatum*, have higher rates of oxygen consumption than starving snails and show a transient twofold increase in ammonia production 8–48 h after feeding (Crisp *et al.*, 1981). On the other hand, Stickle (1971) reported that *Thais lamellosa* increased ammonia excretion during starvation as a result of protein utilization from the foot. *Thais lamellosa, Thais emarginata,* and *Littorina sitchana* transferred from full seawater to half-seawater media showed a transient increase in ammonia excretion as they became acclimated to the diluted medium (Emerson, 1969); transfer of these snails to hypersaline seawater decreased the rate of ammonia production.

The uric acid content of a variety of feeding prosobranchs ranged between 0.4 and about 10 mg/g dry weight, with most in the range of 0.5–2.5 mg/g dry weight (Duerr, 1967). This uric acid content was very much lower than that found in helicid land pulmonates but similar to that found in some freshwater pulmonates. With littorines (*Littorina planaxis*), after 66–70 days of starvation, the free amino acid levels in the tissues fell about 60%, the uric acid content rose from 0.62 to 4.29 mg/g dry weight, and the lipid content fell about 40%, whereas the total carbohydrate and protein contents remained fairly constant (Duerr, 1967; Emerson and Duerr, 1967). Uric acid accumulation accounted for only a third of the total nitrogen lost from the intracellular free amino acid pool and protein; presumably the remaining two thirds of the nitrogen was lost as ammonia (not measured). In studies with *Tegula funebralis,* Peterson and Duerr (1969) found increased uric acid accumulation in the tissues of animals subjected to hyperosmotic stress.

The freshwater and terrestrial ampullarids have been studied more intensely than the other prosobranchs. During estivation, Little (1968) found an increase in the kidney hemolymph uric acid levels of both *Pomacea depressa* and *Pomacea lineata*. *P. lineata* contained 2 to 10 times more uric acid and could estivate for much longer periods than *P. depressa*. When placed in water, most of the uric acid accumulated in the kidney was excreted whereas much of the uric acid in the tissues remained "firmly bound." Little (1968) considers these pomacids to be uricotelic when estivating. During development, the eggs of *Pomacea paludosa* accumulate 10 times more "waste" nitrogen than the eggs of the more aquatic snail *Marisa cornuarietis* (Sloan, 1964). Because pomacid eggs showed a fourfold increase in uric acid content and the marisid eggs showed no increase, Sloan (1964) concludes that the pomacids become partially uricotelic during the developmental process, whereas the more aquatic marisids remain ammonotelic throughout.

Earlier studies (Saxena, 1952, 1955) indicated that *P. globosa* could accumulate large amounts of uric acid in the nephridia and other tissues during estivation. This uric acid in the nephridia was eliminated during periods of activity. More recent studies on *P. globosa* have been summarized by Swami and Reddy (1978). When actively feeding, these snails excrete ammonia and urea and can accumulate substantial amounts of both urea and ammonia in their tissues. With 90 days of estivation, the tissue levels of ammonia remained high (50 $\mu$mol/g wet tissue), the urea levels fell about 3- or 4-fold to about 3.5 $\mu$mol/g wet tissue, and the uric acid content increased 8- to 10-fold to about 60 $\mu$mol/g wet tissue. With arousal, the tissue levels of ammonia and urea rose and some of each was excreted, whereas uric levels remained high in both the mantle and hepatopancreas and decreased in the foot. Because the foot comprises the bulk of the tissue in these animals, a good deal of the uric acid was excreted after arousal. During estivation glycogen levels decrease about 80% and total protein decreases 30–40%. Swami and Reddy (1978) suggest a switch from the ammonotelic–ureotelic mode, to uricotelism during estivation to detoxify ammonia and minimize water loss. Actual biosynthesis of uric acid has not been evaluated in *Pila* or any other prosobranch. Although Horne and Boonkoon (1970) report a complete set of urea cycle enzymes in two other prosobranch species, no urea biosynthesis was detected in *Pila ampullaria* in a [$^{14}$C]bicarbonate-tracer experiment (Horne and Barnes, 1970). These data would seem to indicate an absence or at least only a low level of *de novo* urea biosynthesis in *Pila*. Swami and Reddy (1978) suggest that the urea formed could be derived from both uricolysis and arginase action on dietary or other arginine and that the accumulation of uric acid must result from inhibition of the uricolytic sequence. These ideas on the origins of urea, uric acid, and uricolysis need substantiation; the data indicating very high tissue osmotic pressure (500 mosmol) and ammonia concentrations (50–70 m$M$) in both active and estivating snails should be reviewed and confirmed.

Overall it appears that the availability of food, stage of the reproductive cycle, salinity of the bathing medium, and nature of the habitat have a marked effect on nitrogen excretion in prosobranchs just as they do in bivalves. Among species, these factors, rather than size and metabolic rate, may be better indicators of the actual rate of nitrogen loss. Although uric acid tends to accumulate in the tissues of some snails during periods of xeric stress or starvation, as Duerr (1967) points out, the modest ability to accumulate uric acid seems to be more species specific or phylogenetic and may not be specifically linked to the xeric habitat.

## C. Pulmonate Gastropods

The excretion or loss of nitrogenous "wastes" by pulmonates has been studied in some detail. The results indicate rather complex patterns of ammonotelism, ureotelism, and uricotelism among the various species. The discussion here is in parts: first, reviewing the so-called uricotelic terrestrial pulmonate snails, then the slugs and the aquatic pulmonates.

### 1. Terrestrial Pulmonate Snails

Original studies on the uricotelic nature of *H. pomatia, H. aspersa,* and *O. lactea* has led to many of the generalizations concerning the uricotelic nature of terrestrial pulmonates (Baldwin and Needham, 1934; Needham, 1935; Florkin, 1966; Campbell and Bishop, 1970). Between 90 and 100% of the nitrogen in the excreta lost from the kidney is in the form of uric acid, guanine, and xanthine in both feeding and estivating snails (Jezewska et al., 1963; Jezewska and Sawicka, 1968; Campbell et al., 1972; Speeg and Campbell, 1968b). Little or no urea and/or ammonium ion nitrogen and only a small amount of other nonprotein nitrogen has been detected in these excreta or in the kidney. Hypoxanthine and adenine are found in the form of nucleotides in the tissues of these snails and apparently do not appear in the hemolymph as free bases or as excretory products in the kidney or excreta (Speeg and Campbell, 1968b; Jezewska, 1968). Similar results have been obtained for *S. oblongus* by Tramell and Campbell (1970b), for *Cepaea nemordia, Cepaea veindobonensis,* and *Heligona arbustorum* by Jezewska (1969), for *H. aspersa* by Clark and Rudolph (1979) and Campbell et al. (1972), and for a carnivorous snail, *Euglandina rosea,* by Badman (1971). No xanthine was detected in the excreta of *E. rosea,* and guanine was not found in the excreta of either *H. arbustorium* or *C. vindobonensis.* The general ratio of uric acid:xanthine:guanine in the excreta of *O. lactea* and *H. pomatia* varied with estivation, feeding, and time of the year (Speeg and Campbell, 1968a; Campbell et al., 1972; Jezewska, 1969). During short-term estivation in *O. lactea,* the relative amounts of xanthine and guanine in the excreta remained fairly constant whereas the bulk of the increase in purine accumulation was a result of the large increase in uric acid (Speeg and Campbell, 1968a).

The purines in the excreta of *O. lactea, H. pomatia,* and *H. aspersa* arise by *de novo* purine biosynthesis (see Section IV, L). Pulse-labeling studies of *H. aspersa* hepatopancreas tissue using [$^{14}$C]formate indicate rapid initial labeling of guanine and hypoxanthine with smaller amounts of label in xanthine and uric acid (Clark and Rudolph, 1979). The extremely high initial specific activity of hypoxanthine and guanine when compared to both xanthine and adenine would seem to indicate shunting of the bulk *de novo* synthesized purine nucleotides toward the large adenine nucleotide pool, thus resulting in a greater isotope dilution. Appearance of the purines in the excreta may result from subsequent breakdown of the adenine nucleotides to hypoxanthine, xanthine, and uric acid. These results are in general agreement with those of Speeg and Campbell (1968a), who used [$^{14}$C]glycine tracers with *O. lactea*. Although the metabolic controls regulating uric acid production have not been studied in these molluscs, the suggestion (Campbell and Bishop, 1970) that these snails have a "gouty" purine metabolism resulting from an absence of the phosphoribosyl pyrophosphate transferases required for adenine, guanine, and hypoxanthine salvage, in addition to purine nucleotide pool size regulation, has been shown to be unlikely with the discovery of these enzymes in the tissues of *H. pomatia* by Barankiewicz and Jezewska (1973, 1976).

The rate of purine accumulation by feeding and estivating snails (*O. lactea*) was about 2.8 and 0.7 $\mu$atom purine N/g total weight/day, respectively (Speeg and Campbell, 1968b). Using [$^{14}$C]glycine tracers, the rate of production in *O. lactea* was estimated to be about 2.7 $\mu$atom purine N/g total weight/day, which is in agreement with the accumulation rate of feeding animals. The 75% reduction in purine production during estivation may be a reflection of overall reduction in the metabolic rate during estivation in snails (Horne, 1973a; Schmidt-Nielsen et al., 1971; Machin, 1975).

All of these snails have the capacity for *de novo* urea biosynthesis and urea production from arginine (dietary or other sources), but not from uric acid breakdown. Although small amounts of ammonia and urea appear in the tissues and blood of these snails (Campbell and Speeg, 1968a; Trammell and Campbell, 1970b; Weiser and Schuster, 1975), little or none appeared in the excreta. The absence of urea in the excreta was attributed to the presence of reasonably high tissue urease levels. In testing this idea, Speeg and Campbell (1968b) found that these snails (*O. lactea*) release volatile ammonia and that injection of urea and arginine resulted in a strong stimulation of volatile ammonia production a few hours later. Glutamate injection also stimulated ammonia production although less dramatically. Ammonia was released from the total surface of the animals through the shell. Injection of acetohydroxamate (urease inhibitor) blocked $^{14}$CO$_2$ production from injected [$^{14}$C]ureidocitrulline, [$^{14}$C]guanidinoarginine, and [$^{14}$C]urea, and caused an accumulation of urea in the tissues and blood (Speeg and Campbell, 1969). In a survey (Loest, 1979a,b), feeding snails from several species that produce reasonably large amounts of volatile ammonia had reasonably high urease activity, but not all species with high urease activity produced much volatile ammonia. In any event, if the urea formed is hydrolyzed by this urease, then the special role for urea biosynthesis in ammonia detoxification is unlikely. Urea may act as a carrier molecule for nitrogen transport to a particular organ, such as the mantle or lung, for urease action and ammonia release. A special buffering role in the calcification process has been postulated for this volatile ammonia production from urea (Campbell and Boyan, 1976; Campbell and Speeg, 1969).

Because the uric acid and other purines that accumulate during feeding and estivation are released in bulk from time to time, single analyses of the kidney contents or excreta and the lack of measurements of volatile ammonia production have given the impression that these snails are totally uricotelic. In the well-controlled studies on uric acid and ammonia production by Speeg and Campbell (1968a,b), the volatile ammonia production rate in estivating snails was about 0.31 and 0.58 $\mu$mol ammonia/g weight/day for *H. aspersa* and *O. lactea,* respectively, and uric production was about 0.7 $\mu$atom purine N/g/day. Therefore, volatile ammonia production would account for 45% of the average daily nitrogen loss during estivation. If ammonia production remains at the same rate during feeding, then uric acid would account for about 80% of the average daily nitrogen loss during feeding. One can conclude that these snails tend to be mainly uricotelic when feeding and show combined uricotelism–ammonotelism during estivation. The only commonality between uric acid and gaseous ammonia excretion would seem to be water conservation.

In contrast to these terrestrial pulmonates, certain snails, for example *B. dealbatus,* accumulate large amounts of urea during starvation–estivation (Horne, 1971, 1973a,b). High urea levels have also been reported in some desert snails at certain times of the year (Haggag and Fouad, 1968). Active snails (*B. dealbatus*) contain about 0.9 $\mu$mol of urea, 9.3 $\mu$mol of uric acid, and 6.8 $\mu$mol of ammonia/g wet tissue. With estivation for periods of up to a year in atmospheres of high relative humidity, the urea content of the snail increases with time to over 250 $\mu$mol/g wet tissue and the uric acid content increases two- to threefold. With return to a wetted surface, the snails undergo arousal and void the accumulated urea and uric acid within a few days. During estivation the rate of respiration drops to 16% of the resting rate within 3 days and remains at this level. The R.Q. is about 0.82 with extended estivation (120 days), indicating utilization of protein reserves. After 70 days of estivation, there is little or no decline in lipid or water content but an 85 and 21% loss of carbohydrate reserves and body protein, respectively. This loss of protein nitrogen is more or less balanced by the increase in urea nitrogen, and no nitrogen is lost as volatile ammonia.

During this transition from the ammonotelic–uricotelic pattern of the active, feeding snails (*B. dealbatus*) to the ureotelic pattern of the estivating snails, there is a sharp increase in the level of all the urea cycle enzymes and a 10-fold increase of $^{14}C$ incorporation into urea from $^{14}CO_2$ (Horne, 1971, 1973b). At the same time, the small amount of urease in feeding snails seems to be lost or inhibited in estivating snails. Therefore, the urea accumulated during estivation results from an activation of the urea cycle biosynthetic pathway and a supression or inactivation of the urease activity. The $LD_{50}$ for ammonium acetate injected into active snails was about 16 $\mu$mol/g wet weight, indicating that these snails were somewhat less tolerant to ammonium ion than the aquatic pulmonate snail *B. glabrata* (Meyer and Becker, 1980). Because the tissue level of ammonium ion-ammonia (6.8 $\mu$mol/g wet weight) in both estivating and active snails is below the $LD_{50}$, Horne (1973a) concludes that the enhanced urea biosynthesis during estivation is probably an ammonia-detoxifying process. The increased concentration of urea may add to the osmotic pressure within the tissues and could play a role in reducing water loss by evaporation during estivation.

### 2. Slugs

Until recently, slugs were considered uricotelic because of the relatively large amount of purines in their excreta (Florkin and Bricteux-Grégoire, 1972). Jezeweska (1969) and Horne and Beck (1979) found large amounts of uric acid, xanthine, and guanine with lesser amounts of hypoxanthine and adenine in excreta of three species of slugs. The relative amount of uric acid and xanthine varied with the species although uric acid usually accounted for about half and xanthine about 15% or less of total purines in the excreta. However, with actively feeding slugs (*L. flavus*), about 60% of the nitrogen was excreted as urea and the remaining 40% was lost as purine. Ammonia levels in the excreta were below detectable limits. Although Loest (1979a) found no volatile ammonia production by four species of slugs (*Philomycus carolinianus, Veronicella floridana, Deroceras laeve,* and *Limax maximus*), all species absorbed volatile ammonia from the air on the wetted mucus that coats the surface of the animals.

These data indicate that the slugs may be ureotelic to some degree. Horne (1977b) found a complete set of urea cycle enzymes in the tissues of *L. flavus* and a two- to threefold increase of $^{14}C$ labeling of urea from [$^{14}C$]bicarbonate during fasting. With 16 days of fasting the protein, carbohydrate, and lipid contents declined 34, 25, and 19%, respectively, indicating that protein was used as a major energy source. However, during fasting the levels of the urea cycle enzymes and glutamine synthetase did not increase significantly, and there was very little change in the ammonia or amino acid levels in the tissues. Injection of ammonia, ornithine, aspartate, and arginine stimulated urea production to only a slight extent. Horne (1977b) concludes that the urea excreted is produced *de novo* by the urea cycle and that the increased production during starvation results from increased substrate availability resulting from protein breakdown rather than by increased amounts or activities of the individual urea cycle enzymes. In a separate study, Horne and Beck (1979) found that purine production in terms of synthesis rate ([$^{14}$C]glycine tracer) and actual yield in the excreta decreases (50%) rather than increases during starvation. From these studies, it appears that slugs have a combined ureotelic–uricotelic nitrogen excretory pattern. During starvation or estivation, ureotelism is enhanced by increased *de novo* urea biosynthesis and decreased uric acid production.

## 3. Freshwater Pulmonates

Nitrogen excretion by snails in the species *B. glabrata* has received quite a bit of attention, because they can serve as vectors for schistosomiasis (Becker and Schmale, 1975, 1978). When feeding in water, ammonia production (3.84 $\mu$mol/g wet weight/day) is about four times the urea production (0.98 $\mu$atom urea N/g wet weight/day), whereas when starved, infected, or placed in a humid environment to stimulate an "estivating" condition, the ammonia production falls only slightly (22%) or remains constant while urea excretion increases four- to fivefold, resulting in an overall two- to threefold increase in nitrogen excretion after 5 days. Uric acid levels in the kidney and tissues, when measured, were very low or below detectable limits. Following starvation or parasite infections, hemolymph levels of free amino acids declined two- to fourfold and urea increased four- to fivefold to about 2.4 m$M$, whereas ammonia remained fairly constant at 80–100 $\mu M$ (Stanislowski et al., 1979). Tracer studies (Table VI) using [$^{14}$C]bicarbonate with hepatopancreas tissue incubations indicated a 14- to 40-fold enhancement of urea biosynthesis in the tissues of infected, starved, and "estivating" snails when compared to tissues from fed snails (Schmale and Becker, 1977). Additionally, these investigators found higher levels of urea cycle enzymes in the tissues of stressed snails than in fed snails. The increased nitrogen loss during these metabolic stresses resulted from an increased urea biosynthesis through activation of the urea cycle pathway.

This change from ammonotelism to a combined ammonotelism–ureotelism during periods of physiological stress may be an attempt to detoxify the amino acids or ammonia produced during the stress period. As a corollary to the ammonia production experiments, Becker's group and others performed a series of dietary and ammonotoxicity experiments.

TABLE VI

Urea Biosynthesis in the Hepatopancreas Tissue of *Biomphalaria glabrata:* Variation in the Incorporation of [14C] from [14C]Bicarbonate into Urea[a]

| Physiological pretreatment of snails | dpm in urea/g of tissue[c] |
|---|---|
| Fed snails | 9307 ± 8262    (11)[d] |
| Starving snails[b] | 127,858 ± 58,259    (8) |
| Infected snails (shistosome) | 233,244 ± 181,646 (12) |
| Snails kept dry in 95% humidity[b] | 352,520 ± 129,009   (7) |

[a] Data are from Schmale and Becker (1977).

[b] Starving and dry snails were held for 5 days.

[c] Urea was determined as urease-labile urea and counted as released [14C]CO_2.

[d] Number in parentheses indicates number of trials.

Snails fed test diets high in lipid or carbohydrate and low in protein showed reduced hemolymph urea concentrations compared to starved snails or snails on high-protein diets (Stanislowski and Becker, 1979). Snails fed these enriched diets have elevated hemolymph glucose levels compared to starved or infected snails (Stanislowski and Becker, 1979; Christie et al., 1974). Snails raised under axenic conditions using the diet developed by Machado and Vierra (1979) excreted more or less equal amounts of urea and ammonia nitrogen, whereas the small amount of uric acid in the kidney (average 56.2 $\mu$g/g dry kidney) accounted for only about 4% (average) of the nitrogen in the excreta. This axenic diet developed by Machado and Vierra (1979) may be excessively rich in protein because the amount and pattern of compounds in the excreta mimicked that of stressed snails or snails fed high-protein diets (equally balanced between ammonia and urea), rather than the ammonotelic pattern of naturally fed snails.

To test the premise that the increased urea production during stress might be an ammonia detoxification mechanism, Meyer and Becker (1980) placed snails in water containing between 5 and 35 m$M$ ammonium chloride. With 35 m$M$ $NH_4Cl$, hemolymph urea increased 4-fold to 1.42 m$M$ and ammonia rose 25-fold to 2.3 m$M$. Under these conditions, urea excretion increased 2- to 3-fold to 3.5 $\mu$atom urea N/g wet weight/day and [14C] incorporation from [14C]bicarbonate into urea by hepatopancreas tissue from these animals increased about 8-fold over control animals. Although the snails seemed relatively tolerant of high blood ammonia levels, the results seemed to indicate some role for urea production in ammonia detoxification.

With *L. stagnalis* and other lymneids, ammonia makes up 45–70% of the nitrogen excreted (4–7 $\mu$mol/g wet weight/day) with urea, uric acid, and some unidentified compounds comprising the balance (Freidl, 1975; Duerr, 1967, 1968). Freidl (1975) reported excretion of small amounts of free amino acids (ninhydrin-positive material) and proteinaceous substances (Folen-phenol reagent positive) in addition to these compounds. This proteinaceous material was also excreted or lost by *B. glabrata* (Machado and Vierra, 1979) and may be mucus (Wilson, 1968). Addition of starch or sugar to the diet reduced the amount of nitrogen excreted (ammonia and urea) and caused large increases in hemolymph glucose concentrations (Freidl, 1971, 1975; Veldhuijzen, 1975). With starvation–estivation for 3–5 months under dry pond conditions, the total amount of nitrogen retained in the tissues of *Bakerilymnaea cockerelli* as ammonia, urea, and uric acid increased from 3.5 to 35–37 $\mu$mol N/g wet weight (Newman and Thomas, 1975). After 5 months, 77% of this N was in the form of uric acid, with ammonia and urea accounting for 5 and 17%, respectively. In a separate experiment Newman and Thomas (1975) reported a large and unexplained increase in stored urea with snails held 30 days. Unfortunately, they did not measure nitrogen lost to the environment nor did they account for the change in relative amounts of ammonia, urea, and uric acid. Storage of uric acid in the tissues of active, nonestivating *L. stagnalis* and *L. palustris* has been noted by Dogterom (1980) and Duerr (1967), respectively. The lymneids probably have a complete set of urea cycle enzymes (see Section IV, D on biosynthesis) and may use urea biosynthesis to detoxify ammonia to some extent when starving or estivating. Although more studies are needed on the lymneids, they would appear to be ammonotelic when actively feeding and may show a nitrogen excretion pattern similar to *B. glabrata* during estivation–starvation.

Another freshwater snail, *Lanistes baltemia,* is ammonotelic for the most part, but apparently excretes large amounts of urea and small amounts of uric acid from time to time (Haggag and Fouad, 1968).

In conclusion, freshwater pulmonate snails are mainly ammonotelic when actively feeding in water. Smaller amounts of nitrogen (30% of total) may be lost as urea, uric acid, and free amino acids. These snails apparently have little or no urease so that all urea formed, whether from dietary arginine, *de novo* synthesized arginine, or possibly uricolysis in some cases, is lost as urea and not catabolized to ammonia. Protein loss, through mucus production, may account for a considerable fraction of the nitrogen loss in some species. During estivation, starvation, or schistosome infection, the freshwater pulmonates hold the amount of ammonia in the hemolymph and excreta fairly constant and increase nitrogen output by increasing the urea production. This increased urea production, causing a switch from ammonotelism to a combined ammonotelism–ureotelism, results from an increased *de novo* urea biosynthesis through activation of the urea cycle. Because the snails can tolerate very high non-physiological ammonium ion levels in the blood and bathing media, the induction of urea biosynthesis by the high ammonia levels is puzzling.

# References

Abe, S., and Kaneda, T. (1975a). Studies on the effect of marine products on cholesterol metabolism in rats. X. Isolation of β-homobetaine from oyster and betaine contents in oyster and scallop. *Nippon Suisan Gakkaishi* **41**, 467–471.

Abe, S., and Kaneda, T. (1975b). Studies on the effect of marine products on cholesterol metabolism in rats. XI. Isolation of a new betaine, Ulvaline from a green algae *Monostroma nitidum* and its depressing effect on plasma cholesterol levels. *Nippon Suisan Gakkaishi* **41**, 567–571.

Addink, A. D. F., and Veenhof, P. R. (1975). Regulation of mitochondrial matrix enzymes in *Mytilus edulis* L. *Proc. Eur. Symp. Mar. Biol., 9th, Oban, Scot.* pp. 109–119.

Ahmad, T. A., and Chaplin, A. E. (1979). Seasonal variation in the anaerobic metabolism of the mussel *Mytilus edulis. Comp. Biochem. Physiol. B* **64B**, 351–356.

Aiello, E. (1965). The fate of serotonin in the cell of the mussel *Mytilus edulis. Comp. Biochem. Physiol.* **14**, 71–82.

Aikawa, T. (1966). Adenosine aminohydrolyase from the clam, *Meretrix meretrix lusoria* (Gmelin). *Comp. Biochem. Physiol.* **17**, 271–284.

Aikawa, T., Umemori, Y., and Ishida, S. (1967). Effects of adenosine on action potentials in the oyster heart, with special reference to the activity of adenosine aminohydrolase. *Comp. Biochem. Physiol.* **21**, 579–586.

Aikawa, T., Umemori-Aikawa, Y., and Fisher, J. R. (1977). Purification and properties of the adenosine deaminase from the midgut gland of a marine bivalved mollusc, *Atrina* spp. *Comp. Biochem. Physiol. B* **58B**, 357–364.

Allen, J. A., and Garrett, M. R. (1971a). Taurine in marine invertebrates. *Adv. Mar. Biol.* **9**, 205–253.

Allen, J. A., and Garrett, M. R. (1971b). The excretion of ammonia and urea by *Mya arenaria* L. (Mollusca: Bivalva). *Comp. Biochem. Physiol.* **39**, 633–642.

Allen, J. A., and Garrett, M. R. (1972). Studies on taurine in the euryhaline bivalve, *Mya arenaria. Comp. Biochem. Physiol. A* **41A**, 307–317.

Allen, K. (1961). The effect of salinity on the amino acid concentration in *Rangia cuneata* (Pelecypoda). *Biol. Bull. (Woods Hole, Mass.)* **121**, 419–424.

Allen, K., and Awapara, J. (1960). Metabolism of sulfur amino acids in *Mytilus edulis* and *Rangia cuneata. Biol. Bull. (Woods Hole, Mass.)* **118**, 173–182.

Allen, W. V., and Kilgore, J. (1975). The essential amino acid requirements of the red abalone, *Haliotis rufescens. Comp. Biochem. Physiol. A* **50A**, 771–775.

Amende, L. M., and Pierce, S. K. (1978). Hypotaurine: the identity of an unknown ninhydrin-positive compound co-eluting with urea in amino acid extracts of bivalve tissue. *Comp. Biochem. Physiol. B* **59B**, 257–266.

Amende, L. M., and Pierce, S. K. (1980). Free amino acid mediated volume regulation of isolated *Noetia ponderosa* red blood cells: control by $Ca^{2+}$ and ATP. *J. Comp. Physiol.* **138**, 291–298.

Anderson, J. W. (1975). The uptake and incorporation of glycine by the gills of *Rangia cuneata* (Mollusca; Bivalvia) in response to variations in salinity and sodium. *In* "Physiological Ecology of Estarine Organisms" (F. J. Vernberg, Ed.), pp. 239–258. University of South Carolina Press, Columbia, South Carolina.

Anderson, J. W., and Bedford, W. B. (1973). The physiological response of the estuarine clam, *Rangia cuneata* (Gray), to salinity. II. Uptake of glycine. *Biol. Bull. (Woods Hole, Mass.)* **144**, 229–247.

Andrews, T. R., and Reid, R. G. B. (1972). Ornithine cycle and uricotelic enzymes in four bivalve molluscs. *Comp. Biochem. Physiol. B* **42B**, 475–491.

Aoki, T., and Oya, H. (1979). Glutamine-dependent carbamoyl-phosphate synthetase and control of pyrimidine biosynthesis in the parasitic helminth *Schistosoma mansoni*. *Comp. Biochem. Physiol. B* **63B**, 511–515.

Aoki, T., Oya, H., Mari, M., and Tatibana, M. (1975). Glutamine-dependent carbamoyl-phosphate synthetase in *Ascaris* ovary and its regulatory properties. *Proc. Jpn. Acad. Sci.* **51**, 733–736.

Arch, J. R. S. and Newsholme, E. A. (1978). Activities and some properties of 5'-nucleotidase, adenosine kinase, and adenosine deaminase in tissues from vertebrates and invertebrates in relation to the control of the concentration and the physiological role of adenosine. *Biochem. J.* **174**, 965–977.

Awapara, J. (1976). The metabolism of taurine in the animal. *In* "Taurine" (R. Huxtable and A. Barbeau, eds.), pp. 1–19. Raven, New York.

Awapara, J., and Campbell, J. W. (1964). Utilization of $C^{14}O_2$ for the formation of some amino acids in three invertebrates. *Comp. Biochem. Physiol.* **11**, 231–235.

Awapara, J., Campbell, J. W., and Peck, E. (1963). Formation of amino acids from $C^{14}O_2$ and D-glucose-U-$C^{14}$ in invertebrates. *Fed. Proc., Fed. Am. Soc. Exp. Biol.* **22**, 533.

Badman, D. G. (1971). Nitrogen excretion in two species of pulmonate land snails. *Comp. Biochem. Physiol. A* **38A**, 663–673.

Baginski, R. M. (1978). Evidence for the importance of altered protein metabolism during high salinity acclimation. *Am. Zool.* **18**, 616.

Baginski, R. M., and Pierce, S. K. (1975). Anaerobiosis: a possible source of osmotic solute for high-salinity acclimation in marine molluscs. *J. Exp. Biol.* **62**, 589–598.

Baginski, R. M., and Pierce, S. K. (1977). The time course of intracellular free amino acid accumulation of tissues of *Modiolus demissus* during high salinity adaptation. *Comp. Biochem. Physiol. A* **57A**, 407–412.

Baginski, R. M., and Pierce, S. K. (1978). A comparison of amino acid accumulation during high salinity adaptation with anaerobic metabolism in the ribbed mussel, *Modiolus demissus*. *J. Exp. Zool.* **203**, 419–428.

Baker, P. F., and Potashner, S. J. (1973). The role of metabolic energy in the transport of glutamate by invertebrate nerve. *Biochim. Biophys. Acta* **318**, 123–139.

Baldwin, E. (1935). Problems of nitrogen catabolism in invertebrates. III. Arginase in the invertebrates, with a new method for its determination. *Biochem. J.* **29**, 252–262.

Baldwin, E., and Needham, J. (1934). Problems of nitrogen catabolism in invertebrates. I. The snail (*Helix pomatia*). *Biochem. J.* **28**, 1372–1392.

Ballantyne, J. S., Hochachka, P. W., and Mommsen, T. P. (1981). Studies on the metabolism of the migratory squid, *Loligo opalescens:* enzymes of tissues and heart mitochondria. *Mar. Biol. Lett.* **2**, 75–85.

Bamford, D. R., and Campbell, E. (1976). The effect of environmental factors on the absorption of L-phenylalanine by the gill of *Mytilus edulis. Comp. Biochem. Physiol. A* **53A,** 295–299.

Bamford, D. R., and McCrea, R. (1975). Active absorption of neutral and basic amino acids by the gill of the common cockle, *Cerastoderma edule. Comp. Biochem. Physiol. A* **50A,** 811–817.

Barankiewicz, J., and Jezewska, M. M. (1973). Purine nucleoside: orthophosphate ribosyltransferase activity in the hepatopancreas of *Helix pomatia* (Gastropoda). *Comp. Biochem. Physiol. B* **46B,** 177–186.

Barankiewicz, J., and Jezewska, M. M. (1976). Inosine–guanosine and adenosine phosphorylase activities in hepatopancreas of *Helix pomatia* (Gastropoda). *Comp. Biochem. Physiol. B* **54B,** 239–242.

Baret, R., Mourgue, M., Broc, A., and Charmot, J. (1965). Étude comparative de la desamidination de l'acide γ-guanidinobutyrique et de l'arginine par l'hépatopancréas ou le foie de divers invertébres. *C. R. Seances Soc. Biol. Ses Fil.* **159,** 2446–2450.

Baret, R., Girard, S., and Riou, D. (1972). Sur certaines propriétés des arginases du tissu hépatopancréatique d'*Helix pomatia* Lin. et d'*Helix aspersa* Müll. *Biochimie* **54,** 421–430.

Bargoni, N., and Sisini, A. (1962). Sul metabolismo della D- ed L-serina nell'epatopancreas di *Murex trunculus. Sóc. Ital. Biol. Sper.* **38,** 905–906.

Bartberger, C. A., and Pierce, S. K. (1976). Relationship between ammonia excretion rates and hemolymph nitrogenous compounds of a euryhaline bivalve during low salinity adaptation. *Biol. Bull. (Woods Hole, Mass.)* **150,** 1–14.

Bayne, B. L., ed. (1976). "Marine Mussels: Their Ecology and Physiology." Cambridge Univ. Press, London and New York.

Bayne, B. L., and Scullard, C. (1977a). Rates of nitrogen excretion by species of *Mytilus* (Bivalvia: Mollusca). *J. Mar. Biol. Assoc. U.K.* **57,** 355–369.

Bayne, B. L., and Scullard, C. (1977b). An apparent specific dynamic action in *Mytilus edulis* L. *J. Mar. Biol. Assoc. U.K.* **57,** 371–378.

Bayne, R. A., and Freidl, F. E. (1968). The production of externally measurable ammonia and urea in the snail, *Lymnaea stagnalis jugularis* Say. Comp. Biochem. Physiol. **25,** 711–717.

Becker, W., and Schmale, H. (1975). The nitrogenous products of degradation—ammonia, urea, and uric acid—in the hemolymph of the snail *Biomphalaria glabrata. Comp. Biochem. Physiol. A* **51A,** 407–411.

Becker, W., and Schmale, H. (1978). The ammonia and urea excretion of *Biomphalaria glabrata* under different physiological conditions: starvation, infection with *Schistosoma mansoni*, dry keeping. *Comp. Biochem. Physiol. B* **59B,** 75–79.

Bedford, J. J. (1969). The soluble amino acid pool in *Siphonaria zelandica:* its composition and the influence of salinity changes. *Comp. Biochem. Physiol.* **29,** 1005–1014.

Bedford, J. J. (1971a). Osmoregulation in *Melanopsis trifasciata* Gray 1843. II. The composition of the haemocoelic fluid. *Physiol. Zool.* **45,** 261–269.

Bedford, J. J. (1971b). Osmoregulation in *Melanopsis trifasciata* Gray 1843. III. The intracellular nitrogenous compounds. *Comp. Physiol. Biochem. A* **40A,** 899–910.

Bedford, J. J. (1971c). Osmoregulation in *Melanopsis trifasciata*. IV. The possible control of intracellular isosmotic regulation. *Comp. Biochem. Physiol. A* **40A,** 1015–1027.

Bedford, J. J. (1973). Osmotic relationships in freshwater mussel, *Hydrella menziesi* Gray (Lamellebranchia: Unionidae). *Arch. Int. Physiol. Biochim.* **81,** 819–831.

Bender, A. E., and Krebs, H. A. (1950). The oxidation of various synthetic α-amino acids by mammalian D-amino acid oxidase, L-amino acid oxidase of cobra venom and the L- and D-amino acid oxidases of *Neurospora crassa*. *Biochem. J.* **46**, 210–219.

Bergeron, J. P., and Alayse-Ganet, A. M. (1981). Aspartate transcarbamylase de la coquille Saint-Jacque *Pecten maximus* L. (Mollusque: Lamellibranche): méthod de dosoge et variations de l'activité dans le manteau et la gonade. *J. Exp. Mar. Biol. Ecol.* **50**, 99–117.

Bird, M. I., and Nunn, P. B. (1979). Glycine formation from L-threonine in intact isolated rat liver mitochondria. *Biochem. Soc. Trans.* **7**, 1276–1277.

Bishop, S. H. (1976). Nitrogen metabolism and excretion: regulation of intracellular amino acid concentrations. *In* "Estuarine Processes" (M. Wiley, ed.), Vol. 1, pp. 414–431. Academic Press, New York.

Bishop, S. H., Burcham, J. M., and Greenwalt, D. E. (1980). $\alpha,\alpha$-Iminodiproprionic acid synthesis in a bivalve mollusc (*Modiolus demissus*). *Fed. Proc. Fed. Am. Soc. Exp. Biol.* **39**, 2085.

Bishop, S. H., Greenwalt, D. E., and Burcham, J. M. (1981). Amino acid cycling in ribbed mussel tissues subjected to hyperosmotic shock. *J. Exp. Zool.* **215**, 277–287.

Blaschko, H., and Hawkins, J. (1952). D-Amino-acid oxidase in the molluscan liver. *Biochem. J.* **52**, 306–310.

Blaschko, H., and Himms, J. M. (1955). D-Glutamic acid oxidase in cephalopod liver. *J. Physiol. (London)* **128**, 7P.

Blaschko, H., and Hope, D. B. (1956). The oxidation of L-amino acids by *Mytilus edulis*. *Biochem. J.* **62**, 335–339.

Blaschko, H., and Levine, W. G. (1960). A comparative study of hydroxyindole oxidases. *Br. J. Pharmacol.* **15**, 625–633.

Blaschko, H., and Milton, H. S. (1960). Oxidation of 5-hydroxytryptamine and related compounds by *Mytilus* gill plates. *Br. J. Pharmacol.* **15**, 42–46.

Bricteux-Grégoire, S., and Florkin, M. (1962). Les excreta azotes de l'escargot "*Helix pomatia*" et leur origine métabolique. *Arch. Int. Physiol. Biochim.* **70**, 496–506.

Bricteux-Grégoire, S., and Florkin, M. (1964). Researche des enzymes du cycle de l'ureogenese chez l'escargot *Helix pomatia* L. *Comp. Biochem. Physiol.* **12**, 55–60.

Bricteux-Grégoire, S., Duchâteau-Bosson, G., Jeuniaux, C., and Florkin, M. (1964a). Constituents osmotique actifs des muscles adducteurs de *Mytilus edulis* adaptée a l'eau de mer ou a l'eau saumâtre. *Arch. Int. Physiol. Biochim.* **72**, 116–123.

Bricteux-Grégoire, S., Duchâteau-Bosson, G., Jeuniaux, C., and Florkin, M. (1964b). Constituents osmotique actifs des muscles adducteurs d'*Ostrea edulis* adaptée a l'eau de mer ou a l'eau saumâtre. *Arch. Int. Physiol. Biochim.* **72**, 267–275.

Bricteux-Grégoire, S., Duchâteau-Bosson, G., Jeuniaux, C., and Florkin, M. (1964c). Constituents osmotique actifs des muscles adducteurs de *Gryphaea angulata* adaptée a l'eau de mer ou a l'eau saumâtre. *Arch. Int. Physiol. Biochim.* **72**, 835–842.

Brothers, V. M., Tsubota, S. I., Germeraad, S. E., and Fristrom, J. W. (1978). The *rudimentary* locus of *Drosophila melanogaster*: partial purification of a carbamyl phosphate synthase-aspartate transcarbamylase-dehydroorotase complex. *Biochem. Genet.* **6**, 321–332.

Bryant, C. (1965). The metabolism of the digestive glands of two species of marine gastropod (*Melanerita melanotragus* and *Austrocochlea obtusa*). *Comp. Biochem. Physiol.* **14**, 223–230.

Bryant, C., Hines, W. J. W., and Smith, M. J. H. (1964). Intermediary metabolism in some terrestrial molluscs (*Pomatias, Helix,* and *Cepaea*). *Comp. Biochem. Physiol.* **11**, 147–153.

Burcham, J. M., Greenwalt, D. E., and Bishop, S. H. (1980). Amino acid metabolism in euryhaline bivalves: the L-amino-acid oxidase from ribbed mussel gill tissue. *Mar. Biol. Lett.* **1**, 329–340.

Caldwell, P. C., and Lea, T. J. (1975). Some effects of ouabain on the transport of amino acids into squid giant axons. *J. Physiol.* (*London*) **245**, 91–92.

Caldwell, P. C., and Lea, T. J. (1978). Glycine fluxes in squid giant axons. *J. Physiol.* (*London*) **278**, 1–25.

Campbell, J. W. (1966). A comparative study of molluscan and mammalian arginases. *Comp. Biochem. Physiol.* **18**, 179–199.

Campbell, J. W. (1973). Nitrogen excretion. *In* "Comparative Animal Physiology" (C. L. Prosser, ed.), Vol. 1, pp. 279–316. Saunders, Philadelphia, Pennsylvania.

Campbell, J. W., and Bishop, S. H. (1970). Nitrogen metabolism in molluscs. *In* "Comparative Biochemistry of Nitrogen Fixation" (J. W. Campbell, ed.), Vol. 1, pp. 103–206. Academic Press, New York.

Campbell, J. W., and Boyan, B. D. (1976). On the acid-base balance of gastropod molluscs. *In* "The Mechanisms of Mineralization in the Invertebrates and Plants" (N. Watabe and K. M. Wilbur, eds.), pp. 109–133. Univ. of South Carolina Press, Columbia.

Campbell, J. W., and Speeg, K. V. (1968a). Arginine biosynthesis and metabolism in terrestrial snails. *Comp. Biochem. Physiol.* **25**, 3–32.

Campbell, J. W., and Speeg, K. V. (1968b). Tissue distribution of enzymes of arginine biosynthesis in terrestrial snails. *Z. Vgl. Physiol.* **61**, 164–175.

Campbell, J. W., and Speeg, K. V. (1969). Ammonia and the biological deposition of calcium carbonate. *Nature* (*London*) **224**, 725–726.

Campbell, J. W., and Vorhaben, J. E. (1979). The purine nucleotide cycle in *Helix* hepatopancreas. *J. Comp. Physiol.* **129**, 137–144.

Campbell, J. W., Drotman, R. B., McDonald, J. A., and Trammel, P. R. (1972). Nitrogen metabolism in terrestrial invertebrates. *In* "Nitrogen Metabolism and the Environment" (J. W. Campbell and L. Goldstein, eds.), pp. 11–54. Academic Press, New York.

Cardot, J. (1974). La monoamine-oxydise du système nerveux et du coeur chez le mollusque *Helix pomatia*. *C. R. Seances Soc. Biol. Ses Fil.* **168**, 471–474.

Cardot, J. (1979). les monoamines chez les mollusques. I. Les catécholamines: biosynthèse, mese en place et inactivation. *J. Physiol.* (*Paris*) **75**, 689–713.

Cardot, J. (1980). Hydroxylation de la phénylalanine et biosynthèse de catécholamines par le tissue nerveux du mollusque *Helix pomatia*. *J. Physiol.* (*Paris*) **76**, 3A.

Casola, L., Giuditta, A., and Rocca, E. (1964). Barbiturate inhibition of D-aspartate oxidoreductase: mechanism of action and structural requirements. *Arch. Biochem. Biophys.* **107**, 57–61.

Castille, F. L., and Lawrence, A. L. (1980). Uptake of amino acids and hexoses from sea water by octopod hatchlings. *Physiologist* **21**, 18.

Cavallini, D., Scandura, R., Dupre, S., Federici, G., Santoro, L., Ricci, G., and Barro, D. (1976). Alternative pathways of taurine biosynthesis. *In* "Taurine" (R. Huxtable and A. Barbeau, eds.), pp. 59–66. Raven, New York.

Cavallini, D., Federici, G., Dupre, S., Caunella, C., and Scandurra, R. (1979). Ambiguities in the enzymology of sulfur containing compounds. *Pure Appl. Chem.* **52**, 147–152.

Chambers, J. E., McCorkle, F. M., Carroll, J. W., Hertz, J. R., Lewis, L., and Yarbough, J. D. (1975). Variation in enzyme activities of the american oyster (*Crassostrea virginica*) relative to size and season. *Comp. Biochem. Physiol. B* **51B**, 145–150.

Chaplin, A. E., and Loxtan, J. (1976). Tissue differences in the response of the mussel *Mytilus edulis* to experimentally induced anaerobiosis. *Biochem. Soc. Trans.* **4**, 437–441.

Chen, C., and Awapara, J. (1969a). Intracellular distribution of enzymes catalyzing succinate production from glucose in *Rangia* muscle. *Comp. Biochem. Physiol.* **30**, 727–737.

Chen, C., and Awapara, J. (1969b). Effect of oxygen on the end-products of glycolysis in *Rangia cuneata*. *Comp. Biochem. Physiol.* **31**, 395–401.

Chetty, C. S. R., Naidu, R. C., and Swami, K. S. (1979). Changes in glutamine levels during starvation and aestivation in the indian apple snail *Pila globosa* (Swainson). *Experientia* **35**, 179–180.

Christensen, H. N. (1969). Some special kinetic problems of transport. *Adv. Enzymol. Relat. Areas Mol. Biol.* **32**, 1–20.

Christensen, H. N. (1975). "Biological Transport." Benjamin, New York.

Christie, J. D., and Michelson, E. D. (1975). Transaminase levels in the digestive gland-gonad of *Schistosoma mansoni*-infected *Biomphalaria glabrata*. *Comp. Biochem. Physiol. B* **50B**, 233–236.

Christie, J. D., Foster, W. B., and Stauber, L. A. (1974). The effect of parasitism and starvation on carbohydrate reserves of *Biomphalaria glabrata*. *J. Invertebr. Pathol.* **23**, 297–302.

Chubb, J., and Huxtable, R. J. (1978). Transport and biosynthesis of taurine in the stressed heart. *In* "Taurine and Neurological Disorders" (A. Barbeau and R. J. Huxtable, eds.), pp. 161–178. Raven, New York.

Clark, S. W., and Rudolph, F. B. (1979). Purine biosynthesis in *Helix aspersa:* metabolic fate of labeled precursors. *Comp. Biochem. Physiol. B* **63B**, 369–371.

Collicutt, J. M., and Hochachka, P. W. (1977). The anaerobic oyster heart: Coupling of glucose and aspartate fermentation. *J. Comp. Physiol. A.* **115**, 147–157.

Conway, A. F., Black, R. E., and Morrell, J. B. (1969). Uric acid synthesis in embryos of the pulmonate pond snail, *Limnaea palustris:* evidence for a unique pathway. *Comp. Biochem. Physiol.* **30**, 793–802.

Cooley, L., Crawford, D. R., and Bishop, S. H. (1976). Urease from the lugworm, *Arenicola cristata*. *Biol. Bull. (Woods Hole, Mass.)* **151**, 96–107.

Corrigan, J. J. (1969). D-Amino acids in animals. *Science* **164**, 142–149.

Crisp, M., Gill, C. W., and Thompson, M. C. (1981). Ammonia excretion by *Navaorius reticulatus* and *Buccinum undatum* (Gastropoda: Prosobranchia) during starvation and after feeding. *J. Mar. Biol. Assoc. U.K.* **61**, 381–390.

Crowe, J. H. (1981). Transport of exogenous substrate and cell volume regulation in bivalve molluscs. *J. Exp. Zool.* **215**, 363–370.

Crowe, J. H., Dickson, K. A., Otto, J. L., Calon, R. D., and Farley, K. D. (1977). Uptake of amino acids by the mussel *Modiolus deuissus*. *J. Exp. Zool.* **202**, 323–332.

D'Aniello, A., and Giuditta, A. (1977). Identification of D-aspartic acid in the brain of *Octopus vulgaris* Lam. *J. Neurochem.* **29**, 1053–1057.

D'Aniello, A., and Giuditta, A. (1978). Presence of D-aspartate in squid axoplasm and in other regions of the cephalopod nervous system. *J. Neurochem.* **31**, 1107–1108.

D'Aniello, A., and Rocca, E. (1972). D-Aspartate oxidase the hepatopancreas of *Octopus vulgaris* Lam. *Comp. Biochem. Physiol. B* **41B**, 625–633.

D'Aniello, D., Palescandolo, R., and Scardi, V. (1975). The distribution of the D-aspartate oxidase activity in cephalopoda. *Comp. Biochem. Physiol. B* **50B**, 209–210.

Deffner, G. G. J. (1961a). The dialyzable free organic constituents of squid blood: A comparison with nerve axoplasm. *Biochim. Biophys. Acta* **47**, 378–388.

Deffner, G. G. J. (1961b). Chemical investigations of the giant nerve fibers of the squid. V. Quanternary ammonium ions in axoplasm. *Biochim. Biophys. Acta* **50**, 555–564.

Delauny, H. (1931). L'excrétion azotée des invertébrés. *Biol. Rev. (Cambridge Philos. Soc.)* **6**, 265–301.

Delauny, H. (1934). Le métabolisme de l'ammoniaque d'après les recherches relatives aux invertébrés. *Ann. Physiol. Physicochim. Biol.* **10**, 695–724.

Delluva, A. M., Markley, K., and Davis, R. E. (1968). The absence of gastric urease in germ-free animals. *Biochim. Biophys. Acta* **151**, 646–650.

de Vooys, C. G. N, and de Zwaan, A. (1978). The rate of oxygen consumption and ammonia excretion by *Mytilus edulis* after various periods of exposure to air. *Comp. Biochem. Physiol. A* **60A**, 343–347.

de Zwaan, A. (1977). Anaerobic energy metabolism in bivalve molluscs. *Oceanogr. Mar. Biol.* **15**, 103–187.

de Zwaan, A., and van Marrewijk, J. A. (1973). Anaerobic glucose degradation in the sea mussel, *Mytilus edulis* L. *Comp. Biochem. Physiol. B* **44B**, 429–439.

de Zwaan, A., and Wijsman, T. C. M. (1976). Anaerobic metabolism in bivalvia (Mollusca) characteristics of anaerobic metabolism. *Comp. Biochem. Physiol. B* **54B**, 313–324.

de Zwaan, A., de Bont, A. M. T., and Kluytmans, J. H. F. M. (1975). Metabolic adaptations on the aerobic–anaerobic transition in the sea mussel, *Mytilus edulis* L. *Proc. Eur. Symp. Mar. Biol. 9th, Oban, Scot.* pp. 121–138.

Dixon, M., and Kenworthy, P. (1967). D-Aspartate oxidase of kidney. *Biochim. Biophys. Acta* **146**, 54–76.

Dixon, N. E., Gazzola, C., Blakeley, R. L., and Zirner, B. (1975). Jackbean urease (E. C. 3.5.1.5). A metalloenzyme. A simple biological role for nickel. *J. Am. Chem. Soc.* **97**, 4131–4133.

Dogterom, A. A. (1980). The effect of the growth hormone of the freshwater snail *Lymanaea stagnalis* on biochemical composition and nitrogenous wastes. *Comp. Biochem. Physiol. B* **65B**, 163–167.

Duerr, F. G. (1967). The uric acid content of several species of prosobranch and pulmonate snails as related to nitrogen excretion. *Comp. Biochem. Physiol.* **22**, 333–340.

Duerr, F. G. (1968). Excretion of ammonia and urea in seven species of marine prosobranch snails. *Comp. Biochem. Physiol.* **26**, 1051–1059.

DuPaul, W. D., and Webb, K. L. (1970). The effect of temperature on salinity-induced changes in the free amino acid pool of *Mya arenaria*. *Comp. Biochem. Physiol.* **32**, 785–801.

DuPaul, W. D., and Webb, K. L. (1974). Salinity induced changes in the alanine and aspartic aminotransferase activity in three marine bivalve molluscs. *Arch. Int. Physiol. Biochim.* **82**, 817–822.

Ellington, W. R. (1981). Energy metabolism during hypoxia in the isolated, perfused ventricle of the Whelk, *Busycon contrarium* Conrad. *J. Comp. Physiol. B.* **142**, 457–464.

Ellis, L. L., Burcham, J. M., and Bishop, S. H. (1981). Control of intracellular glycine levels in gills of the ribbed mussel, *Modiolus demissus*. *Fed. Proc. Fed. Am. Soc. Exp. Biol.* **40**, 1684.

Emerson, D. N. (1969). Influence of salinity on ammonia excretion rates and tissue constituents of euryhaline invertebrates. *Comp. Biochem. Physiol.* **29**, 1115–1133.

Emerson, D. N., and Duerr, F. G. (1967). Some affects of starvation in the intertidal *Littorina planaxis* (Philippi, 1847). *Comp. Biochem. Physiol.* **20**, 45–53.

Ericson, L. E. (1960a). Transmethylation in *Anodonta cygnea*. *Nature (London)* **185**, 465–466.

Ericson, L. E. (1960b). Betaine Homocysteine-Methyl-Transferases. III. The methyl donor specificity of the transferase isolated from pig liver. *Acta Chem. Scand.* **14**, 2127–2134.

Falany, C. N., and Freidl, F. E. (1981). Amino acid transamination in the freshwater clams

*Anodonta couperiana* and *Popenaias buckleyi*. *Comp. Biochem. Physiol. B* **68B**, 119–123.

Fellman, J. H., Roth, E. S., and Fujita, T. S. (1978). Taurine is not metabolized to isethionate in mammalian tissue. *In* "Taurine and Neurological Disorders" (A. Barbeau and R. J. Huxtable, eds.), pp. 19–23. Raven, New York.

Fellman, J. H., Roth, E. S., Avedovech, N. A., and McCarthy, K. D. (1980). The metabolism of taurine to isethionate. *Arch. Biochem. Biophys.* **204**, 560–567.

Fields, J. H. A., and Hochachka, P. W. (1981). Purification and properties of alanopine dehydrogenase from the adductor muscle of the oyster, *Crassostrea gigas* (Mollusca, Bibalvia). *Eur. J. Biochem.* **114**, 615–621.

Fields, J. H. A., Eng, A. K., Ramsden, W. D., Hochachka, P. W., and Weinstein, B. (1980). Alanopine and strombine are novel imino acids produced by a dehydrogenase found in the adductor muscle of the oyster *Crassostrea gigas*. *Arch. Biochem. Biophys.* **201**, 110–114.

Fisher, H. F. (1973). Glutamate dehydrogenase-ligand complexes and their relationship to the mechanism of the reaction. *Adv. Enzymol. Relat. Areas Mol. Biol.* **39**, 369–417.

Flashner, M. I. S., and Massey, J. (1974). Regulatory properties of the flavoprotein L-lysine monooxygenase. *J. Biol. Chem.* **249**, 2587–2592.

Flohé, L., and Günzler, W. A. (1976). Glutathione-dependent enzymatic oxidoreduction reactions. *In* "Glutathione: Metabolism and Function" (I. M. Arias and W. B. Jakoby, eds.), pp. 17–35. Raven, New York.

Florkin, M. (1966). Nitrogen metabolism. *In* "Physiology of Mollusca" (K. M. Wilbur and C. M. Yonge, eds.), Vol. 2, pp. 309–351. Academic Press, New York.

Florkin, S., and Bricteux-Grégoire, S. (1972). Nitrogen metabolism in molluscs. *In* "Chemical Zoology" (M. Florkin and B. T. Scheer, eds.), Vol. 7, pp. 301–348. Academic Press, New York.

Freidl, F. E. (1971). Hemolymph glucose in the freshwater pulmonate snail *Lymnaea stagnalis:* basal values and an effect of ingested carbohydrate. *Comp. Biochem. Physiol. A* **39A**, 605–610.

Freidl, F. E. (1974). Nitrogen excretion by freshwater pulmonate snail, *Lymnaea stagnalis jugularis* Say. *Comp. Biochem. Physiol. A* **49A**, 617–622.

Freidl, F. E. (1975). Nitrogen excretion in *Lymnaea stagnalis:* an effect of ingested carbohydrate. *Comp. Biochem. Physiol. A* **52A**, 377–379.

Freidl, F. E. (1979). Some aspects of amino acid catabolism in the fresh water pulmonate snail *Lymnaea stagnalis*. *Malacologia* **18**, 595–604.

Freidl, F. E., and Bayne, R. A. (1966). Ureogenesis in the snail *Lymnaea stagnalis jugularis*. *Comp. Biochem. Physiol.* **17**, 1167–1173.

Fujimoto, D., and Adams, E. (1965). D-Serine dehydrase of earthworm tissues. *Biochim. Biophys. Acta* **105**, 596–599.

Fyhn, H. J. (1976). A note on the hyperosmotic regulation in the brackish-water clam *Rangia cuneata*. *J. Comp. Physiol. A.* **107**, 159–167.

Gäde, G. (1980). Biological role of octopine formation in marine molluscs. *Mar. Biol. Lett.* **1**, 121–135.

Gainey, L. F. (1978a). The response of the Corbiulidae (Mollusca: Bivalvia) to osmotic stress: the organismal response. *Physiol. Zool.* **51**, 68–78.

Gainey, L. F. (1978b). The response of the Corbiulidae (Mollusca: Bivalvia) to osmotic stress: the cellular response. *Physiol. Zool.* **51**, 79–91.

Gainey, L. F., and Greenberg, M. J. (1977). The physiological basis of the molluscan species-abundance curve: a speculation. *Mar. Biol. (Berlin)* **40**, 41–49.

Gaston, S., and Campbell, J. W. (1966). Distribution of arginase activity in mollusks. *Comp. Biochem. Physiol.* **17**, 259–270.

Gershenfeld, H. M. (1973). Chemical transmission in invertebrate central nervous systems and neuromuscular junctions. *Physiol. Rev.* **53**, 1–119.

Gibbs, K. L., and Bishop, S. H. (1977). Adenosine triphosphate-activated adenylate deaminase from marine invertebrate animals. Properties of the enzyme from lugworm (*Arenicola cristata*) body-wall muscle. *Biochem. J.* **163**, 511–516.

Gibbs, K. L., Reiss, P. M., Ribnik, L. R., and Bishop, S. H. (1976). AMP deaminase from marine invertebrates. *Am. Zool.* **16**, 211.

Gilbertson, D. E., and Schmid, L. S. (1975). Free amino acids in the hemolymph of five species of fresh water snails. *Comp. Biochem. Physiol. B* **51B**, 201–203.

Gilles, R. (1969). Effect of various salts on the activity of enzymes implicated in amino-acid metabolism. *Arch. Int. Physiol. Biochim.* **77**, 441–464.

Gilles, R. (1970). Intermediary metabolism and energy production in some invertebrates. *Arch. Int. Physiol. Biochim.* **78**, 313–326.

Gilles, R. (1972). Osmoregulation in three molluscs: *Acanthochitona discrepans* (Brown), *Glycymeris glycymeris* (L.) and *Mytilus edulis* (L.) *Biol. Bull.* (*Woods Hole, Mass.*) **142**, 25–35.

Gilles, R. (1974). Métabolisme des acides aminés et control du volume cellulaire. *Arch. Int. Physiol. Biochim.* **82**, 423–589.

Gilles, R. (1979). Intracellular organic effectors. *In* "Mechanisms of Osmoregulation in Animals" (R. Gilles, ed.), pp. 111–154. Wiley, New York.

Glahn, P. E., Manchow, P., and Roche, J. (1955). L-Amino acid oxidases of the hepatopancreas of lamellibranchs. *C. R. Seances Soc. Biol. Ses Fil.* **149**, 509–513.

Glynn, B. P., and Johnson, D. B. (1981). γ-Glutamyltranspeptidase in echinoderms and marine molluscs. *Comp. Biochem. Physiol. B* **68B**, 361–362.

Goldman, J. E., and Schwartz, J. H. (1977). Metabolism of ($^3$H) serotonin in the marine mollusc, *Aplysia californica*. *Brain Res.* **136**, 77–88.

Gorzkowski, B. (1969). Utilization of $^{14}$C-labelled purine precursors for uric acid synthesis in *Helix pomatia*. *Acta Biochim. Pol.* **16**, 1963–2000.

Gould, R. M., and Cottrell, G. A. (1974). Putrescine in Molluscs: identification and occurrence in neurons and other tissues. *Comp. Biochem. Physiol. B* **48B**, 591–597.

Greenberg, D. M. (1962). Pyrroline-5-carboxylate reductase. *In* "Preparation and Assay of Enzymes" (S. P. Colowick and N. O. Kaplan, eds.), Methods in Enzymology, Vol. 5, pp. 959–964. Academic Press, New York.

Greenwalt, D. E. (1981). Role of amino acids in cell volume control in the ribbed mussel: Alanine and proline metabolism. Ph.D. Thesis, Iowa State Univ., Ames.

Greenwalt, D. E., and Bishop, S. H. (1980). Effect of aminotransferase inhibitors in the pattern of free amino acid accumulation in isolated mussel hearts subjected to hyperosmotic stress. *Physiol. Zool.* **53**, 262–269.

Guchhait, R. B., Bourgeois, J. G., and Bueding, F. (1980). Biogenic amine metabolism in *Biomphalaria glabrata*. I. Catechol-*o*-methyltransferase activity. *Int. J. Neurosci.* **11**, 17–23.

Haggag, G., and Fouad, Y. (1968). Comparative study of nitrogenous excretion in terrestrial and fresh-water gastropods. *Z. Vgl. Physiol.* **57**, 428–431.

Hammen, C. S. (1966). Carbon dioxide fixation in marine invertebrates. V. Rate and pathway in the oyster. *Comp. Biochem. Physiol.* **17**, 289–296.

Hammen, C. S. (1968). Aminotransferase activities and amino acid excretion of bivalve molluscs and brachiopods. *Comp. Biochem. Physiol.* **26**, 697–705.

Hammen, C. S., and Wilbur, K. M. (1959). Carbon dioxide fixation in marine invertebrates. I. The main pathway in the oyster. *J. Biol. Chem.* **234,** 1268–1271.

Hammen, C. S., Hanlon, D. P., and Lum, S. C. (1962). Oxidative metabolism of *Lingula. Comp. Biochem. Physiol.* **5,** 185–191.

Hammen, C. S., Miller, H. F., and Geer, W. H. (1966). Nitrogen excretion of *Crassostrea virginica. Comp. Biochem. Physiol.* **17,** 1199–1200.

Hanley, M. R. (1975). The identification of dansyl sarcosine and its occurrence in molluscs. *Experientia* **31,** 881–882.

Hanlon, D. P. (1975). The distribution of arginase and urease in marine invertebrates. *Comp. Biochem. Physiol. B* **52B,** 261–264.

Hanson, J. A., and Dietz, T. H. (1976). The role of free amino acids in cellular osmoregulation in the freshwater bivalve *Ligumia subrostrata. Can. J. Zool.* **54,** 1927–1931.

Harbison, G. R., and Fisher, J. R. (1973a). Purification, properties and temperature dependence of the adenosine deaminase from a poikilotherm (bay scallop). *Arch. Biochem. Biophys.* **154,** 84–95.

Harbison, G. R., and Fisher, J. R. (1973b). Comparative studies on the adenosine deaminases of several bivalved molluscs. *Comp. Biochem. Physiol. B* **46B,** 283–293.

Harbison, G. R., and Fisher, J. R. (1974). Substrate dependent apparent activation energies of the adenosine deaminases from bivalved molluscs. *Comp. Biochem. Physiol. B* **47B,** 27–32.

Hartman, W. J., Clark, W. G., Gyr, S. D., Jordon, A. L., and Leibold, R. A. (1960). Pharmacologically active amines and their biogenesis in the *Octopus. Ann. N.Y. Acad. Sci.* **90,** 637–649.

Henry, R. P., and Mangum, C. P. (1981a). Salt and water balance in the oligohaline clam, *Rangia cuneata.* I. Anisosmotic extracellular regulations. *J. Exp. Zool.* **211,** 1–10.

Henry, R. P., and Mangum, C. P. (1981b). Salt and water balance in the oligohaline clam, *Rangia cuneata.* III. Reduction of the free amino acid pool during low salinity adaptation. *J. Exp. Zool.* **211,** 25–32.

Henry, R. P., Mangum, C. P., and Webb, K. L. (1981). Salt and water balance in the oligohaline clam, *Rangia cuneata.* II. Accumulation of intracellular free amino acids during high salinity adaptation. *J. Exp. Zool.* **211,** 11–24.

Hochachka, P. W., Moon, T. W., Mustafa, T., and Storey, K. B. (1975). Metabolic sources of power for mantle muscle of a fast swimming squid. *Comp. Biochem. Physiol. B* **52B,** 151–158.

Hochachka, P. W., French, C. J., and Meredith, J. (1978). Metabolic and ultrastructural organization in *Nautilus* muscles. *J. Exp. Zool.* **205,** 51–62.

Hope, D. B., and Horncastle, K. C. (1967). The oxidation of lysine and oxalysine by *Mytilus edulis,* identification of the products formed in the presence and absence of catalase. *Biochem. J.* **102,** 910–916.

Hope, D. B., Horncastle, K. C., and Aplin, R. T. (1967). The dimerization of Δ-piperidine-2-carboxylic acid. *Biochem. J.* **105,** 663–667.

Horne, F. B. (1971). Accumulation of urea by a pulmonate snail during aestivation. *Comp. Biochem. Physiol. A* **38A,** 565–570.

Horne, F. B. (1973a). The utilization of foodstuffs and urea production by a land snail during estivation. *Biol. Bull. (Woods Hole, Mass.)* **144,** 321–330.

Horne, F. B. (1973b). Urea metabolism in an estivating terrestrial snail Bulimulus dealbatus. *Am. J. Physiol.* **224,** 781–787.

Horne, F. B. (1977a). Ureotelism in the slug, *Limax flavus* Linne. *J. Exp. Zool.* **199,** 227–232.

Horne, F. B. (1977b). Regulation of urea biosynthesis in the slug, *Limax flavus* Linne. *Comp. Biochem. Physiol. B* **56B**, 63–69.

Horne, F. B., and Barnes, G. (1970). Reevaluation of urea biosynthesis in prosobranch and pulmonate snails. *Z. Vgl. Physiol.* **69**, 452–457.

Horne, F. B., and Beck, S. (1979). Purine production during fasting in the slug, *Limax flavus* Linne. *J. Exp. Zool.* **209**, 309–316.

Horne, F., and Boonkoon, V. (1970). The distribution of the ornithine cycle enzymes in twelve gastropods. *Comp. Biochem. Physiol.* **32**, 141–153. -

Hoskin, F. C. G., and Brande, M. (1973). An improved sulphur assay applied to a problem of isethionate metabolism in squid axon and other nerves. *J. Neurochem.* **20**, 1317–1327.

Hoskin, F. C. G., and Kordik, E. R. (1977). Hydrogen sulfide as a precursor for the synthesis of isethionate in the squid giant axon. *Arch. Biochem. Biophys.* **180**, 583–586.

Hoskin, F. C. G., Pollock, M. L., and Prusch, R. D. (1975). An improved method for the measurement of $^{14}CO_2$ applied to a problem of cysteine metabolism in squid nerve. *J. Neurochem.* **25**, 445–449.

Hoyaux, J., Gilles, R., and Jeuniaux, C. (1976). Osmoregulation in molluscs of the intertidal zone. *Comp. Biochem. Physiol. A* **53A**, 361–365.

Huggins, A. K., and Woodruff, G. N. (1968). Histamine metabolism in invertebrates. *Comp. Biochem. Physiol.* **26**, 1107–1111.

Huggins, A. J., Rick, T., and Kerkut, G. A. (1967). A comparative study of the intermediary metabolism of L-glutamate in muscle and nervous tissue. *Comp. Biochem. Physiol.* **21**, 23–30.

Huxtable, R. (1976). Metabolism and function of taurine in the heart. *In* "Taurine" (R. Huxtable and A. Barbeau, eds.), p. 99–119. Raven, New York.

Huxtable, R. J. (1978). Regulation of taurine in the heart. *In* "Taurine and Neurological Disorders" (A. Barbeau and R. J. Huxtable, eds.), pp. 5–17. Raven, New York.

Ishida, S., Umemori, Y., and Aikawa, T. (1969). Production and decomposition of adenosine in the oyster. *Comp. Biochem. Physiol.* **28**, 465–469.

Ivanovici, A. M., Ranier, S. F., and Wadley, V. A. (1981). Free amino acids in three species of mollusc: Responses to factors associated with reduced salinity. *Comp. Biochem. Physiol. A* **70A**, 17–22.

Jacobsen, J. G., and Smith, L. H. (1968). Biochemistry and physiology of taurine and taurine derivatives. *Physiol. Rev.* **48**, 424–511.

Jarry, B. (1976). Isolation of a multifunctional complex containing the first three enzymes of pyrimidine biosynthesis in *Drosophila melanogaster*. *FEBS Lett.* **70**, 71–75.

Jefferies, H. P. (1972). A stress syndrome in the hard clam, *Mercenaria mercenaria*. *J. Invertebr. Pathol.* **20**, 242–251.

Jezewska, M. M. (1968). The presence of uric acid, xanthine, and guanine in the hemolymph of the snail *Helix pomatia* (Gastropoda). *Bull. Acad. Pol. Sci., Ser. Sci. Biol.* **16**, 73–76.

Jezewska, M. M. (1969). The nephridial excretion of guanine, xanthine and uric acid in slugs (Limacedae) and snails (Helicidae). *Acta Biochim. Pol.* **16**, 313–320.

Jezewska, M. M., and Sawicka, T. (1968). Purine and pyrimidine compounds in the albumen gland of *Helix pomatia* (Gastropoda) during hibernation and in the feeding periods. *Bull. Acad. Pol. Sci., Ser. Sci. Biol.* **16**, 197–201.

Jezewska, M. M., Gorzkowski, B., and Heller, J. (1963). Nitrogen compounds in snail *Helix pomatia* excretion. *Acta Biochim. Pol.* **10**, 55–65.

Jezewska, M. M., Gorzkowski, B., and Heller, J. (1964). Utilization of (1-$^{14}$C) glycine in purine biosynthesis in *Helix pomatia*. *Acta Biochim. Pol.* **11**, 135–138.

Johnson, A. G., and Utter, F. M. (1973). Electrophoretic variants of aspartate aminotransferase of the bay mussel, *Mytilus edulis* (Linnaeus, 1958). *Comp. Biochem. Physiol. B* **44B,** 317–323.

Johnson, A. G., Utter, F. M., and Niggol, K. (1972). Electrophoretic variants of aspartate aminotransferase and adductor muscle proteins in the native oyster (*Ostrea lurida*). *Anim. Blood Groups Biochem. Genet.* **3,** 109–113.

Jones, M. E. (1980). Pyrimide nucleotide biosynthesis in animals: genes, enzymes, and regulation of UMP biosynthesis. *Ann. Rev. Biochem.* **49,** 253–279.

Jorgensen, N. O. G. (1980). Uptake of glycine and release of primary amines by the polychaete *Nereis virens* (Sars) and the mud snail *Hydrobia neglecta* Muus. *J. Exp. Mar. Biol. Ecol.* **47,** 281–297.

Kampa, L., and Peisach, J. (1980). Purification and characterization of hyroxyindole oxidase from gills of *Mytilus edulis. J. Biol. Chem.* **255,** 595–601.

Kaplowitz, N. (1980). Physiological significance of glutathione *s*-transferase. *Am. J. Physiol.* **239,** G439–G444.

Kasschau, M. R. (1975a). The relationship of free amino acids to salinity changes and temperature-salinity interactions in the mud-snail, *Nassarius obsoletus. Comp. Biochem. Physiol. A* **51A,** 301–308.

Kasschau, M. R. (1975b). Changes in concentration of free amino acids in larval stages of the trematode, *Himasthla quissetensis* and its intermediate host, *Nassarius obsoletus. Comp. Biochem. Physiol. B* **51B,** 273–280.

Kerkut, G. A., Rick, J. T., and Huggins, A. K. (1969). The intermediary metabolism *in vivo* of glucose and acetate by ganglia from *Helix aspersa* and the effects of amphetamine. *Comp. Biochem. Physiol.* **28,** 765–770.

Kluytmans, J. H., de Bont, A. M. T., Janus, J., and Wijsman, T. C. M. (1977). Time dependent changes and tissue specificities in the accumulation of anaerobic fermentation products in the sea mussel *Mytilus edulis* L. *Comp. Biochem. Physiol. B* **58B,** 81–87.

Kluytmans, J. H., van Graft, M., Janus, J., and Pieters, H. (1978). Production and excretion of volatile fatty acids in the sea mussel *Mytilus edulis* L. *J. Comp. Physiol. A.* **123,** 163–167.

Kluytmans, J. H., Zandee, D. I., Zurburg, W., and Pieters, H. (1980). The influence of seasonal changes on energy metabolism in *Mytilus edulis* (L). III. Anaerobic energy metabolism. *Comp. Biochem. Physiol. B* **67B,** 307–315.

Kochakian, C. D. (1976). Taurine and related sulfur compounds in the reproductive tract of marine animals. *In* "Taurine" (R. Huxtable and A. Barbeau, eds.), pp. 375–378. Raven, New York.

Kolenbrander, H. M., and Berg, C. P. (1967). Role of urocanic acid in the metabolism of L-histidine. *Arch. Biochem. Biophys.* **119,** 110–118.

Kondo, H., Anada, H., Ohsana, K., and Ishimoto, M. (1971). Formation of sulfo-acetaldehyde from taurine in bacterial extracts. *J. Biochem (Tokyo)* **69,** 621–623.

Konosu, S., and Hayashi, T. (1975). Determination of β-alanine betaine and glycine betaine in some marine invertebrates. *Nippon Suisan Gakkaishi* **41,** 743–746.

Kostenko, M. A., Smolikhina, T. I., Gakhova, E. N., and Tretyak, N. N. (1979). Input of $^{14}$C-leucine and $^3$H-uridine into mollusc nervous tissue as related to the action of ion active transport system and to changes caused by pronase treatment (in Russ.). *Tsitologiya* **21,** 586–593.

Krebs, H. A., Hims, R., Lund, P., Halliday, D., and Read, W. W. C. (1978). Sources of ammonia for mammalian urea synthesis. *Biochem. J.* **176,** 733–737.

Lange, R. (1963). The osmotic function of amino acids and taurine in the mussel, *Mytilus edulis. Comp. Biochem. Physiol.* **10,** 173–179.

Lange, R. (1970). Isomotic intracellular regulation and euryhalinity in marine bivalves. *J. Exp. Mar. Biol. Ecol.* **5**, 170–179.

Lange, R. (1972). Some recent work on osmotic, ionic and volume regulations in marine animals. *Oceanogr. Mar. Biol.* **10**, 97–136.

Lee, P. C., Fisher, J. R., and Ma, P. F. (1973). Immunochemical studies of adenosine deaminases from several vertebrates and a mollusc. *Comp. Biochem. Physiol. B* **46B**, 483–486.

Lee, T. W., and Campbell, J. W. (1965). Uric acid synthesis in the terrestrial snail, *Otala lactea*. *Comp. Biochem. Physiol.* **15**, 457–468.

Levinton, J. S., and Koehn, R. K. (1976). Population genetics of mussels. *In* "Marine Mussels" (B. L. Bayne, ed.), pp. 357–384. Cambridge Univ. Press, London and New York.

Lewin, J., Eckman, J. E., and Ware, G. N. (1979). Blooms of surf-zone diatoms along the coast of the Olympic Peninsula, Washington. XI. Regeneration of ammonium in the surf environment by the pacific razor clam *Siliqua patula*. *Mar. Biol. (Berlin)* **52**, 1–9.

Linton, S. N., and Campbell, J. W. (1962). Studies on urea cycle enzymes in the terrestrial snail, *Otala lactea*. *Arch. Biochem. Biophys.* **97**, 360–369.

Little, C. (1968). Aestivation and ionic regulation in two species of *Pomacea* (Gastropoda, Prosobranchia). *J. Exp. Biol.* **48**, 569–585.

Livingstone, D. R., and Bayne, B. L. (1977). Responses of *Mytilus edulis* L. to low oxygen tension: anaerobic metabolism of the posterior adductor muscle and mantle tissues. *J. Comp. Physiol. B.* **114**, 143–155.

Livingstone, R. D., Widdows, J., and Fieth, P. (1979). Aspects of nitrogen metabolism of the common mussel *Mytilus edulis:* adaptation to abrupt and fluctuating changes in salinity. *Mar. Biol. (Berlin)* **53**, 41–55.

Loest, R. A. (1979a). Ammonia volatilization and absorption by terrestrial gastropods: a comparison between shelled and shell-less species. *Physiol. Zool.* **52**, 461–469.

Loest, R. A. (1979b). Ammonia forming enzymes and calcium carbonate deposition in terrestrial pulmonates. *Physiol. Zool.* **52**, 470–483.

Lombardini, J. B., Pang, P. K. T., and Griffith, R. W. (1979). Amino acids and taurine in intracellular osmoregulation in marine animals. *Occas. Pap. Calif. Acad. Sci. No.* **134**, 160–169.

Lowenstein, J. M. (1972). Ammonia production in muscle and other tissues: the purine nucleotide cycle. *Physiol. Rev.* **52**, 382–414.

Lowenstein, J. M., and Goodman, M. N. (1978). The purine nucleotide cycle in skeletal muscle. *Fed. Proc., Fed. Am. Soc. Exp. Biol.* **37**, 2308–2312.

Lum, S. C., and Hammen, C. S. (1964). Ammonia excretion of *Lingula*. *Comp. Biochem. Physiol.* **12**, 185–190.

Lynch, M. P., and Wood, L. (1966). Effects of environmental salinity on free amino acids of *Crassostrea virginica* Gmelin. *Comp. Biochem. Physiol.* **19**, 783–790.

McAdoo, D. J., Ilifee, T. M., Price, C. H., and Novak, R. A. (1978). Specific glycine uptake by identified neurons of *Aplysia californica*. II. Biochemistry. *Brain Res.* **154**, 41–51.

McDonald, J. A., and Campbell, J. W. (1980). Invertebrate urease: effects of sulfhydryl reagents and heavy metals on the *Otala* enzyme. *Comp. Biochem. Physiol. B* **66B**, 215–222.

McDonald, J. A., Speeg, K. V., and Campbell, J. W. (1972). Urease: a sensitive and specific radiometric assay. *Enzymologia* **42**, 1–9.

McDonald, J. A., Vorhaben, J. E., and Campbell, J. W. (1980). Invertebrate urease: Purification and properties of the enzyme from a land snail, *Otala lactea*. *Comp. Biochem. Physiol. B* **66B**, 225–231.

Machado, C. R., and Vierra, E. C. (1979). Nitrogenous compounds in the excreta of the germ free snail *Biomphalaria glabrata* (Mollusca: Planorbidae). *Comp. Biochem. Physiol. A* **61A**, 961–964.

Machin, J. (1975). Water relationships. *In* "Pulmonates" (V. Fretter and J. Peake, eds.), Vol. 1, pp. 105–163. Academic Press, New York.

McIntyre, T. M., and Curthoys, N. P. (1980). The interorgan metabolism of glutathione. *Int. J. Biochem.* **12**, 545–551.

McManus, D. P., and James, B. L. (1975a). Tricarboxylic acid cycle enzymes in the digestive gland of *Littorina saxatilis rudis* (Maton) and in the daughter sporocysts of *Microphallus similus* (Jag.) (Digenea: Microphallididae). *Comp. Biochem. Physiol. B* **50B**, 491–495.

McManus, D. P., and James, B. L. (1975b). Anaerobic glucose metabolism in the digestive gland of *Littorina saxatilis rudis* (Maton) and in the daughter sporocysts of *Microphallus similis* (Jag.) (Digenea: Microphallidae). *Comp. Biochem. Physiol. B* **51B**, 293–297.

McManus, D. P., and James, B. L. (1975c). Pyruvate kinases and carbon dioxide fixating enzymes in the digestive gland of *Littorina saxatilis rudis* (Maton) and in the daughter sporocysts of *Microphallus similis* (Jag.) (Digenea: Microphallidae). *Comp. Biochem. Physiol. B* **51B**, 299–306.

Malanga, C. J., and Aiello, E. L. (1972). Succinate metabolism in the gills of the mussels *Modiolus demissus* and *Mytilus edulis*. *Comp. Biochem. Physiol. B* **43B**, 795–806.

Mangum, C. P., Dykens, J. A., Henry, R. P., and Polites, G. (1978). The excretion of $NH_4^+$ and its ouabain sensitivity in aquatic annelids and molluscs. *J. Exp. Zool.* **203**, 151–157.

Meijer, A. J., Gimpel, J. A., Deleeuw, G. A., Tager, J. M., and Williamson, J. R. (1975). Role of anion translocation across the mitochondrial membrane in the regulation of urea synthesis from ammonia by isolated rat hepatocytes. *J. Biol. Chem.* **250**, 7728–7738.

Meijer, A. J., Gimpel, J. A., Deleeuw, G. A., Tischler, M. J., Tager, J. M., and Williamson, J. R. (1978). Interrelationships between gluconeogenesis and ureogenesis in isolated hepatocytes. *J. Biol. Chem.* **253**, 2308–2320.

Meister, A. (1973). On the enzymology of amino acid transport. *Science* **180**, 33–39.

Mermel, L., Guchhait, R. B., Bourgeois, J. G., and Bueding, E. (1981). Biogenic amine metabolism in *Biomphalaria glabrata*—II. Monoamine oxidase activity. *Comp. Biochem. Physiol. C* **69C**, 227–234.

Mettrick, D. F., and Telford, J. M. (1965). The histamine content and histidine decarboxylase activity of some marine and terrestrial animals from the West Indies. *Comp. Biochem. Physiol.* **16**, 547–559.

Meyer, R., and Becker, W. (1980). Induced urea production in *Biomphalaria glabrata,* a snail host of *Schistosoma mansoni*. *Comp. Biochem. Physiol. A* **66A**, 673–677.

Meyer, R. A., and Terjung, R. L. (1979). Differences in ammonia and adenylate metabolism in contracting fast and slow muscle. *Am. J. Physiol.* **237**, C111–C118.

Michalek-Moricca, H. (1965). Comparative investigation on tryosine metabolism in animals. *Acta Biochim. Pol.* **12**, 167–177.

Mirolli, M. (1968). Decarboxylation of 5-hydroxytryptophane in ganglia of *Busycon canaliculatum* (L.) treated with reserpine. *Comp. Biochem. Physiol.* **24**, 847–854.

Mommsen, T. P., Ballantyne, J., MacDonald, J., Gosline, J., and Hochachka, P. W. (1981). Analogs of red and white muscle in squid mantle. *Proc. Natl. Acad. Sci. U.S.A.* **78**, 3274–3278.

Mori, M., and Tatibani, M. (1975). Purification of homogeneous glutamine-dependent carbamyl phosphate synthetase from ascites hepatoma cells as a complex with aspartate transcarbamylase and dihydroorotase. *J. Biol. Chem.* **78**, 239–242.

Nakazawa, T., Hori, K., and Hayaishi, D. (1972). Studies on monooxygenases. V. Manifestation of amino acid oxidase activity by L-lysine monooxygenase. *J. Biol. Chem.* **247**, 3439–3444.

Needham, J. (1935). Problems of nitrogen catabolism in invertebrates—II. Correlation between uricotelic metabolism and habitat in the phylum Mollusca. *Biochem. J.* **29**, 238–251.

Negus, M. R. S. (1968). Oxygen consumption and amino acid levels in *Hydrobia ulvae* (Pennant) in relation to salinity and behaviour. *Comp. Biochem. Physiol.* **24**, 317–325.

Newman, K. C., and Thomas, R. E. (1975). Ammonia, urea and uric acid levels in active and estivating snails *Bakerilymnaea cockerelli*. *Comp. Biochem. Physiol. A* **50A**, 109–112.

Norris, E. R., and Benoit, G. J. (1945). Studies on trimethylamine oxide. I. Occurrence of trimethylamine oxide in marine organisms. *J. Biol. Chem.* **158**, 433–438.

Norton, R. S. (1979). Identification of mollusc metabolites by natural-abundance [13]C NMR studies of whole tissue and tissue homogenates. *Comp. Biochem. Physiol. B* **63B**, 67–72.

Olomucki, A., Thoai, N. V., and Roche, J. (1960). Dégradation de l'ornithine et de la lysine chez *Limnaea stagnalis* L. *In* "Biochemie Comparée des Acides Aminés Basiques" (J. Roche, ed.), No. 67, pp. 171–179. Colloq. Int. CNRS, Paris.

Orive, E., Berjon, A., and Fernandez-Otero, M. P. (1980). Metabolism of nutrients during intestinal absorption in *Helix pomatia* and *Arion empiricorum* (Gastropoda: Pulmonata). *Comp. Biochem. Physiol. B* **66B**, 155–158.

Osborne, N. N. (1971). Occurrence of GABA and taurine in the nervous systems of the dogfish and some invertebrates. *Comp. Gen. Pharmacol.* **2**, 433–438.

Osborne, N. N. (1976a). The uptake of ([3]H) choline by snail (*Helix pomatia*) nervous tissue in vitro. *J. Neurochem.* **27**, 517–522.

Osborne, N. N. (1976b). Dopamine metabolism and inactivation in the gastropod brain. *In* "Neurobiology of Invertebrates: Gastropod Brain" (J. Salanki, ed.), pp. 141–161. Akadémiai Kiadó, Budapest.

Osborne, N. N., Briel, G., and Neuhoff, V. (1971). Distribution of GABA and other amino acids in different tissues of the gastropod mollusc *Helix pomatia* including *in vitro* ex periments with [14]C-glucose and [14]C-glutamic acid. *Int. J. Neurosci.* **1**, 265–272.

Osborne, N. N., Hiripi, L., and Neuhoff, V. (1975). The *in vitro* uptake of biogenic amines by snail (*Helix pomatia*) nervous tissue. *Biochem. Pharmacol.* **24**, 2141–2148.

Osborne, N. N., Schröder, H. U., and Neuhoff, V. (1978). The accumulation of DL-glutamate by the central nervous system of the snail *Helix pomatia*. *Brain Res.* **152**, 543–553.

Otto, J., and Pierce, S. K. (1981a). An interaction of extra- and intracellular osmoregulatory mechanisms in the bivalve mollusc *Rangia cuneata*. *Mar. Biol.* (Berlin) **61**, 193–198.

Otto, J. and Pierce, S. K. (1981b). A critical level of blood $Ca^{2+}$ is required by bivalves for amino acid regulation in hyperosmotic salinities. *Mar. Biol.* **61**, 193–198.

Pamatmat, M. M. (1978). Oxygen uptake and heat production in a metabolic conformer (*Littorina irrorata*) and a metabolic regulator (*Uca pugnax*). *Mar. Biol.* (Berlin) **48**, 317–325.

Pamatmat, M. M. (1979). Anaerobic heat production of bivalves (*Polymesoda caroliniana* and *Modiolus demissus*) in relation to temperature, body size, and duration of anoxia. *Mar. Biol.* (Berlin) **53**, 223–229.

Péquignat, E. (1973). A kinetic and autoradiographic study of the direct assimilation of amino acids and glucose by organs of the mussel *Mytilus edulis*. *Mar. Biol.* (Berlin) **19**, 227–244.

Peraino, C., and Pitot, H. C. (1963). Ornithine-δ-transaminase in the rat. I. Assay and some general properties. *Biochim. Biophys. Acta* **73**, 222–231.

Peterson, M. B., and Duerr, F. G. (1969). Studies on osmotic adjustment in *Tegula funebralis* (Adams 1854). *Comp. Physiol. Biochem.* **28**, 633–644.

Philley, J. C. (1978). Anaerobic uptake and utilization of glycine-2-$^{14}$C by the estaurine bivalve *Rangia cuneata* at three salinities. *Comp. Biochem. Physiol. B* **61B**, 565–569.

Pierce, S. K. (1971). A source of solute for volume regulation in marine mussels. *Comp. Biochem. Physiol. A* **38A**, 619–692.

Pierce, S. K., and Amende, L. M. (1981). Control mechanisms of amino acid-mediated cell volume regulation in salinity-stressed molluscs. *J. Exp. Zool.* **215**, 247–257.

Pierce, S. K., and Greenberg, M. J. (1972). The nature of cellular volume regulation in marine bivalves. *J. Exp. Biol.* **57**, 681–692.

Pierce, S. K., and Greenberg, M. J. (1973). The initiation and control of free amino acid regulation of cell volume in salinity stressed marine bivalves. *J. Exp. Biol.* **59**, 435–446.

Pierce, S. K., and Greenberg, M. J. (1976). Hypoosmotic cell volume regulation in marine bivalves: the effects of membrane potential change and metabolic inhibition. *Physiol. Zool.* **45**, 417–424.

Porembska, Z., Gorykowski, B., and Jezewska, M. (1966). Utilization of ($^{14}$C) orotate in the biosynthesis of pyrimidines in *Helix pomatia* and *Celerio euphorbiae. Acta Biochim. Pol.* **13**, 107–111.

Porembska, Z., Gasiorowska, I., and Mochnacka, I. (1968). Isolation of arginase and guanidinobutyrate ureohydrolase from hepatopancreas of *Helix pomatia. Acta Biochim. Pol.* **15**, 171–181.

Potts, W. T. W. (1954). The inorganic composition of *Mytilus edulis* and *Anodonta cygnaea. J. Exp. Biol.* **31**, 376–385.

Potts, W. T. W. (1958). The inorganic and amino acid composition of some lamellibranch muscles. *J. Exp. Biol.* **35**, 749–764.

Potts, W. T. W. (1967). Excretion in the molluscs. *Biol. Rev. (Cambridge Philos. Soc.)* **42**, 1–41.

Price, C. H., Coggenshall, R. E., and McAdoo, D. J. (1978). Specific glycine uptake by identified neurons of *Aplysia californica*. I. Autoradiography. *Brain Res.* **154**, 25–40.

Price, C. H., McAdoo, D. J., Farr, W., and Okuda, R. (1979). Bidirectional axonal transport of free glycine in identified neurons R3-R14 of *Aplysia. J. Neurobiol.* **10**, 551–571.

Prota, G., Ortonne, J. P., Voulot, C., Khatchadourean, C., Nardi, G., and Palumbo, A. (1981). Occurrence and properties of tyrosinase in the ejected ink of cephalopods. *Comp. Biochem. Physiol. B* **68B**, 415–419.

Rahmen, S. A., and Decker, P. (1966). Comparative study of the urease in the rumen wall and rumen content. *Nature (London)* **209**, 618–619.

Ratner, S. (1973). Enzymes of arginine and urea synthesis. *Adv. Enzymol. Relat. Areas Mol. Biol.* **39**, 1–90.

Rava, R., Spinosi, G., and Brunetti, A. (1981). Puriicazione parziole e caratterizzazione di una D-aminoacido ossidasi da epatopancreas di *Octopus vulgaris. Boll. Soc. Ital. Biol. Sper.* **58**, 111–117.

Rawls, J. M. (1979). The enzymes for *de novo* pyrimidine biosynthesis in *Drosophila melanogaster:* their localization, properties and expression during development. *Comp. Biochem. Physiol. B* **62B**, 207–216.

Razet, P., and Dagobert, C. (1968). Recherches des enzymes de la chaine de l'uricolyse chez les mollusques terrestres et dulçagnicoles. *Arch. Sci. Physiol.* **22**, 173–181.

Read, K. R. H. (1962). Transamination in certain tissue homogenates of the bivalve molluscs *Mytilus edulis* L. and *Modiolus modiolus* L. *Comp. Biochem. Physiol.* **7**, 15–22.

Read, K. R. H. (1963). Thermal inactivation of preparations of aspartic/glutamic transaminase from species of bivalved molluscs from the sublittoral and intertidal zones. *Comp. Biochem. Physiol.* **9**, 161–180.

Reddy, S. R. R., and Baby, T. B. (1976). The inhibition of arginase from the hepatopancreas of terrestrial snail by amino acids. *Arch. Int. Physiol. Biochim.* **84**, 759–766.

Reddy, S. R. R., and Campbell, J. W. (1968). A low molecular weight arginase in the earthworm. *Biochim. Biophys. Acta* **159**, 557–560.

Reddy, S. R. R., and Campbell, J. W. (1970). Molecular weights of arginase from different species. *Comp. Biochem. Physiol.* **32**, 499–509.

Reddy, Y. S., and Swami, K. S. (1975). On the significance of enhanced glutamine synthetase and its regulation during aestivation in *Pila globosa*. *Curr. Sci.* **44**, 191–192.

Reddy, Y. S., and Rao, R. V., and Swami, K. S. (1974). Probable significance of urea and uric acid accumulation during aestivation in the gastropod, *Pila globosa* (Swainson). *Indian J. Exp. Biol.* **12**, 454–456.

Reiss, P. M., Pierce, S. K., and Bishop, S. H. (1977). Glutamate dehydrogenase from tissues of the ribbed mussel *Modiolus demissus*. ADP activation and possible physiological significance. *J. Exp. Zool.* **202**, 253–258.

Reite, O. B. (1972). Comparative physiology of histamine. *Physiol. Rev.* **52**, 778–919.

Rice, M. A., Wallis, K., and Stephens, G. C. (1980). Influx and net flux of amino acids into larval and juvenile European flat oysters, *Ostrea edulis* (L.). *J. Exp. Mar. Biol. Ecol.* **48**, 51–59.

Riley, R. T. (1976). Changes in the total protein, lipid carbohydrate, and extracellular body fluid free amino acids of the Pacific oyster, *Crassostrea gigas,* during starvation. *Proc. Natl. Shellfish. Assoc.* **65**, 84–90.

Riley, R. T. (1980). The effect of prolonged starvation on the relative free amino acid composition of the extracellular body fluids and protein bound amino acids in the oyster *Crassostrea gigas*. *Comp. Biochem. Physiol. A* **67A**, 279–281.

Roberts, N. R., Coelho, R. R., Lowry, D. H., and Crawford, E. J. (1958). Enzyme activities of giant squid axoplasm and axon sheath. *J. Neurochem.* **30**, 109–115.

Robertson, J. D. (1965). Studies on the chemical composition of muscle tissue. III. The mantle muscle of cephalopod molluscs. *J. Exp. Biol.* **42**, 153–175.

Rocca, E., and Ghiretti, F. (1968). Purification and properties of D-glutamic acid oxidase from *Octopus vulgaris* Lam. *Arch. Biochem. Biophys.* **77**, 336–349.

Roche, J., Glahn, P. E., Mandon, P. L., and Thoai, N. V. (1959). Sur une nonvelle L-amino acid oxydase, activable par le magnesium. *Biochim. Biophys. Acta* **35**, 111–122.

Rodrick, G. E. (1979). Selected enzyme activities in *Mya arenaria* hemolymph. *Comp. Biochem. Physiol. B* **62B**, 313–316.

Roesijadi, G., and Anderson, J. W. (1979). Condition index and free amino acid content of *Macoma inquinata* exposed to oil-contaminated marine sediments. *In* "Marine Pollution: Functional Responses" (W. B. Vernberg, A. Calabrese, F. P. Thurberg, and F. J. Vernbeg, eds.), pp. 69–83. Academic Press, New York.

Rosenfeld, M. G., and Leiter, E. H. (1977). Isolation and characterization of a mitochondrial D-amino acid oxidase from *Neurospora crassa*. *Can. J. Biochem.* **55**, 66–74.

Rowsell, E. V., Carnie, J. A., Wahbi, S. D., Al-Tai, A. H., and Rowsell, K. V. (1979). L-Serine dehydrase and L-serine-pyruvate aminotransferase activities in different animal species. *Comp. Biochem. Physiol. B* **63B**, 543–555.

Saijo, Y., and Mitamura, O. (1971). Regeneration of nutrients in the waters of a coastal oyster bed. *In* "The Ocean World" (M. Uda, ed.), pp. 242–248. Jpn. Soc. Promotion Sci., Tokyo.

Sansone, G., Biordi, A., and Noviello, L. (1978). Free amino acids in fluids and tissues of *Mytilus galloprovincialis* in relation to the environment. Their behavior as an index of normality of metabolism. *Comp. Biochem. Physiol. A* **61A**, 133–139.

Sato, M., Sato, Y., and Tsuchiya, Y. (1977). Studies of the extractives of Molluscs—II. α-Iminodipropionic acid isolated from squid muscle extracts. *Nippon Suisan Gakkaishi* **43**, 1441–1443.

Sato, M., Sato, Y., and Tsuchiya, Y. (1978). D-α-Iminopropioacetic acid isolated from the adductor muscle of scallop. *Nippon Suisan Gakkaishi* **44**, 247–250.

Sato, M., Sato, Y., and Tsuchiya, Y. (1979). L-Threo-β-hydroxyaspartic acid isolated from the muscle of short-necked clam. *Nippon Suisan Gakkaishi* **45**, 636–638.

Saxena, B. B. (1952). Uricotelism in the common Indian apple snail, *Pila globosa* (Swainson). *Nature (London)* **170**, 1024.

Saxena, B. B. (1955). Physiology of excretion in the common Indian apple snail, *Pila globosa* (Swainson). Part I. *J. Anim. Morphol. Physiol.* **2**, 87–95.

Schirch, L., and Gross, T. (1968). Serine hydroxymethyltransferase. Identification as the threonine and allothreonine aldolases. *J. Biol. Chem.* **243**, 5651–5655.

Schmale, H., and Becker, W. (1977). Studies on the urea cycle of *Biomphalaria glabrata* during normal feeding activity, in starvation and with infection of *Schistosoma mansoni*. *Comp. Biochem. Physiol. B* **58B**, 321–330.

Schmidt-Nielsen, K., Taylor, C. R., and Shkolnik, A. (1971). Desert snails: problems of heat, water and food. *J. Exp. Biol.* **55**, 385–398.

Schoffeniels, E., and Gilles, R. (1972). Ionoregulation and osmoregulation in molluscs. *In* "Chemical Zoology" (M. Florkin and B. T. Scheer, eds.), pp. 393–420. Academic Press, New York.

Schriver, C. R., and Wilson, O. H. (1964). Possible locations for a common gene product in membrane transport of imino-acids and glycine. *Nature (London)* **202**, 92–93.

Schwartz, J. H., Eisenstadt, M. K., and Cedar, H. (1975). Metabolism of acetylcholine in the nervous system of *Aplysia californica*. Source of choline and its uptake by intact nervous tissue. *J. Gen. Physiol.* **65**, 255–273.

Shieh, H. S. (1968). The characterization and incorporation of radioactive bases into scallop phospholipids. *Comp. Biochem. Physiol.* **27**, 533–541.

Shumway, S. E., and Youngson, A. (1979). The effects of fluctuating salinity on the physiology of *Modiolus demissus* (Dillwyn). *J. Exp. Mar. Biol. Ecol.* **40**, 167–181.

Shumway, S. E., Gabbott, P. A., and Youngson, A. (1977). The effect of fluctuating salinity on the concentrations of free amino acids and ninhydrin-positive substances in the adductor muscle of eight species of bivalve molluscs. *J. Exp. Mar. Biol. Ecol.* **29**, 131–150.

Simpson, J. W., and Awapara, J. (1965). Biosynthesis of glucose from pyruvate-2-$^{14}$C and aspartate-3-$^{14}$C in a mollusc. *Comp. Biochem. Physiol.* **15**, 1–6.

Simpson, J. W., Allen, K., and Awapara, J. (1959). Free amino acids in some aquatic invertebrates. *Biol. Bull. (Woods Hole, Mass.)* **117**, 371–381.

Sloan, W. C. (1964). The accumulation of nitrogenous compounds in terrestrial and aquatic eggs of prosobranch snails. *Biol. Bull. (Woods Hole, Mass.)* **126**, 302–306.

Smith, E. L., Austen, B. M., Blumenthal, K. M., and Nyc, J. F. (1975). Glutamate dehydrogenase. *In* "The Enzymes" (P. D. Boyer, ed.), 3rd ed., Vol. 11, pp. 294–368. Academic Press, New York.

Sollock, R. L., Vorhaben, J. E., and Campbell, J. W. (1979). Transaminase reactions and glutamate dehydrogenase in gastropod hepatopancreas. *J. Comp. Physiol. A.* **129**, 129–135.

Speeg, K. V., and Campbell, J. W. (1968a). Formation and volitilization of ammonia gas by terrestrial snails. *Am. J. Physiol.* **214**, 1392–1402.

Speeg, K. V., and Campbell, J. W. (1968b). Purine biosynthesis and excretion in *Otala* (=*Helix*) *lactea:* An evaluation of the nitrogen excretory potential. *Comp. Biochem. Physiol.* **26,** 579–595.

Speeg, K. V., and Campbell, J. W. (1969). Arginine and urea metabolism in terrestrial snails. *Am. J. Physiol.* **216,** 1003–1012.

Spitzer, J. (1937). Physiologisch okologesche Untersuchungen über den Exkretsloffwechsel der Mollusken. *Zool. Jahrb., Abt. Allg. Zool. Physiol. Tiere* **57,** 457–496.

Srna, R. F., and Baggaley, A. (1976). Rate of excretion of ammonia by the hard clam *Mercenaria mercenaria* and the american oyster *Crassostrea virginica. Mar. Biol. (Berlin)* **36,** 251–258.

Strange, K. B., and Crowe, J. H. (1979a). Acclimation to successive short term salinity changes by the bivalve *Modiolus demissus.* I. Changes in hemolymph osmotic concentration, hemolymph ion concentration and tissue water content. *J. Exp. Zool.* **210,** 221–226.

Strange, K. B., and Crowe, J. H. (1979b). Acclimation to successive short term salinity changes by the bivalve *Modiolus demissus.* II. Nitrogen metabolism. *J. Exp. Zool.* **210,** 227–236.

Stanislowski, E., and Becker, W. (1979). Influences of semi-synthetic diets, starvation, and infection with *Schistosoma mansoni* (Trematoda) on the metabolism of *Biomphalaria glabrata* (Gastropoda). *Comp. Biochem. Physiol. A* **63A,** 527–533.

Stanislowski, E., Becker, W., and Müller, G. (1979). Alterations of the free amino acid content in the hemolymph of *Biomphilaria glabrata* (Pulmonata) in starvation and after infection with *Schistosoma mansoni* (Trematoda). *Comp. Biochem. Physiol. B* **63B,** 477–482.

Stephens, G. C. (1972). Amino acid accumulation and assimilation in marine organisms. *In* "Nitrogen Metabolism and the Environment" (J. W. Campbell and L. Goldstein, eds.), pp. 155–184. Academic Press, New York.

Stewart, M. G. (1978a). The uptake and utilization of dissolved amino acids by the bivalve *Mya arenaris* (L.). *Physiol. Behav. Mar. Org., Proc. Eur. Symp. Mar. Biol., 12th, Stirling, Scot., 1977* pp. 165–176.

Stewart, M. G. (1978b). Kinetics of neutral amino-acid transport of isolated gill tissue of the bivalve *Mya arenaria* (L.). *J. Exp. Mar. Biol. Ecol.* **32,** 39–52.

Stewart, M. G. (1979). Absorption of dissolved organic nutrients by marine invertebrates. *Oceanogr. Mar. Biol.* **17,** 163–192.

Stewart, M. G. (1981). Kinetics of dipeptide uptake by the mussel *Mytilus edulis. Comp. Biochem. Physiol. A* **69A,** 311–315.

Stewart, M. G., and Bamford, D. R. (1975). Kinetics of alanine uptake by the gills of the soft shelled clam *Mya arenaria. Comp. Biochem. Physiol. A* **52A,** 67–74.

Stewart, M. G., and Bamford, D. R. (1976). The effect of environmental factors on the absorption of amino acids by isolated gill tissue of the bivalve *Mya arenaria* (L.). *J. Exp. Mar. Biol. Ecol.* **24,** 205–212.

Stewart, M. G., and Dean, R. C. (1980). Uptake and utilization of amino acids by the shipworm *Bankia gouldi. Comp. Biochem. Physiol. B* **66B,** 443–450.

Stewart, M. G., Gosling, E., and Dean, R. C. (1979). Skin digestion in bivalve molluscs and its relation to amino acid absorption from sea water. *In* "Animals and Environmental Fitness" (R. Gilles, ed.), pp. 157–158. Pergammon, Oxford.

Stickle, W. B. (1971). The metabolic effects of starving *Thais lamellosa* immediately after spawning. *Comp. Biochem. Physiol. A* **40A,** 627–634.

Stickle, W. B., and Duerr, F. G. (1970). The effects of starvation on the respiration and major nutrient stores of *Thais lamellosa. Comp. Biochem. Physiol.* **33,** 689–695.

Stokes, T., and Awapara, J. (1968). Alanine and succinate as end-products of glucose degradation in the clam, *Rangia cuneata*. *Comp. Biochem. Physiol.* **25**, 883–892.

Storey, K. B., and Storey, J. M. (1978). Energy metabolism in the mantle muscle of the squid, *Loligo pealeii*. *J. Comp. Physiol. A.* **123**, 169–175.

Storey, K. B., Fields, J. H. A., and Hochachka, P. W. (1978). Purification and properties of glutamate dehydrogenase from mantle muscle of the squid *Loligo pealeii*. Role of the enzyme in energy production from amino acids. *J. Exp. Zool.* **205**, 111–118.

Struck, J., and Sizer, I. W. (1960). Oxidation of L-α-amino acids by chicken microsomes. *Arch. Biochem. Biophys.* **90**, 22–30.

Suyama, M., and Kobayashi, H. (1980). Free amino acids and quaternary ammonium bases in mantle muscle of squids. *Nippon Suisan Gakkaishi* **46**, 1261–1264.

Swami, K. S., and Reddy, Y. S. (1978). Adaptive changes to survival during aestivation in the gastropod snail, *Pila globosa*. *J. Sci. Ind. Res.* **37**, 144–157.

Swinehart, J. H., Crowe, J. H., Giannin, A. P., and Rosenbaum, D. A. (1980). Effects of divalent cations on amino acid and divalent cation fluxes in gills of the bivalve mollusc, *Mytilus californianus*. *J. Exp. Zool.* **212**, 389–396.

Tabor, C. W., and Tabor, H. (1976). 1, 4-Diaminobutane (putrescine), spermidine, and spermine. *Annu. Rev. Biochem.* **45**, 285–306.

Thoai, N. V., Glahn, P. E., Hedegaard, J., Manchon, P., and Roche, J. (1954). Sur un aspect nouveau du metabolisme de la L-histidine. *Biochim. Biophys. Acta* **15**, 87–94.

Tramell, P. R., and Campbell, J. W. (1970a). Carbamyl phosphate synthesis in a land snail, *Strophocheilus oblongus*. *J. Biol. Chem.* **245**, 6634–6641.

Tramell, P. R., and Campbell, J. W. (1970b). Nitrogenous excretory products of the giant South American land snail, *Strophocheilus oblongus*. *Comp. Biochem. Physiol.* **32**, 569–571.

Tramell, P. R., and Campbell, J. W. (1971). Carbamyl phosphate synthesis in invertebrates. *Comp. Biochem. Physiol. B* **40B**, 395–406.

Tramell, P. R., and Campbell, J. W. (1972). Arginine and urea metabolism in the South American land snail, *Strophocheilus oblongus*. *Comp. Biochem. Physiol. B* **42B**, 439–449.

Treherne, J. E. (1980). Neuronal adaptations to osmotic and ionic stress. *Comp. Biochem. Physiol. B* **67B**, 455–463.

Trytek, R. E., and Allen, W. V. (1980). Synthesis of essential amino acids by bacterial symbionts in the gills of the shipworm *Bankia setacea* (Tryon). *Comp. Biochem. Physiol. A* **67A**, 419–427.

Umemori, Y. (1967). Ribonucleotide phosphohydrolases in the clam *Meretrix meretrix lusoria* (Gmelin). *Comp. Biochem. Physiol.* **20**, 635–639.

Umemori-Aikawa, Y. (1971). An alkaline phosphatase in the clam *Meretrix meretrix lusoria* (Gmelin) with affinities for nucleotides. *Comp. Biochem. Physiol. B* **40B**, 347–358.

Umemori-Aikawa, Y., and Aikawa, T. (1974). Adenosine deaminase from the clam, *Tapes phillippinarum*: Partial purification and detection of activity on disc electrophoresis. *Comp. Biochem. Physiol. B* **49B**, 353–359.

Umiastowski, J. (1964). AMP aminohydrolase in the muscles of some vertebrates and invertebrates. *Acta. Biochim. Pol.* **11**, 459–464.

Veldhuijzen, J. P. (1975). Effects of different kinds of foods, starvation, and restart of feeding on the haemolymph-glucose of the pond snail *Lymnaea stagnalis*. *Neth. J. Zool.* **25**, 88–101.

Virkar, R. A., and Webb, K. L. (1970). Free amino acid composition of the soft-shell clam *Mya arenaria* in relation to salinity of the medium. *Comp. Biochem. Physiol.* **32**, 775–783.

von Brand, T., McMahon, P., and Nolan, M. O. (1957). Physiological observations of starvation and dessication of the snail *Australorbis glabratus*. *Biol. Bull.* (*Woods Hole, Mass.*) **113,** 89–102.

von Wachtendonk, D., and Kaeppler, M. (1977). Amino acids and biogenic amines in the smooth muscles and hemolymphs of different molluscs. *Comp. Biochem. Physiol. C* **56C,** 19–24.

Waite, J. H., and Wilbur, K. M. (1976). Phenoloxidase in the periostracum of the marine bivalve *Modiolus demissus* Dillwyn. *J. Exp. Zool.* **195,** 359–368.

Watts, D. M. (1971). The effects of Digenea on the free amino acid pool of *Littorina littorea*. *Parasitology* **62,** 361–366.

Watts, J. A., and Pierce, S. K. (1978). A correlation between the activity of divalent cation activated adenosine triphosphatase in the cell membrane and low salinity tolerance of the ribbed mussel, *Modiolus demissus demissus*. *J. Exp. Zool.* **204,** 49–56.

Weinreich, D. (1979). γ-Glutamylhistamine: a major product of histamine metabolism in ganglia of the marine mollusk, *Aplysia californica*. *J. Neurochem.* **32,** 363–369.

Weinreich, D., and Rubin, L. (1981). Irreversible inhibitors of histidine decarboxylase in *Aplysia* ganglia: a tool for the identification of histaminergic synapes. *Comp. Biochem. Physiol. C* **69C,** 383–385.

Weinreich, D., and Yu, Y. T. (1977). The characterization of histidine decarboxylase and its distribution in nerves, ganglia and in single neuronal cell bodies of the CNS of *Aplysia californica*. *J. Neurochem.* **28,** 361–369.

Weinreich, D., Weiner, C., and McCaman, R. E. (1975). Endogenous levels of histamine in single neurons from the CNS of *Aplysia californica*. *Brain Res.* **84,** 341–345.

Weiser, W. (1980). Metabolic end products in three species of marine gastropods. *J. Mar. Biol.* **60,** 175–180.

Weiser, W., and Schuster, M. (1975). The relationship between water content, activity, and free amino acids in *Helix pomatia* L. *J. Comp. Physiol. B.* **98,** 169–181.

Whiteley, H. R. (1960). The distribution of the formate activating enzyme and other enzymes involving tetrahydrofolic acid in animal tissues. *Comp. Biochem. Physiol.* **1,** 227–247.

Wickes, M. A., and Morgan, R. P. (1976). Effects of salinity on three enzymes involved in amino acid metabolism from the american oyster, *Crassostrea virginica*. *Comp. Biochem. Physiol. B* **53B,** 339–343.

Widdows, J., Bayne, B. L., Livingstone, D. R., Newell, R. I. E., and Donkin, P. (1979). Physiological and biochemical responses of bivalve molluscs to exposure to air. *Comp. Biochem. Physiol. A* **62A,** 301–308.

Wijsman, T. C. M., de Bont, A. M. T., and Kluytmans, J. H. F. M. (1977). Anaerobic incorporation from 2, 3-$^{14}$C-succinic acid into citric acid cycle intermediates and related compounds in the sea mussel *Mytilus edulis* L. *J. Comp. Physiol. B.* **114,** 167–175.

Wilson, R. A. (1968). An investigation into the mucus produced by *Lymnaea truncatula*, the snail host of *Fasciola hepatica*. *Comp. Biochem. Physiol.* **24,** 629–633.

Wong, D. T., Fuller, R. W., and Malloy, B. B. (1973). Inhibition of amino acid transaminases by L-cycloserine. *Adv. Enzyme Regul.* **11,** 139–154.

Wright, S. H., and Stephens, G. C. (1977). Characteristics of influx and net flux of amino acids in *Mytilus californianus*. *Biol. Bull.* (*Woods Hole, Mass.*) **152,** 295–310.

Wright, S. H., and Stephens, G. C. (1978). Removal of amino acid during a single passage of water across the gill of marine mussels. *J. Exp. Zool.* **205,** 337–352.

Wright, S. H., Johnson, T. L., and Crowe, J. H. (1975). Transport of amino acids by isolated gills of the mussel, *Mytilus californianus* Conrad. *J. Exp. Biol.* **62,** 313–325.

Wright, S. H., Becker, S. A., and Stephens, G. C. (1980). Influence of temperature and unstirred layers on the kinetics of glycine transport in isolated gills of *Mytilus californianus*. *J. Exp. Zool.* **214**, 27–35.

Wu, J. Y. (1976). Purification, characterization and kinetic studies of GAD and GABA-T from mouse brain. *In* "GABA in Nervous System Function" (E. Roberts, T. N. Chase, and D. B. Tower, eds.), pp. 2–56. Raven, New York.

Yoneda, T. (1967). Cysteamine as a metabolite in marine mollusk, *Mytilus edulis*. *Hokkaido Daigaku Suisangakubu Kenkyu Iho* **18**, 88–94.

Yoneda, T. (1968). Enzymatic oxidation of cysteamine to taurine in marine mollusc, *Mytilus edulis*. *Hokkaido Daigaku Suisangakubu Kenkyu Iho* **19**, 140–146.

Yoshida, T., and Kikuchi, G. (1972). Comparative study on major pathways of glycine and serine catabolism in vertebrate livers. *J. Biochem. (Tokyo)* **72**, 1503–1516.

Zaba, B. N., de Bont, A. M. T., and de Zwaan, A. (1978). Preparation and properties of mitochondria from tissues of the sea mussel, *Mytilus edulis* L. *Int. J. Biochem.* **9**, 191–197.

Zammit, V. A., and Newsholme, E. A. (1976). The maximum activities of hexokinase, phosphorylase, phosphofructokinase, glycerol phosphate dehydrogenases, lactate dehydrogenase, octopine dehydrogenase, phosphoenol pyruvate carboxykinase, nucleoside diphosphatekinase, glutamate–oxaloacetate transaminase and arginine kinase in relation to carbohydrate utilization in muscles from marine invertebrates. *Biochem. J.* **160**, 447–462.

Zandee, D. I., Nijkamp, H. J., Roosheroe, I., deWaart, J., Sedee, J. W., and Vonk, H. J. (1958). Transaminations in invertebrates. *Arch. Int. Physiol. Biochim.* **66**, 220.

Zandee, D. I., Kluytmans, J. H., Zurburg, H., and Pieters, H. (1980). Seasonal variations in biochemical composition of *Mytilus edulis* with reference to energy metabolism and gametogenesis. *Neth. J. Sea Res.* **14**, 1–29.

Zeman, G. H., Myers, P. R., and Dalton, T. K. (1975). Gamma-aminobutyric acid uptake and metabolism in *Aplysia dactylomela*. *Comp. Biochem. Physiol. C* **51C**, 291–299.

Zurburg, W., and de Zwaan, A. (1981). Role of amino acids in anaerobiosis and osmoregulation in bivalves. *J. Exp. Zool.* **215**, 315–326.

Zurburg, W., and Ebberink, R. H. (1981). The anaerobic energy demand of *Mytilus edulis*. Organ specific differences in ATP-supplying processes and metabolic routes. *Mol. Physiol.* **1**, 153–164.

Zurburg, W., and Kluytmans, J. H. (1980). Organ specific changes in energy metabolism due to anaerobioses in the sea mussel *Mytilus edulis* (L.). *Comp. Biochem. Physiol. B* **67B**, 317–322.

# 7

# Lipids: Their Distribution and Metabolism

**Peter A. Voogt**

Laboratory of Chemical Animal Physiology
The State University of Utrecht
3508 TB Utrecht
The Netherlands

## I. Introduction

To the many articles dealing with lipid composition and lipid distribution in molluscs a progressively increasing number is added each year. The data reported in these recent articles are usually obtained with highly improved and much more reliable methods and instruments, providing us with more details than most of the older work. As a result the scientific value of the latter studies has decreased considerably; therefore, they will only be taken into account in this chapter if still relevant. Another reason for paying less attention to these older data is that they have been reviewed extensively before (Voogt, 1972a). In that review it was written: "Attention paid to each of the five classes within the phylum of Mollusca is quite different. Most papers are concerned with bivalves and gastropods, while only few deal with cephalopods and still fewer with Amphineura. Nothing seems to be known about lipid composition in scaphopods" (p. 245). In this respect hardly anything has changed and the actual situation is still the same as that described for 1972. It was suggested then that possible causes for this strongly unequal interest might be the abundance or the economic value of several representatives of bivalves, gastropods, and cephalopods, as well as the minute dimensions of scaphopods. To these possible causes another two may be added: First, there is

THE MOLLUSCA, VOL. 1
Metabolic Biochemistry
and Molecular Biomechanics

an increasing interest in the origin of lipids, especially in tracing lipid components in the food chain. By virtue of their abundance, gastropods and bivalves are important links in the food chain, the filter-feeding bivalves in particular being important because of their potency in accumulating lipids from previous links. A second cause is the increasing interest in the problem of marine-water pollution. The sessile bivalves, which cannot escape from unfavorable conditions, may be indicators for environment pollution and very attractive for studies on the biological removal of pollutants from the environment as well as on the tolerance for and the detoxification of harmful substances in animals.

Although knowledge of lipids in molluscs has increased considerably, there is still little insight as to how molluscs manage to achieve and to maintain a characteristic composition of lipids, despite the often quite different lipid composition of their diet. Studies on this homeostasis are badly needed.

The foregoing emphasized the great difference in the knowledge about each of the classes of molluscs. A similar situation exists with respect to the different classes of lipids. Whereas a rather detailed knowledge about sterols has been built up, enabling Bergmann (1949, 1962) as the first, and some others after him, to review these data, our knowledge about lipids belonging to other classes is still fragmentary.

Progress in lipid research has been strongly determined by progress in separation and identification techniques. However, the evolution in instrumentation will not be discussed here.

The structure and biosynthesis of naturally occurring lipids will be dealt with only briefly in this chapter.

## A. Sterols

Sterols may occur in free form or esterified to fatty acids. The parent nucleus of sterols is formed by cholestane (I), which is also the basis for their nomenclature. Probably their main function is in regulating the viscosity and thus permeability of cell membranes. This essential role likely requires a specific sterol composition. Reviews of the structure and distribution of sterols have been given by Bergmann (1949, 1962), Austin (1970), Idler and Wiseman (1972), Voogt (1972a), and Goad (1976).

The biosynthesis of sterols can be divided into three parts. The first part is the formation of squalene (II) from acetyl-CoA. In this route the irreversible conversion of 3-hydroxy-3-methyl-glutaryl CoA (HMG-CoA) into mevalonate, catalyzed by the enzyme HMG-CoA reductase, is found to be in vertebrates the rate-limiting and regulatory step in sterol biosynthesis. The enzyme has a short half-life (in the order of some hours), and recently it has been shown that in liver the activity of the enzyme is regu-

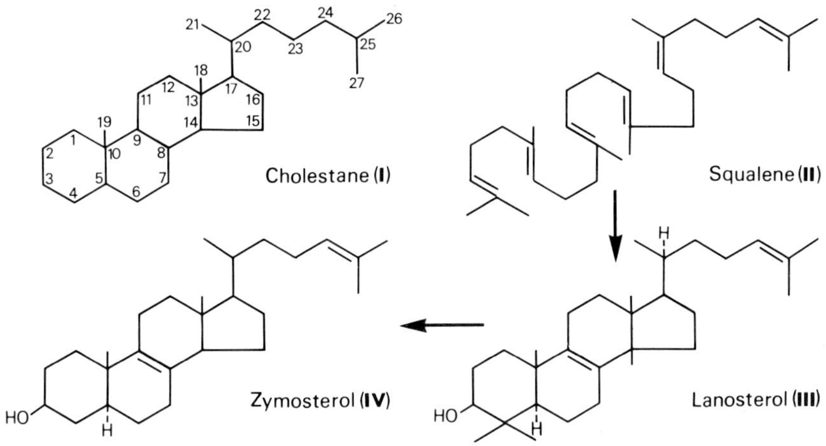

Cholestane (I)

Squalene (II)

Zymosterol (IV)

Lanosterol (III)

lated by a bicyclic phosphorylation system (Brown et al., 1979; Hunter and Rodwell, 1980; Ingebritsen et al., 1981 Arebalo et al., 1981; Shah, 1981).

In the second part of sterol biosynthesis squalene is cyclized to form lanosterol (III), a triterpene. In the third part, lanosterol is demethylated to form zymosterol (IV), which is believed to be the precursor of all other sterols.

## B. Fatty Acids

Fatty acids may occur in free form, but mostly they are present in esterified form. They are found in sterol esters, acylglycerols, and phospholipids. Most fatty acids possess straight chains, but considerable amounts of branched-chain fatty acids may be present. The monobranched fatty acids comprise the iso- and the anteiso fatty acids. The common multibranched or isoprenoid fatty acids are 4,8,12-trimethyltridecanoic acid, 2,6,10,14-tetramethylpentadecanoic (pristanic) acid, and 3,7,11,15-tetramethylhexadecanoic (phytanic) acid.

Generally double bonds in unsaturated fatty acids are methylene interrupted. However, a number of dienoic fatty acids are known in which the double bonds are non-methylene interrupted (Paradis and Ackman, 1975).

The fatty acid composition in organisms is influenced by several external conditions. Among these are temperature (Lewis, 1962) and diet. So fatty acid composition in storage organs will vary with the diet, and that in the membranes with the temperature. As a result the fatty acid composition of a whole animal at a certain moment will give but little information. For this general reason, fatty acid compositions will not be given here.

Fatty acids—for example, the non-methylene-interrupted ones, have been used in food web studies (Paradis and Ackman, 1977). Certain (unsaturated) fatty acids or combinations of them are indicative of the natural habitat in which an animal lives (Ackman, 1964, 1967; Lovern, 1964). So eicosapentenoic acid (20:5 $\omega$ 3) and docosahexenoic acid (22:6 $\omega$ 3) are characteristic for marine lipids.

## C. Phospholipids

This group is very heterogeneous, but all compounds have in common that they contain phosphorus. Depending on the alcohol used for esterification, they are divided into glycerophospholipids and sphingolipids. The first group—with glycerol as the alcohol—comprises phosphatidic acid and lipids derived from it by esterification of the phosphate residue with serine, ethanolamine, or choline leading to phosphatidyl-serine, and so on.

The main constituent of the sphingolipids, ceramide, is built up from the amino alcohol sphingosine (V) and a fatty acid linked to the amino group by a carbon–nitrogen bond. Ceramide can be esterified by its terminal hydroxyl group to phosphorylcholine, forming sphingomyelin; to phosphorylethanolamine (2-aminoethyl phosphate, VI); or to 2-amino-ethyl phosphonic acid (VII). Phospholipids containing the latter component are also called phosphonolipids.

$$H_3C-(CH_2)_{12}-CH=CH-\overset{\overset{\displaystyle OH}{|}}{CH}-\underset{\underset{\displaystyle NH_2}{|}}{CH}-CH_2OH$$

Sphingosine

(V)

$$NH_2-CH_2-CH_2-O-PO_3H_2$$

Phosphorylethanolamine
(2-aminoethyl phosphate)

(VI)

$$NH_2-CH_2-CH_2-\overset{\overset{\displaystyle O}{\|}}{\underset{\underset{\displaystyle OH}{|}}{P}}-OH$$

2-aminoethyl
phosphonic acid

(VII)

## D. Lipids Containing Ether Bonds

These lipids are encountered in both neutral and polar lipids. They are not generally distinguished as a separate lipid class, but are classed with the triacylglycerols and the phospholipids. Instead of a fatty acid in acyl linkage, a hydrocarbon in either linkage is present. Usually, the alkyl chain is at the 1 position of glycerol, but dialkyl ethers are also known.

Alkyl ether bonds are synthesized by microsomal enzymes from fatty alcohols and glyceraldehyde 3-phosphate or, to a lesser extent, dihydroxyacetone phosphate.

Glyceryl ethers in which the alkyl chain is $\alpha,\beta$-unsaturated are often called plasmalogens. They are derived from the normal alkylglycerols by desaturation of the alkyl chain.

The biochemistry, biological function, and biosynthesis of alkylglycerols has been reviewed by Snyder (1969).

## E. Glycolipids

These lipids, often called cerebrosides, consist of ceramide and one or more sugars (neutral sugars or amino sugars) linked to the terminal hydroxyl group of sphingosine.

## II. Distribution of Lipids

### A. Amphineura

This class has been the subject of but few investigations, which moreover are all concerned with the subclass Polyplacophora. Most data on lipid content and function in chitons have been obtained by Giese and his co-workers. They showed lipids to be important reserves in *Katharina tunicata* (Giese and Araki, 1962), varying in amount during the reproductive cycle (Giese and Hart, 1967). The gonads and particularly the ovaries, turned out to be very rich in lipid (Giese, 1966; Lawrence and Giese, 1969). Lipids in ovaries comprised 80% neutral lipids; in testes the corresponding value amounted to 60%. These sex differences were not found in the lipid composition of the other organs. Lawrence and Giese (1969) further showed that the foot, mantle, and digestive gland contribute lipid reserves for normal metabolic use and for gametogenesis.

In a study on the interorgan transport of lipids—the promising first in this field—Allen (1977) showed that the blood plasma of *Cryptochiton stelleri* contained alkyldiacylglycerols, triacylglycerols, and fatty acids, in addition to phospholipids and sterols. He obtained evidence that fatty acids are the principal transport form of acyl lipid and calculated from their rate of turnover in plasma that this process may support some 10% of the energy demand.

### 1. Sterols

Toyama and Tanaka (1953) proved the sterols of *Liolophura japonica* to consist of $\Delta^7$-cholestenol with some minor unidentified components. This has been found since then for several other chitons (Voogt, 1972a).

Recent analysis of the sterols of *Lepidochitona* sp. (Voogt and van Rheenen, 1974), *Chaetopleura apiculata* (Idler and Wiseman, 1971), and *L. japonica* (Teshima and Kanazawa, 1972) gave percentages of $\Delta^7$-cholestenol varying from 89 to 93%, with $\Delta^5$-cholestenol as the second most important sterol. Sterols of an unidentified chiton species even contained 24.7% $\Delta^5$-cholestenol, in addition to 75.0% $\Delta^7$-cholestenol

(Idler and Wiseman, 1971). Idler et al. (1978) reported the sterols of *Placophorella velate* and *Ischnochiton* sp. to contain 64.2 and 86% $\Delta^7$-cholestenol, respectively, and the amount of cholestanol present to surpass that of $\Delta^5$-cholestenol. In a scrutinous analysis of the sterols of *L. japonica*, Teshima et al. (1982) identified 20 compounds most of them present in small or even trace amounts; $\Delta^7$-cholestenol made up 95%, whereas cholesterol and the other $\Delta^5$-sterols together were less than 1%. These data show that in chitons the sterol spectrum is simple, with $\Delta^7$-cholestenol as the main sterol, setting them aside in the phylum Mollusca in which $\Delta^5$-sterols are common. They also show that in the same animal $\Delta^7,\Delta^5$-, and saturated sterols are present together, which may be indicative for sterol metabolism.

## 2. Fatty Acids

Since Takagi and Toyama (1956) reported for *L. japonica* a ratio of di- and triunsaturated fatty acids to tetra- and pentaunsaturated ones larger than for other aquatic animals, only one article on fatty acids in chitons has appeared. Johns et al. (1980), in a food web investigation, studied in detail the fatty acid composition of various organs of *Ponerplax costata*. The major polyunsaturated fatty acids were 20:4 $\omega$ 6 and 20:5 $\omega$ 3, their level being high in the gill and showing seasonal variation in the foot. The levels of 13:1 $\omega$ 9 are high both absolutely and relatively to 18:1 $\omega$ 7.

## 3. Lipids Containing Ether Bonds

Thompson and Lee (1965) showed ether bonds to be present in neutral lipids (4.1%) and phospholipids (7.9%) of *K. tunicata*. In neutral lipids the alkyl group consisted of 15:0, 16:0, 18:0, and 18:1; in the phospholipids 14:0 was present as well. The content of 18:1 far exceeded that in the representatives of the other classes. Isay et al. (1976) reported that 1-alkyliacylglycerols are characteristic for marine animals, their concentration in all phyla being about 5%. Within the Mollusca their concentration generally is lower than 5%, but in the chiton *Ischnochiton hakodadensis* the concentration amounted to 8.4%, being much higher than in the other classes.

## B. Gastropoda

### 1. Prosobranchia

Within this subclass great differences are observed with reference to the nature of reserves. Glycogen is stored in the foot of the archeogastropod *Haliotis cracheroidii* (Webber, 1970). Possibly it is converted into lipids and subsequently transferred to the gonads, which are rich in lipid. This holds particularly for ovaries, which indicates that lipids are the nutritive storage product of the eggs. According to Webber (1970), the high lipid:carbohydrate ratio in the ovaries may be associated with the marine habitat of *Haliotis*. Also Giese (1966) mentioned the storage of large amounts of glycogen in archeogastropods. In *Turbo sarmaticus*, lipid values are low and correlated with temperature, suggesting that lipids are involved in normal maintenance (McLachlan and Lombard, 1980).

Holland et al. (1975) reported high lipid contents in the veliger larvae of several *Littorina* species. Particularly neutral lipids were used during development and metamorphosis.

Lambert and Dehnel (1974) showed the gonads and digestive gland of the carnivorous neogastropod *Thais lamellosa* to be rich in lipids, which were used during winter. Stickle (1975) showed that in the female *T. lamellosa*, lipid loss was correlated with oocyte production. In both sexes of *Thais haemastoma* (Belisle and Stickle, 1978), lipid concentrations were strongly correlated with the reproductive cycle, contents in the testes (6.37–11.32% of dry weight) being about half those in the ovaries. On starvation *Littorina littorea* utilized both carbohydrates and lipids (Holland et al., 1975). Under these conditions *T. lamellosa* did not use lipids, but polysaccharides and proteins (Stickle, 1971); in the males, even lipogenesis occurred. This and the effect of starvation on respiration in this animal might indicate that metabolism in carnivorous prosobranchs is "lipid oriented" (Stickle and Duerr, 1970).

In the freshwater prosobranch *Melania scabra* (Muley, 1975), a reciprocal relationship was observed between the synthesis of lipids and utilization of proteins, and vice versa. Also in two other freshwater prosobranchs, *Viviparus bengalensis* and *Acrostoma variabile*, lipids were used during breeding (Chatterjee and Ghose, 1973).

**a. Archeogastropoda.** (I) STEROLS. Sterols in archeogastropods have been studied rather extensively. The older data have been reviewed previously (Voogt, 1972a); results of recent analyses are given in Table I. From these data it can be concluded that in Haliotidae sterol composition is very simple and that in Trochidae the greatest number of components is present, none of them reaching high values, however. It can be further concluded that in this order cholesterol is predominant, making up 80–100% of the total sterols. In general desmosterol is present, attaining considerable percentages only in Acmaeidae. Because this sterol is an intermediate in cholesterol synthesis, its presence may be indicative for sterol-synthesizing capacity. Brassicasterol and 22-dehydrocholesterol, both with a double bond at C-22, are rather common, but reach higher values only in Trochidae. Campes-

**TABLE I**

**Percentage Composition of Sterols in Prosobranch Gastropods**

| Molluscs | % Composition of sterols[a] | | | | | | | | | | | | | | Reference |
|---|---|---|---|---|---|---|---|---|---|---|---|---|---|---|---|
| | A | B | C | D | E | F | G | H | I | J | K | L | M | N | |
| Order Archeogastropoda | | | | | | | | | | | | | | | |
| Family Haliotidae | | | | | | | | | | | | | | | |
| Haliotis gurneri | 1 | 6 | | | 93 | | | | tr[b] | | | | | | Teshima and Kanazawa (1972) |
| Haliotis discus hannai | | | | | 100 | | | | | | | | | | Hayashi and Yamada (1972) |
| H. discus hannai | | tr | | | 97.5 | tr | tr | | | | | | | | Joh and Kim (1974) |
| Family Acmaeidae | | | | | | | | | | | | | | | |
| Amacea testudinalis alvea | 0.1 | 0.3 | | | 97.0 | 2.0 | tr | tr | 0.6 | | | | tr | | Idler and Wiseman (1971) |
| Cellana nigrolineata | | | | | 98 | 2 | | | | | | | | | Teshima and Kanazawa (1972) |
| C. nigrolineata | | | | | 86 | 10 | tr | 2 | | | | | | | Teshima and Kanazawa (1972) |
| Patelloida saccharina | | | | | 100 | | | | | | | | | | Teshima and Kanazawa (1972) |
| Patella coerulea | | | | | 69.5 | 18 | | | 2.5 | 0.5 | 2.0 | 7.5 | | 2.5 | Voogt (1971a) |
| Family Trochidae | | | | | | | | | | | | | | | |
| Chlorostoma argyrostoma lischkei | | | | | 92.5 | | | | 2.8 | | 2.3 | | | 2.5 | Hayaski and Yamada (1974) |
| Monodonta turbinata | | | | | 80 | 8 | | | 2.7 | 0.4 | 1.7 | 2.3 | | 4.5 | Voogt (1971a) |
| Monodonta labio | | | | | 99 | tr | tr | | | | | | | | Teshima and Kanazawa (1972) |
| Teculatus maximus | | 4 | | | 79 | 3 | tr | 6 | tr | 1 | 8 | tr | | | Teshima and Kanazawa (1972) |
| Trochus maculatus | | 2 | | | 82 | 5 | | 7 | | | 5 | | | | Teshima and Kanazawa (1972) |
| Trochus niloticus | | 2 | | 4 | 85 | 3 | | 3 | | tr | 3 | | | | Ando et al. (1979) |
| Family Turbinidae | | | | | | | | | | | | | | | |
| Astralium haemastragum | | | | | 78 | 3 | | 7 | | | 8 | 4 | | | Teshima and Kanazawa (1972) |
| Turbo cornutus | | | | | 99 | | tr | | tr | | tr | tr | | | Teshima and Kanazawa (1972) |
| Family Neritidae | | | | | | | | | | | | | | | |
| Nerita argus | | 1 | 2 | | 96 | tr | tr | 1 | | | 1 | | | | Ando et al. (1979) |
| Ritena plicata | | 1 | 1 | | 96 | tr | | 2 | | | 1 | | | | Ando et al. (1979) |
| Order Mesogastropoda | | | | | | | | | | | | | | | |
| Family Viviparidae | | | | | | | | | | | | | | | |
| Viviparus fasciatus | 0.3 | 2 | 1 | | 72 | | 3 | 4.5 | 2 | 2.5 | 10 | 1 | | 2 | Voogt (1971a) |

| | | | | | | | | | | | | | |
|---|---|---|---|---|---|---|---|---|---|---|---|---|---|
| **Family Littorinidae** | | | | | | | | | | | | | |
| *Littorina littorea* | tr | tr | 67 | 15 | 4.5 | 4 | 4 | | 3 | 2 | | | Voogt (1971a) |
| *L. littorea* | 0.9 | 1.7 | 56.0 | 29.5 | 4.3 | 1.9 | 1.8 | | | 0.6 | 1.6 | 1.4 | Idler and Wiseman (1971) |
| *Littorina coccinea* | | tr | 96 | 1 | 2 | tr | | | tr | | | | Ando et al. (1979) |
| *Littorina scabra* | | tr | 99 | 1 | tr | tr | | | tr | | | | Ando et al. (1979) |
| **Family Calyptraeidae** | | | | | | | | | | | | | |
| *Crepidula fornicata* | 3.2 | 7 | | 58 | 4 | 1.5 | 8 | 2.5 | 3 | | | | Voogt (1971b) |
| *C. fornicata* | 2.8 | 2.8 | 0.4 | 74.0 | 4.4 | 2.8 | 3.9 | | 5.9 | | 3.0 | 11.5 | Idler and Wiseman (1971) |
| **Family Naticidae** | | | | | | | | | | | | | |
| *Natica catena* | 3.4 | 9.5 | 1 | 52 | 12 | | 7.5 | 2 | 4 | | | 8 | Voogt (1971b) |
| *Lunatia groenlandica* | 0.2 | 0.4 | tr | 98.0 | 0.1 | 0.4 | | | 0.4 | | 0.1 | | Idler and Wiseman (1971) |
| **Family Strombidae** | | | | | | | | | | | | | |
| *Conomurex luhuanus* | | 5 | | 64 | 9 | | 2 | tr | 10 | | | | Teshima and Kanazawa (1972) |
| **Family Cerithidae** | | | | | | | | | | | | | |
| *Cerithium echinatus* | | 4 | | 77 | 7 | 5 | tr | | 6 | tr | | | Ando et al. (1979) |
| *Rhinoclavis fasciatus* | | 2 | tr | 63 | 11 | 9 | 6 | | 8 | 1 | | | Ando et al. (1979) |
| **Family Cypraeidae** | | | | | | | | | | | | | |
| *Peribolus arabica* | tr | 4 | 2 | 80 | 4 | tr | 4 | 2 | 5 | tr | | | Teshima and Kanazawa (1972) |
| *Ravitrona captserpentis* | | | | 100 | | | | | | | | | Teshima and Kanazawa (1972) |
| *Monetaria obvelata* | | 2 | | 86 | 5 | 3 | tr | | 4 | | | | Ando et al. (1979) |
| *Monetaria moneta* | | 1 | | 74 | 8 | 8 | 2 | | 4 | 3 | | | Ando et al. (1979) |
| **Family Cymatidae** | | | | | | | | | | | | | |
| *Charonia sauliae* | tr | 1 | | 42 | 4 | 5 | | tr | 4 | | | | Teshima and Kanazawa (1972) |
| *Cymatidae sp.* | | 6 | 35 | 52 | 22 | 16 | | | 4 | | | | Teshima and Kanazawa (1972) |
| *Pterotrachea coronata* | 14.3 | | tr | 69.7 | 12.8 | | 2.6 | | | | | 0.6 | Kanazawa et al. (1976) |
| **Order Neogastropoda** | | | | | | | | | | | | | |
| **Family Muricidae** | | | | | | | | | | | | | |
| *Urosalpinx cinereus* | tr | 0.5 | | 98.5 | tr | 1.0 | | | | | | | Idler and Wiseman (1971) |
| *Murex trunculus* | 0.4 | 2.4 | 1.2 | 72.2 | 5.0 | 1.1 | 3.2 | | | | | | Sica (1980) |
| *Murex brandaris* | tr | tr | 3.0 | 58.8 | 3.9 | 3.8 | 2.4 | 2.8 | 5 | | | 7 | Voogt (1972b) |
| *Murex asianus* | tr | 5 | tr | 56 | 17 | 12 | 3 | tr | 5 | 1 | | | Teshima and Kanazawa (1972) |
| *Purpura bronni* | | 11 | | 86 | 4 | | | | | | | | Teshima and Kanazawa (1972) |
| *Purpura lapillus* | 1.2 | 3.9 | 4 | 77 | 4.5 | 2 | | 1 | 0.5 | | | 4.5 | Voogt (1972b) |
| *Purpura haemastoma* | 0.9 | 5.8 | 1.1 | 75.1 | 5.8 | 0.6 | 2.6 | | | | | 5.4 | Sica (1980) |

*(Continued)*

337

**TABLE I** (*Continued*)

% Composition of sterols[a]

| Molluscs | A | B | C | D | E | F | G | H | I | J | K | L | M | N | Reference |
|---|---|---|---|---|---|---|---|---|---|---|---|---|---|---|---|
| *Drupa rubscidaeus* | | tr | | | 80 | | 3 | 8 | | | 9 | | | | Ando et al. (1979) |
| *Drupa elata* | | | | | 73 | | tr | 22 | 5 | | | | | | Ando et al. (1979) |
| **Family Buccinidae** | | | | | | | | | | | | | | | |
| *Buccinum undatum*[c] | 1.5 | 1.1 | tr | | 94.4 | 2.0 | 0.3 | tr | 0.7 | | | | | | Idler and Wiseman (1971) |
| *B. undatum*[d] | 2.8 | 2.0 | 0.5 | | 83.0 | 2.0 | 4.5 | 0.5 | 2.0 | 0.3 | 0.6 | | | 2.3 | Voogt (1972c) |
| **Family Neptunidae** | | | | | | | | | | | | | | | |
| *Neptunea decemcostrate*[c] | 0.6 | 1.3 | 0.3 | | 93.5 | | 0.7 | 1.1 | 2.5 | | tr | tr | | | Idler and Wiseman (1971) |
| *Neptunea antiqua*[d] | 2.0 | 3.0 | 0.5 | | 80.0 | 1.0 | 6.0 | 2.0 | 2.5 | 1.0 | 2.5 | | | 1.5 | Voogt (1972c) |
| *Busycon canaliculatum* | 0.2 | 0.5 | 0.1 | | 85.5 | | 0.8 | 8.1 | 0.4 | 0.4 | 3.7 | | 0.7 | 3.7 | Idler and Wiseman (1971) |
| **Family Rapidae** | | | | | | | | | | | | | | | |
| *Coralliobis violacea* | | | | | 65 | | 9 | 22 | 2 | | | | | 2 | Ando et al. (1979) |
| **Family Fasciolariidae** | | | | | | | | | | | | | | | |
| *Peristomia nassatula* | | 1 | | | 87 | | 9 | 1 | | 1 | 2 | | | | Ando et al. (1979) |
| *Pleoroploca trapezium* | | | | | 83 | | 9 | 3 | | | 4 | | | | Teshima and Kanazawa (1972) |
| **Family Conidae** | | | | | | | | | | | | | | | |
| *Lithoconus pardus* | | 2 | | 3 | 80 | | 6 | 7 | 1 | | 1 | tr | | | Ando et al. (1979) |
| *Conus taitensis* | | tr | | tr | 80 | | 15 | 3 | | | | | | | Ando et al. (1979) |
| *Puncticulis pulicarium* | | 1 | | 2 | 87 | | 3 | 3 | tr | 1 | 2 | 1 | | | Ando et al. (1979) |
| *Rhombus imperialis* | | tr | tr | tr | 71 | | 12 | 8 | | 6 | 2 | | | | Ando et al. (1979) |
| **Family Terebridae** | | | | | | | | | | | | | | | |
| *Terebra maculata* | | 2 | | 5 | 74 | | 8 | 3 | 3 | 4 | 1 | | | | Ando et al. (1979) |
| *Terebra guttata* | | tr | | 2 | 85 | | 4 | 3 | 3 | 3 | tr | | | | Ando et al. (1979) |
| *Terebra sublaia* | | tr | | tr | 71 | | 9 | 8 | 5 | 7 | tr | | | | Ando et al. (1979) |

[a] A, 22-trans-24-norcholesta-5,22-dienol; B, 22-trans-cholesta-5,22-dienol; C, 22-cis-cholesta-5,22-dienol; D, cholestanol; E, cholesterol; F, desmosterol; G, 24-methylcholesta-5,22-dienol; H, 24-methylcholesterol; I, 24-methylenecholesterol; J, 24-ethylcholesta-5,22-dienol; K, 24-ethylcholesterol; L, fucosterol; M, 28-isofucosterol; N, other sterols.

[b] tr, Trace amounts.

[c] Muscle.

[d] Whole animal.

terol or its C-24 epimer and $\beta$-sitosterol are important only in Trochidae. Cholestanol was observed only in coral reef molluscs.

(II) FATTY ACIDS. Investigations of fatty acid compositions of some *Haliotis* species (Shimma and Taguchi 1964; Bannatyne and Thomas, 1969) and of *Chlorostoma argyrostoma lischkei* (Hayashi and Yamada, 1974) showed the main fatty acids to be, in order of decreasing abundancy, 16:0, 18:1, 20:4, 20:5, and 22:5. Similar results were obtained by Johns et al. (1980) for several Australian gastropods. The relative amounts of polyenoic acids show strong variations: Low contents (5.9%) of these acids—accompanied by very high contents of dienoic ones (32.5%)—were found in *Patella coerulea* (Tibaldi, 1966), whereas extremely high contents (43.2%) were found in *Turbo cornutus* (Shimma and Taguchi, 1964).

Visceral lipids appear to be richer in monoenoic acids and less rich in polyenoic ones than flesh lipids (Hayashi and Yamada, 1972, 1974). Gardner and Riley (1972) found a branched C-21 fatty acid to be the main one in phospholipids of *Patella vulgata*, whereas this acid held only the third position in triglycerides.

Vaver et al. (1972) encountered in *Acmaea palloda* an extremely high content of diollipids, the ratio of ethylene glycol:glycerol being 7:3. This ratio was much higher than that in a bivalve and a holothurian studied at the same time.

III. PHOSPHOLIPIDS. Phospholipids, and particularly glycerolphosphatides, of archeogastropods have been relatively little studied in the last 10 years. Rouser et al. (1963) discovered a 2-aminoethyl phosphonic acid containing sphingolipid of which De Koning (1966a,b) showed 6% to be present in the lipids of *Haliotis midae*. He also showed that the fatty acid composition in this ceramide-2-aminoethyl phosphonate (CAEP) was quite different from that in other lipid classes by the strong preference for saturated fatty acids (75% of the total). Hori et al. (1976b) reported the presence of CAEP and the absence of ceramide-aminoethyl phosphate in seven archeogastropod species. They (Hori et al., 1967a) suggested that sphingomyelin and CAEP might partly replace each other. An analog, ceramide monomethyl-aminoethyl phosphonate, (CMAEP) was found in *Monodonta labio* (Hori and Arakawa, 1969).

A new sphingolipid, *N*-palmitoylsphingosyl(C-16)-2-*N*-methylaminoethyl phosphonate, was isolated from the viscera of *T. cornutus* (Hayashi et al., 1969a). Matsuura et al. (1973) showed molecular species of this compound to contain normal fatty acids and dihydroxy long-chain bases. The mixture turned out to be rather complex, as in CMAEP several species were composed of a 2-hydroxy fatty acid with 16 or 17 carbon atoms linked to a trihydroxy(phyto)sphingosine with different numbers of carbon atoms. Main components were (Hayashi and Matsuura, 1973) 2-hydroxyhexadecanoyl(C-18)-phytosphingosine (46%), 2-hydroxyhexadecanoyl(C-19)phytosphingosine (17%), and 2-hy-droxyhexadecanoyl(C-22)dehydrophytosphingosine (27%). Two forms of the C-22 species—namely, docosa-4,15-sphingadienine and 4-hydro-oxydocosa-15-sphingenine—have been reported (Hayashi et al., 1975). More data are needed before generalizations can be made.

IV. LIPIDS CONTAINING ETHER BONDS. Little is known about glyc-eryl ethers in archeogastropods (for older data, see Voogt, 1972a). Hayashi and Yamada (1972) reported the presence of ethers in visceral lipids of *Haliotis discus hannai*. The alkyl chain ranged from 14:0 to 10:0, 18:0 being the main one (34%). Further, the alkyl chains 16:1, 18:1, and 20:1 were present.

V. GLYCOLIPIDS. Hayashi et al. (1969b) isolated from *T. cornutus* a glycolipid containing sphingosine bases, fatty acids, and the sugars glu-cose, galactose, fucose, and glucosamine.

Later (Matsubara and Hayashi, 1981), these glycolipids turned out to be composed of five types. The numbers 1–3 represent mono-, di-, and trigalactosylceramide, respectively. Number 4 possibly is identical with tetragalactosylceramide. The fifth contained fucose, galactose, glucose, and N-acetylglucosamine in the molar ratio 1:2:1:1.

In glycolipid 1 the fatty acid was mainly palmitic acid; the long-chain base consisted mainly of C-18:1 sphingosine, whereas on acid hydroly-sis galactose and 2-N-methylaminoethyl phosphonic acid were set free. The molar ratio was 1:1:1:1 (Hayashi and Matsuura, 1971). Matsuura (1977) proposed for this compound the structure 1-O[6'-O-(N-methyl-aminoethyl-phosphonyl)galactopyranosyl]ceramide. This compound is closely related to galactosylceramide, which is present in large amounts. Both compounds do contain the combinations of dihydroxy long-chain bases with nonhydroxy fatty acids, and of dihydroxy long-chain bases with hydroxy fatty acids. From the muscle of *T. cornutus* the same compound was isolated. A minor component was 1-O[6'-O-(aminoethylphosphonyl)galactosyl]ceramide. The main fatty acids were 16:0 (53.3%) and 2-hydroxypalmitic acid (14.6%). The long-chain bases were composed mainly of C-22:2 (36.6%), C-18:1 (14.6%), and C-18:2 (11.3%). Trihydroxy bases were minor components (Hayashi and Mat-suura, 1978). The same compounds were isolated from *M. labio* (Mat-suura, 1979). A novel pentaglycosylceramide has been isolated from *Haliotis japonica* (Matsubara and Hayashi, 1982) with the structure

$\alpha$Fuc-(1 $\rightarrow$ 3)-$\alpha$GalNAc-(1 $\rightarrow$ 3)-$\alpha$Fuc-(1 $\rightarrow$ 2)-$\beta$Gal-(1 $\rightarrow$ 4)-$\beta$Glc-1-ceramide. The main fatty acid and long-chain base were 16:0 and 16:1 (hexadecasphing-4-enine), respectively.

**b. Mesogastropods.** I. STEROLS. Most studies on mesogastropods are concerned with sterols. Recent data are listed in Table I. This table shows that sterol composition in mesogastropods is more complex than in archeogastropods. Cholesterol still is the main sterol but it seldom reaches the percentages common in archeogastropods. Sterols with double bonds at C-22 are common. Striking are the high values of desmosterol in some species, the possible significance of which will be discussed later. Sterol composition in *Charonia tritonis* (Teshima et al. 1979a) is interesting: Cholesterol (29.5%) and 24-methylcholesterol (16.7%) are the main sterols, but also considerable amounts (34.4%) of $\Delta^7$-sterols are present among which are cholest-7-enol (5.4%), 24-methylcholest-7,22-dienol (6.7%), and 24-methylcholest-7-enol (15.5%). As the triton predates on the seastar *Acanthaster planci* (with $\Delta^7$-sterols), these latter sterols in *Charonia* probably have a dietary origin.

II. FATTY ACIDS. These have been little studied in mesogastropods. Johns et al. (1980) found the main fatty acid in *Telescopium telescopium* and *Terebralia palustris* to be 16:0, followed by 18:1 $\omega$ 9, respectively, and 20:4 $\omega$ 6 + 20:5 $\omega$ 3. This is in accordance with the situation in archeogastropods, in which, 16:0 is generally the main fatty acid for whole animals. Gardner and Riley (1972) studied the fatty acid composition in triglycerides, phospholipids, and sterol esters of *Crepidula fornicata* and found the main fatty acid in phospholipids to be 20:1 (12.2%), 16:0 (11.7%), and 18:1 (9.2%).

Non-methylene-interrupted fatty acids in *L. littorea* and *Lunatia triseriata* were studied (Ackman and Hooper, 1973), as well as the distribution of saturated and isoprenoid fatty acids in *L. littorea* (Ackman et al., 1971).

III. PHOSPHOLIPIDS. In contrast to Archeogastropoda, in which only CAEP has been found, Mesogastropoda contain, besides CAEP, also ceramide-2-aminoethyl phosphate (Hori et al., 1966, 1967b).

**c. Neogastropoda.** I. STEROLS. Results of recent investigations are given in Table I. Cholesterol, the main sterol, reaches values between those of archeogastropods and mesogastropods, ranging from 70 to 95% with a high concentration of values between 80 and 90%. Only in the genus *Murex* low contents of cholesterol are found. C-26 Sterols and 22-*cis*-cholesta-5,22-dienol seem to be absent, whereas cholestanol is present only in gastropods from coral reefs (Ando et al., 1979). 24-Ethylcholesta-5,22-dienol is more common, whereas fucosterol and isofucosterol are less encountered in this order than in the former two.

II. FATTY ACIDS. Fatty acids in *Murex brandaris* were studied by Calzolari et al. (1971); 16:0 and 18:1 were the main acids whereas 20:5 was only fifth in importance. In phospholipids of *Neptunea antiqua* the main acids were 20:1 and 16:0, followed by 17B (Gardner and Riley, 1972). These authors had also found a branched fatty acid (21B) to be the main acid in phospholipids of *P. vulgata*. In flesh of *M. brandaris* (Hayashi and Yamada, 1974) the main acids were 20:1, 20:5, and 16:0, whereas in flesh of *Neptunea intersculpta,* and three *Buccinum* species, the main acids were 20:5, 16:0, and 18:1 (Hayashi and Yamada, 1975). In agreement with former observations, visceral lipids had much higher contents of monoenoic acids than that of flesh.

III. PHOSPHOLIPIDS. Hori et al. (1967a) showed the presence of ceramide-2-aminoethyl phosphonate (CAEP) in *Purpura bronni* and *Pugilina ternatana,* whereas ceramide-aminoethyl phosphate is absent in these animals. Unfortunately no further data have become available since then to confirm the resemblance between the archeogastropods and neogastropods in this respect.

IV. LIPIDS WITH ETHER BONDS. Our knowledge of these lipids in neogastropods seems to be confined to two articles. Rapport and Alonzo (1960) studied these lipids in *Busycon canaliculatum,* and Thompson and Lee (1965) did so in *T. lamellosa.* No generalizations can be made.

### 2. Opisthobranchia

Little is known about lipid distribution in this subclass.

**a. Sterols.** Results of recent analyses of opisthobranch sterols are summarized in Table II. They clearly show that cholesterol, the main sterol, generally constitutes somewhat less than half of the total sterols, and that 22-*trans*-cholesta-5,22-dienol is often present in relatively high amounts. Desmosterol is always present, mostly in rather high and in the case of *Lamellidoris* in very high percentage, which may be indicative for sterol metabolism.

**b. Fatty Acids.** Studies are confined to some rather old ones: on *Aplysia kurodai* (Tanaka and Toyama, 1959) and on *Aplysia fasciata* and

**TABLE II**

**Percentage Composition of Sterols in Opisthobranchia and Pulmonata**

| Molluscs | % Composition of sterols[a] | | | | | | | | | | | | | Reference |
|---|---|---|---|---|---|---|---|---|---|---|---|---|---|---|
| | A | B | C | D | E | F | G | H | I | J | K | L | M | |
| Subclass Opisthobranchia | | | | | | | | | | | | | | |
| Order Pleurocoela | | | | | | | | | | | | | | |
| Aplysia depilans | 0.5 | | 2.6 | 87.4 | 8.3 | | | 0.4 | | | | | 1.1 | Voogt and Van Rheenan (1973a) |
| Order Pteropoda | | | | | | | | | | | | | | |
| Spiratella helicina | 3.7 | 4.6 | 1.0 | 45.6 | 9.8 | 4.2 | 3.5 | 8.1 | | 12.0 | 1.8 | 5.7 | | Idler and Wiseman (1971) |
| Order Acoela | | | | | | | | | | | | | | |
| Archidoris tuberculata | 3.3 | 11.1 | 0.9 | 58.1 | 10.0 | 12.5 | 2.0 | tr[b] | 1.0 | 1.2 | | | | Voogt (1973a) |
| Archidoris montereyensis | 1.4 | 3.6 | tr | 60.4 | 6.6 | 3.0 | 0.3 | 21.2 | | 2.7 | tr | 0.8 | tr | Idler et al. (1978) |
| Lamellidoris bilamellata | 0.9 | 9.0 | | 39.9 | 42.0 | 3.5 | | 3.7 | 0.1 | | | 0.6 | 0.5 | Voogt (1973a) |
| Peltodoris atromaculata | 0.3 | 2.7 | 0.6 | 44.0 | 14.3 | 2.3 | 2.0 | 2.2 | 4.1 | 22.6 | 1.3 | 0.4 | 3.3 | Voogt (1973a) |
| Dendronotus frondosus | 7.4 | 16.8 | 2.0 | 43.0 | 17.3 | 3.2 | 3.0 | 1.3 | | 0.9 | | 0.3 | 5.1 | Voogt (1973a) |
| Subclass Pulmonata | | | | | | | | | | | | | | |
| Order Basommatophora | | | | | | | | | | | | | | |
| Lymnaea stagnalis | 0.7 | 1.0 | | 88.0 | 1.1 | 0.9 | 2.5 | | 1.1 | 1.9 | | | 3.1 | Voogt (1972e) |
| Planorbarius corneus | 0.3 | 1.6 | | 88.0 | | 2.0 | 3.5 | | 1.0 | 2.4 | | | 0.6 | Voogt (1972e) |
| Melampus lineatus | 1.0 | 1.9 | tr | 85.1 | | 3.7 | 2.2 | 1.9 | | 3.1 | | | | Idler and Wiseman (1971) |
| Order Stylommatophora | | | | | | | | | | | | | | |
| Cepaea nemoralis | | | | 93.1 | | 1.5 | 2.8 | | | 2.6 | | | | Van der Horst and Voogt (1972) |
| Arion rufus | 0.5 | | | 78.7 | 7.8 | 6.1 | 2.6 | | 2.1 | 1.6 | | | 0.5 | Voogt (1972d) |
| Arianta arbustorum | 1.1 | | | 83.7 | 4.6 | 1.0 | 4.0 | | 0.6 | 1.6 | | | 3.2 | Voogt (1972d) |
| Succinea putris | 1.7 | | | 83.8 | 5.0 | tr | 3.7 | | 0.6 | 4.4 | | | | Voogt (1972d) |

[a] A, 22-trans-24-norcholesta-5,22-dienol; B, 22-trans-cholesta-5,22-dienol; C, 22-cis-cholesta-5,22-dienol; D, cholesterol; E, desmosterol; F, 24-methylcholesta-5,22-dienol; G, 24-methylcholesterol; H, 24-methylenecholesterol; I, 24-ethylcholesta-5,22-dienol; J, 24-ethylcholesterol; K, fucosterol; L, 28-isofucosterol; M, other sterols.

[b] tr, Trace amounts.

*Pleurobranchaea meckeli* (Tibaldi, 1966). In the latter species the main fatty acids were 16:0, 20:2, and 22:2.

## 3. Pulmonata

**a. Basommatophora.** Little is known about economics and distributions of lipids in this order. In the freshwater limpet *Ancylus fluviatilis,* lipid reserves are stored in the midgut gland. Stores are larger in summer than in winter, although feeding takes place also in winter. During starvation lipid is used first, followed by carbohydrates (Streit, 1978). In *Bulinus africanus* the main reserve seems to be protein; during estivation lipid is metabolized preferentially over carbohydrates, whereas under starvation conditions the opposite takes place (Heeg, 1977).

I. STEROLS. Data on sterol composition are given in Table II. Cholesterol is predominant; 22-*cis*-cholesta-5,22-dienol, sterols with $\Delta^{24(28)}$ double bond, and desmosterol are not very common.

II. FATTY ACIDS. Lipids of *Siphonaria diemenensis* (Johns et al., 1980) and *Heterogen longispira* (Weerasinghe and Obara, 1974a) had very high contents of polyunsaturated fatty acids: 60 and 61.2%, respectively. In the former species, 20:4 ω 6 + 20:5 ω 3 were the main components; in the latter species 20:4 and 22:6 were. Monoenoic acids had low concentrations: 12–19% and 21.7%, respectively.

III. PHOSPHOLIPIDS. Liang and Strickland (1969) studied phospholipids of *Lymnaea stagnalis* in detail. Phosphatidylethanolamine was the major type (30.7%), but CAEP was also present (7.5%). In both the choline- and ethanolamine-containing types, glyceryl ethers and plasmalogens were present. Similar results were obtained for *H. longispira* (Weerasinghe and Obara, 1974b). The unusually high amount of phosphatidylinositol (38%) seems to be characteristic of *Heterogen* sp. Plasmalogens were chiefly of the ethanolamine type; CAEP was present.

**b. Stylommatophora.** Metabolism in stylommatophora seems to be carbohydrate oriented. As a result lipids are not very important as reserves, though in *Cryptozona ligulata* lipid amounts decrease during estivation (Krupanidhi et al., 1978). In *Semperula maculata* lipid content increases during pregametogenesis, is maximal when the ovotestes are fully gravid, and strongly decreases after the liberation of gametes (Nanaware and Varute, 1976). These authors suggest that lipids may be converted into sugars. Catalan et al. (1977) showed great variations of the levels of lipids during the development of *Arion empiricorum*.

I. STEROLS. Older data regularly mentioned the presence of $\Delta^{5,7}$-sterols in stylommatophoran snails, often in rather high levels (see Voogt, 1972a). In recent investigations small amounts of $\Delta^{5,7}$-sterols were found. The results are given in Table II. Sterol compositions are simple, cholesterol being by far predominant. Some tendencies started in Basommatophora are even more pronounced. So 22-dehydrocholesterol, both cis and transforms, is absent, as are compounds with the $\Delta^{24(28)}$ double bond.

II. FATTY ACIDS. Thiele and Kröber (1963) analyzed the fatty acid composition in various lipid classes of *Helix pomatia*. In free fatty acids the main ones, in order of decreasing abundancy, were 20:4, 20:2, and 18:2, whereas in the complex lipids there was a preference for the saturated fatty acids 16:0 and 18:0. In total lipids of *Cepaea nemoralis* (Van der Horst, 1970), the main fatty acids were 20:4 ω 6, 18:2 ω 6, 20:2, and 18:0, which are the same as given by Thiele and Kröber (1963) for the free fatty acids of *H. pomatia*. Fatty acids of *C. nemoralis* are mainly unsaturated, saturated ones constituting only 26.88% of the total fatty acids. A similar value (20%) was reported for *Euhadra herklotsi* (Takagi and Toyama, 1958). Mainly saturated fatty acids were present in *Arianta arbustorum* (45.3%) and *Succinea putris* (55.5%) (Van der Horst and Voogt, 1969a,b), the main fatty acids being 18:0, 18:1, 16:0, 20:1 in *A. arbustorum*, and 18:1, 16:0, 18:0, 18:2, in *S. putris*. The same fatty acids were predominant in storage lipids of *H. pomatia* (Thiele and Kröber, 1963).

III. PHOSPHOLIPIDS. Phospholipids constitute about 50% of the total lipids in Stylommatophora. In *C. nemoralis* they amount even to 65% (Van der Horst et al., 1973). The authors carefully studied the composition of the phospholipids in *C. nemoralis* and determined the fatty acid pattern in each phospholipid class. Phosphatidylcholine was the major class (47%), whereas CAEP made up only 6.8%. Fatty acids in phosphatidylcholine and diphosphatidylglycerol were highly unsaturated (80 and 83.5%, respectively), whereas fatty acids in CAEP were mainly saturated (54%), with 16:0 as the major one. Van der Horst et al. (1973) showed that each phospholipid class had its own fatty acid profile.

IV. LIPIDS CONTAINING ETHER BONDS. Thompson and Hanahan (1963) showed phospholipids of terrestrial slugs to be rich in 1-alkylglyceryl ethers. In *Arion ater* phospholipids constituted 40% of total lipids and consisted of about 26 mol % glyceryl ethers. The alkyl side chain consisted mainly of 16:0 (86–89%), whereas unsaturated side chains were absent. In *Ariolimax* glyceryl ethers were mainly present in phosphatidylcholine (49 mol 15%). Side chains were different for ethers isolated from triglycerides and the various phospholipid classes, but with a strong preference for 16:0 and 18:0.


## C. Bivalvia

Great differences are known in the lipid contents of bivalves. In several *Donax* species lipid contents are about 4.5% of dry weight (Balasubramanian et al., 1979), whereas in *Siliqua patula* lipids make up 42% of the dry weight (Lewin et al., 1979). Mane and Nagabhushanam (1975) attributed the low content of lipid in several bivalves to their sedentary way of living and, related to this, their capacity to survive anaerobic conditions. Although in general, glycogen is the most suitable substrate for anaerobic energy metabolism, some authors point to the role of lipids under these conditions (Zs.-Nagy and Galli, 1977; Van er Horst, 1974; Oudejans and Van der Horst, 1974a).

Seasonal changes, which are difficult to distinguish from changes related to the reproductive cycle, have been observed frequently (Ansell, 1974; Ando et al., 1976; Balasubramanian et al., 1979; De Moreno et al., 1976a, 1980; Desai et al., 1979; Nagabhushanam and Deshmukh, 1974; Nagabhushanam and Dhamne, 1977; Nagabhushanam and Talikhedkar, 1977; Nagabhushanam and Mane, 1978; Pollero et al., 1979; Swift et al., 1980). From the data on *Mytilus viridis* (Mane and Nagabhushanam, 1975), an inverse relationship between the lipid content of digestive gland and gonad can be deduced. Ansell (1974) mentioned that there was only little indication of any net utilization of lipid during winter. Zandee et al. (1980) suggested that in *Mytilus edulis* lipids are the main source of energy production in growing mussels. From autumn until spring lipids may be saved for gametogenesis.

Starvation experiments showed lipid to be the most important reserve in *Crassostrea gigas* (Riley, 1976). In *Paphia laterisulca,* however, lipid contents remained constant after starvation for 12 days, glycogen contents decreased, and water contents increased (Nagabhushanam and Dhamne, 1977). From the foregoing it may provisionally be concluded that lipids in bivalves are multifunctional, and that in different species one or some of these functions can be more noticeable.

In the neutral lipids of *Modiolus difficilis,* the alcohols to which fatty acids are esterified (ethylene glycol and glycerol) were present in the ratio

8:9, respectively, whereas in a gastropod this ratio amounted to 7:3 (Vaver et al., 1972).

## 1. Sterols

Sterols of bivalves have been reviewed by Toyama (1958), Bergmann (1962), and Voogt (1972a). Here only data that have not been reviewed before or are necessary to make the picture more complete will be discussed. They are summarized in Table III. These data show that cholesterol is the main sterol in the Protobranchia, constituting more than half but in the other orders only about one third of the total sterols. Brassicasterol (24-methylcholesta-5,22-dienol) is the second sterol, particularly in Filibranchia where it makes up about 20% of the total sterols. Extremely high levels of campesterol or 22,23-dihydrobrassicasterol (the two epimers of 24-methylcholesterol) are encountered in representatives of the genuses *Tridacne* and *Hipponps*. In Filibranchia the levels of 22-*trans*-cholesta-5,22-dienol are somewhat higher than in the other orders. In all filibranchian species investigated poriferasterol is found. Desmosterol is more common in filibranchia than in other orders; the reverse holds for fucosterol and 28-isofucosterol.

## 2. Fatty Acids

Many articles have been devoted to fatty acid composition in bivalves. The older ones have been reviewed previously (Voogt, 1972a), but since that time several new ones have appeared (Calzolari et al., 1971; Paradis and Ackman, 1977; Gardner and Riley, 1972; Pollero et al., 1979). These data confirm that in bivalves the typical marine unsaturated fatty acids 20:5 ω 3 and 22:6 ω 3 are predominant. It is remarkable that in the freshwater species of *Anodonta* (Gardner and Riley, 1972), the same predominance was found. The data also show that the levels of these two fatty acids undergo seasonal variations, their level being high in the cold season. There are indications that these variations are related to differences in the diet. Pollero et al. (1979) suggest that in *Chlamys tehuelcha*, levels of 20:5 ω 3 are due to increased ingestion of diatoms, whereas the 22:6 ω 3 may be derived from dinoflagellates. In the razor clam *S. patula* the main unsaturated fatty acids are 20:5 ω 6 and 22:6 ω 3 (Lewin et al., 1979). In the freshwater species *Diplodom patagonicus*, 20:4 ω 6 is predominant (Pollero et al., 1981), which was also often found in gastropods.

## 3. Phospholipids

Glycerol phospholipids of *Corbicula sandai* are rich in ethanolamine (Hori and Itasaka, 1961). This also holds for phospholipids in *Tapes japonica*, in which high percentages of phosphatidylethanolamine, etha-

## TABLE III
## Percentage Composition of Sterols in Bivalvia

| Molluscs | A | B | C | D | E | F | G | H | I | J | K | L | M | Reference |
|---|---|---|---|---|---|---|---|---|---|---|---|---|---|---|
| | | | | | | | | % Composition of sterols[a] | | | | | | |
| **Order Protobranchia** | | | | | | | | | | | | | | |
| Nucula sp. | 0.3 | 8.6 | 0.3 | 63.2 | | 10.7 | 5.4 | 5.9[b] | | 5.6 | | | | Idler and Wiseman (1971) |
| Solemya velum | | | | 100 | | | | | | | | | | Idler and Wiseman (1971) |
| **Order Filibranchia** | | | | | | | | | | | | | | |
| Atrina pectinata | 1 | 16 | 1 | 26 | | 23 | 1 | 21 | 2 | 6 | 1 | | | Teshima and Kanazawa (1972) |
| Atrina fragilis | 3.6 | 11.2 | 7.8 | 31.1 | 0.9 | 23.2 | 1.8 | 7.3 | 2.7 | 5.1 | | | 5.4 | Voogt (1975a) |
| Limaria fragilis | | 3 | | 60 | | 14 | 5 | 4 | 5 | 9 | | | | Ando et al. (1979) |
| Arca nivea | tr[c] | 8 | 1 | 36 | | 24 | 10 | 12 | 3 | 6 | | | | Ando et al. (1979) |
| Mytilus edulis | 5.9 | 8.6 | 0.9 | 58.7 | 8.1 | 9.6 | 0.6 | | 6.4 | 1.2 | | | | Idler and Wiseman (1971) |
| M. edulis | 2 | 25 | 4 | 30 | 4 | 23 | | 12 | 2 | | | | | Teshima and Kanazawa (1972) |
| M. edulis | 5.0 | 9.0 | 1.0 | 34.0 | 10.5 | 21.5 | 3.0 | 9.0 | 3.0 | 3.5 | 0.3 | | 0.3 | Voogt (1975a) |
| Ostrea edulis | 7.5 | 11.6 | 3.6 | 33.2 | | 24.7 | 2.6 | 9.5 | 1.5 | 3.4 | 1.7 | | 0.4 | Voogt (1975a) |
| Crassostrea virginica | 4.6 | 7.5 | 1.1 | 34.1 | | 28.1 | 1.7 | 8.1 | 1.5 | 3.7 | 0.1 | 8.3 | 1.2 | Idler and Wiseman (1971) |
| C. virginica | 4.0 | 10.2 | | 34.0 | 0.2 | 15.6 | 3.7 | 12.6 | 2.0 | 3.7 | 1.5 | 4.6 | | Teshima et al. (1980) |
| **Order Eulamellibranchia** | | | | | | | | | | | | | | |
| Anodonta cygnea | tr | 4.2 | 1.9 | 47.4 | | 18.5 | 8.5 | 8.5[b] | | 10.2 | | | 0.3 | Voogt (1975b) |
| A. cygnea | 2 | 2 | | 42 | | 24 | | 25[b] | | | 7 | | | Popov et al. (1981) |

| | A | B | C | D | E | F | G | H | I | J | K | L | M | Reference |
|---|---|---|---|---|---|---|---|---|---|---|---|---|---|---|
| Pseudoanodonta complanata | | | 4 | 48 | 12 | 26 | | | | 10 | | | 3 | Popov et al. (1981) |
| Unio timidus | | | 2 | 52 | 16 | 27 | | | | | | | 3 | Popov et al. (1982) |
| Arctica islandica | 6.7 | 1.5 | 11.3 | 35.8 | 21.6 | 10.9 | 1.7 | 3.6 | | | 3.5 | | 0.4 | Voogt (1975b) |
| A. islandica | 5.5 | 4.3 | 5.5 | 44.5 | 7.0 | 21.6 | 1.7 | 4.2 | 3.3 | | | | 2.4 | Idler and Wiseman (1971) |
| Cardium edule | 4.2 | 0.7 | 6.9 | 24.9 | 22.0 | 13.0 | 5.0 | 8.2 | 5.2 | 8.1 | | | 2.1 | Voogt (1975b) |
| Tridacna maxima | tr | tr<sup>b</sup> | 2 | 17 | 8 | 16 | 5 | 36 | 3 | | | tr | 16 | Ando et al. (1979) |
| Tridacna crocea | 1 | | 4 | 22 | 17 | 13 | 3 | 44 | tr | | | tr | 1 | Teshima et al. (1974) |
| Tridacna noae | tr | | 1 | 12 | 5 | 10 | tr | 65 | | | | tr | 4 | Teshima et al. (1974) |
| Tridacna squamosa | tr | | 3 | 24 | 11 | 13 | 3 | 39 | 1 | | | tr | 7 | Teshima et al. (1974) |
| Hipponps hipponps | tr | | 3 | 29 | 14 | 17 | 4 | 34 | tr | | | tr | 3 | Teshima et al. (1974) |
| Tapes philippinarum | 4 | 1 | 6 | 35 | 14 | 9 | | tr | 4 | 8 | | 3 | | Teshima and Kanazawa (1972) |
| Tapes japonica | 2.7 | | 7.2 | 35.4 | 15.7 | 25.8 | | | | | | | 14.2 | Yasuda (1970) |
| Spisula solidissima | 3.9 | 1.9 | 2.9 | 68.0 | 8.5 | 5.4 | 4.8 | | | | | | 2.4 | Idler and Wiseman (1971) |
| Spisula sachalinensis | 3.0 | | 10.2 | 48.6 | 13.8 | 11.5 | 11.9 | | | | | | | Joh and Kim (1976) |
| Mactra sulcataria | 3.0 | | 6.7 | 39.6 | 14.1 | 19.4 | 10.5 | 4.3 | | | | | | Joh and Kim (1976) |
| Mya arenaria | 2.4 | 0.5 | 4.0 | 59.0 | 12.3 | 10.8 | 4.1 | | | | | | 1.6 | Idler and Wiseman (1971) |
| M. arenaria | 3.4 | 0.6 | 9.5 | 38.1 | 13.2 | 16.6 | 2.0 | 3.9 | | | | | 1.0 | Voogt (1975b) |

[a] A, 22-trans-24-norcholesta-5,22-dienol; B, 22-trans-cholesta-5,22-dienol; C, 22-cis-cholesta-5,22-dienol; D, cholesterol; E, desmosterol; F, 24-methylcholesta-5,22-dienol; G, 24-methylcholesterol; H, 24-methylenecholesterol; I, 24-ethylcholesta-5,22-dienol; J, 24-ethylcholesterol; K, fucosterol; L, 28-isofucosterol; M, other sterols.

[b] Combined values for the sterols unified by this symbol.

[c] tr, Trace amounts.

nolamine, plasmalogen, and sphingoethanolamine are present (Yasuda, 1967). Zama et al. (1960) and Zama (1963), on the other hand, reported low percentages of phosphatidylethanolamine in phospholipids of *Pecten yessoensis, Chlamys nipponensis,* and *Mactra sacchalinensis.* Shieh (1968) analyzed the fatty acid composition of the various phospholipids of *Placopecten magellanicus* and found in all cases 16:0, 20:5, and 22:6 to be the main fatty acids. Ceramide-aminoethyl phosphonic acid (CAEP) was demonstrated to be present in *C. sandai* by Hori et al. (1964b). Since then its presence has been mentioned for several bivalves (Sugita et al., 1968; Quin, 1965; Higashi and Hori, 1968). In *Crassostrea virginica,* very high levels of phosphonates are present, 25–35% of it phospholipids (Quin and Shelburne, 1969; Sampugna et al., 1972). Swift (1977) showed that during starvation and preparation for reproduction phosphonate bonds are conserved at the expense of phosphodiester bonds. This was also found for the individual tissues with the exception of the adductor muscle, in which the phosphonate : phosphodiester ratio remained constant during starvation but had increased after spawning.

## 4. Lipids with Ether Bonds

Isay et al. (1976) assumed alkyl ethers to be characteristic for marine lipids, occurring in about the same percentages in all phyla. Within the Mollusca, alkyl ethers contribute less than 5% of the lipids.

The alkyl chains of glyceryl ethers of *Protothaca staminea* were analyzed by Thompson and Lee (1965). The main ones were 16:0 and 18:0 in neutral and phospholipids, and 18:0 and a branched 18:0 in plasmalogens.

## 5. Glycolipids

Hori et al. (1964a) found the glycolipids of *C. sandai* to consist of four groups, all of them containing phosphorus. Since then each group has been studied more or less extensively, and fatty acid composition as well as sugar components have been determined (Hori et al., 1968, and references therein).[1]

The glycolipids of oyster have also been studied extensively. Nakazawa (1959) identified a new glycolipid, isolated some years before (Akiya and Nakazawa (1955), to be *O*-(14-methyl-4-pentadecenoyl)choline-*N*-*O*-[L-fucopyranosyl-(1→4)-D-glucopyranosyl-(1→4)-D-glycopyranosyl]lactyl-taurate. Hayashi and Matsubara (1966) found the glycolipid of oyster gills to be a mucolipid containing fatty acid, sphingosine, and neutral and amino sugars. This glycolipid resembles that of *Corbicula* as described by

---

[1] See note added in proof, p. 369.

Hori et al. (1964a). In oyster mantle a glycolipid was present different from that in gill (Hayashi and Matsubara, 1969).

## D. Scaphopoda

Only one article seems to have appeared concerning scaphopods. Idler et al. (1978) gave the following sterol composition for *Dentalium entate:* 22-*trans*-24-norcholesta-5,22-dienol (0.5%), 22-*trans*-cholesta-5,22-dienol (2.2%), cholesterol (89.5%), brassicasterol (1.9%), 24-methyl-cholesterol (2.6%), 24-methylenecholesterol (1.4%), and 24-ethylcholesterol (1.6%).

## E. Cephalopoda

Phospholipids were the main lipids in the squids *Loligo vulgaris* and *Illex illecebrosus coindetti* (Shchepkin et al., 1976). Triacylglycerol levels were high in the hepatopancreas. This, together with the presence of high levels of fatty acids in the fin of *Loligo,* may indicate a high lipid metabolism in this species. The differences in lipid distribution in the two species may be related to differences in their ways of living. In accordance with these data Boucaud-Camou (1971) showed triacylglycerols to be the reserve material in the hepatopancreas of *Sepia officinalis*. Besides triacylglycerols, alkoxydiacyl-glycerols and ceramides were present.

### 1. Sterols

Sterols of cephalopods have been little studied and include five species (Voogt, 1973b; Idler et al., 1978; Sica, 1980). Cholesterol is by far predominant in cephalopods (89–95%), and only small amounts of C-28 and C-29 sterols are present. Within molluscs such simple sterol patterns have been encountered only in archeogastropods and neogastropods.

### 2. Fatty Acids

Recently no reports on fatty acids in cephalopods seem to have appeared. Older data reviewed previously (Voogt, 1972a) showed that in species studied, 22:6 is predominant—in flesh as much as 37.1% of the total—followed by 16:0 and 20:5. Thus fatty acid composition in cephalopods shows a pattern that is characteristic for marine lipids.

### 3. Phospholipids

Zama (1963) studied the phospholipids in muscle and liver of *Ommastrephes sloani pacificus* and in the liver of *Octopus dofleini*. Lipid distribution in the liver of the two species was quite different, but also within one species there were differences related to the sex of the animal and to the season. The presence of CAEP and sphingomyelin was reported for *Polypus vulgaris* (Hori et al., 1967a). Phospholipids of *Octopus vulgaris* (De Koning, 1972) were composed of 13% CAEP, 42% phosphophatidyl-choline, 30% phosphatidylethanolamine, and small amounts of other glyc-erol- and sphingophosphatides. The major fatty acids in them were 16:0, 22:6, 20:5, and 18:0.

### 4. Lipids with Ether Bonds

These lipids have been studied mainly in the genus *Octopus*. High amounts of glyceryl ethers were found in the hepatopancreas of *Octopus rugosus, O. dofleini,* and *Octopus* sp. (Karnovsky et al., 1946; Thompson and Lee, 1965; Isay et al., 1976). In the alkyl side chains of 1-alkylglyceryl ethers from *O. dofleini* (Thompson and Lee, 1965) and *Loligo* sp. (Lewis, 1966), 16:0 was far predominant, followed by 18:0. In the plasmalogens, which were present in both neutral lipids and phospholipids of *O. dofleini* (Thompson and Lee, 1965), 18:0 was predominant.

### F. Conclusions

Im amphineurans, several gastropods, and in cephalopods, consider-able amounts of lipids, particularly triacylglycerols, can be stored in the hepatopancreas as reserve material. In some gastropods and in bivalves the metabolism seems to be more carbohydrate oriented, and glycogen is stored. Lipids can be used during starvation and are important, even in bivalves, during gametogenesis.

### 1. Sterols

From the data on sterol composition in molluscs some tendencies can be indicated: The Amphineura form a separate group by their characteris-tic $\Delta^7$-sterols. Within the prosobranchs, the archeogastropods show sim-ple sterol patterns, the C-27 sterols constituting $94 \pm 7\%$ ($n = 19$) of the total sterols. Cholesterol makes up 91% of the sterols, the corresponding value in mesogastropods and neogastropods being 72.6 and 79%, respec-tively. In the Opisthobranchia, *Aplysia* has a simple pattern, with high contents of cholesterol (87%). In the order Acoela, cholesterol consti-tuted only $48.5 \pm 8.6\%$ ($n = 6$) of the total sterols, with high levels of des-mosterol. The Pulmonata are rich in cholesterol ($85.8 \pm 4.5\%$ of total sterols, $n = 7$). In this order till now the $\Delta^{24(28)}$ double bond has not been found.

In Bivalvia the Protobranchia are rich in cholesterol ($81.6 \pm 26\%$, $n = 2$), whereas Filibranchia and Eulamellibranchia are poor in choles-terol ($37.7 \pm 11.7\%$, $n = 10$, and $36.3 \pm 15.0\%$, $n = 16$, respectively).

Cephalopoda have rather simple sterol patterns; cholesterol is the main sterol (91.8 ± 2.7%, $n = 5$). Anticipating the next paragraph, it may be assumed here that sterols in molluscs are partly of exogenous origin and partly endogenous. Because of their function in cell membranes, the composition of sterols in these membranes probably will be characteristic. This was actually shown for the seastar *Asterias rubens* (Voogt, 1982). Sterol composition in the reserve organs of this starfish was strongly influenced by the diet. This might hold good as well for molluscs. In analyses of whole animals, great differences in sterol composition may be found for different geographic localities and depending on the amount of stored sterols for different times in the year. Therefore, it is recommended that the reserve organs be excluded when sterol compositions in molluscs are determined in future.

## 2. Fatty Acids

Fatty acid compositions of total animals will be of little value for the reason indicated under sterols (Section II,F,1). Fatty acids in triacylglycerols will be strongly influenced by the diet, and of course some typical fatty acids herein can be used for food web studies. Fatty acid compositions in phospholipids are likely influenced by temperature as indicated before, thus lists of fatty acids present at a certain moment in molluscs have only limited value. Therefore, it would be advisable to undertake studies to determine the unique contribution of the molluscs in achieving a certain fatty acid pattern in phospholipids, and to unravel the mechanisms by which they maintain or adapt it.

## III. Metabolism of Lipids

In comparison with the knowledge about the distribution of lipids in the phylum Mollusca, that about metabolism of these lipids is very limited. Only a few investigators have worked in this field and still fewer have studied the subject in detail. As a result, our insight into the metabolic pathways is still limited. Voogt has systematically studied the capacity of synthesizing sterols in the several taxa of the molluscs, and as a spin-off found that all species investigated are able to synthesize fatty acids from acetate (Voogt, 1972a). He concluded that probably all molluscs can synthesize fatty acids. The individual data will not be given here.

General conclusions concerning the capacity of synthesizing sterols were not allowed at that time, and therefore these data will be used here to strengthen the evidence. Steroid metabolism has been studied in a number of species. These endocrinological aspects will not be dealt with.

## A. Amphineura

Voogt (1970) and Voogt and Van Rheenen (1974) showed that *Lepido-chitona cinerea* is able to synthesize sterols from mevalonate, but sterols were not further identified. Teshima and Kanazawa (1973, 1976), using the same precursor in *L. japonica,* found incorporation of radioactivity in $\Delta^7$-cholestenol only, whereas cholesterol was unlabeled. Radioactive cholesterol administered to *L. japonica* was converted into $\Delta^7$-cholestenol, mainly via cholestanol (Teshima and Kanazawa, 1978). These results point to the mechanisms by which chitons attain and maintain their characteristic sterol pattern: On the one hand, they can synthesize $\Delta^7$-cholestenol *de novo;* on the other, this sterol can be obtained by bioconversion of the corresponding dietary $\Delta^5$-sterol. This situation strongly resembles that in starfish. As cholestanol is an intermediate, the bioconversion pathway may be the same as that proposed for starfish (Voogt, 1982).

## B. Gastropoda

### 1. Prosobranchia

**a. Archeogastropoda.**  Voogt (1968b) showed that *P. coerulea* and *Monodonta turbinata* are able to synthesize sterols from acetate, but did not identify which sterols had been synthesized. It is notable that these species had considerable amounts of desmosterol (Table I). Biosynthesis of sterols from mevalonate was also observed in *Haliotis gurneri* (Teshima and Kanazawa, 1974). Collignon-Thiennot et al. (1973) isolated radiolabeled cholesterol from *P. vulgata* after injection with [1-$^{14}$C]acetate, [5-$^{14}$C]mevalonate, [3-$^3$H]-$\beta$-sitosterol, or [3-$^3$H]fucosterol. This means that *Patella* can synthesize cholesterol *de novo* from acetate or mevalonate, but also can form cholesterol by modification (dealkylation) of dietary 24-alkylsterols. In dealkylation first a double bond between C-24 and C-28 is introduced; then the alkyl group is split off under the concomitant formation of a $\Delta^{24}$double bond. Thus 24-methylenecholesterol, fucosterol, and desmosterol seem to be intermediates in the dealkylation process. It is notable that desmosterol is also an intermediate in the *de novo* synthesis of cholesterol.

**b. Mesogastropoda.**  *L. littorea* and *Viviparus fasciatus,* mesogstropods with similar feeding habits to archeogastropods, are able to synthesize (not further identified) sterols from acetate (Voogt, 1969). This was also found for the filter feeder *C. fornicata* (Voogt, 1971b). Further, Walton and Pennock (1972) observed synthesis of sterols from mevalonate in

*L. littorea*. However, hardly any radioactivity (8 dpm/mg and 7 dpm/mg) was found in 2 successive years in the sterols of the carnivorous species *Natica cataena* after the injection of [1-$^{14}$C]acetate. When [2-$^{14}$C]mevalonate was used as the precursor, sterol biosynthesis was beyond doubt (Voogt, 1971b). From this it may be concluded that *N. cataena* is able to synthesize sterols but that probably this synthesis is strictly regulated, the limiting and controlled reaction being between acetate and mevalonate. It is tempting to suppose that this will be the reaction catalyzed by HMG-CoA reductase.

**c. Neogastropoda.** The differences in the sterol composition of *Urosalpinx cinereus* and its prey, which is oysters and other bivalves, were interpreted by Kind and Goldberg (1953) as indirect evidence for the *de novo* synthesis of sterols by *Urosalpinx*. However, modification of the dietary sterols, as discussed already (Collignon-Thiennot et al., 1973), would suffice to explain the observed differences in sterol patterns. No or hardly any radioactivity from [1-$^{14}$C]acetate was incorporated into the sterols of *Buccinum undatum, N. antiqua, Purpura lapillus,* and *M. brandaris* (Voogt, 1967b, 1972b,c). When [2-$^{14}$C]mevalonate was used, radioactivity was recovered in the sterols of *B. undatum, P. lapillus,* and *M. brandaris*. Sterol synthesis in *B. undatum* from mevalonate was also found by Teshima and Kanazawa (1976), and in *Nucella lapillus* by Walton and Pennock (1972). In contrast, a small but probably significant incorporation of radioactivity was found in the free and particularly in the esterified sterols of *Nassa* sp. after administration of [1-$^{14}$C]acetate (Voogt and Van Rheenen, 1973b). Apart from the latter finding, the situation in these carnivorous species is identical to that described for *N. cataena* (Mesogastropoda) and the considerations given there may hold good here as well. In addition, it may be noticed that desmosterol is only encountered in small amounts in this order (Table I). Khalil and Idler (1976) studied further the synthesis of sterols in *B. undatum*. Lanosterol, but not cycloartenol, was converted to cholesterol, indicating that the *de novo* synthesis proceeds via lanosterol. 24-Methylenecholesterol and β-sitosterol were dealkylated to cholesterol.

## 2. Opisthobranchia

The herbivorous slug *Aplysia depilans* showed an active biosynthesis of sterols from acetate (Voogt and Van Rheenen, 1973a). The same was found for *Lamellidoris bilamellata* (Voogt, 1973a). Low but probably significant radioactivities were obtained in the sterols of *Dendronotus frondosus* and *Archidoris tuberculatum* (Voogt, 1973a). Generally high or very high levels of desmosterol are present (Table II). The differences in the intensity of sterol synthesis observed in this order may be related to the feeding habits of the species investigated.

## 3. Pulmonata

**a. Basommatophora.** *L. stagnalis, Lymnaea peregra,* and *Planorbarius corneus* are able to synthesize 3β-sterols from acetate (Voogt, 1968a). De Souza and De Oliveira (1976) showed high esterification and high-hydrolytic activities in the hemolymph of *Biomphalaria glabrata.* Cholesterol was mainly esterified to saturated and polyunsaturated fatty acids. This system may be important in regulating the level of free sterols. Esterification of β-sitosterol was observed in *Omphalius pfeifferi,* which animal was also able to convert β-sitosterol by dealkylation into cholesterol (Teshima et al., 1979b).

**b. Stylommatophora.** I. STEROLS.  The first direct proof for the biosynthesis of sterols in stylommatophoran snails was given by Addink and Ververgaert (1963) for *H. pomatia.* Later this capacity was demonstrated for several species: *Arion rufus* (Voogt, 1967a) *Ariolimax californicus* (Gottfried and Dorfman, 1969), *S. putris,* and *A. arbustorum* (Voogt and Van der Horst, 1972), *C. nemoralis* (Van der Horst and Voogt, 1972).

Thus it can be concluded that all gastropods investigated are able to synthesize sterols from mevalonate and that phytophagous species can also synthesize them from acetate, whereas in carnivorous species sterol synthesis from acetate proceeds extremely slowly or cannot be detected. It was supposed that synthesis here is regulated at the HMG-CoA reductase level. In some species dealkylation of 24-alkylsterols has been demonstrated, which points to the utilization of dietary sterols and enables the animal to maintain a characteristic sterol composition. It may be that dealkylation is a more common process in gastropods than is known at this moment.

II. FATTY ACIDS.  Fatty acid metabolism in *C. nemoralis* has been studied intensively by Van der Horst. Fatty acid composition did not show appreciable seasonal changes, and dietary influences were not detectable (Van der Horst and Zandee, 1973). Fatty acids were synthesized at high rates (Van der Horst, 1973), palmitic acid being the main end product of *de novo* synthesis. Longer fatty acids were formed by elongation, and saturated fatty acids were actively converted into unsaturated ones. One of the most remarkable outcomes was that linoleic acid (18:2 ω 6) is rapidly synthesized from acetate and thus is not essential for this animal. As a consequence, high levels of this fatty acid, and especially of its following product, arichidonic acid (20:4 ω 6), are present. Linolenic acid (18:3 ω 3), on the other hand, is an essential fatty acid. Fatty acid biosynthesis was continued under anoxia and was fully comparable with that under normoxic conditions (Oudejans and Van der Horst, 1974a). This prompts the question of the anaerobic formation of acetyl-CoA. Van der Horst (1974) showed that after short exposure to anoxia little radioactivity from acetate was incorporated into unsaturated fatty acids; with longer incubation times this amount increased, whereas under prolonged conditions of anoxia the biosynthesis of saturated fatty acids was stimulated. Van der Horst et al. (1974) also studied lipid synthesis during hibernation. Several precursors turned out to be usable for lipid synthesis, which was markedly reduced by not essentially different from that in nonhibernating animals. Excessive ingestion of oleic acid had little effect on the fatty acid composition of phospholipids but profound effects on the composition of triacylglycerols (Oudejans and Van der Horst, 1974b). This means that fatty acid composition in membrane lipids is highly characteristic, in contrast to that of reserve lipids. When linoleic and linolenic acids were provided via the diet, the major part was degraded. Incorporation of linoleic acid into phospholipids was lower than that of linolenic acid, which may be correlated to the *de novo* synthesis of the former acid. Fatty acids incorporated into phospholipids were intensively changed, in contrast with those incoporated into triglycerides (Van der Horst and Oudejans, 1976).

## C. Bivalvia

### 1. Sterols

Data on the capacity of synthesizing sterols in bivalves are controversial. Fagerlund and Idler (1960) reported the incorporation of [2-$^{14}$C]acetate in sterols of *Mytilus californianus* and *Saxidomus giganteus*. *Saxidomus* converted [11,14-$^{14}$C]squalene into a monounsaturated $\Delta^5$-sterol with only traces of radioactivity present in 24-methylenecholesterol (Fagerlund and Idler, 1961b). *Saxidomus* was also able to convert cholesterol into 24-methylenecholesterol and to introduce unsaturation into the side chain (Fagerlund and Idler, 1961a). Indications were obtained that biosynthesis and metabolism of cholesterol and of 24-methylenecholesterol are related (Idler et al., 1964; Tamura et al., 1964). Voogt (1975a) did not recover radioactivity in the sterols of *Ostrea edulis* 6.5 h after injection with [1-$^{14}$C]acetate but 7 days after the injection the sterols had a specific radioactivity of 265 dpm/mg, which after four recrystallizations had increased to 348 dpm/mg. On the other hand, Salaque et al. (1966) did not find radioactivity in the sterols of *Ostrea gryphea* after the administration of labeled mevalonate or methionine. Recently Teshima and Patterson (1981a) showed that *C. virginica* is able to synthesize cholesterol, desmosterol, isofucosterol, and 24-methylenecholesterol from acetate.

The route probably proceeds via lanosterol.[2] No radioactivity was found in the sterols of *M. edulis* after the injection of [1-14C]acetate or [2-14C]-acetate, even after very long incubation times (Voogt, 1975a). Neither was radioactivity from mevalonate incorporated in the sterols (Walton and Pennock, 1972). On the other hand, Teshima and Kanazawa (1974) reported that *M. edulis* was able to synthesize from mevalonate the sterols: cholesterol, 22-dehydrocholesterol, desmosterol, and 24-methyl-enecholesterol. This result is fully comparable with that of Teshima and Patterson (1981a) for *C. virginica*.

In comparing the sterol composition in wild and cultivated oysters, Berenberg and Patterson (1981) got indications that *C. virginica* is able to synthesize cholesterol or to dealkylate phytosterols. Saliot and Barbier (1973) observed that in *O. edulis*, fucosterol is converted into cholesterol via desmosterol. In *M. edulis*, desmosterol is converted into cholesterol, 22-dehydrocholesterol, and 24-methylenecholesterol (Teshima and Kana-zawa, 1973). Teshima et al. (1979b) observed that in *M. edulis* β-sito-sterol is esterified or can be converted to β-sitostanol.

In conclusion, *de novo* biosynthesis of sterols in *Anisomyaria* has been shown by Fagerlund and Idler (1960) for *M. californianus* and *S. gigan-teus*, by Teshima and Patterson (1981a) for *C. virginica*, by Teshima and Kanazawa (1974) for *M. edulis*, and with some reservations by Voogt (1975a) for *O. edulis*. No biosynthesis was found by Walton and Pennock (1972) and by Voogt (1975a) for *M. edulis*, and by Salaque et al. (1966) for *O. gryphea*. Several investigators have pointed to the possibilities of forming alkylsterols from cholesterol via desmosterol and sterols with a $\Delta^{24(28)}$ double bond, whereas the reverse route can also be followed.

Fewer data are available on sterol biosynthesis in Eulamellibranchia. *Cyprina islandica* did not synthesize sterols from [1-14C]acetate or [2-14C]mevalonate (Voogt, 1975b). No radioactivity was found in the sterols of *Mya arenaria* 4.5 h after the injection of [1-14C]acetate but a small amount (67 dpm/mg) was recovered in them after 120 h. In the sterols of *Anodonta cygnea* a small amount of radioactivity was recovered 120 h after the injection of [1-14C]acetate (Voogt, 1975b) and 14 days after the injection of [2-14C]acetate (Popov et al., 1981). Reproducibly, *Cardium edule* did not synthesize sterols from [1-14C]acetate, whereas it did from [2-14C]acetate (Voogt, 1975b); no sterols were formed from [2-14C]mevalon-ate (Walton and Pennock, 1972). A slight incorporation of radioactivity from [2-14C]-mevalonate was observed in the hepatopancreas of *Tapes philippinarum* (Teshima and Kanazawa, 1976). These data indicate that in

---

[2] Information on the possible steps after lanosterol was obtained from an analysis of the 4α-methyl sterols (Teshima and Patterson, 1981b). The authors suggest that the synthesis of sterols may proceed in the sequence: 4,4′,14-trimethyl $\Delta^{8,24}$ sterol (lanosterol) $\twoheadrightarrow$ 4α-methyl-$\Delta^{8(14)}$ sterol $\rightarrow$ 4α-methyl-$\Delta^{8(9)}$ sterol $\rightarrow$ 4α-methyl-$\Delta^7$-sterol $\rightarrow$ $\Delta^7$-sterol $\rightarrow$ $\Delta^5$-sterol. Regulation points in the biosynthesis may be the removal of the 4α-methyl groups and the isomerization of the $\Delta^{8(14)}$ to the $\Delta^{8(9)}$ bond.

principle Eulamellibranchia can synthesize sterols, but the degree to which they use this capacity seems to be even more limited than in *Anisomyaria*. The different incorporation of radiolabel from [1-¹⁴C]acetate and [2-¹⁴C]acetate in the sterols of *Cardium* might indicate that in the second case the methyl group ultimately has been used only for alkylation of dietary sterols and not for *de novo* synthesis.[3]

## 2. Fatty Acids

Data giving direct evidence about fatty acid metabolism in Bivalvia are largely absent. From his studies using [1-¹⁴C]acetate, Shieh (1968) concluded that fatty acids of *P. magellanicus* likely are of dietary origin. From feeding experiments it was concluded that *C. virginica* and *O. edulis* are able to convert quantitatively or qualitatively unusual fatty acids to maintain the species-oriented fatty acid composition (Watanabe and Ackman, 1974). De Moreno et al. (1976b, 1977) showed elongation and further desaturation of administered polyunsaturated fatty acids in *Mesodesma mactroides*. Palmitic and stearic acids administered were incorporated into diacyl- and triacyl-glycerols, and were elongated and desaturated. [1-¹⁴C]Acetate was taken up from the medium and incorporated in several fatty acids from 12:0 up to 20:3. Obviously the end product of *de novo* fatty acid synthesis is 14:0.

Vassalo (1973) found that [1-¹⁴C]acetate was incorporated in the lipids of *Chlamys hericia*. Initially radioactivity was found in the hepatopancreas; later it was also found in the gonads. Radioactivity in the hepatopancreas was recovered in phospholipids and triacylglycerols. In the testes radioactivity was mainly in triacylglycerols, in the ovaries in hydrocarbons. Shieh (1968) found incorporation of choline, ethanolamine, and serine in phospholipids of *P. magellanicus*. However, methionine was not utilized. Itasaka et al. (1969) observed incorporation of ³²P into all types of phospholipids present in *Hyriopsis schlegelii*. The hepatopancreas of this animal was capable of synthesizing CAEP.

## D. Cephalopoda

### 1. Sterols

Incorporation of [1-¹⁴C]acetate into sterols was not observed in *S. officinalis* (Zandee, 1967) and *Eledone aldrovandi* (Voogt, 1973b). A small incorporation of this precursor was found in the sterols of *O. vulgaris*

---

[3] Popov et al. (1981) showed that after the injection of [4-¹⁴C]cholesterol in *A. cygnea* radioactivity could be recovered in the sterols with 27, 28, or 29 carbon atoms and eventually Δ²²-bond in addition to the Δ⁵-bond. This is the first pertinent data on the capacity of alkylating sterols in Eulamellibranchia. These authors also showed that methionine can be used for this alkylation. These processes are rather slow, this in contrast to the introduction of Δ²²-bond. The injection of an excess of cholesterol resulted in the restoration of the original sterol composition by, among others, alkylation and Δ²² dehydrogenation.

(Voogt, 1973b). The author observed a distinct incorporation of [2-[14]C]mevalonate into the sterols of *S. officinalis*. Thus incorporation of [1-[14]C]acetate is limited, but mevalonate is incorporated well. This resembles what has been found in Neogastropoda. Because both groups are carnivorous, the diet may provide them with sufficient cholesterol leading to a feedback inhibition at the HMG-CoA reductase level, as discussed before.

## E. Conclusions

Sterol composition seems to be species oriented when storage organs are excluded. This composition will be a dynamic equilibrium between the processes that provide the animal with sterols,—namely, biosynthesis *de novo* and resorption eventually followed by modification of dietary sterols and those processes that eliminate sterols from an animal. There is an increasing knowledge about the distribution of sterol-synthesizing capacity within the phylum Mollusca. Uncertainties have remained, particularly with reference to Neogastropoda and Bivalvia. Studies are needed to test the hypothesis of the regulation of sterol biosynthesis at the HMG-CoA reductase level. Further, the sterols synthesized should be identified to get a better insight into the qualitative aspects of sterol synthesis. With some exceptions, little is known about the fate and utilization of exogenous sterols. In particular, data on the quantitative aspects of sterol supply, and the relative contribution of synthesis *de novo* and of the diet are completely absent. The same holds with respect to the processes that remove sterols from the animal. Studies to fill these gaps are badly needed. Similar remarks can be made for the other lipid classes.

## References

Ackman, R. G. (1964). Structural homogeneity in unsaturated fatty acids of marine lipids. A review. *J. Fish. Res. Board Can.* **21**, 247–254.

Ackman, R. G. (1967). Characteristics of the fatty acid composition and biochemistry of some fresh-water fish oils and lipids in comparison with marine oils and lipids. *Comp. Biochem. Physiol.* **22**, 907–922.

Ackman, R. G., and Hooper, S. N. (1973). Nonmethylene interrupted fatty acids in lipids of shallow water marine invertebrates. A comparison of 2 mollusks *Litterina littorea* and *Lunatia triseriata* with the sand shrimp *Crangon septemspinosus*. *Comp. Biochem. Physiol. B* **46B**, 153–165.

Ackman, R. G., Hooper, S. N., and Ke, P. J. (1971). The distribution of saturated and isoprenoid fatty acids in the lipids of three species of molluscs, *Littorina littorea, Crassostrea virginica,* and *Venus mercenaria. Comp. Biochem. Physiol. B.* **39B**, 579–587.

Addink, A. D. F., and Ververgaert, P. H. J. T. (1963). Biosynthesis of cholesterol and fatty acids in a snail, *Helix pomatia* L., after administration of 1-[14]C-acetate. *Arch. Int. Physiol. Biochim.* **71**, 797–801.

Akiya, S., and Nakazawa, Y. (1955). New glycolipid in oyster. I. Its isolation and purification. *Yakugaku Zasshi* **75**, 1332–1334.

Allen, W. V. (1977). Interorgan transport of lipids in the blood of the gumboot chiton *Cryptochiton stelleri* (Middendorff). *Comp. Biochem. Physiol. A* **57A,** 41–46.

Ando, T., Mouëza, M., and Ceccaldi, H. J. (1976). Variations des lipides et des stérols chez *Donax trunculus* L. (Mollusque, Lamelli branche) durant les mois d'automne et d'hiver. *C. R. Seances Soc. Biol. Ses Fil.* **170,** 149–153.

Ando, T., Kanazawa, A., Teshima, S., and Miyawski, H. (1979). Sterol components of coral-reef molluscs. *Mar. Biol. (Berlin)* **50,** 169–173.

Ansell, A. D. (1974). Seasonal changes in biochemical composition of the bivalve *Lima hians* from the Clyde Sea area. *Mar. Biol. (Berlin)* **27,** 115–122.

Arebalo, R. E., Hardgrave, J. E., and Scallen, T. J. (1981). The *in vivo* regulation of rat liver 3-hydroxyl-3-methylglutaryl Coenzyme A-reductase. Phosphorylation of the enzyme as an early regulatory response following cholesterol feeding. *J. Biol. Chem.* **256,** 571–574.

Austin, J. (1970). The sterols of marine invertebrates and plants. *Adv. Steroid Biochem. Pharmacol.* **1,** 73–96.

Balasubramanian, T., Sumitra-Vijayaraghavan, and Krishna Kumari, L. (1979). Energy content of the wedge clam, *Donax incarnatus* Gmelin. *Indian J. Mar. Sci.* **8,** 193–195.

Bannatyne, W. R., and Thomas J. (1969). Fatty acid composition of New Zealand shellfish lipids. *N.Z. J. Sci.* **12,** 207–212.

Belisle, B. W., and Stickle, W. B. (1978). Seasonal patterns in the biochemical constituents and body component indexes of the muricid gastropod, *Thais haemastoma. Biol. Bull. (Woods Hole, Mass.)* **155,** 259–272.

Berenberg, C. J., and Patterson, G. W. (1981). The relationship between dietary phytosterols and the sterols of wild and cultivated oysters. *Lipids* **16,** 276–278.

Bergmann, W. (1949). Comparative biochemical studies on the lipids of marine invertebrates, with special reference to the sterols. *J. Mar. Res.* **8,** 137–176.

Bergmann, W. (1962). Sterols: their structure and distribution. *In* "Comparative Biochemistry" (M. Florkin and H. S. Mason, eds.), Vol. 3, pp. 103–162. Academic Press, New York.

Boucaud-Camou, E. (1971). Constituants lipidiques du foie de *Sepia officinalis. Mar. Biol. (Berlin)* **8,** 66–69.

Brown, M. S., Goldstein, J. L., and Dietschy, J. M. (1979). Active and inactive form of 3-hydroxy-3-methylglutarylcoenzyme A reductase in the liver of the rat. Comparison with the rate of cholesterol synthesis in different physiological states. *J. Biol. Chem.* **254,** 5144–5149.

Calzolari, C., Cerma, C., and Stancher, B. (1971). Applicazione della gas-cromatografia nella determinazione degli acidi grassi di alcuni Gastropodi e Lamellibranchi dell'alto Adriatico durante un ciclo annuale. *Riv. Ital. Sostanze Grasse* **48,** 605–611.

Catalan, R. E., Castillon, M. P., and Rallo, A. (1977). Lipid metabolism during development of the mollusc *Arion empiricorum.* Distribution of lipids in midgut gland, genitalia and foot muscle. *Comp. Biochem. Physiol. B* **57B,** 73–79.

Chatterjee, B., and Ghose, K. C. (1973). Seasonal variation in stored glycogen and lipid in the digestive gland and genital organs of two freshwater prosobranchs. *Proc. Malacol. Soc. London* **40,** 407–412.

Collignon-Thiennot, F., Allain, J.-P., and Barbier, M. (1973). Existence de deux voies de biosynthèse du cholestérol chez un mollusque: la patelle *Patella vulgata* L. (Gastéropodes, Prosobranches, Archéogastéropodes). *Biochimie* **55,** 579–582.

De Koning, A. J. (1966a). Phospholipids of marine origin. I. The hake (*Merluccius capensis*). *J. Sci. Food Agric.* **17,** 112–117.

De Koning, A. J. (1966b). Phospholipids of marine origin. IV. The abalone (*Haliotis midae*). *J. Sci. Food Agric.* **17,** 460–464.

De Koning, A. J. (1972). Phospholipids of marine origin. VI. The octopus (*Octopus vulgaris*). *J. Sci. Food Agric.* **23,** 1471–1475.

De Moreno, J. E. A., Moreno, V. J., and Brenner, R. R. (1976a). Lipid metabolism of the Yellow Clam, *Mesodesma mactroides:* I. Composition of the lipids. *Lipids* 11, 334–340.

De Moreno, J. E. A., Moreno, V. J., and Brenner, R. R. (1976b). Lipid metabolism of the Yellow Clam, *Mesodesma mactroides:* 2—Polyunsaturated fatty acid metabolism. *Lipids* 11, 561–566.

De Moreno, J. E. A., Moreno, V. J., and Brenner, R. R. (1977). Lipid metabolism in the Yellow Clam, *Mesodesma mactroides:* 3—Saturated fatty acids and acetate metabolism. *Lipids* 12, 804–808.

De Moreno, J. E. A., Pollero, R. J., Moreno, V. J., and Brenner, R. R. (1980). Lipids and fatty acids of the mussel (*Mytilus platensis* d'Orbigny) from South Atlantic waters. *J. Exp. Mar. Biol. Ecol.* 48, 263–276.

Desai, K., Hirani, G., and Nimavat, D. (1979). Studies on the Pearl Oyster, *Pinctada fucata* (Gould). Seasonal biochemical changes. *Indian J. Mar. Sci.* 8, 49–50.

De Souza, A. M. F., and De Oliveira, D. N. G. (1976). Esterification of cholesterol and cholesterol ester hydrolysis by the hemolymph of the mollusk *Biomphalaria glabrata*. *Comp. Biochem. Physiol. B* 53B, 345–347.

Fagerlund, U. H. M., and Idler, D. R. (1960). Marine sterols, VI. Sterol biosynthesis in molluscs and echinoderms. *Can. J. Biochem. Physiol.* 38, 997–1002.

Fagerlund, U. H. M., and Idler, D. R. (1961a). Marine sterols. VIII. *In vivo* transformation of the sterol side chain by a clam. *Can. J. Biochem. Physiol.* 39, 505–509.

Fagerlund, U. H. M., and Idler, D. R. (1961b). Marine sterols. IX. Biosynthesis of 24-methylene-cholesterol in clams. *Can. J. Biochem. Physiol.* 39, 1347–1355.

Gardner, D., and Riley, J. P. (1972). The component fatty acids of the lipids of some species of marine and freshwater molluscs. *J. Mar. Biol. Assoc. U.K.* 52, 827–838.

Giese, A. C. (1966). Lipids in the economy of marine invertebrates. *Physiol. Rev.* 46, 244–298.

Giese, A. C., and Araki, G. (1962). Chemical changes with reproductive activity of the chitons, *Katharina tunicata* and *Mopalia hindsii*. *J. Exp. Zool.* 151, 259–267.

Giese, A. C., and Hart, M. A. (1967). Seasonal changes in component indices and chemical composition in *Katharina tunicata*. *J. Exp. Mar. Biol. Ecol.* 1, 34–46.

Goad, L. J. (1976). The steroids of marine algae and invertebrate animals. *Biochem. Biophys. Perspect. Mar. Biol.* 3, 213–318.

Gottfried, H., and Dorfman, R. J. (1969). The occurrence of *in vivo* cholesterol biosynthesis in an invertebrate, *Ariolimax californicus*. *Gen. Comp. Endocrinol., Suppl.* No. 2, 590–593.

Hayashi, A., and Matsubara, T. (1966). Biochemical studies on marine shellfish lipid. I. Glycolipid of Oyster Gills. *J. Fac. Science and Technology, Kinki University* 1, 25–37.

Hayashi, A., and Matsubara, T. (1969). Biochemical studies on marine shellfish lipid. II. Glycolipid of oyster mantle. *J. Biochem. (Tokyo)* 65, 503–511.

Hayashi, A., and Matsuura, F. (1971). Isolation of a new sphingophosphonolipid containing galactose from the viscera of *Turbo cornutus*. *Biochim. Biophys. Acta* 248, 133–136.

Hayashi, A., and Matsuura, F. (1973). 2-Hydroxy fatty acid—and phytosphingosine containing ceramide 2-*N*-methylaminoethylphosphonate from *Turbo cornutus*. *Chem. Phys. Lipids* 10, 51–65.

Hayashi, A., and Matsuura, F. (1978). Characterization of aminoalklphosphonyl cerebrosides in muscle tissue of *Turbo cornutus*. *Chem. Phys. Lipids* 22, 9–24.

Hayashi, A., Matsubara, T., and Matsuura, F. (1969a). Biochemical studies on the lipids of *Turbo cornutus*. I The conjugated lipids of viscera (Part I). *Yukagaku* 18, 118–123.

Hayashi, A., Matsuura, F., and Matsubara, T. (1969b). Isolation and characterization of a new sphingolipid containing 2-*N*-methylaminoethylphosphonic acid from the viscera of *Turbo cornutus*. *Biochim. Biophys. Acta* 176, 208–210.

Hayashi, A., Matsubara, T., and Matsuura, F. (1975). Characterization of docosa-4,15 sphingadienine and 4-hydroxydocosa-15-sphingenine in sphingophosphonolipids from *Turbo cornutus* by gaschromatography–mass spectrometry. *Chem. Phys. Lipids* **14**, 102–105.

Hayashi, K., and Yamada, M. (1972). Studies on the lipids of shellfish—I. On the visceral lipid composition of abalone, *Haliotis discus hannai* (Ino). *Nippon Suisan Gakkaishi* **38**, 255–263.

Hayashi, K., and Yamada, M. (1974). Studies on the lipids of shellfish—III. On the fatty acid and sterol compositions of a purple and lischke's tegula top shell snail. *Hokkaido Daigaku Suisangakubu Kenkyu Iho* **25**, 247–255.

Hayashi, K., and Yamada, M. (1975). Studies on the lipids of shellfish—IV. On the fatty acid compositions of five species of snails from Toyama bay. *Hokkaido Daigaku Suisangakubu Kenkyu Iho* **26**, 176–181.

Heeg, J. (1977). Oxygen consumption and the use of metabolic reserves during starvation and aestivation in *Bulinus* (Physopsis) *africanus* (Pulmonata : Planorbidae). *Malacologia* **16**, 549–560.

Higashi, S., and Hori, T. (1968). Studies on sphingolipids of freshwater mussel spermatozoa. *Biochim. Biophys. Acta* **152**, 568–575.

Holland, D. L., Tantanasiriwong, R., and Hannant, P. J. (1975). Biochemical composition and energy reserve in the larvae and adults of the four British periwinkles *Littorina littorea, L. littoralis, L. saxatilis* and *L. neritoides. Mar. Biol.* (*Berlin*) **33**, 235–239.

Hori, T., and Arakawa, I. (1969). Isolation and characterization of new sphingolipids containing *N,N*-acylmethylaminoethylphosphonic acid and *N*-acylaminoethylphosphonic acid from the mussel, *Corbicula sandai. Biochim. Biophys. Acta* **176**, 898–900.

Hori, T., and Itasaka, O. (1961). Lipids of fresh-water mussels. IV. Complex lipids from *Corbicula sandai. Seikagaku* **33**, 169–173.

Hori, T., Itasaka, O., and Hashimoto, T. (1964a). Biochemistry of shellfish lipid. I. Glycolipid of *Corbicula sandai. J. Biochem.* (*Tokyo*) **55**, 1–10.

Hori, T., Itasaka, O., Inoue, H., and Yamada, K. (1964b). Structural components of the pyridine-insoluble sphingolipid from *Corbicula sandai,* and the distribution in other species. *J. Biochem.* (*Tokyo*) **56**, 477–479.

Hori, T., Itasaka, O., Inoue, H., Gamo, M., and Arakawa, I. (1966). Biochemistry of shellfish lipids. IV. Purification and characterization of a new phosphosphingolipid in the pond-snail, *Heterogen longispira. Jpn. J. Exp. Med.* **36**, 85–89.

Hori, T., Itasaka, O., Sugita, M., and Arakawa, I. (1967a). Distribution of ceramide-2-aminoethylphosphonate in nature and its quantitative correlation to sphingomyelin. *Shiga Daigaku Kyoikugakubu Kiyo, Shizen Kagaku* **17**, 23–26.

Hori, T., Arakawa, I., and Sugita, M. (1967b). Distribution of ceramide-2-aminoethylphosphonate and ceramide aminoethylphosphate (sphingoethanolamine) in some aquatic animals. *J. Biochem.* (*Tokyo*) **62**, 67–70.

Hori, T., Itasaka, O., and Kamimura, M. (1968). Biochemistry of shellfish lipids. VIII. Occurrence of ceramide mono- and dihexoside in Corbicula, *Corbicula sandai. J. Biochem.* (*Tokyo*) **64**, 125–128.

Hori, T., Sugita, M., Kanbayashi, J., and Itasaka, O. (1977a). Studies on glycosphingolipids of fresh-water bivalves. III. Isolation and characterization of a novel globoside containing mannose from spermatozoa of the fresh-water bivalve, *Hyriopsis schlegelii. J. Biochem.* (*Tokyo*) **81**, 107–114.

Hori, T., Takeda, H., Sugita, M., and Itasaka, O. (1977b). Studies on glycosphingolipids of fresh-water bivalves. IV. Structure of a branched globoside containing mannose from spermatozoa of the fresh-water bivalve, *Hyriopsis schlegelii. J. Biochem.* (*Tokyo*) **82**, 1281–1285.

Hori, T., Sugita, M., Ando, S., Kuwahara, M., Kumauchi, K., Sugie, E., and Itasaka, O.

(1981). Characterization of a novel glycosphingolipid, ceramide nonasaccharide, isolated from spermatozoa of the fresh-water bivalve, *Hyriopsis schlegelii*. *J. Biol. Chem.* **256**, 10979–10985.

Hunter, C. F., and Rodwell, V. W. (1980). Regulation of vertebrate liver HMG-CoA reductase via reversible modulation of its catalytic activity. *J. Lipid Res.* **21**, 399–405.

Idler, D. R., and Wiseman, P. (1971). Sterols of molluscs. *Int. J. Biochem.* **2**, 516–528.

Idler, D. R., and Wiseman, P. (1972). Molluscan sterols: A review. *J. Fish. Res. Board Can.* **29**, 385–398.

Idler, D. R., Tamura, T., and Wainai, T. (1964). Seasonal variations in the sterol, fat and unsaponifiable components of scallop muscle. *J. Fish. Res. Board Can.* **21**, 1035–1042.

Idler, D. R., Khalil, M. W., Brooks, C. J. W., Emonds, C. G., and Gilbert, J. D. (1978). Studies of sterols from marine molluscs by gas chromatography and mass spectrometry. *Comp. Biochem. Physiol. B* **59B**, 163–168.

Ingebritsen, T. S., Parker, R. A., and Gibson, D. M. (1981). Regulation of liver hydroxy-methylglutaryl-CoA reductase by a bicyclic phosphorylation system. *J. Biol. Chem.* **256**, 1138–1144.

Isay, S. V., Makarchenko, M. A., and Vaskovsky, V. E. (1976). A study of glycerylethers—I. Content of $\alpha$-glycerylethers in marine invertebrates from the Sea of Japan and tropical regions of the Pacific Ocean. *Comp. Biochem. Physiol. B* **55B**, 301–305.

Itasaka, O., Hori, T., and Sugita, M. (1969). Biochemistry of shellfish lipids. XI. Incorporation of ($^{32}$P) orthophosphate into ceramide ciliatine (2-aminoethylphosphonic acid) of the freshwater mussel, *Hyriopsis schlegelii*. *Biochim. Biophys. Acta* **176**, 783–788.

Joh, Y. G., and Kim, Y. K. (1976). The origin of molluscs sterol (1). The sterol composition of bivalves and snails. *Bull. Korean Fish. Soc.* **9**, 185–193.

Johns, R. B., Nichols, P. D., and Perry, G. J. (1980). Fatty acid components of nine species of mollusks of the littoral zone from Australian waters. *Comp. Biochem. Physiol. B* **65B**, 207–214.

Kanazawa, A., Teshima, S. I., and Ceccaldi, H. J. (1976). Chemical composition of some Mediterranean macroplanktonic organisms. 2. Sterols, *Tethys* **8**, 323–326.

Karnovsky, M. L., Rapson, W. S., and Black, M. (1946). South African fishproducts. Part XXIV—The occurrence of $\alpha$-glycerylethers in the unsaponifiable fractions of natural fats. *J. Soc. Chem. Ind., London, Trans. Commun.* **65**, 425–428.

Khalil, M. W., and Idler, D. R. (1976). Steroid biosynthesis in the whelk, *Buccinum undatum. Comp. Biochem. Physiol. B* **55B**, 239–242.

Kind, C. A., and Goldberg, M. H. (1953). Sterols of marine mollusks. III. Further observations on component sterols. *J. Org. Chem.* **18**, 203–206.

Krupanidhi, S., Reddy, V. V., and Najdu, B. P. (1978). Organic composition of tissues of the snail *Cryptozona ligulata* (Ferussac) with special reference to aestivation. *Indian J. Exp. Biol.* **16**, 611–612.

Lambert, P., and Dehnel, P. A. (1974). Seasonal variations in biochemical composition during the reproductive cycle of the intertidal gastropod *Thais lamellosa* (Gastropoda, Prosobranchia). *Can. J. Zool.* **52**, 305–318.

Lawrence, J. M., and Giese, A. C. (1969). Changes in the lipid composition of the chiton, *Katharina tunicata*, with the reproductive and nutritional state. *Physiol. Zool.* **42**, 353–360.

Lewin, J., Chen, C.-H., and Hruby, T. (1979). Blooms of surf-zone diatoms along the coast of the Olympic Peninsula Washington. X. Chemical composition of the surf diatom *Chaetoceros armatum* and its major herbivore, the Pacific razor clam *Siliqua patula*. *Mar. Biol.* (*Berlin*) **51**, 259–265.

Lewis, R. W. (1962). Temperature and pressure effects on the fatty acids of some marine ectotherms. *Comp. Biochem. Physiol.* **6**, 75–89.

Lewis, R. W. (1966). Studies of the glyceryl ethers of the stomach oil of Leach's petrol *Oceanodroma leucorhoa* (Viellot). *Comp. Biochem. Physiol.* **19**, 363–377.

Liang, C. R., and Strickland, K. P. (1969). Phospholipid metabolism in the molluscs. I. Distribution of Phospholipids in the water snail *Lymnaea stagnalis*. *Can. J. Biochem.* **47**, 85–89.

Lovern, J. A. (1964). The lipids of marine organisms. *Oceanogr. Mar. Biol.* **2**, 169–191.

McLachlan, A., and Lombard, H. W. (1980). Seasonal variations in energy and biochemical components of an edible gastropod, *Turbo sarmaticus* (Turbinidae). *Aquaculture* **19**, 117–125.

Mane, U. H., and Nagabhushanam, R. (1975). Body distribution and seasonal changes in the biochemical composition of the estuarine mussel, *Mytilus viridis* at Ratnagiri. *Riv. Idrobiol.* **14**, 163–175.

Matsubara, T., and Hayashi, A. (1981). Structural studies on glycolipid of shellfish. III. Novel glycolipids from *Turbo cornutus*. *J. Biochem.* (*Tokyo*) **89**, 645–650.

Matsubara, T., and Hayashi, A. (1982). Structural studies on glycolipid in shellfish. IV. A novel pentaglycosylceramide from abalone, *Haliotis japonica*. *Biochim. Biophys. Acta* **711**, 551–553.

Matsuura, F. (1977). Phosphonosphingoglycolipid, a novel sphingolipid from the viscera of *Turbo cornutus*. *Chem. Phys. Lipids* **19**, 223–242.

Matsuura, F. (1979). The identification of aminoalkylphosphonyl cerebrosides in the marine gastropod, *Monodonta labio*. *J. Biochem.* (*Tokyo*) **85**, 433–441.

Matsuura, F., Matsubara, T., and Hayashi, A. (1973). Identification of molecular species of ceramide 2-*N*-methylaminoethyl phosphonates containing normal fatty-acids and dihydroxy long chain bases from *Turbo cornutus*. *J. Biochem.* (*Tokyo*) **74**, 49–57.

Muley, E. V. (1975). Seasonal variations in bio-chemical composition of the freshwater prosobranch, *Melania scabra*. *Riv. Idrobiol.* **14**, 199–208.

Nagabhushanam, R., and Deshmukh, R. S. (1974). Seasonal changes in body component indices and chemical composition in the estuarine clam, *Meretrix meretrix*. *Indian J. Fish.* **21**, 531–542.

Nagabhushanam, R., and Dhamne, K. P. (1977). Seasonal variations in biochemical constituents of the clam, *Paphia laterisulca*. *Hydrobiologia* **54**, 209–214.

Nagabhushanam, R., and Mane, U. H. (1978). Seasonal variation in the biochemical composition of *Mytilus viridis* at Ratnagiri on the west coast of India. *Hydrobiologia* **57**, 69–72.

Nagabhushanam, R., and Talikhedkar, P. M. (1977). Seasonal variations in proteins, fat and glycogen of the wedge clam *Donax cuneatus*. *Indian J. Mar. Sci.* **6**, 85–87.

Nakazawa, Y. (1959). Studies on the new glycolipids in oyster. V. On the nitrogenous components and structure of the glycolipids. *J. Biochem.* (*Tokyo*) **46**, 1579–1585.

Nanaware, S. G., and Varute, A. T. (1976). Biochemical studies on the reproductive organs of a land pulmonate, *Semperula maculata* (Tompleton, 1858; Semper, 1885) during seasonal breeding-aestivation cycle: I. Biochemical seasonal variations in proteins and lipids. *Veliger* **19**, 96–106.

Oudejans, R. C. H. M, and Van der Horst, D. J. (1974a). Aerobic–anaerobic biosynthesis of fatty acids and other lipids from glycolytic intermediates in the pulmonate land snail *Cepaea nemoralis* (L.) *Comp. Biochem. Physiol. B* **47B**, 139–147.

Oudejans, R. C. H. M., and Van der Horst, D. J. (1974b). Effect of excessive fatty acid ingestion upon composition of neutral lipids and phospholipids of snail *Helix pomatia* L. *Lipids* **9**, 798–803.

Paradis, M., and Ackman, R. G. (1975). Occurrence and chemical structure of nonmethylene-interrupted dienoic fatty acids in American oyster *Crassostrea virginica*. *Lipids* **10**, 12–16.

Paradis, M., and Ackman, R. G. (1977). Potential for employing the distribution of anomalous non-methylene-interrupted dienoic fatty acids in several marine invertebrates as part of food web studies. *Lipids* **12**, 170–176.

Pollero, R. J., Ré, M. E., and Brenner, R. R. (1979). Seasonal changes of the lipids of the mollusc *Chlamys tehuelcha*. *Comp. Biochem. Physiol.* **64A**, 257–263.

Pollero, R. J., Brenner, R. R., and Gros, E. G. (1981). Seasonal changes in lipid and fatty acid composition of the freshwater mollusk, *Diplodom patagonicus*. *Lipids* **16**, 109–113.

Popov, S., Stoilov, I., Marehov, N., Kovachev, G., and Andreev, St. (1981). Sterols and their biosynthesis in some fresh-water bivalves. *Lipids* **16**, 663–669.

Quin, L. D. (1965). The presence of compounds with a carbon-phosphorus bond in some marine invertebrates. *Biochemistry* **4**, 324–330.

Quin, L. D., and Shelburne, F. A. (1969). An examination of marine animals for the presence of carbon-bound phosphorus. *J. Marine Res.* **27**, 73–84.

Rapport, M. M., and Alonzo, N. F. (1960). The structure of plasmalogens. V. Lipids of marine invertebrates. *J. Biol. Chem.* **235**, 1953–1956.

Riley, R. T. (1976). Changes in the total protein, lipid, carbohydrate, and extracellular body fluid free amino acids of the Pacific oyster, *Crassostrea gigas*, during starvation. *Proc. Nat. Shellfish. Assoc.* **65**, 84–90.

Rouser, G., Kritchevsky, G., Heller, D., and Lieber, E. (1963). Lipid composition of beef brain, beef liver, and the sea anemone; two approaches to quantitative fractionation of complex lipid mixtures. *J. Am. Oil. Chem. Soc.* **40**, 425–454.

Salaque, A., Barbier, M., and Lederer, E. (1966). Sur la biosynthèse des stérols de l'huître (*Ostrea gryphea*) et de l'oursin (*Paracentrotus lividus*). *Comp. Biochem. Physiol.* **19**, 45–51.

Saliot, A., and Barbier, M. (1973). Stérols en solution dans l'eau de mer: Leur utilisation par les invertebres marins. *J. Exp. Mar. Biol. Ecol.* **13**, 207–214.

Sampugna, J., Johnson, L., Bachman, K., and Keeney, M. (1972). Lipids of *Crassostrea virginica*. I. Preliminary investigations of aldehyde and phosphorous containing lipids in oyster tissue. *Lipids* **7**, 339–343.

Shah, S. N. (1981). Modulation *in vitro* of 3-hydroxy-3-methylglutaryl Coenzyme A Reductase in brain microsomes: Evidence for the phosphorylation and dephosphorylation associated with inactivation and activation of the enzyme. *Arch. Biochem. Biophys.* **211**, 439–446.

Shchepkin, V. Y., Shul'man, G. Y., and Sigayeva, T. G. (1976). Tissue lipids in Mediterranean squids of different ecology. *Gidrobiol. Zh.* **12**, 76–79.

Shieh, H. S. (1968). The characterization and incorporation of radioactive bases into scallop phospholipids. *Comp. Biochem. Physiol.* **27**, 533–541.

Shimma, Y., and Taguchi, H. (1964). A comparative study on fatty acid composition of shellfish. *Nippon Suisan Gakkaishi* **30**, 153–160.

Sica, C. (1980). Sterols from some mollusks. *Comp. Biochem. Physiol. B* **65B**, 407–410.

Snyder, F. (1969). The biochemistry of lipids containing ether bonds. *Prog. Chem. Fats Other Lipids* **10**, 287–335.

Stickle, W. B. (1971). The metabolic effects of starving *Thais lamellosa* immediately after spawning. *Comp. Biochem. Physiol. A* **40A**, 627–634.

Stickle, W. B. (1975). The reproductive physiology of the intertidal prosobranch *Thais lamellosa* (Gmelin). II. Seasonal changes in biochemical composition. *Biol. Bull.* (*Woods Hole, Mass.*), **148**, 448–460.

Stickle, W. B., and Duerr, F. G. (1970). The effects of starvation on the respiration and major nutrient stores of Thais lamellosa. *Comp. Biochem. Physiol.* **33**, 689–695.

Streit, B. (1978). Changes in protein, lipid, and carbohydrate content during starvation in the freshwater limpet *Ancylus fluviatilis* (Basommatophora). *J. Comp. Physiol. B* **123**, 149–154.

Sugita, M., Arakawa, I., Hori, T., and Sawada, Y. (1968). Shellfish lipids. X. Fatty acid components of ceramide 2-aminoethylphosphonate. *Seikagaku* **40**, 158–162.

Swift, M. R. (1977). Phosphono-lipid content of the oyster, *Crassostrea virginica*, in three physiological conditions. *Lipids* **12**, 449–451.

Swift, M. R., White, D., and Ghassemieh, M. B. (1980). Distribution of neutral lipids in the tissues of the oyster *Crassostrea virginica*. *Lipids* **15**, 129–132.

Takagi, T., and Toyama, Y. (1956). Fatty oils of aquatic invertebrates. XVII. Fatty oils from echinoderms and mollusks with a particular reference to their sterol components. *Mem. Fac. Eng., Nagoya Univ.* **8**, 177–182.

Takagi, T., and Toyama, Y. (1958). Fatty oils of snail, *Euhadra herklotsi* (Martens). *Mem. Fac. Eng., Nagoya Univ.* **10**, 84–87.

Tamura, T. Truscott, B., and Idler, D. R. (1964). Sterol metabolism in the oyster. *J. Fish. Res. Board Can.* **21**, 1519–1522.

Tanaka, T., and Toyama, Y. (1959). Fatty oils of aquatic invertebrates. XXI. Fatty oil of the sea hare and its sterol components. *Mem. Fac. Eng., Nagoya Univ.* **11**, 182–186.

Teshima, S.-I., and Kanazawa, A. (1972). Comparative study on the sterol composition of marine mollusks. *Nippon Suisan Gakkaishi* **38**, 1299–1304.

Teshima, S.-I., and Kanazawa, A. (1973). Biosynthesis of 7-cholestenol in the chiton, *Liolophura japonica*. *Comp. Biochem. Physiol. B* **44B**, 881–887.

Teshima, S.-I., and Kanazawa, A. (1974). Biosynthesis of sterols in abalone, *Haliotis gurneri*, and mussel, *Mytilus edulis*. *Comp. Biochem. Physiol. B* **47B**, 555–561.

Teshima, S.-I., and Kanazawa, A. (1976). Comparison of the sterol-synthesizing ability in some marine invertebrates. *Kagoshima Daigaku Suisangakubu Kiyo* **25**, 33–39.

Teshima, S.-I., and Kanazawa, A. (1978). Bio conversion of cholesterol to 7-cholestenol in a chiton. *Nippon Suisan Gakkaishi* **44**, 1265–1268.

Teshima, S.-I., and Patterson, G. W. (1981). Sterol biosynthesis in the oyster, *Crassostrea virginica*. *Lipids* **16**, 234–239.

Teshima, S-I., and Patterson, G. W. (1981b). Identification of 4α-methyl sterols in the oyster, *Crassostrea virginica*. *Comp. Biochem. Physiol. B* **69B**, 175–181.

Teshima, S.-I., Kanazawa, A., and Ando, T. (1974). Sterols of the killer clams, Mollusca Pelecypoda. *Kagoshima Daigaku Suisangakubu Kiyo* **23**, 106–110.

Teshima, S.-I., Kanazawa, A., Hyodo, S., and Ando, T. (1979a). Sterols of the triton *Charonia tritonis*. *Comp. Biochem. Physiol. B* **64B**, 225–228.

Teshima, S.-I., Kanazawa, A., and Miyawaki, H. (1979b). Metabolism of β-sitosterol in the mussel and the snail. *Comp. Biochem. Physiol. B* **63B**, 323–328.

Teshima, S.-I., Patterson, G. W., and Dutky, S. R. (1980). Sterols of the oyster, *Crassostrea virginica*. *Lipids* **15**, 1004–1011.

Teshima, S-I., Kanazawa, A., Yamada, I., and Kimura, S. (1982). Sterols of the chiton (Liolophura japonica). A new $C_{30}$ sterol, (24Z)-24-propylidenecholest-7-enol and other minor sterols. *Comp. Biochem. Physiol. B* **71B**, 373–378.

Thiele, O. W., and Kröber, G. (1963). Die Lipide der Weinbergschnecke (*Helix pomatia* L.), III. Uber die Fettsäuren der acetonlöslichen Lipide. *Hoppe-Seyler's Z. Physiol. Chem.* **334**, 63–70.

Thompson, G. A., Jr., and Hanaban, D. J. (1963). Identification of α-glyceryl ether phospholipids as major lipid constituents in two species of terrestrial slug. *J. Biol. Chem.* **238**, 2628–2631.

Thompson, G. A., Jr., and Lee, P. (1965). Studies of the α-glyceryl ether lipids occurring in molluscan tissues. *Biochim. Biophys. Acta* **98**, 151–159.

Tibaldi, E. (1966). Ricerche preliminari sugli acidi grassi di alcune specie di molluschi marini. *Atti Accad. Naz. Lincei, Cl. Sci. Fis., Mat. Nat., Rend.* **40,** 921–925.

Toyama, Y. (1958). Die Sterine der fetton Öle von wirbellosen Wassertieren. *Fette, Seifen, Anstrichm.* **60,** 909–915.

Toyama, Y., and Tanaka, T. (1953). Sterols and other unsaponifiable substances in the lipids of shell-fishes, Crustacea and echinoderms. XII. Occurrence of $\Delta^7$-cholestenol as a major component of the sterol mixture of chiton. *Bull. Chem. Soc. Jpn.* **26,** 497–499.

Van der Horst, D. J. (1970). Investigation of the synthesis and distribution of fatty acids in the lipids of the snail *Cepaea nemoralis* (L.). I. The fatty acid composition of the total lipids. *Neth. J. Zool.* **20,** 433–444.

Van der Horst, D. J. (1973). Biosynthesis of saturated and unsaturated fatty acids in the pulmonate land snail, *Cepaea nemoralis* (L.). *Comp. Biochem. Physiol. B* **46B,** 551–560.

Van der Horst, D. J. (1974). *In vivo* biosynthesis of fatty acids in the pulmonate land snail *Cepaea nemoralis* (L.). under anoxic conditions. *Comp. Biochem. Physiol. B* **47B,** 181–187.

Van der Horst, D. J., and Oudejans, R. C. H. M. (1976). Fate of dietary linoleic and linolenic acids in the land snail, *Cepaea nemoralis* (L.). *Comp. Biochem. Physiol. B* **55B,** 167–170.

Van der Horst, D. J., and Voogt, P. A. (1969a). Investigation of the fatty acid composition of the snail, *Succinea putris* L. *Comp. Biochem. Physiol.* **31,** 763–769.

Van der Horst, D. J., and Voogt, P. A. (1969b). Investigation of the fatty acid composition of the snail, *Arianta arbustorum.* *Arch. Int. Physiol. Biochim.* **77,** 507–514.

Van der Horst, D. J., and Voogt, P. A. (1972). Biosynthesis and composition of sterols and sterol esters in the land snail, *Cepaea nemoralis* (L.) (Gastropoda, Pulmonata, Stylommatophora). *Comp. Biochem. Physiol. B* **42B,** 1–6.

Van der Horst, D. J., and Zandee, D. I. (1973). Invariability of the composition of fatty acids and other lipids in the pulmonate land snail *Cepaea nemoralis* (L.) during an annual cycle. *J. Comp. Physiol.* **85,** 317–326.

Van der Horst, D. J., Kingma, F. J., and Oudejans, R. C. H. M. (1973). Phospholipids of the pulmonate land snail *Cepaea nemoralis* (L.). *Lipids* **8,** 759–765.

Van der Horst, D. J., Oudejans, R. C. H. M., Meyers, J. A., and Testerink, G. J. (1974). Fatty acid metabolism in hibernating *Cepaea nemoralis* (Mollusca: Pulmonata). *J. Comp. Physiol.* **91,** 247–256.

Vassalo, M. T. (1973). Lipid storage and transfer in the scallop *Chlamys hericia* Gould. *Comp. Biochem. Physiol. A.* **44A,** 1169–1175.

Vaver, V. A., Pisareva, N. A. and Bergelson, L. D. (1972). Diol lipids. XXI. The high ethyleneglycol content of marine invertebrate lipids. *Chem. Phys. Lipids* **8,** 82–86.

Voogt, P. A. (1967a). Biosynthesis of 3β-sterols in a snail, *Arion rufus* L., from 1-$^{14}$C-acetate. *Arch. Int. Physiol. Biochim.* **75,** 492–500.

Voogt, P. A. (1967b). Investigations on the capacity of synthesizing 3β-sterols in Mollusca. I—Absence of 3β-sterol synthesis in a whelk, *Buccinum undatum* L. *Arch. Int. Physiol. Biochim.* **75,** 809–815.

Voogt, P. A. (1968a). Investigations of the capacity of synthesizing 3β-sterols in Mollusca. II—Study on the biosynthesis of 3β-sterols in some representatives of the order Basommatophora. *Comp. Biochem. Physiol.* **25,** 943–948.

Voogt, P. A. (1968b). Investigations of the capacity of synthesizing 3β-sterols in Mollusca. III—The biosynthesis of 3β-sterols in some archeogastropods. *Arch. Int. Physiol. Biochim.* **76,** 721–730.

Voogt, P. A. (1969). Investigations of the capacity of synthesizing 3β-sterols in Mollusca. IV—The biosynthesis of 3β-sterols in some mesogastropods. *Comp. Biochem. Physiol.* **31,** 37–46.

Voogt, P. A. (1970). Investigations into the capacity of synthesizing sterols and into sterol composition in the phylum mollusca. Ph.D. Thesis, Univ. of Utrecht, Utrecht.

Voogt, P. A. (1971a). Sterols of some prosobranchs. *Arch. Int. Physiol. Biochim.* **79**, 391–400.

Voogt, P. A. (1971b). Investigations of the capacity of synthesizing 3β-sterols in Mollusca. V—The biosynthesis and composition of 3β-sterols in the mesogastropods *Crepidula fornicata* and *Natica cataena*. *Comp. Biochem. Physiol. B* **39B**, 139–149.

Voogt, P. A. (1972a). Lipid and sterol components and metabolism in Mollusca. *In* "Chemical Zoology" (M. Florkin, and B. T. Scheer, eds.), Vol. 7, pp. 245–300. Academic Press, New York.

Voogt, P. A. (1972b). Investigations of the capacity of synthesizing 3β-sterols in Mollusca. VI—The biosynthesis and composition of 3β-sterols in the neogastropods *Purpura lapillus* and *Murex brandaris*. *Comp. Biochem. Physiol. B* **41B**, 831–841.

Voogt, P. A. (1972c). Investigations of the capacity of synthesizing 3β-sterols in Mollusca. VII—The biosynthesis and composition of 3β-sterols in the neogastropods *Buccinum undatum* and *Neptunea antiqua*. *Neth. J. Zool.* **22**, 59–71.

Voogt, P. A. (1972d). Analysis of the sterols of the pulmonates *Arion rufus, Arianta arbustorum*, and *Succinea putris*. *Neth. J. Zool.* **22**, 489–496.

Voogt, P. A. (1972e). A gaschromatographic analysis of the sterols of *Lymnaea stagnalis* and *Planorbarius corneus*. *Arch. Int. Physiol. Biochim.* **80**, 697–704.

Voogt, P. A. (1973a). Investigations of the capacity of synthesizing 3β-sterols in Mollusca. XI—Biosynthesis and composition of 3β-sterols in some Acoela (Opisthobranchia). *Int. J. Biochem.* **4**, 479–488.

Voogt, P. A. (1973b). Investigations of the capacity of synthesizing 3β-sterols in Mollusca. X—Biosynthesis and composition of 3β-sterols in Cephalopoda. *Arch. Int. Physiol. Biochim.* **81**, 401–407.

Voogt, P. A. (1975a). Investigations of the capacity of synthesizing 3β-sterols in Mollusca. XIII—Biosynthesis and composition of sterols in some bivalves (Anisomyasia). *Comp. Biochem. Physiol. B* **50B**, 499–504.

Voogt, P. A. (1975b). Investigations of the capacity of synthesizing 3β-sterols in Mollusca. XIV—Biosynthesis and composition of sterols in some bivalves (Eulamellibranchia). *Comp. Biochem. Physiol. B* **50B**, 505–510.

Voogt, P. A. (1982). Steroid metabolism and nutrition in echinoderms. *In* "Echinoderm Nutrition" (M. Jangoux and J. M. Lawrence, eds.), Chap. 18. Balkema, Rotterdam.

Voogt, P. A., and Van der Horst, D. J. (1972). Investigations of the capacity of synthesizing 3β-sterols in Mollusca. VIII—The biosynthesis of 3β-sterols in the pulmonates *Succinea putris* and *Arianta arbustorium*. *Arch. Int. Physiol. Biochim.* **80**, 293–297.

Voogt, P. A., and Van Rheenen, J. W. A. (1973a). Sterols and sterol biosynthesis in the slug *Aplysia depilans*. *Experientia* **29**, 1070–1071.

Voogt, P. A., and Van Rheenen, J. W. A. (1973b). Investigations of the capacity of synthesizing 3β-sterols in Mollusca. IX—The biosynthesis and the composition of 3β-sterols in *Nassa*. *Neth. J. Sea Res.* **6**, 409–416.

Voogt, P. A., and Van Rheenen, J. W. A. (1974). Investigations of the capacity of synthesizing 3β-sterols in Mollusca. XII—Composition and biosynthesis of 3β-sterols in *Lepidochitona cinerea*. *Comp. Biochem. Physiol. B* **47B**, 131–137.

Walton, M. J., and Pennock, J. F. (1972). Some studies on the biosynthesis of ubiquinone, isoprenoid alcohols, squalene and sterols by marine invertebrates. *Biochem. J.* **127**, 471–479.

Watanabe, T., and Ackman, R. G. (1974). Lipids and fatty acids of the American (*Crassostrea virginica*) and European flat (*Ostrea edulis*) oysters from a common habitat, and after one feeding with *Dicrateria inornata* or *Isochrysis galbana*. *J. Fish. Res. Board Can.* **31**, 403–409.

Webber, H. H. (1970). Changes in metabolic composition during the reproductive cycle of the abalone *Haliotis cracheroidii* (Gastropoda:Prosobranchiata). *Physiol. Zool.* **43**, 213–231.

Weerasinghe, L., and Obara, T. (1974a). Characteristics of pond snail (*Heterogen longispira*) lipids. *Lebensm.-Wiss. Technol.* **7**, 208–210.

Weerasinghe, L., and Obara, T. (1974b). Phospholipids of pond snail (*Heterogen longispira*). *Lebensm.-Wiss. Technol.* **7**, 211–213.

Yasuda, S. (1967). Studies on the lipids of Japanese little neck, *Tapes japonica* Deshayes. II. Composition of phospholipids. *Yukagaku* **16**, 596–600.

Yasuda, S. (1970). Lipids of the Japanese little neck, *Tapes japonica*. IV. Sterols. *Yukagaku* **19**, 1014—1019.

Zama, K. (1963). Studies on the phospholipids of aquatic animals. *Mem. Fac. Fish., Hokkaido Univ.* **11**, 1–72.

Zama, K., Hatano, M., and Igarashi, H. (1960). The phosphatide of aquatic animals. XXIII. The phosphatides of Mollusca. *Nippon Suisan Gakkaishi* **26**, 917–920.

Zandee, D. I. (1967). The absence of cholesterol synthesis in *Sepia officinalis* L. *Arch. Int. Physiol. Biochim.* **75**, 487–491.

Zandee, D. I., Kluytmans, J. H., Zurburg, W., and Pieters, H. (1980). Seasonal variations in biochemical composition of *Mytilus edulis* with reference to energy metabolism and gametogenesis. *Neth. J. Sea Res.* **14**, 1–29.

Zs.-Nagy, I., and Galli, C. (1977). On the possible role of unsaturated fatty acids in the anaerobiosis of *Anodonta cygnea* L. (Mollusca, Pelecypoda). *Acta Biol. Acad. Sci. Hung.* **28**, 123–131.

## Note Added in Proof[4]

A second group of glycolipids studied by Hori and co-workers were those isolated from spermatozoa of the freshwater bivalve, *Hyriopsis schlegelii*. They consisted of three compounds called I, II, and III.

Glycolipid I (making up about 60% of the total glycolipid fraction) was identified to be a tetraglycosyl ceramide, with the structure $\beta$GlcNAc-(1→2)-$\beta$Man-(1→3)-$\beta$Man-(1→4)-$\beta$Glc-(1→1)-ceramide (Hori et al., 1977a). This glycolipid is unique in its sugar chain because it contains mannose instead of galactose. The glycolipid contained saturated fatty acids ranging in chain length from $C_{16}$ to $C_{21}$, with 18:0 and 16:0 making up 51.8 and 22.7%, respectively. The main long-chain base (97%) was found to be octadecasphing-4-enine.

Glycolipid II (about 30% of the total glycolipid fraction) was identified as a pentaglycosyl ceramide with the structure $\beta$GlcNAc-(1→2)-$\beta$Man-(1→3)-[$\beta$Xyl-(1→2)]-$\beta$Man-(1→4)-$\beta$Glc-(1→1)-ceramide (Hori et al., 1977b). The sugar chain differs from that of glycolipid I in that a branched xylose is linked to it. Fatty acid and long-chain base composition of lipid II was nearly identical to that of glycolipid I.

Glycolipid III was identified as a nonaglycosyl ceramide with the structure 3-$O$-Me-$\alpha$-Fuc-(1→2)-3-$O$-Me-$\beta$-Xyl-(1→4)-[3-$O$-Me-$\alpha$GalNAc-(1→3)]-$\alpha$Fuc-(1→4)-$\beta$GlcNAc-(1→2)-$\alpha$Man-(1→3)-[$\beta$Xyl-(1→2)]-$\beta$Man-(1→4)-$\beta$Glc-(1→1)-ceramide (Hori et al., 1981). This sugar chain is structurally related to that of glycolipid II in that three 3-$O$-methyl sugars are linked to the chain of the latter lipid by an internal fucose moiety. Fatty acid and long-chain base composition of lipid III was nearly identical with those of lipids I and II.

[4] See Section II,C,5, p. 350.

# 8

# Molluscan Collagen and Its Mechanical Organization in Squid Mantle

JOHN M. GOSLINE
ROBERT E. SHADWICK

Department of Zoology
University of British Columbia
Vancouver, British Columbia V6T 2A9
Canada

## I. Introduction

Collagen is the principle structural protein in multicellular animals. It is a major component of materials such as the body wall connective tissues of most invertebrates; vertebrate skins; internal organs such as arteries, veins, and guts; and skeletal materials such as bone, tendon, ligaments, and cartilage. In addition, loose connective tissues containing collagen tie

371

THE MOLLUSCA, VOL. 1
Metabolic Biochemistry
and Molecular Biomechanics

together all the other soft internal organs of most animals. As might be expected for such an important protein, a great deal is known about its chemistry, structure, and properties, but virtually everything that is known is based on vertebrate (primarily mammalian) collagen. However, there is a limited but growing body of evidence indicating that invertebrate "tissue" collagens (mesodermally derived collagens) are similar to, and evolutionarily related to, vertebrate collagens (Adams, 1978). Indeed, Mathews (1980) suggests that the collagen gene arose very early in metazoan evolution and has been altered only slightly in the last 700 million years. Thus it should be possible to make useful inferences about molluscan tissue collagens on the basis of our understanding of vertebrate collagen. On the other hand, invertebrate "secreted" or epidermally derived collagens (e.g., nematode and annelid cuticle collagens and bivalve byssus collagen) are probably distinct from the tissue collagens, although they are remarkably similar in structure and properties. This chapter will first review the literature on molluscan tissue collagens with reference to our current understanding of the structure and chemistry of vertebrate collagen. We will then look at the mechanical organization of collagen fibers in squid mantle. This will involve the discussion of some previously unpublished data on the mechanical properties of squid collagen, and an analysis of the organization and functional properties of the mantle connective tissue system in the mechanics of swimming. A discussion of collagen in bivalve byssus will be found in Chapter 4.

## II. Chemistry and Structure of Collagen

There are many recent reviews on all aspects of collagen structure and chemistry (Viidik and Vuust, 1980; Bornstein and Sage, 1980; Bornstein and Traub, 1979; Ramachandran and Reddi, 1976; Traub and Piez, 1971), and the reader should consult these articles for more specific information and detailed references. We will restrict ourselves to a very general review of the collagen literature, with specific reference to the limited information on molluscan collagen and to areas of possible future research interest.

Collagen functions as an extracellular protein fiber with a high modulus of elasticity (stiffness) and great tensile strength. These extracellular fibers can be identified by several common characteristics.

1. Collagen fibers usually exhibit a 640-Å axial periodicity when observed in the electron microscope.
2. Large-angle X-ray diffraction patterns exhibit characteristic reflec-

tions that arise from the helical arrangement of the polypeptide chains in collagen fibers.

3. The amino acid composition of collagen is about one-third glycine and is rich in the imino acids proline and hydroxyproline.

4. With heating, collagen fibers shrink to about one-third their original length and become rubbery.

These structural and chemical characteristics and the functional mechanical properties that arise from them can be traced directly to the molecular level. This section on chemistry and structure will attempt to develop an understanding of the relationships between the chemistry of collagen and the various levels of structure that arise in the collagen fiber from the chemical information coded in the sequence of the collagen polypeptides.

## A. Tropocollagen

When macroscopic collagen fibers are placed in weak acid solutions (pH about 3), many of the cohesive forces that hold the constituent molecules together into the fibrous form are weakened, and it is possible to isolate a single molecule, called tropocollagen, that is the basic building block of collagen fibers. This molecule has a molecular weight of about 300,000. Solution studies on the intact or "native" tropocollagen molecule indicate that it is a very elongated, rodlike molecule with a length of about 2800 Å and a diameter of about 15 Å (see Fig. 1). Under denaturing conditions tropocollagen can be disrupted to yield three polypeptide chains, each of about 100,000 MW and containing somewhat more than 1000 amino acid residues. These individual polypeptide chains are referred to as $\alpha$ chains. Denatured tropocollagen can also yield dimers ($\beta$ collagen) and trimers ($\gamma$ collagen), where the disordered peptide chains are linked together by covalent cross-links that were introduced following the assembly of the tropocollagen molecule (see Fig. 1).

Invertebrate tissue collagens appear to fit this general pattern, but there are insufficient data available at present to verify all the details. Kimura and Matsuura (1974) report that tropocollagen from the skin of *Octopus vulgaris* has a molecular weight of 240,000 as determined by sedimentation analysis, whereas the tropocollagen of the lobster *Panulirus japonicus* has a molecular weight of 270,000. The $\alpha$ chains from the collagen of both of these animals comigrate with the $\alpha$ chains from shark collagen on SDS–polyacrylamide gels, and thus apparently have a molecular weight of 95,000. The differences in apparent molecular weights of the native tropocollagen molecules remain unexplained, but may well arise from the unusual sedimentation of these anisotropic molecules.

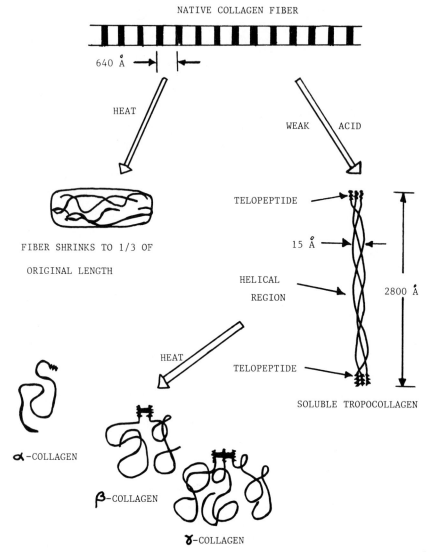

NATIVE COLLAGEN FIBER

640 Å

HEAT

WEAK    ACID

FIBER SHRINKS TO 1/3 OF

ORIGINAL LENGTH

TELOPEPTIDE

15 Å

HELICAL
REGION

2800 Å

HEAT    TELOPEPTIDE

SOLUBLE TROPOCOLLAGEN

α-COLLAGEN

β-COLLAGEN

γ-COLLAGEN

**Fig. 1.** Molecular organization of collagen. When native collagen fibers are heated they shrink to about one-third of their length and become rubbery. This shrinkage is due to the collapse of helical molecules in the fiber. When native fibers are extracted in weak acid, some of the constituent molecules (called tropocollagen) will go into solution. These molecules maintain their native helical conformation, and they exist as long, rodlike molecules about 2800 Å long and 15 Å in diameter. When soluble tropocollagen is heated it denatures to form random coils. Tropocollagen is made from three polypeptide chains, and if there are no intramolecular cross-links, then heating will yield three $\alpha$ chains. If intramolecular cross-links are present, then the denaturation will yield dimers ($\beta$ collagen) and even some trimers ($\gamma$ collagen).

## B. Tropocollagen Structure

A variety of evidence that is too extensive to be reviewed here (see Bornstein and Traub, 1979; Fraser and McRae, 1973; Traub and Piez, 1971) indicates that the basic molecular conformation that characterizes the rodlike tropocollagen molecule is helical. Each of the three $\alpha$ chains exists in a left-handed polyglycine II-type helix, with a pitch of approximately 10 Å and about three amino acids per turn. These three helical peptides are then twisted around each other in a right-handed superhelix with a pitch of around 100 Å. This coiled-coil structure is very compact (i.e., only 15 Å in diameter) because the individual peptide helices always have a glycine residue on the face pointing toward the center of the super-helix. Because glycine has only a hydrogen for its side chain, the three individual helices can be packed very tightly in the superhelix. This packing arrangement requires that glycine must occupy every third position in the polypeptide sequence, and we will see shortly that this is the dominant feature observed for collagen sequences.

The peptides in this coiled coil are held together by interchain hydrogen bonds, with at least one hydrogen bond per tripeptide. In addition, it has been suggested that bridges involving tightly bound water molecules may also stabilize the structure. Although virtually all features of collagen structure have been worked out for vertebrate collagen, the limited X-ray data available for invertebrate collagens indicate that the molecular conformation is exactly the same (Traub and Salem, 1972).

## C. Collagen Types

Recent research on vertebrate collagens indicates that there are several chemically distinct types of collagen (see Bornstein and Sage, 1980). Type I collagen is found widely distributed in the vertebrate body, existing in most structural tissues such as bone, skin, artery, and tendon. This collagen type is composed of two different types of $\alpha$ chains, designated $\alpha1(I)$ and $\alpha2(I)$. Type I tropocollagen contains two $\alpha1$ chains and a single $\alpha2$ chain, which together form the characteristic triple helix. Thus type I collagen is formed from two distinct gene products. Type II collagen, found in cartilage, is produced from yet another gene and is composed of three identical $\alpha1(II)$ polypeptide chains. Type III collagen is often found in the same tissues as type I collagen, but it is produced on yet another gene and is composed of three identical $\alpha1(III)$ polypeptides. Type IV collagen contains two different types of $\alpha$ chains and is found in basement membranes. In addition, there is recent evidence for other collagen types. Clearly there has been considerable evolution of the collagen gene, but the various collagen types share all the important chemical and structural characteristics.

Nordwig and Hayduk (1969) report that the tropocollagen from the sea anemone (*Actinia*) and the fluke (*Fasicola*) consists of three identical α chains. Kimura and Matsuura (1974) report that the collagen from the foot of the abalone *Haliotis discus* contains three identical α chains, whereas the collagen from the skin of *O. vulgaris* and the skin of the squid *Todarodes pacificus* contains two different types of α chain in both cases. More recently, Schmut et al. (1980) report that the collagen from the skin of the snail *Helix pomatia* contains three identical α chains. There are clearly not sufficient data to establish any pattern for the molluscs, but it is tempting to speculate that there are differences between gastropods and cephalopods. However, it is likely that, as in the vertebrates, there are differences between tissues within any group.

## D. Amino Acid Composition and Peptide Sequences

The amino acid composition provides several important clues about the chemistry of collagen. Perhaps the most striking and consistent feature of the composition is the glycine content. In all tissue collagens studied, glycine makes up almost exactly one-third of the total. In addition, proline and the unusual imino acid hydroxyproline are found in very large amounts, being typically of the order of 200 residues per 1000. The proline and hydroxyproline residues are believed to contribute to the stability of the coiled-coil structure described previously.

Table I shows amino acid composition data for bovine types I and II collagen as well as the composition for collagen from the muscular foot of the gastropod *H. discus,* and the composition of the α1 and α2 chains of the collagen from the skin of the squid *T. pacificus.* Additional amino acid composition data are available in Williams (1960) and Rigby and Mason (1967) for gastropods, and in Isemura et al. (1973) and Kimura and Matsuura (1974) for cephalopods. In all the collagens shown in Table I, the glycine content is very close to being exactly one-third of the total composition, ranging from 327 to 335 residues per 1000. This composition is consistent with the proposed coiled-coil structure of tropocollagen, where every third amino acid in the individual α chains should be glycine. Indeed, a fair amount of sequence data are now available for selected vertebrate collagens (see Fietzek and Kuhn, 1976; Bornstein and Traub, 1979), and in all cases glycine always occupies the third position. In fact, the best way to regard the sequence of collagen is as $(Gly-X-Y)_n$. The imino acids, proline and hydroxyproline, also occupy characteristic positions in this repeating tripeptide. Again, from the work on vertebrate collagen,

TABLE I
The Amino Acid Composition of Several Different Collagens[a]

| Amino acid | Bovine | | Abalone[d] (Haliotis discus) | Squid skin (Todarodes pacificus)[d] | |
| | Type 1(I)[b] | Type 1(II)[c] | | α1 Chain | α2 Chain |
|---|---|---|---|---|---|
| 3-Hyp | 1 | 2 | 1 | 2 | 2 |
| 4-Hyp | 85 | 91 | 80 | 75 | 71 |
| Asp | 45 | 43 | 59 | 53 | 64 |
| Thr | 16 | 22 | 17 | 24 | 20 |
| Ser | 34 | 26 | 58 | 44 | 35 |
| Glu | 77 | 87 | 96 | 87 | 82 |
| Pro | 135 | 129 | 102 | 105 | 105 |
| Gly | 327 | 333 | 334 | 331 | 335 |
| Ala | 120 | 102 | 91 | 106 | 78 |
| Cys | — | — | — | — | — |
| Val | 18 | 17 | 16 | 18 | 24 |
| Met | 7 | 11 | 11 | 13 | 10 |
| Ile | 9 | 9 | 12 | 12 | 24 |
| Leu | 21 | 26 | 26 | 26 | 32 |
| Tyr | 4 | 1 | 6 | 5 | 2 |
| Phe | 12 | 14 | 7 | 11 | 8 |
| Hyl | 5 | 23 | 14 | 11 | 25 |
| Lys | 32 | 15 | 8 | 13 | 11 |
| His | 3 | 2 | 1 | 5 | 8 |
| Arg | 50 | 51 | 60 | 61 | 64 |

[a] Data are expressed as residues per 1000.
[b] Rauterberg and Kuhn (1971).
[c] Miller and Lunde (1973).
[d] Kimura and Matsuura (1974).

prolines are usually found in the X position, whereas hydroxyproline is virtually always found in the Y position. The observation that hydroxyproline always occupies the Y position arises from the fact that the enzyme that hydroxylates proline to hydroxyproline, following the synthesis of the individual α chains, is specific for amino acid residues in this position. Thus the sequence $(Gly\text{-}Pro\text{-}Hyp)_n$ is perhaps an even better description of collagen in general, and synthetic polypeptides with this sequence take on structures that are apparently identical to native collagen. In addition, it appears that the number of these (Gly-Pro-Hyp) tripeptides relative to the number of tripeptides where the X and Y positions are occupied by other amino acids strongly affects the thermal stability of the tropocollagen molecule.

When tropocollagen molecules in solution or when macroscopic collagen fibers are heated, they denature. This denaturation involves the collapse of the highly extended rodlike molecules into essentially random coils. The temperature $(T_s)$ at which macroscopic collagen fibers denature, as indicated by a shrinkage to about one-third of their original length, is usually 20–30°C higher than the temperature $(T_d)$ at which individual tropocollagen molecules denature. This difference undoubtedly exists because tropocollagen molecules in a large-scale structure like a fiber are stabilized by many additional forces. Josse and Harrington (1964) pointed out a positive correlation between the content of Pro + Hyp and the denaturation temperature $(T_d)$ of various collagens, and they suggested that the imino acids directly increase the stability of the coiled-coil structure. Rigby (1968) made the interesting observation that the denaturation temperature $(T_d)$ is very close to the upper limit of the environmental temperature for the animal from which the collagen is obtained. For example, $T_d$ values for mammalian collagen and for the collagen of parasites that live in mammals fall in the range of 38–40°C. $T_d$ for the body wall collagen of *Helix* is about 30°C, and $T_d$ is about 14°C for cod skin collagen. Pikkarainen et al. (1968) report that $T_d$ for the body wall collagen of the squid *Loligo,* obtained in the Mediterranean, is about 25°C. Thus it appears that evolution ensures the stability of the relatively labile tropocollagen molecule by matching the content and presumably the specific placement of imino acids in collagen sequences to the normal range of environmental temperatures experienced by the animal. There does not appear to be any similar relationship between environmental temperature and the shrinkage temperature $(T_s)$ of the macroscopic fibers, because the fibers are always more stable than the individual tropocollagen molecules from which they are formed.

One other characteristic feature of the amino acid composition of collagen is the presence of hydroxylysine. As with hydroxyproline, the lysines are hydroxylated after the individual $\alpha$ chains are synthesized, but before the triple-helical structure is assembled, and the hydroxylysines always occupy the Y position in the repeating tripeptide. Hydroxylysines are important both in cross-linking and in the attachment of carbohydrate chains to collagen (see later). The hydroxylation of the lysines in invertebrate collagens is often higher than that in vertebrate types I and III collagens (Nordwig and Hayduk, 1969), but the levels are considerably lower than those seen for type IV (basement membrane collagen; 49 residues per 1000 of Hyl). The levels are about the same as those seen for type II (cartilage) collagen, and this may reflect the increased interaction between collagen and carbohydrates seen for invertebrate tissues as well as in vertebrate cartilage.

Finally, detailed sequence studies on vertebrate collagen reveal that although tropocollagen is a triple-helical structure along its entire length, there are small regions at each end of the molecule that do not contain the characteristic (Gly-X-Y) repeat and are therefore not helical. These regions, called telopeptides, are about 15 residues in length and are believed to be involved in intra- and intermolecular cross-links. They are, in fact, just the remnants of very much larger nonhelical regions that are part of a precursor molecule to the tropocollagen. This precursor molecule, called procollagen, has a molecular weight of the order of 450,000, and the nonhelical regions are thought to aid in the assembly of the helical region or possibly to prevent the molecules from forming fibers intracellularly. At any rate, these large nonhelical regions are enzymatically removed to yield tropocollagen as the molecule is released from the cell that has synthesized it. All that is known about this aspect of collagen biochemistry is based on work with vertebrate collagens, and it should be very interesting indeed to see if the same pattern is exhibited in the lower animals as well.

## E. Structure of the Collagen Fiber

The most characteristic feature of the collagen fiber is the 640-Å axial periodicity that can be observed in the electron microscope and in low-angle X-ray diffraction patterns. This axial repeat is a direct reflection of the pattern by which individual tropocollagen molecules are assembled to form fibers, a pattern that is based on precise overlapping of parallel tropocollagen molecules. Because tropocollagen is roughly four times longer than the axial periodicity, it was originally thought that the adjacent molecules were shifted one-quarter of their length relative to each neighboring molecule, referred to as the "quarter-stagger" model (Schmitt et al., 1953). More recently, Hodge and Petruska (1963) realized that the 2800-Å length of the tropocollagen molecule was greater than four times the 640-Å axial repeat, and they suggested a modification of the quarter-stagger arrangement. Figure 2 summarizes the features of these packing arrangements. The Hodge-Petruska model (Fig. 2B), which is now generally accepted, has five overlap regions of 640 Å, with the fifth region being occupied by only a short (~240 Å) region of the C-terminus of the tropocollagen molecule and a 400-Å gap.

Although this overlap pattern provides an adequate description of the relationship between immediately adjacent molecules, it should not be taken as a complete description of an entire collagen fiber. The collagen fiber is probably made up of microfibrils that are brought into precise register to maintain the axial periodicity at the level of the entire fiber. One of several proposals currently being investigated has a sheet of five tropocollagen molecules (i.e., one repeat of the Hodge-Petruska overlap arrangement) rolled to form a cylindrical microfibril with a diameter of about

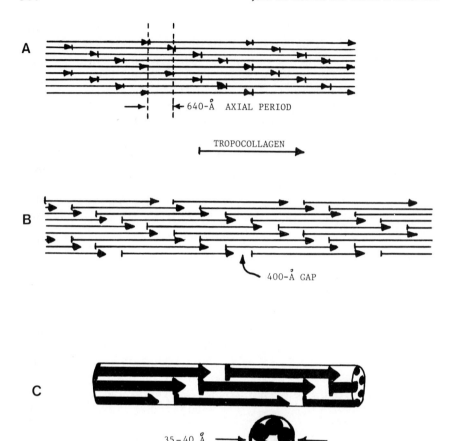

**Fig. 2.** The organization of the tropocollagen molecules in collagen fibers. (**A**) The "quarter-stagger" model, in which the tropocollagen molecules are arranged end to end and adjacent molecules overlap each other by one-quarter of their length. (**B**) The Hodge-Petruska model, a modification of the quarter-stagger model, in which there is a 400-Å gap between the ends of the tropocollagen molecules. (**C**) A microfibrillar form of the Hodge-Petruska model, in which one lateral repeat (i.e., five tropocollagen molecules) is rolled to form a cylindrical fiber about 40 Å wide.

35–40 Å (see Fig. 2C). The microfibrils are then packed in a highly ordered lattice to form larger scale fibers. More detailed discussions of the various possibilities are presented in Bornstein and Traub (1979) and Parry and Craig (1979).

Although there are some fairly firm ideas about the organization of the collagen fiber, all that we know is based on vertebrate collagens. However, most of the collagens in the invertebrates show the same or similar axial periodicity, and it seems very likely that the basic organization is the same for all tissue collagens. There is, however, some variation in the size of the axial periodicity measured for some invertebrate collagens (Meek, 1966; Travis et al., 1967), and the origins of these differences, if they are real, are unknown. Figure 3 shows collagen fibers from the dorsal aorta of *Octopus dofleini*. The axial periodicity, measured at 600–640 Å, is quite obvious, as are finer striations within the major period. The squid mantle collagen periodicity is 680 Å (Hunt et al., 1970).

One of the very interesting and important properties of soluble tropocollagen is its ability to aggregate spontaneously to form fibers that are identical to the native fibers seen in intact tissues (Schmitt et al., 1953). In addition, under special conditions the tropocollagen molecules will aggregate without overlap in perfect register to form segment long spacing (SLS) structures (see Fig. 4). In the electron microscope SLS collagen exhibits many fine striations that represent the distribution of polar and nonpolar amino acid residues along the length of the tropocollagen molecule. That is, darkly stained regions represent areas rich in polar amino acids, and light regions represent areas rich in nonpolar residues. Detailed analyses of these SLS striations reveal 640-Å groupings that undoubtedly explain the tendency of these molecules to aggregate in the normal fiber pattern (see Bornstein and Traub, 1979). The banding patterns for SLS collagens from the sea anemone *Actinia* and the liver fluke *Fasicola* have been shown to be identical to those of calf skin collagen (Nordwig and Hayduk, 1969), and this indicates that the distribution of polar and nonpolar regions along the tropocollagen molecule has remained constant throughout the evolution of most of the metazoan animals. It would be surprising, therefore, if there were any major differences in the arrangement of molecules in the fibers of different animal types. Unfortunately, nothing is known about the organization of a molluscan collagen at this level, but if any differences are present, this sort of analysis will reveal it.

Finally, at the light microscope level, most vertebrate collagens are seen to take on a wavy or crimped appearance (Diamant et al., 1972). The appearance varies from tissue to tissue. In tendon, for example, the fibers have a tight crimp with a very small crimp angle relative to the long axis of the tendon. In skin and artery the collagen fibers follow a more sinuous pattern. It has been suggested that these differences arise, in part at least, from the presence of the different collagen types (see Bornstein and Traub, 1979). Invertebrate collagens also show these sorts of differences in macroscopic form. For example, the octopus aorta collagen (Fig. 3) is obviously like its mammalian analog, as it follows a very wavy pattern. On the other hand, the collagen fibers in squid mantle appear to be per-

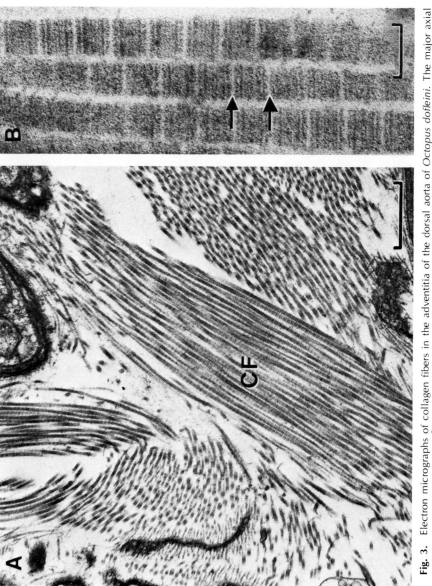

**Fig. 3.** Electron micrographs of collagen fibers in the adventitia of the dorsal aorta of *Octopus dofleini*. The major axial periodicity is approximately 640 Å, as indicated by the arrows in (**B**). Finer striations are visible within the major period. (**A**) Bar = 1 μm. (**B**) Bar = 0.1 μm. CF, Collagen fibers.

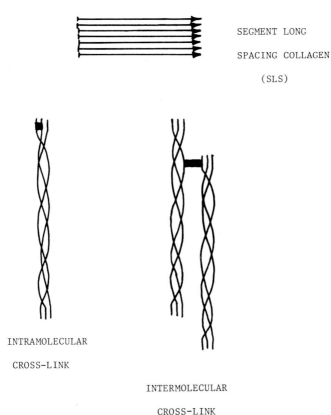

SEGMENT LONG

SPACING COLLAGEN

(SLS)

INTRAMOLECULAR

CROSS-LINK

INTERMOLECULAR

CROSS-LINK

**Fig. 4.** Segment long spacing (SLS) collagen aggregates are assemblies of collagen molecules aligned in perfect register with no overlap. Stained SLS structures reveal detailed banding patterns that indicate the distribution of polar and nonpolar residues along the tropocollagen molecule. Intramolecular cross-links form in the nonhelical telopeptide region between chains within a single tropocollagen molecule. Intermolecular cross-links form between adjacent molecules, joining the telopeptide region of one molecule to an appropriately placed lysine in the helical region of another molecule.

fectly straight (see Section IV). It is tempting to speculate that these differences in morphology are due to different collagen types, but a great deal more research on molluscan collagens is obviously needed.

## F. Cross-Linking

In addition to the weak noncovalent forces that hold the tropocollagen molecules together into the characteristic overlap pattern, strong covalent cross-links are present that can tie molecules together into a mechanically rigid structure (see Wainwright et al., 1976). At present, everything that is known about the cross-link chemistry is based on studies with vertebrate collagen. There are intramolecular cross-links between the three $\alpha$ chains of tropocollagen molecules, and there are intermolecular cross-links that link adjacent tropocollagen molecules to one another (see Fig. 4). Most of the cross-links are based on lysines and on lysine-derived aldehydes. The short nonhelical telopeptide regions at either end of the tropocollagen molecule provide the major site for cross-link formation. Lysines in these regions are oxidized to aldehydes and then interact with lysines in adjacent regions of neighboring molecules to form intermolecular cross-links. Detailed sequence studies indicate that there are lysines strategically placed at exactly the places required by the overlap pattern for cross-link formation. That intramolecular cross-links are present in molluscan collagen is indicated by the isolation of $\beta$ and $\gamma$ collagens. Because the high tensile strength of collagen fibers is due to the presence of intermolecular cross-links (Bailey, 1968), we can infer that molluscan collagen contains intermolecular crosslinks as well (see Section III), but nothing is known about the chemistry of the cross-links.

## G. Carbohydrate Interactions

Finally, collagen is not just a protein, it is a glycoprotein. Some of the hydroxylysines carry a glucose–galactose disaccharide, and the carbohydrate content varies considerably in different tissues. Vertebrate skin collagen (types I and III) contains only about 0.4% carbohydrate; cartilage (type II) collagen, about 4%; and basement membrane (type IV) collagen, about 12% (Bornstein and Traub, 1979). Studies on molluscan collagens (Kimura, 1972a,b) indicate that the carbohydrate content of collagen from the abalone, from the squid, and from the octopus are at a level of about 3–4%. This is much higher than most vertebrate collagens but about the same level as cartilage collagen. The functional significance of the increased carbohydrate levels is at present unknown. Of the roughly 15 residues per 1000 of hydroxylysine in molluscan collagens, about 40% carry carbohydrate units, and 95% of the carbohydrate units are the glucosylgalactosylhydroxylysine type. The remaining 5% are simply galactosylhydroxylysine (Kimura, 1972b; Stemberger and Nordwig, 1974). In addition, it is known that the glycosylated hydroxylysines in vertebrate collagens are always found in a characteristic sequence (Gly-X-Hyl-Gly-Y-Arg), and Isemura et al. (1973) have shown that the glycosylated hydroxylysines from cuttlefish skin collagen fall in exactly the same sequences. This interesting information indicates that along with all the other similarities (in fact homologies) between vertebrate and invertebrate tissue collagens, the specific enzymes that add carbohydrate units during collagen biosynthesis are also very similar. This provides additional support for the notion that invertebrate tissue collagens are related directly to vertebrate collagens.

## III. Mechanical Properties of Squid Mantle Collagen

To date, no one has determined the mechanical properties of any isolated invertebrate collagen. Because of the similarities in chemistry and structure, it has always been assumed that vertebrate and invertebrate collagen have similar mechanical properties. We have recently carried out tests on sheets of collagen isolated from squid mantle, and we find that it is virtually identical in mechanical properties to vertebrate tendon. The results for squid collagen will be presented in the context of our general understanding of the mechanical properties of vertebrate collagens. More detailed reviews can be found in Kastelic and Baer (1980), Wainwright et al. (1976), Viidik (1973), and Harkness (1968).

From a mechanical point of view, collagen can be considered as a tensile material; that is, a material designed to carry large loads applied parallel to the long axis of the fiber and resist extension resulting from these loads. The tensile properties arise at the molecular level from the parallel arrangement of the polypeptide chains in the tropocollagen molecules and from the higher level organization of the molecules parallel to the long axis of the fiber. To a first approximation, collagen fibers are stiff and strong because applied loads are carried by covalent bonds in this highly ordered structure. Tissues such as tendon and ligament are flexible because they are composed of many very thin fibers ($\sim$0.1 $\mu$m diameter) that are not glued together by a stiff matrix, as is the case in rigid composites like fiberglass and bone. Thus the individual thin collagen fibers bend independently, and the structure as a whole is very flexible. For example, a tendon of modest length ($\sim$10 cm long) can support a tensile load of about 50,000 times its own weight, and yet the same tendon will bend easily under its weight. The flexibility of the collagen fiber is probably as important in allowing structures like tendons to flex, as are the tensile properties of stiffness and strength in transmitting forces and in limiting the distension of tissues like skin and artery at high loads.

Figure 5 shows the tensile properties of outer tunic collagen from the mantle of the squid *Loligo opalescens*. Longitudinal strips of mantle tissue taken from the midventral region ($\sim$1 $\times$ 7 cm) were first split to isolate the outer half of the mantle. Then most of the muscle tissue was scraped off with a razor blade, to yield a fairly pure sheet of isolated colla-

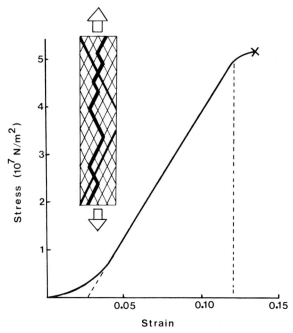

**Fig. 5.** The mechanical properties of squid mantle collagen. The modulus of elasticity in the linear region is $5.4 \times 10^8$ N/m². The region of linear extension covers a range of about 8%, and the specimen breaks at about 13% extension. Elastic energy storage in the linear region, as indicated by the area under the stress/strain curve bounded by the dashed lines, is 2000 J/kg. The inset diagram shows the pattern of collagen fibers in the test sample.

gen. This sheet was then clamped and glued with cyanoacrylate cement into specimen grips and extended at a strain rate of about 2%/sec on a constant strain-rate tensile testing machine. The initial length and width of the test specimen were measured with a millimeter rule. The thickness of the collagen sheet was measured from frozen sections of the mantle taken immediately adjacent to the location of the test specimen. In *L. opalescens* the outer tunic is about 20 $\mu$m thick and is composed of six layers of collagen fibers.

Perhaps the most characteristic tensile property of collagen fibers is the high tensile strength, where tensile strength is defined as the stress ($\sigma$ = force per unit cross-sectional area) at which the sample fails. For a large number of different kinds of vertebrate collagen, tensile strength falls in the range of $2 \times 10^7$ to $1.4 \times 10^8$ N/m², with a value of about $8 \times 10^7$ N/m² being a good average (Wainwright et al., 1976). For comparison, the tensile strength of mild steel is about $4 \times 10^8$ N/m², or only about five times greater than that of collagen. The tensile strength of the squid collagen shown in Fig. 5 is about $5 \times 10^7$ N/m², and a duplicate test yielded a value of $4.8 \times 10^7$ N/m². Similar tests run on isolated outer tunic from the squid *Nototodarus sloani* (tunic thickness 50–60 $\mu$m) range between $7 \times 10^7$ and $1.2 \times 10^8$ N/m². We certainly do not have sufficient data at this time to determine if there are real differences between the two squid, but it should be clear that the values are in line with what is expected for a collagen.

The tensile modulus of elasticity ($E \times \Delta\sigma/\Delta\epsilon$, where $\epsilon$ is the strain and is equal to the change in length divided by the initial length) is typically in the range of $10^9$ N/m² for vertebrate collagen fibers (Wainwright et al., 1976). The tensile modulus, as derived from the slope of the linear region of Fig. 5, is $5.4 \times 10^8$ N/m² for *Loligo* tunic collagen. The tensile modulus of *Nototodarus* tunic collagen is in the range of $7–9.5 \times 10^8$ N/m². Again, squid collagen is very similar to vertebrate collagen.

The shape of the stress/strain curve is also quite similar to that seen for vertebrate tendons. There is typically a low-modulus "toe" region at low extensions before the curve becomes linear and exhibits the characteristic modulus of $10^9$ N/m². In tendon, this low-modulus region is due to the straightening of the slightly crimped collagen fibers. In the squid tunic, this region probably arises from a different fiber morphology. The collagen fibers in the tunic of the squid mantle are arranged as a crossed-fiber array, with fibers running at 27° to the long axis of the animal (see Section IV for more details). The inset diagram in Fig. 5 shows how the collagen fibers run in the test samples that were used to obtain the data reported here. No single fiber runs the entire length of the sample; in fact, no two fibers run the entire length with only a single join. Therefore the high-modulus region must correspond to a transfer of load at many junction points, as suggested by the zigzag bold line. The need to transfer load at many junction points is a fundamental difference between tendon and the squid tunic. We know nothing of the linkages between the collagen fibers at these crossover points, but the high tensile strength and stiffness of the structure suggests that these linkages are very strong. Perhaps the collagen fibers weave between the larger scale fibers represented in this diagram.

Regardless of the exact morphology, it seems likely that this structure will be a bit looser than the parallel collagen fibers of a tendon, and thus that estimates of extension at failure may be too large and the modulus values may be too small. The low-modulus region of vertebrate tendon can extend up to strains as small as 0.02 (2% extension) and as large as 0.15 (15% extension), depending on the crimp angle. The collagen fibers in squid tunic appear to be quite straight in both the light and electron microscopes (Ward and Wainwright, 1972; Otwell and Giddings, 1980), but the rotation of fibers in the crossweave allows the low-modulus region to extend up to strains of about 0.04. The range of the high-modulus region in vertebrate tendon is typically 5–6% beyond the toe region. In the squid tunic it appears to be about 8% (i.e., from strain 0.04 to about 0.12). This apparently greater extensibility of squid collagen probably reflects some slippage of the collagen fibers at the junction points in the crossweave, rather than a difference in the properties of the individual fibers. However, it is very difficult to quantify this slippage, and we are only able to say that the extensibility of squid collagen is similar to that of vertebrate collagen.

Finally, because collagen fibers are both stiff and reasonably extensible, it is possible to store large amounts of elastic energy in the deformation of these fibers. The energy storage per unit volume can be calculated from the area under a stress/strain curve, and the energy storage by squid tunic collagen, as indicated by the area bounded by the stress/strain curve and the dashed lines in Fig. 5, is of the order of 2.4 $J/cm^3$, or about 2000 J/kg. This value is quite typical of vertebrate tendons, and as pointed out by Alexander and Bennet-Clark (1977), the storage of elastic energy in materials like collagen is much more efficient than in activated muscle (the energy storage in active muscle is maximally 5 J/kg). The use of collagen fibers to store elastic energy in the squid mantle will be considered in the next section.

## IV. Mechanical Organization of the Collagen in Squid Mantle

### A. Morphology

The collagen fiber system in squid mantle was described first by Ward and Wainwright (1972), and more recently by Otwell and Giddings (1980) and Bone et al. (1981). The mantle is a muscular tissue, composed of alternating layers of circular and radially oriented fibers (see Fig. 6A). These muscle layers are encased and constrained by two connective tissue fiber systems. One system, the inner and outer tunics, lines the inner and outer surfaces of the bullet-shaped mantle structure. These tunics are in intimate contact with the muscular tissue of the mantle, and as described by Ward and Wainwright, the tunics are robust layers of collagen fibers, with the collagen being arranged in 6–10 layers of crossed-helical fibers. The layers formed alternating right- and left-handed helices, with a very regular fiber angle of 27° relative to the long axis of the mantle (see Fig. 6B).

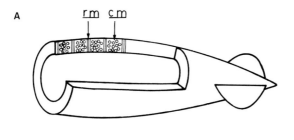

**Fig. 6.**   Organization of the squid mantle, after Ward and Wainwright (1972). **(A)** The mantle is composed of alternating layers of circular (cm) and radial (rm) muscle. These muscle layers are encased between two collagen tunics that line the inner and outer surfaces of the mantle structure. **(B)** The outer tunic is composed of several layers of collagen fibers arranged as right- and left-handed helices that spiral around the mantle at a fiber angle of 27° relative to the long axis of the mantle.

The second system of fibers runs through the thickness of the mantle muscle tissue (intermuscular fibers) in several different orientations. One set of intermuscular fibers (IM-1) can be seen in longitudinal sections of the mantle (Fig. 7A) running from the outer tunic to the inner tunic, with an angle of about 30° relative to the long axis of the animal (Ward and Wainwright, 1972). In sections cut tangentially to the surface of the mantle (i.e., at right angles to the longitudinal sections; Fig. 7B), these fibers appear not to run parallel to the long axis of the animal but run at a small angle (~10–15°) relative to this axis (Bone et al., 1981). Thus these fibers follow a steep "spiral" through the muscle tissue. A second set of intermuscular fibers (IM-2) can be seen in transverse sections that pass through layers of radially oriented muscles (Fig. 7C). These fibers also run from the outer tunic to the inner tunic and are oriented at about 55° relative to the mantle surface. They are confined to the plane of the radial muscle bands (Bone et al., 1981). The fibers in both the inner and outer tunics and in IM-1 and IM-2 all appear to follow very straight paths. There is however, a third set of intermuscular fibers (IM-3) that can be seen running parallel to the circular muscles. These fibers are not straight but appear coiled or buckled (Bone et al., 1981).

The fibers of the inner and outer tunics can be identified as collagen (*a*)

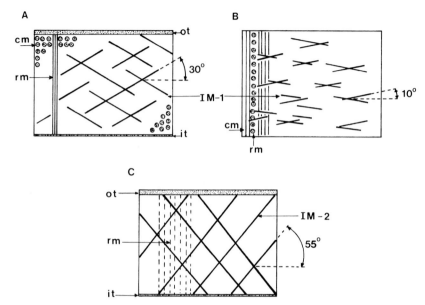

**Fig. 7.** The organization of intermuscular fibers in squid mantle. In addition to the collagen tunics, there are several sets of fibers that run within the muscle layers. (**A**) In longitudinal sections, the fibers in system IM-1 are seen to run from the outer tunic to the inner tunic at an angle of about 30° relative to the long axis. (**B**) In sections cut tangentially to the surface of the mantle, the fibers in IM-1 run at an angle of 10–15° relative to the long axis. (**C**) Intermuscular fiber system IM-2 can be seen in transverse sections that run through a layer of radial muscle. These fibers run from the outer tunic to the inner tunic at an angle of about 55° relative to the surface of the mantle. ot, Outer tunic; it, inner tunic; cm, circular muscle; rm, radial muscle.

on the basis of their mechanical properties (Section III), (*b*) because they show the thermal shrinkage expected for collagen (Ward and Wainwright, 1972), and (*c*) because the tunics contain hydroxyproline (Otwell and Giddings, 1980). The intermuscular fibers cannot be identified unequivocally as collagen at this time. In fact, Bone et al. (1981) report that some of the intermuscular fibers stain with histological stains that are supposed to be specific for rubber-like "elastic" fibers. However, we have observed that the fibers in IM-1 are highly birefringent, and exhibit a birefringence, $B = 1.3 \times 10^2$. This is just the birefringence we observe for rat tail tendon collagen fibers. "Elastic" fibers, on the other hand, are typically isotropic, and this difference in optical properties reflects the basic differences in molecular organization of tensile materials like collagen and rubbery materials like "elastic" fibers. The parallel arrangement of the molecular elements in collagen makes collagen fibers birefringent. The random molecular conformation characteristic of all rubbery materials makes them optically isotropic (Aaron and Gosline, 1980). We have not yet been able to measure the birefringence of the fibers in systems IM-2 and IM-3, probably because these fibers are seen only in sections that run parallel to a set of muscle fibers, and the very strong birefringence of the myofibrils masks that of the connective tissue fibers. Alternately, these fibers may not be birefringent and may indeed be "elastic." However, Otwell and Giddings (1980) report that the hydroxyproline content of the mantle muscle layer of *Loligo pealei* is about 0.08% of the wet weight. Cephalopods do not contain elastin (Sage and Gray, 1977). Cephalopods do, however, contain "elastic" fibers (Chapter 9, this volume), but these "elastic" fibers do not contain hydroxyproline. Squid mantle collagen contains 90 residues per 1000 Hyp (Hunt et al., 1970). Thus the presence of collagen in the region of the mantle where the intermuscular fibers are seen is certain, and on the basis of the hydroxyproline content we estimate that about 1% of the mantle tissue volume is occupied by collagen. Note that this figure does not include the collagen present in the tunics. Judging from the histological sections, a collagen content of 1% seems reasonable, but such visual estimates are not nearly accurate enough to preclude the possibility that fiber sets IM-2 and IM-3 are noncollagenous.

Ward and Wainwright (1972) suggest that the function of the two tunic layers is to prevent the mantle from increasing in length during the jet power stroke (i.e., contraction of the circular muscles) and thus to facilitate the refilling of the mantle cavity by the contraction of the radial muscles. The mantle tissue is envisioned as a hydrostatic system, and all the shape change brought about by the shortening of the circular muscles is focused into extending the radial muscles (i.e., increasing the thickness of the mantle wall) because the tunics will not allow the mantle to get longer. The mantle cavity refills when the radial muscles contract, thinning the mantle wall and thus increasing the circumference of the structure. The 27° fiber angle of the tunics was seen as a compromise between having stiff fibers parallel to the long axis that could rigidly resist longitudinal extensions, and having a system that could contract circumferentially without buckling when the circular muscles contract. The function of the intermuscular fibers remains unclear at present, but both Ward and Wainwright (1972) and Bone et al. (1981) suggest that these various fiber systems might contribute to the refilling of the mantle cavity following the jet power stroke by storing elastic energy that could be paid back as an elastic recoil during the refilling phase of the jet cycle. It is this possibility, that the intermuscular fibers provide an elastic mechanism for mantle refilling, that we will consider quantitatively in the following pages.

## B. Elastic Properties of Squid Mantle

If the intermuscular fibers are capable of storing elastic energy, then we expect that the tissue as a whole will exhibit elastic properties that correspond to the elastic distortion of these fibers. Accordingly, we have carried out mechanical tests on whole mantle to determine (*a*) if the structure is stiff enough to store reasonable amounts of elastic energy, and (*b*) if the resilience or efficiency of energy storage is high enough to make energy storage a practical possibility. It is quite easy to make tensile measurements on the mantle tissue, but the mantle is loaded in compression when it is deformed by the contraction of the circular muscles. Therefore, the data presented here are for the constant strain-rate (quasi-static) compression and for the dynamic compression of samples of the mantle of the squid *L. opalescens,* loaded parallel to the circular muscles. The test specimens were cut as shown in the inset to Fig. 8, with the circumferen-

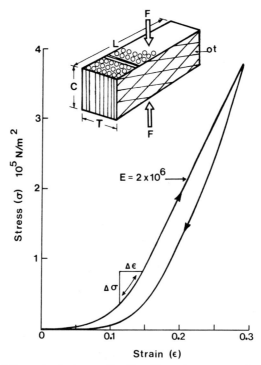

**Fig. 8.** Quasi-static mechanical properties of squid mantle compressed in the direction indicated in the inset diagram. The $\delta\sigma$ and $\delta\epsilon$ symbols indicate the range of dynamic measurements carried out on this sample. The dimensions of the sample were L = 1 cm, T = 0.24 cm, and C = 0.26 cm. ot, Outer tunic; F, the applied compressive force. The elastic modulus (*E*) in the linear region is $2 \times 10^6$ N/m².

tial dimension of the specimen (i.e., in the direction of application of the compressive load) being only slightly greater than the mantle thickness so as to prevent buckling of the sample. In the constant strain-rate studies, the tissue was compressed at a strain rate of the order of 0.01/sec (1%/sec). This strain rate is much slower than the peak rates that must occur in the normal jet cycle (~100%/sec), and thus we refer to this as a quasi-static test. In addition we carried out dynamic tests by applying a sinusoidal compression at frequencies in the range of 0.05–30 Hz. The dynamic loading of the mantle tissue *in vivo* probably corresponds to a loading frequency in the range of 1–3 Hz.

Figure 8 shows the results of the quasi-static tests. Once the tissue gets beyond an initial low modulus, the stress increases linearly with strain, yielding a modulus of elasticity of $2 \times 10^6$ N m$^2$. We do not know if the "slack" region that extends up to strains of about 0.10 in this test reflects real properties of the tissue, a low-modulus region of the constituent fibers, or loose ends created in the isolation of the specimen from the intact structure. The linear region, however, undoubtedly represents the true tensile properties of the fibrous network. By comparison of the area under the load and the unload curves (see the arrows), it is possible to determine the resilience of the structure, which turns out to be about 75%. This means that 75% of the energy put into deforming the sample is recoverable to power the refilling phase of the jet cycle. The dynamic experiments show much the same thing. The sample was preloaded and then sinusoidally compressed with a peak-to-peak strain amplitude of approximately 0.04, as indicated in Fig. 8. The resilience was computed according to Ferry (1970) by calculating the energy storage and the energy dissipated per cycle from the storage ($E'$) and loss ($Ee$) moduli. At frequencies of about 0.5–2 Hz the resilience is approximately 70%. As the frequency is increased the resilience drops gradually, becoming 62% at 5 Hz, 59% at 12 Hz, and 53% at 30 Hz. Thus, under dynamic loading conditions that approximate the normal frequency range *in vivo*, the resilience is about 70%. The combination of reasonably high stiffness and resilience makes the storage of elastic energy in the mantle tissue at least a feasible proposition. For comparison, the stiffness and resilience of the mammalian artery wall, a structure that is known to store elastic energy and thus facilitate the flow of blood, are almost exactly the same (Bergel, 1961).

## C. Calculated Strain-Energy Storage in the Intermuscular Fiber Systems

We will now attempt to determine if these elastic properties can be explained in terms of intermuscular collagen fibers, and to do this we must first calculate the level of strain that these fibers experience during a normal jet cycle. Ward (1972) measured the dimension changes of the mantle of *Lolliguncula brevis* during the jet cycle, and observed that the mantle thickness increased by about 18% and that the midwall circumference decreased by a similar amount. These are average values based on the analysis of 30 jet cycles, and thickness changes as large as 30% were observed. Taking these values as an indication of the dimension changes, we have attempted to calculate the strain levels that would occur in intermuscular fiber systems IM-1 and IM-2 during the jet power stroke (i.e., contraction of the circular muscles). We have not made any attempt to analyze the strains in system IM-3 because these fibers lie parallel to the circular muscles and can only be further buckled by the contraction of the circular muscles. Figure 9A summarizes the calculations for system IM-1. The diagram represents a longitudinal section through the mantle tissue that has increased in thickness by 18%. Assuming that the mantle does not get longer, the collagen fiber aligned at an initial fiber angle of 30° will be strained by 0.047 or extended by 4.7%, and the fiber angle will be increased to about 34°. Note that in this calculation we have ignored the fact that the fibers in IM-1 spiral through the muscle tissue at a steep angle (Fig. 7), and that this spiral orientation will reduce the net strain on the collagen fiber. However, as long as the spiral angle (i.e., the angle seen in tangential sections, Fig. 7B) is small, this approximation is reasonable. For a 30% increase in thickness the fibers in IM-1 will be extended very close to the failure strain for a collagen fiber (see Fig. 5), and we suspect that the reason that these fibers lie at an angle of about 30° is to match the thickness changes of the mantle precisely to the extensibility of the material from which the fibers are constructed (i.e., collagen). Similar calculations for system IM-2 are shown in Fig. 9B. In this case the fibers run at an initial angle of 50° relative to the surface of the mantle, and the thickness of the mantle increases by 18% whereas the width decreases by 18%. In this situation, the fiber strain will also be 0.047 or a 4.7% extension, and the fiber angle will increase to about 60°. Recall that Bone et al. (1981) observed a fiber angle of 55°, but it is not known if this represents a relaxed or contracted value. Therefore we have chosen values that roughly span their value. Again, it is clear that at this fiber angle, the fiber, if it is collagen, is loaded to about 50% of its breaking extension, and as with system IM-1, if the mantle thickness change approaches 30% and the lateral contraction increases proportionally, then the fibers will be extended right to the point of failure. Again, the orientation of the fibers seems to be perfectly matched to the properties of collagen.

From the foregoing calculations we can estimate that an 18% increase in thickness creates about a 4.7% extension in the fibers in both IM-1 and

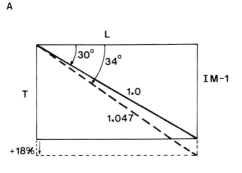

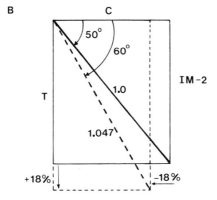

**Fig. 9.** Calculated strains in intermuscular fiber systems. (**A**) System IM-1; (**B**) system IM-2. The mantle length (L) is assumed to remain constant. Thickness (T) increases by 18%, whereas the circumference (C) decreases by a similar amount. In each diagram the solid diagonal line represents an unstrained, intermuscular collagen fiber (length = l) in the relaxed mantle. The dashed line represents the length and orientation of this fiber after the mantle has contracted. The new fiber length (i.e., 1.047) indicates that the intermuscular collagen fiber is extended to a strain of 0.047 in this example.

IM-2. Thus if compression, as created either by the contraction of the circular muscles or by the application of an external load in our test apparatus, causes an equal increase in thickness (i.e., if the tunics do actually prevent the mantle from getting longer), then we expect that the ratio of the compressive strain on the whole specimen to the tensile strain in the intermuscular fibers will be about 4:1. Ward (1972) observed, and we can confirm, that the length change of the mantle during a jet cycle is less than about ±1%. Therefore we should be able to predict the compressive stiffness of the mantle on the basis of the calculated intermuscular collagen content of 1%. Because the stiffness of squid collagen is about $7 \times 10^8$

N/m² (Section III), and the collagen makes up only 1% of the total tissue volume, we would, as a first estimate, predict a stiffness of about $7 \times 10^6$ N/m². However, the collagen fibers are only strained to one-quarter of the strain of the whole tissue, so the modulus should be reduced by a factor of four to $1.75 \times 10^6$ N/m². The observed stiffness is $2 \times 10^6$ N/m². Considering the numerous errors possible in this analysis, it is remarkable that these two values are so close, and thus we conclude that the elastic properties of the squid mantle arise virtually entirely from the distortion of the intermuscular collagen fibers.

We still cannot exclude the possibility that system IM-2 is made from rubber-like "elastic" fibers. However, the precise matching of the orientation of these fibers with the properties of collagen make this possibility unlikely. Further, if the fibers in IM-2 are "elastic" fibers, then they can only contribute a negligible amount to the stiffness of, and hence the elastic energy storage by, the mantle structure. From Section III, we observed that collagen fibers can store approximately 2000 J/kg at extensions of the order of 8%. Rubber-like "elastic" materials can also store similar amounts of elastic energy, but they must be strained to extensions of the order of 150–200%. At the small extensions seen for the intermuscular fibers (maximally 8%), rubber-like "elastic" proteins with properties similar to elastin or resilin (Gosline, 1980) will only be able to store about 6 J/kg. Thus it is unlikely that "elastic" fibers play a major role in squid mantle. Indeed, for "elastic" fibers to replace the energy storage system provided by a collagen network occupying 1% of the volume, the mantle would have to be made of solid rubber, leaving no room for the muscles that are obviously needed to power the system.

We have clearly demonstrated that the squid mantle is an elastic structure and that the intermuscular collagen fibers provide an efficient elastic energy storage system that can antagonize the circular muscles and account, in part at least, for the refilling of the mantle cavity. However, we do not know what portion of the total energy required for refilling is provided by this system. Such questions are really beyond the scope of this chapter, and will require more detailed information on the energetics of the jet cycle.

### REFERENCES

Aaron, B. B., and Gosline, J. M. (1980). Optical properties of single elastin fibres indicate random protein conformation. *Nature* (*London*) **287**, 865–867.

Adams, E. (1978). Invertebrate collagens. *Science* **202**, 591–598.

Alexander, R. Mc. N., and Bennet-Clark, H. C. (1977). Storage of elastic strain energy in muscle and other tissues. *Nature* (*London*) **265**, 114–117.

Bailey, A. J. (1968). Intermediate labile intermolecular crosslinks in collagen fibres. *Biochim. Biophys. Acta* **160**, 447–453.

Bergel, D. H. (1961). Dynamic elastic properties of the arterial wall. *J. Physiol.* (*London*) **156**, 458–469.

Bone, Q., Pulsford, A., and Chubb, A. C. (1981). Squid mantle muscle. *J. Mar. Biol. Assoc. U.K.* **61**, 327–342.

Bornstein, P., and Sage, H. (1980). Structurally distinct collagen types. *Annu. Rev. Biochem.* **49**, 957–1003.

Bornstein, P., and Traub, W. (1979). Chemistry and biology of collagen. *In* "The Proteins" (H. Neurath, R. L. Hill, and C. Boeder, eds.), 3rd ed., Vol. 4, pp. 412–632. Academic Press, New York.

Diamant, J., Keller, A., Baer, E., Litt, M., and Arridge, R. (1972). Collagen: ultrastructure and its relation to mechanical properties as a function of ageing. *Proc. R. Soc. London, Ser. B* **180**, 293–315.

Ferry, J. D. (1970). "Viscoelastic Properties of Polymers," 2nd ed. Wiley, New York.

Fietzek, P. P., and Kuhn, K. (1976). Primary structure of collagen. *Int. Rev. Connect. Tissue Res.* **7**, 1–60.

Fraser, R. D. B., and McRae, T. P. (1973). "Conformation in Fibrous Proteins." Academic Press, New York.

Gosline, J. M. (1980). Elastic properties of rubber-like proteins and highly extensible tissues. *In* "Mechanical Properties of Biological Materials" (J. F. V. Vincent and J. D. Currey, eds.), pp. 331–357. Cambridge Univ. Press, London and New York.

Harkness, R. D. (1968). Mechanical properties of collagenous tissues. *In* "Treatise on Collagen" (B. S. Gould, ed.), Vol. 2A, pp. 249–309. Academic Press, New York.

Hodge, A. J., and Petruska, J. A. (1963). Recent studies with the electron microscope on ordered aggregates of the tropocollagen molecule. *In* "Aspects of Protein Structure" (G. N. Ramachandran, ed.), pp. 289–300. Academic Press, New York.

Hunt, S., Grant, M. E., and Liebovich, S. J. (1970). Polymeric collagen isolated from squid (*Loligo paelii*) connective tissue. *Experientia* **26**, 1204–1205.

Isemura, M., Ikenaka, I., and Matsushima, Y. (1973). Comparative study of carbohydrate protein complexes. I. The structure of glycopeptides from cuttlefish skin collagen. *J. Biochem.* (*Tokyo*) **74**, 11–27.

Josse, J., and Harrington, W. F. (1964). Role of pyrrolidine residues in the structure and stabilization of collagen. *J. Mol. Biol.* **9**, 269–287.

Kastelic, J., and Baer, E. (1980). Deformation in tendon collagen. *In* "Mechanical Properties of Biological Materials" (J. F. V. Vincent and J. D. Currey, eds.), pp. 397–435. Cambridge Univ. Press, London and New York.

Kimura, S. (1972a). Determination of glycosylated hydroxylysines in several invertebrate collagens. *J. Biochem.* (*Tokyo*) **71**, 367–370.

Kimura, S. (1972b). Studies on marine invertebrate collagens. V. Neutral sugar compositions and glycosylated hydroxylysines contents of several collagens. *Nippon Suisan Gakkaishi* **38**, 1153–1161.

Kimura, S., and Matsuura, F. (1974). The chain compositions of several invertebrate collagens. *J. Biochem.* (*Tokyo*) **75**, 1231–1240.

Mathews, M. B. (1980). Coevolution of collagen. *In* "Biology of Collagen" (A. Viidik and J. Vuust, eds.), pp. 193–210. Academic Press, New York.

Meek, G. A. (1966). Electron microscopy of some invertebrate collagens. *Proc. R. Microsc. Soc.* **1**, 100.

Miller, E. J., and Lunde, L. G. (1973). Isolation and characterization of the cyanogen bromide peptides from the $\alpha1(II)$ chains of bovine and human cartilage collagen. *Biochemistry* **12**, 3153–3161.

Nordwig, A., and Hayduk, U. (1969). Invertebrate collagens: isolation, characterization, and phylogenitic aspects. *J. Mol. Biol.* **44,** 161–172.

Otwell, W. S., and Giddings, G. G. (1980). Scanning electron microscopy of squid *Loligo paelei. Mar. Fish. Rev.* **42,** 67–73.

Parry, D. A. D., and Craig, A. S. (1979). Electron microscope evidence for an 80 A unit in collagen fibrils. *Nature (London)* **282,** 213–215.

Pikkarainen, J., Rantanen, J., Vastmaki, K., Lampiaho, A., Kari, A., and Kulonen, E. (1968). On collagens of invertebrates with special reference to *Mytilus edulis. Eur. J. Biochem.* **4,** 555–563.

Ramachandran, C. V., and Reddi, A. H. (1976). "Biochemistry of Collagen." Plenum, New York.

Rauterberg, J., and Kuhn, K. (1971). Acid soluble calf skin collagen, characterization, of the peptides obtained by cyanogen bromide cleavage of its α1-chain. *Eur. J. Biochem.* **19,** 398–415.

Rigby, B. J. (1968). Thermal transitions in some invertebrate collagens and their relation to amino acid content and environmental temperature. *In* "Symposium on Fibrous Proteins" (W. G. Creqther, ed.), pp. 217–225. Butterworth, London.

Rigby, B. J., and Mason, P. (1967). Thermal transitions in gastropod collagen and their correlation with environmental temperature. *Aust. J. Biol. Sci.* **20,** 265–271.

Sage, E. H., and Gray, W. R. (1977). Evolution of elastin structure. *In* "Elastin and Elastic Tissue" (L. B. Sandberg, W. R. Gray, and C. Franzblau, eds.), pp. 291–312. Plenum, New York.

Schmitt, F. O., Gorss, J., and Reich, M. E. (1953). A new particle type in certain connective tissue extracts. *Proc. Natl. Acad. Sci. U.S.A.* **39,** 459–464.

Schmut, O., Roll, P., and Reich, M. E. (1980). Biochemical and electron microscopical investigation of *Helix pomatia* collagen. *Z. Naturforsch.* **35C,** 376–379.

Stemberger, A., and Nordwig, A. (1974). Invertebrate collagens: qualitative and quantitative studies on their carbohydrate moieties. *Hoppe-Seyler's Z. Physiol. Chem.* **355,** 721–724.

Traub, W., and Piez, K. (1971). Chemistry and structure of collagen. *Adv. Protein Chem.* **25,** 243–352.

Traub, W., and Salem, G. (1972). Comparative structural investigation of collagens from various species. *Isr. J. Chem.* **10,** 111–123.

Travis, D. F., Francois, C. J., Bonnar, L. C., and Glimcher, M. J. (1967). Comparative study of the organic matrices of invertebrate mineralized tissues. *J. Ultrastruct. Res.* **18,** 519–550.

Viidik, A. (1973). Functional properties of collagenous tissues. *Int. Rev. Connect. Tissue Res.* **6,** 127–215.

Viidik, A., and Vuust, J., eds. (1980). "Biology of Collagen." Academic Press, New York.

Wainwright, S. A., Biggs, W. D., Currey, J. D., and Gosline, J. M. (1976). "Mechanical Design in Organisms." Arnold, London.

Ward, D. V. (1972). Locomotory function of the squid mantle. *J. Zool.* **167,** 487–499.

Ward, D. V., and Wainwright, S. A. (1972). Locomotory aspects of squid mantle structure. *J. Zool.* **167,** 437–449.

Williams, A. P. (1960). Chemical composition of snail gelatin. *Biochem. J.* **74,** 304–306.

# 9

# Molecular Biomechanics of Protein Rubbers in Molluscs

**ROBERT E. SHADWICK**
**JOHN M. GOSLINE**

Department of Zoology
University of British Columbia
Vancouver, British Columbia V6T 2A9,
Canada

## I. Introduction

Molluscs, like most animals, have a body form in which many structural components are subject to periodic stresses from the external environment or from within the organism. Although some molluscs are encased in rigid exoskeletons, others have internal skeletons, and still others rely entirely on hydraulic mechanisms for support. Regardless of the type of skeleton, there is a requirement for pliant connective tissues that can be bent, stretched, or sheared readily during the course of locomotion or other body functions (Wainwright et al., 1976). Many pliant ma-

THE MOLLUSCA, VOL. 1
Metabolic Biochemistry
and Molecular Biomechanics

terials have the property of long-range elasticity. This means that these materials (*a*) have a modulus (stiffness) that is similar to that of lightly vulcanized rubbers, (*b*) can undergo large extensions when loaded by an external force, and (*c*) recover their initial dimensions rapidly when the force is removed. The work done in deforming such a material can be stored as potential energy by the elastic mechanism, and most of this energy can be recovered as useful work when the material is unloaded.

In the vertebrate body, pliant tissues such as skin, lung, artery wall, and elastic ligaments are actually composite materials that have mechanical properties that depend on the presence of two fibrous proteins: elastin and collagen. The rubber-like protein elastin is responsible for the high extensibility and long-range elasticity under normal physiological stress, whereas the stiff, relatively inextensible collagen fibers are arranged in such a manner that they are recruited to bear load and provide the limiting properties at high extensions (Roach and Burton, 1957; Wainwright et al., 1976; Gosline, 1980). Collagen is an important structural protein in all metazoan animals (for a review of the physical and chemical properties of collagen see Chapter 8, this volume). On the other hand, elastin is found only in the vertebrates (Sage and Gray, 1977, 1980). However, it seems likely that other protein rubbers, similar to elastin, will be present to provide the long-range elasticity of pliant tissues in molluscs and other invertebrates.

Presumptive elastic fibers have been demonstrated histologically in the dermal and body wall connective tissues, among muscles, in suspensory ligaments, around nerve cords, and in the digestive and circulatory systems of many invertebrates (Krumbach, 1935; Nutting, 1951; Mabillot, 1954, 1955; Jullien et al., 1957, 1958; Barnes and Gonor, 1958; Elder and Owen, 1967; Elder, 1966a,b, 1972, 1973; Bone et al., 1981). Although this histological evidence supports the suggestion that elastin analogs have broad distribution and diversity in the invertebrates, to date only three of these protein rubbers, two of which are molluscan, have been isolated and studied in detail: (1) resilin, a component of insect cuticle (Weis-Fogh, 1960); (2) abductin, which occurs in the inner hinge ligament of bivalve molluscs (Kelly and Rice, 1967); and (3) a new elastic protein that has been isolated from octopus arteries (Shadwick and Gosline, 1981) and is called octopus arterial elastomer (OAE).

In this chapter we will first briefly describe the morphology of bivalve hinge ligaments and molluscan blood vessels. We will then examine the current knowledge of the chemistry and network properties of the protein rubbers abductin and OAE, and then follow with a discussion of the functional role of these elastic materials in molluscs.

## II. Distribution and Morphology

The morphology of the hinge ligament in bivalves has been a subject of considerable interest for many years, particularly from a phylogenetic point of view (Bowerbank, 1844; Jackson, 1891; Yonge, 1936, 1973, 1978; Trueman, 1950, 1951, 1953a; Owen et al., 1953; Owen, 1958), although the elastic properties of the hinge were investigated as early as 1884 by Plateau (1884). In general, the ligament is a proteinaceous region of the shell that forms a hinged articulation between the calcareous valves. Like the shell, the hinge ligament is secreted in three distinct layers by the mantle tissue (Beedham, 1958; Owen et al., 1953). In fact, the ligament represents a modified dorsal region of what was originally a single laterally compressed shell (Owen et al., 1953; Yonge, 1978). There is a superficial noncalcareous periostracum, and an outer and inner ligament proper. The outer ligament layer is made of uncalcified quinone-tanned protein, laid down in parallel lamellae that probably represent growth layers. The inner layer is also typically a lamellar structure, composed of a fibrous protein matrix with variable degrees of calcification. In most bivalves calcium carbonate occupies 60–80% of the dry weight, whereas in the Pectinidae the inner ligament is essentially uncalcified (Trueman, 1950, 1951; Beedham, 1958; Galtsoff, 1964; Taylor et al., 1969; Kahler et al., 1976; Yonge, 1978).

The ligament functions as a hinge and keeps the valves in alignment as they articulate. In most bivalves the outer ligament layer lies above the pivotal axis and is stretched when the valves close. At the same time, the inner ligament layer, which is below the pivotal axis, is compressed. The whole ligament is an elastic structure and thus acts as a spring to open the valves when the adductor muscles relax. With adaptive radiation in the bivalves, the hinge ligament has undergone extensive modification. Perhaps the most interesting, from a biomechanical point of view, is the highly specialized ligament that has developed in association with swimming by jet locomotion in scallops (Pectinidae).

In *Pecten* and related genera the roles of the hinge and spring are performed separately by the outer and inner ligament layers, respectively. The outer layer is reduced to a very thin structure that forms a long flexible hinge as it unites the edentulous dorsal margins of the valves (Fig. 1). This layer prevents the valves from rotating relative to each other when the single eccentric adductor muscle contracts (Trueman, 1953a,b). The inner layer is a dark brown pyramid-shaped block of rubbery material, situated centrally, below the hinge line (Fig. 1). This part of the ligament is calcified only at the lateral attachments to the shell, with the central por-

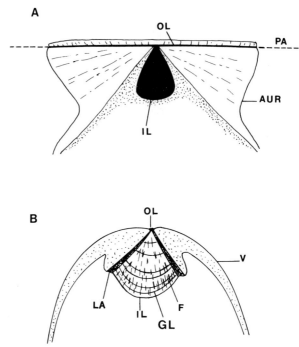

**Fig. 1.** (A) Diagrammatic view of the inside surface of one valve of a scallop (Pectinidae). The outer layer of the ligament (OL) extends across the anterior and posterior auricles (AUR) to form a long hinge line. The inner layer of the ligament (IL) is a pyramid-shaped block of rubber-like protein that lies below the pivotal axis (PA) and is compressed when the valves are closed. (B) Diagrammatic transverse section of a scallop to show the attachment of the hinge ligament at the valves (V). The inner layer is calcified only at the lateral attachment regions (LA), whereas the central portion is composed of protein fibers (F). Growth lines (GL) are seen across the cut surface of the ligament. (Redrawn from Trueman, 1953b.)

tion being composed almost entirely of a fibrous protein-rubber called abductin (Trueman, 1953a,b; Alexander, 1966; Kelly and Rice, 1967). In the scallop the inner ligament provides almost all the elastic recoil necessary to open the valves during the refilling phase of the jet cycle (Trueman, 1953b). Abductin is very efficient at storing elastic energy and causing the valves to "spring" open rapidly, thus allowing the scallop to swim effectively by jetting at 3–4 Hz (Trueman, 1953b; Alexander, 1966; Moore and Trueman, 1971; Morton, 1980).

Early investigators noted the presence of extracellular connective tissue fibers in molluscan blood vessels that had morphological and histological features in common with vertebrate elastic fibers (Dahlgren and

Kepner, 1908; Wetekamp, 1915). These observations have been repeated and extended to include the arteries of several species of gastropods, bivalves, and cephalopods (Jullien et al., 1957, 1958; Elder, 1973; Shadwick and Gosline, 1981; Gosline and Shadwick, 1983). In general, there is an internal "elastic" layer and, in most species, a series of concentric "elastic" lamellae present throughout the muscle layers of the artery wall. These structures are very similar to the internal elastica and medial elastin lamellae, respectively, found in vertebrate arteries. The long-range passive elasticity of the major arteries is a fundamentally important feature of the vertebrate circulation. The tissue elasticity provides an elastic "reservoir" that is inflated with blood when the heart pumps, and that drains by elastic recoil to maintain peripheral blood flow while the heart refills (Taylor, 1964). Based on blood pressure data from several species, it has been suggested that an elastic component is also present in the cardiovascular system of molluscs, and that this is particularly well developed in the closed circulation of cephalopods (Johansen and Martin, 1962; Bourne and Redmond, 1976a,b; Bourne et al., 1978; Wells, 1979; Shadwick, 1980). Recently it has been shown by mechanical tests that the aorta in cephalopods is indeed a distensible elastic tube (Shadwick and Gosline, 1981; Gosline and Shadwick, 1983). It is presumed that the elasticity of mollusc arteries, as in vertebrate arteries, is provided by the histologically identified "elastic" fibers. These fibers have been isolated from the aorta of the large cephalopod, *Octopus dofleini,* and subsequently shown to be composed of a rubber-like protein with elastic properties very much like those of elastin and the bivalve hinge protein abductin.

Figure 2 shows a histological section through the internal elastica of the aorta of *O. dofleini.* The internal elastica forms a 5- to 7-$\mu$m thick layer that lines the vessel lumen. Elastic fibers appear to lie parallel to the long axis in the unpressurized aorta. However, when the vessel is inflated to physiological pressures the internal elastica opens like a trelliswork, revealing many lateral connections. This arrangement allows the aorta to expand in length as well as in circumference under pressure from the heart, with suitable elastic reinforcement in both of these directions (Shadwick, 1981). The elastic fibers in the muscle layers of the aorta are not arranged in the discrete, concentric lamellae as is the case in vertebrate arteries and in other species of molluscs so far studied. Instead, these intermuscular fibers in the octopus aorta form a very fine but mechanically continuous network throughout the artery wall. Variations in the relative abundance and macroscopic organization of the elastic fibers in different species of cephalopods and the mechanical consequences of these variations are discussed by Gosline and Shadwick (1983).

**Fig. 2.**  Light micrograph of a section cut tangentially to the lumenal surface of an unpressurized aorta from *Octopus dofleini*. The internal elastica (IE) is stained by the aldehyde fuchsin technique for elastic fibers. The IE is a lattice-work of rubber-like fibers, predominantly oriented parallel to the long axis of the artery (LA); CM, circular muscle. Bar = 50 $\mu$m.

## III. Chemical Structure

### A. Amino Acid Composition and Peptide Chemistry

Various aspects of the chemistry and structure of abductin and octopus arterial elastomer (OAE) have been investigated, and the results support the idea that these proteins have the basic structural properties necessary to conform to the kinetic theory of rubber elasticity. Basically, this theory requires (*a*) that the material be composed of flexible long-chain random-coil molecules, (*b*) that the molecular chains be joined together by stable cross-links into an isotropic three-dimensional network, and (*c*) that the entire network be kinetically free and in a state of maximal entropy when undeformed. Contrary to our general concept of proteins as having fixed, stable conformations that arise uniquely from each amino acid sequence, a protein rubber must have *no* stable conformation in order to maintain its network mobility and rubber-like elastic properties. How this unusual class of proteins achieves a kinetically free, disordered state under conditions where other proteins have stable secondary and tertiary structures remains an unanswered question at this time.

When viewed under polarized light both abductin and OAE are non-birefringent in the unstressed state (Alexander, 1966; Shadwick and Gosline, 1981). This property is characteristic of rubbers and is consistent with the random-network model (Weis-Fogh, 1960; Treloar, 1975; Aaron and Gosline, 1980). In contrast, the large intrinsic birefringence of collagen fibers is indicative of a highly ordered molecular structure. When protein rubbers are deformed, they become birefringent, presumably because deformation imposes a preferred orientation on the network chains, and this gives rise to optical anisotropy (Weis-Fogh, 1960, 1961; Aaron and Gosline, 1981). We have also observed this phenomenon in stretched abductin and in OAE fibers. Further evidence in support of a random molecular conformation in the molluscan protein rubbers comes from the observation that both appear as homogeneously amorphous materials in the electron microscope, and in addition, no distinct X-ray diffraction patterns are seen in abductin in either the stretched or unstretched state (Kelly and Rice, 1967; Shadwick, 1981).

As yet, we can only begin to make some broad generalizations about the molecular design of the protein rubbers. Amino acid compositions have been reported for hinge proteins from a variety of bivalves (Kelly and Rice, 1967; Thornhill, 1971; Kahler et al., 1976; Sage and Gray, 1977). These data are summarized in Tables I and II, along with the compositions of octopus arterial elastomer, resilin, and elastin. Here, the term abductin refers only to the noncalcified inner hinge ligament of the Pectinacea. Abductin is characterized by extremely high levels of glycine (greater than 60%) and by an unusually large methionine content. Other bivalve hinge proteins shown in Table I have amino acid compositions and elastic properties that are fairly similar to those of abductin. Therefore it seems likely that these proteins are evolutionarily related. However, the inner ligament of the oyster, *Crassostrea virginica,* has an amino acid composition that is quite different from abductin (Kahler et al., 1976); the glycine content is much lower and proline and aspartic acid are present at higher levels. At present, we can not make any definite conclusions about the relationships between the compositions of various hinge proteins and their mechanical properties.

It is evident from the wide variation in composition between the protein rubbers—abductin, OAE, elastin, and resilin—that there must be many amino acid combinations and therefore numerous sequence patterns that will give rise to a kinetically free molecular structure under normal physiological conditions. Abductin, resilin, and elastin all have substantial quantities of small amino acids (see Table II), such that (Gly + Ala + Ser)

TABLE I

The Amino Acid Composition of Various Protein Rubbers

| Amino acid | Molluscan hinge proteins | | | | OAE[c] | Elastin[d] | Resilin[e] |
|---|---|---|---|---|---|---|---|
| | Aquipecten irridians[a] | Spondylus varians[b] | Spisula solidissima[b] | Mytilus edulis[b] | | | |
| Asp | 23 | 6 | 23 | 39 | 91 | 6 | 94 |
| Thr | 10 | 2 | 8 | 14 | 64 | 15 | 20 |
| Ser | 61 | 16 | 22 | 89 | 72 | 12 | 128 |
| Glu | 14 | 3 | 20 | 37 | 121 | 19 | 42 |
| Pro | 10 | 72 | 28 | 47 | 55 | 113 | 75 |
| Gly | 627 | 544 | 629 | 411 | 85 | 313 | 422 |
| Ala | 50 | 75 | 35 | 42 | 72 | 244 | 70 |
| ½Cys | 1 | 1 | 10 | 13 | 7 | — | — |
| Val | 5 | 116 | 9 | 22 | 62 | 128 | 12 |
| Met | 92 | 2 | 132 | 97 | 21 | — | — |
| Ile | 6 | 2 | 22 | 31 | 60 | 18 | 9 |
| Leu | 3 | 2 | 19 | 72 | 74 | 54 | 30 |
| Tyr | 1 | 20 | 2 | 4 | 36 | 19 | 12 |
| Phe | 83 | 108 | 20 | 22 | 42 | 33 | 12 |
| Lys | 8 | 5 | 10 | 34 | 68 | 5 | 8 |
| His | 2 | 1 | 2 | 4 | 21 | 1 | 12 |
| Arg | 4 | 30 | 11 | 22 | 46 | 8 | 46 |
| Hyp | — | — | — | — | — | 9 | — |

[a] Composition (residues per 1000) averaged from the data of Kelly and Rice (1967); Thornhill (1971); Kahler et al. (1976); Sage and Gray (1977).
[b] Kahler et al. (1976).
[c] Shadwick and Gosline (1981).
[d] Sage and Gray (1977).
[e] Andersen (1971).

TABLE II

Summary of Amino Acid Composition Data for Abductin from *Aquipecten irradians,* Octopus Arterial Elastomer (OAE), Elastin, and Resilin[a]

| | Abductin | OAE | Elastin | Resilin |
|---|---|---|---|---|
| Small amino acids (Gly + Ala + Ser) | 738 | 229 | 569 | 620 |
| Polar residues | 123 | 519 | 94 | 362 |
| Nonpolar residues | 250 | 393 | 590 | 208 |
| Average residue weight (daltons) | 79 | 110 | 85 | 89 |

[a] These summary data are based on the amino acid composition data presented in Table I (residues per 1000).

makes up 74, 62, and 57% of these three proteins, respectively. For this reason the average residue weights are quite a bit lower than that observed for most other proteins. It has been suggested that a high proportion of small amino acids may be important in minimizing steric restrictions to the rotational freedom of protein chains, thus permitting a kinetically free network (Gosline, 1980). However, a preponderance of small amino acids alone does not prevent proteins from having stable conformations. For example, glycine makes up about one-third of the amino acid content of collagen, and glycine, alanine, and serine together account for over 85% of the composition of silkworm silk. Both collagen and silk are highly crystalline proteins, with mechanical properties very different from those of a rubber. In addition, the composition of OAE, which is rubber-like, suggests that a large quantity of small amino acids is not an essential requirement for a kinetically free network. Here (Gly + Ala + Ser) comprises only 23% of the total. In fact, the abundance of charged amino acids in OAE may represent an alternative design strategy, in which these charged amino acids are present to reduce the probability that stable secondary structures will form. Consequently the average residue weight for OAE is relatively high (110 daltons), and this high value has a significant effect on the network flexibility, as discussed in Section IV,B.

The foregoing generalizations based on amino acid composition are highly speculative. Information on the sequence patterns in these proteins is necessary before a clear understanding of the molecular design of protein rubbers can be achieved. So far a very limited amount of sequence data are available, and only for elastin (Gray et al., 1973). We cannot yet propose any general sequence properties that give rise to random, kinetically free protein chains, except that very regular repeating sequences, such as those found in collagen and silk, are inappropriate (Gosline, 1980).

It is important to note that rubber-like behavior is exhibited only when the proteins are swollen in water or some other polar solvent. When dehydrated, the protein chains will fold into a rigid, fixed conformation held by intrachain hydrogen bonds, and will become brittle and glasslike (Gosline, 1976). Water presumably disrupts these bonds, imparting kinetic freedom to the network. Thus it is essential that the sequence design of a protein rubber be one that does not allow strong peptide–peptide hydrogen bonds in the presence of water.

Based on the diversity of composition seen among the four protein rubbers, it has been suggested that each arose independently during evolution, and that other invertebrate elastic tissues may well contain protein rubbers of different composition than those already described (Shadwick

and Gosline, 1981). Within the vertebrates there are some striking variations in the amino acid composition of elastin, which Sage and Gray
(1977) attribute to a segment-multiplication mode of evolution. They speculate that this process could give rise to rapid shifts in composition of any
structural protein. Interestingly, additional evidence to support this hypothesis comes from the abductins. Table I shows that there are distinct
differences in the content of some amino acids in abductins from the
closely related *Aquipecten irradians* and *Spondylus varians* (both Pectinacea). Note that there appears to be an almost complete switch from methionine to valine between these two species. Sage and Gray (1977) also
demonstrated that a similar switch from methionine to valine occurs in
abductions within the same genus, and they concluded that this protein
has the potential for rapid evolutionary change in composition and properties.

## B. Cross-Linking

An important requirement of the random-network model for rubber-like
proteins is that the molecular chains be joined by stable cross-links to prevent viscous flow when the material is deformed. Elastin is cross-linked
primarily by desmosine, a special amino acid that is formed by the condensation of four lysine side chains (Thomas et al., 1963). In resilin, the
protein chains are linked through tyrosine residues that are oxidized to
form di- and trityrosine (Andersen, 1964). In neither of these protein rubbers are disulfide bridges important as network cross-links. The structure
of the cross-links in the molluscan protein rubbers is as yet unknown.
However, abductin and OAE are insoluble in all protein solvents tested,
including urea, formic acid, guanidine hydrochloride, and mercaptoethanol. In fact, these proteins are only dissolved by agents that hydrolyze
peptide bonds (Kelly and Rice, 1967; Shadwick and Gosline, 1981). The
high degree of insolubility exhibited by these proteins (also typical of elastin and resilin) indicates that the peptide chains must be held together by
stable covalent cross-links, other than disulfide bridges. Elucidation of
the molecular structure of the cross-links in an insoluble protein is a difficult task because the cross-links may or may not be stable to hydrolysis,
and in addition, it may be virtually impossible to separate and identify native structures from degradation artifacts in the hydrolysate. Kelly and
Rice (1967) found no unusual components in hydrolysates of abduction
that could be regarded as potential cross-link structures. They suggested
that a cross-link component may have remained tightly bound to the ion-
exchange column used in their fractionation procedure. This is known to
occur with aromatic compounds like di- and trityrosine. Andersen (1963)

used special techniques to separate these cross-links from hydrolysates of resilin, and he found that they could be located and identified by their strong fluorescence and UV absorbance. Andersen (1967) also found phenolic compounds in hydrolysates of abductin. One of these compounds was identified as 3,3'-methylenebistyrosine (Tyr-CH$_2$-Tyr) and proposed as a possible cross-link structure. However, the presence of this compound in native abductin has not been established (Thornhill, 1971), and it now seems likely that this compound is a degradation product of acid hydrolysis (Juřicová and Deyl, 1973; Price and Hunt, 1974).

Thornhill (1971) found that dilute solutions of pronase-digested abductin had intense fluorescence, characterized by excitation and emission at 380 and 490 nm, respectively, in acid, and 360 and 450 nm, respectively, in alkali. A brown pigment, released by acid hydrolysis, is apparently responsible for the fluorescence of the protein. Further observations on the UV absorption and fluorescence spectrum led to the conclusion that the chromophore–fluorophore is probably an aromatic compound, but not a phenolic or quininoid one. In addition, the chromophore–fluorophore is located in the enzyme-resistant region of abductin, and this makes it a possible candidate for a cross-linking structure (Thornhill, 1971).

It is interesting to note that fluorescent compounds have been found in many other structural proteins of molluscs (Hunt, 1970; Price and Hunt, 1973, 1974). The fibrous protein from the egg case of the whelk *Buccinum undatum* contains an aromatic fluorophore that is apparently absent from the soluble precursor protein, suggesting that the fluorophore is part of the cross-link structure. At present we know nothing about precursors of the molluscan protein rubbers, nor do we know anything about the biosynthetic mechanisms involved. Presumably cross-links are introduced into the network as the fibers are formed extracellularly from a soluble precursor, as is the case with other structural proteins. Cross-links in abductin may well be associated with the fluorescent components, as they are in resilin and elastin (Andersen, 1964; Thornhill, 1971, 1972; Deyl et al., 1980), although we do not know their structure. We have observed that OAE is fluorescent in the native state, but no attempts have been made to isolate a fluorophore or cross-link structure from this protein.

## IV. Mechanochemistry

### A. Thermodynamics of the Elastic Mechanism

According to the kinetic theory of rubber elasticity, when a rubber is deformed the conformational entropy of the kinetically free chains in the molecular network is decreased because the random-coil polymer molecules are forced to adopt new conformations of lower probability than in the undeformed state. The elastic restoring force arises from the tendency of the molecular chains to return spontaneously to the state of higher entropy, that is to the more probable conformations (Treloar, 1975; Gosline, 1976). Thus, in theory, the elasticity of a rubber-like material is entirely due to a decrease in the conformational entropy of the molecular network, with no contribution from changes in internal energy (i.e., energy stored in the distortion of chemical bonds). These predictions can be tested experimentally by using a thermodynamic analysis based on the following relationship (see Flory, 1953):

$$f = (\partial H/\partial L)_{T,P} - T(\partial S/\partial L)_{T,P} \tag{1}$$

where $f$ is the elastic force, $H$ is the enthalpy, $L$ is the sample length, $S$ is the entropy, $T$ is the absolute temperature, and $P$ is the pressure. This equation states that the elastic force for any elastic solid deformed at constant temperature and pressure arises from enthalpy and entropy changes. In addition, it can be shown that $-(\partial S/\partial L)_{T,P} = (\partial f/\partial T)_{L,P}$. Therefore if a sample of an elastic solid is stretched and held at constant length while the elastic force is measured as a function of temperature, then the slope of the force–temperature plot $(\partial f/\partial T)_{L,P}$ will give the entropic component of the force, and the intercept at 0°K will be the enthalpic component of the force. The theory of rubber elasticity is based on the assumption that all of the elastic force arises from changes in conformational entropy, and the theory states specifically that the internal energy $U$, and not the enthalpy $H$, must remain unchanged when a rubber is deformed at constant volume. That is, $(\partial U/\partial L)_{T,V} = 0$. If the experiment just described can be carried out in such a manner that the volume of the test sample does not change, then $(\partial H/\partial L)_{T,P} = (\partial U/\partial L)_{T,V}$, and the experiment will provide a useful way of testing the prediction that $(\partial U/\partial L)_{T,V} = 0$ and that the elastic force is due entirely to changes in conformational entropy.

For unswollen rubbers, where volume changes due to thermal expansion are small, thermoelastic measurements (as described previously) have shown that the kinetic theory is indeed valid and that the elastic mechanism in these materials is clearly different from that of crystalline solids (Treloar, 1975). The situation with protein rubbers is complicated by the requirement that they must be tested while swollen in water, where there may be large temperature-dependent volume changes due to the movement of water into or out of the network (Gosline, 1976). In this case, $(\partial H/\partial L)_{T,P}$ will not be just the internal energy change associated with the distortion of chemical bonds in the network chains, but will also have a component associated with the mixing of water with the protein chain (Gosline, 1976). For protein rubbers it may be necessary to apply corrections to the thermoelastic data in order to account for the swelling effects, and thus obtain reasonable estimates of $(\partial U/\partial L)_{T,V}$.

Figure 3A shows the results of thermoelastic experiments on small bundles of OAE fibers (10–20 $\mu$m in diameter), isolated from the internal elastica of the octopus aorta. As expected for a rubber, the force required to hold the fibers extended at a given length increases linearly with increasing temperature from 0 to 35°C. An analysis of these data according to Eq. (1) is given in Fig. 3B, where the entropy and enthalpy components of the elastic force are plotted as a function of extension at 20°C. The results show that for OAE tested in water, the enthalpy change with extension $(\partial H/\partial L)_{T,P}$ is large and *negative*. Consequently the decrease in entropy that is associated with the stretching of this material is much larger than the force, so that the thermodynamic relationship of Eq. (1) is satisfied. Similar results have been obtained for elastin stretched in water, as shown in Fig. 3C and D. In this case the isometric force–temperature plots are not linear, but exhibit an increasing slope as the temperature is decreased. Extrapolation of these curves for elastin to obtain the force intercept at 0°K—that is, $(\partial H/\partial L)_{T,P}$—yields a large negative value, and a correspondingly large decrease in entropy $(\partial S/\partial L)_{T,P}$. It has been shown that the volume of water-swollen elastin increases dramatically with decreasing temperature, and that these changes are due to the extreme hydrophobicity of this protein (Gosline, 1978b). It is now apparent that the absorption of water onto the hydrophobic regions of the elastin protein is entirely responsible for the large negative enthalpy change $(\partial H/\partial L)_{T,P}$ obtained when elastin is tested in equilibrium with surrounding water (Weis-Fogh and Andersen, 1970; Gosline, 1978a). Furthermore, thermoelastic data for elastin obtained either at constant volume (Hoeve and Flory, 1958; Grut and McCrum, 1974) or by applying appropriate corrections for swelling changes in water (Mistrali et al., 1971; Dorrington and McCrum, 1977) indicate that the internal energy change $(\partial U/\partial L)_{T,V}$ is close to zero, as expected, and that elastin is a true rubber elastomer (Fig. 4C).

It appears that almost all of the negative enthalpy component observed for the OAE fibers (Fig. 3B) can also be attributed to temperature-dependent swelling of the protein network. Like elastin, OAE fibers absorb water with decreasing temperature, and the coefficient of linear expansion is constant at about $-0.001/°C$ between 0 and 35°C. In other words, the unstrained length $L_0$ is temperature dependent. Therefore the stretched length $L$, which is held constant during the thermo-

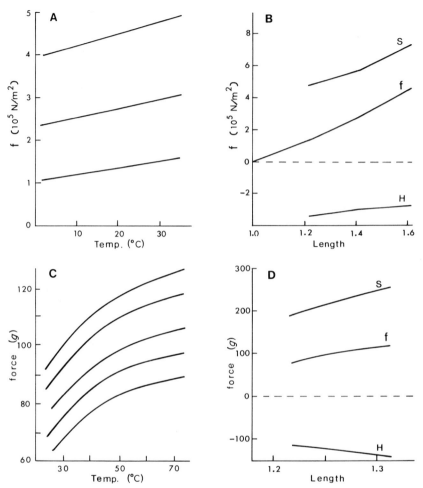

**Fig. 3.** (A) Isometric force–temperature plots for OAE fibers in water, at three different lengths. Values for curves, from top to bottom are $L = 1.66$, $1.44$, and $1.22$, (arbitrary units) respectively. (B) Data from (A) analyzed according to Eq. (1). The force–length curve ($f$) at 20°C is resolved into the corresponding entropy ($S$) and enthalpy ($H$) components. (C) Isometric force–temperature plots for ligamentum nuchae at five different lengths, top to bottom: $L = 1.31$, $1.30$, $1.26$, $1.23$, and $1.22$, respectively. (D) The data from (C) analyzed according to Eq. (1). The entropy ($S$) and enthalpy ($H$) components of the elastic force ($f$) are plotted as a function of extension at 40°C. In (B) and (D), $S$ represents $-T(dS/dL)_{T,P}$ and $H$ represents $(dH/dL)_{T,P}$. [(C) and (D) from Mistrali et al., 1971.]

elastic experiment, represents a decreasing extension ratio ($\lambda = L/L_0$) as the temperature is reduced. To overcome this problem we transformed the force–temperature data from constant length to constant extension ratio, according to the method of Dorrington and McCrum (1977). Assuming that swelling is isotropic, this procedure gives a correction to constant volume, and allows us now to determine the magnitude of the internal energy change that arises from the distortion of chemical bonds when the molecular network is deformed. This analysis is based on the following equation (Flory, 1953):

$$f \simeq (\partial U/\partial L)_{T,V} + T(\partial f/\partial T)_{\lambda,P} \qquad (2)$$

where $-(\partial S/L)_{T,V} \simeq (\partial f/\partial T)_{\lambda,P}$. The results, shown in Fig. 4A, indicate that $(\partial U/\partial L)_{T,V}$ is close to zero, and that the decrease in entropy of the protein chains, $-(\partial S/\partial L)_{T,V}$, accounts for virtually all of the elastic force at low extensions and about 80% of the force at the highest extension. Thus the large enthalpy component shown in Fig. 3B is likely associated with the mixing of water and nonpolar groups in the protein.

Figure 4B shows some results of thermoelastic experiments carried out on abductin from *Pecten maximus* (Alexander, 1966). The swollen volume of abductin changes very little with temperature between 20 and 70°C, so we can regard the enthalpy changes as being a close approximation of the internal energy change. Here too, the internal energy change with extension is very small, and the elastic force is almost entirely due to changes in the conformational entropy of the molecular network.

These studies demonstrate that the molluscan protein rubbers, like elastin, appear to conform to the predictions of the kinetic theory of rubber elasticity. Temperature-dependent swelling changes give rise to a larger enthalpy change with extension in OAE than in abductin, but when analyzed at constant volume the thermodynamic properties of these two protein rubbers are very similar to what is observed for natural rubber (Fig. 4D). Thus we can conclude that the elastic mechanism in abductin and OAE is based on the presence of a kinetically free, random-coil structure.

## B. Mechanical Properties of the Molecular Network

We have seen how the elasticity of a rubber-like material is based on conformational changes of flexible long-chain molecules in contrast to the elasticity of stiff solids such as glass or steel, where elastic energy is stored in the distortion of chemical bonds. The kinetic theory of rubber elasticity also provides a statistical treatment of the polymer chains

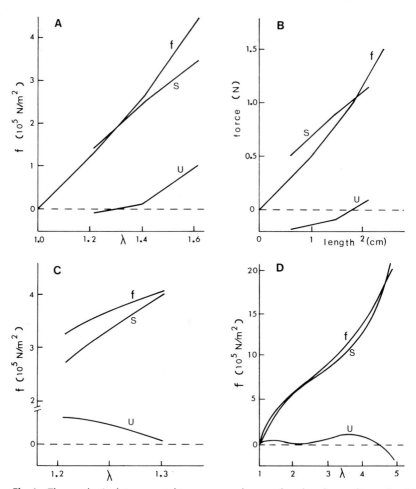

**Fig. 4.** Thermoelastic data corrected to constant volume and analyzed according to Eq. (2). (**A**) OAE fibers, 20°C; (**B**) Abductin, 27°C. (data from Alexander, 1966). (**C**) Elastin, 40°C. (Redrawn from Mistrali et al., 1971.) (**D**) Natural rubber, 20°C. (Redrawn from Anthony et al., 1942.)

that has been used as a model to quantify the network properties of the protein rubbers. This model idealizes each molecular chain between cross-links as a series of $n$ thermally agitated random links, each of length $l$. The orientation in space of any chain link, at any instant in time, is random and independent of the position of any neighboring links. Such random chains would have a high degree of flexibility. Clearly, complete rotational freedom cannot be achieved in real molecules due to steric hindrances. In proteins, rotation of the $\alpha$-carbon single bonds will be restricted to various degrees by the presence of side groups on the chain backbone, whereas the peptide bond is considered to be fixed and planar. In practical application of the theory based on an ideal random molecule, we must define a "functional random link" as a segment of the polymer chain containing enough bonds of partial rotational freedom that together approach the properties of the ideal random link.

When a single random chain is subjected to a small deformation, such that the distance between the ends of the chain $r$ is much less than the fully extended length $nl$, then the entropy change associated with this deformation can be calculated from a Gaussian probability function. Applying the analysis to a whole cross-linked network, the following relationship between uniaxial force and extension of a swollen rubber is obtained (Treloar, 1975).

$$f = NkTv^{1/3}(\lambda - \lambda^{-2}) \tag{3}$$

where $f$ is the nominal stress, defined as the force per unit swollen, unstrained cross-sectional area; $N$ is the number of network chains per unit volume, $k$ the Boltzmann constant, $T$ the absolute temperature, $v$ the volume fraction of the polymer in the swollen sample, and $\lambda$ the extension ratio. Thus the elastic force is directly proportional to temperature and to the number of chains in the network. The quantity $NkT$ is a measure of the material's stiffness and is usually referred to as the elastic modulus, $G$,

$$G = NkT = \rho RT/M_c \tag{4}$$

where $\rho$ is the density of the dry polymer, $R$ the gas constant, and $M_c$ the average molecular weight of the random chains between cross-links in the network. According to Eq. (3), a plot of $f$ versus $(\lambda - \lambda^{-2})$ should yield a straight line with a slope equal to $Gv^{1/3}$. If $v$ is known, then the modulus $G$ can determined.

Figure 5 shows the results of force–extension tests on small bundles of OAE fibers isolated from the octopus aorta. These fibers have long-range, reversible elastic properties and can be extended to well over 100% extension without breaking, but Fig. 5 only shows the data for extension up to $\lambda = 1.5$ (i.e., 50% extension). At these relatively small deformations, Guassian statistics appear to provide an adequate description of the mechanical properties, because the data in Fig. 5 fit a straight line quite well. The elastic modulus $G$, calculated from the slope of the regression line, is $4.7 \times 10^5$ N/m² ($v = 0.40$ at 20°C). Using a value of 1.33 g/cm³ for $\rho$, we estimate that the molecular weight of the chains between cross-links, $M_c$, is about 6900 [Eq. (4)]. Because the OAE protein has an average residue

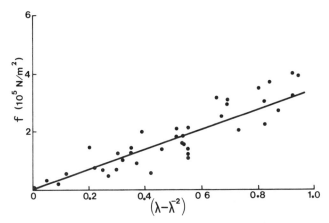

**Fig. 5.** Results of force–extension tests on small bundles of OAE fibers (Shadwick and Gosline, 1981). Up to extensions of $\lambda = 1.5$ [i.e., $(\lambda - \lambda^{-2}) = 1.0$] the data are linear as predicted by the Gaussian statistics for a rubber (Eq. 3). The slope of the line is $3.44 \times 10^5$ N/m$^2$ and the regression coefficient, $r^2 = 0.75$.

weight of 110 daltons, then each chain in the cross-linked network must be about 60 amino acid residues long.

When tested in compression, abductin has a Young's modulus of elasticity $E$ of about $3 \times 10^6$ N/m$^2$ (Trueman, 1953b; Alexander, 1966). Kelly and Rice (1967) reported $E$ for abductin tested in tension as $1.25 \times 10^6$ N/m$^2$. $E$ is defined as the initial slope of the stress–extension curve, and for rubbers at small extensions $E \simeq 3G$. Therefore $G$ for abductin is approximately $7 \times 10^5$ N/m$^2$, which gives this protein rubber nearly the same stiffness as resilin ($G = 6.5 \times 10^5$ N/m$^2$; Weis-Fogh, 1961), and almost twice that of OAE and elastin ($G$ for elastin is $4.1 \times 10^5$ N/m$^2$; Aaron and Gosline, 1981). From Eq. (4), $M_c$ for abductin is about 4600, and from the average residue weight (79 daltons) there must be approximately 60 amino acid residues between cross-links in the abductin network.

As mentioned before, the Gaussian equation, Eq. (3) is applicable to rubbers only at small extensions. At larger extensions, where $r$ (the chain end separation) approaches $nl$ (the extended length of the random chain), the entropy changes more rapidly than predicted by the Gaussian approximation. This results in an increase in the material stiffness with increasing $\lambda$, such that the stress–extension curve departs rapidly from the Gaussian curve. The amount of deformation that is possible before a rubber enters its non-Gaussian region and the shape of the non-Gaussian curve depend directly on the number of "functional random links" in the network

chains, $n$. Thus $G$ characterizes the behavior of a rubber at low extensions, whereas $n$ characterizes the non-Gaussian properties. Figure 6
shows force–extension curves for a lightly cross-linked natural rubber,
and for various protein rubbers. In all cases we see that the curves depart
from the Gaussian plot, Eq. (3), for which $n$ is considered to be infinite. It
is possible to estimate the number of functional random links per chain for
each material by comparing the force–extension curves to theoretical
curves plotted for different values of $n$ (Treloar, 1975). The curve (dotted
line) shown for natural rubber corresponds to the prediction for $n = 75$,
whereas the curves for OAE, resilin, and elastin correspond to about
$n = 5$, 7, and 10 random links, respectively. This means that the number
of amino acids that are necessary to provide one functional random link in
the OAE chains is about 12, compared to about 8 for elastin and 9 for resi-

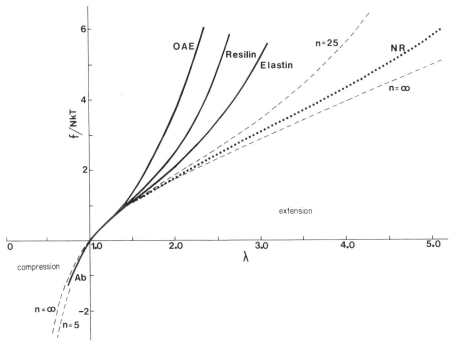

**Fig. 6.** Force–extension curves for rubber networks plotted as the force per chain ($f/NkT$)
against extension ratio $\lambda$. Two theoretical curves are also shown in dashed lines: One is for a
network with 25 random links per chain ($n = 25$) and the other is the Gaussian curve, Eq. (3),
for which $n = \infty$. The data for OAE, resilin, and elastin are best described by curves with 5, 7,
and 10 random links per chain, respectively (solid lines). Extension ratios greater than 1.0 represent elongation, and less than 1.0 represent compression. Dotted curve, Natural rubber,
$n = 75$; Ab, Abductin. (Data from Alexander, 1966.)

lin. Because each amino acid in a peptide allows two single bonds with partial rotational freedom, some 24 points of partial freedom are required for the OAE protein to approach the complete rotational freedom assumed for "ideal" random links, with correspondingly smaller values for elastin and resilin. Thus the protein chains in OAE appear to be less flexible than those in resilin and elastin, and this probably results from the relatively high proportion of amino acids with large side chains in OAE. Compared to natural rubber, all the protein rubbers are very much less flexible, because the average molecular weight of the chains in the natural rubber shown in Fig. 6 is about the same as that for the three protein rubbers, but the number of functional random links is about an order of magnitude greater. The protein rubbers will also fail at lower extensions compared to natural rubber, because the ultimate extension ratio of a rubber will be approximately equal to $n^{1/2}$ (Treloar, 1975).

It is difficult to estimate the non-Gaussian parameter from compression data such as those shown for abductin (Fig. 6), because the range of extensions is rather limited. However, we have estimated that the molecular chains in abductin are about the same length as those in OAE and resilin, but are made of smaller amino acids, which should impose somewhat less steric hindrance to rotational freedom. Therefore we expect that the abductin chains will be more flexible than the other protein rubbers.

## V. Functional Mechanical Properties

### A. The Bivalve Hinge

An important condition of the random-coil model for rubber elasticity is that the molecular chains must be in constant motion due to thermal agitation. This means that the thermal energy that drives the system must be greater than the frictional forces between chains. These frictional forces are the basis of the time-dependent, viscous properties of polymers. As a result of viscous interactions, a certain portion of the energy that goes into the deformation of a rubber network is lost as heat, whereas the rest is stored elastically and can be recovered to do useful work. Thus all polymeric materials have the property of viscoelasticity. The greater the viscous loss, the lower the efficiency of elastic energy storage. For biological tissues that are designed to provide long-range elasticity, such as the molluscan hinge ligament and artery wall, the elastic efficiency is an important consideration.

All bivalves rely on elastic recoil from the hinge ligament to open the valves. In most cases these animals are rather sedentary, and although it may be necessary to have sudden closure of the valves, the rate at which they open is usually of little consequence. However, in the case of the Pectinidae, which swim by rapid adduction of the valves, it is important that the passive reopening be very quick, and that the elastic efficiency of the ligament be high. This has been demonstrated in several studies of the mechanical properties of the hinge ligaments from a variety of bivalves (Trueman, 1953b; Russel-Hunter and Grant, 1962; Kahler et al., 1976), and some of the results are shown in Fig. 7. These are plots of the force required to hold the valves at a given degree of gape during a closing and opening cycle. The area under the closing curve is equal to the energy put into the ligament by the applied load ($E_{in}$), whereas the area under the opening curve is the energy recovered during elastic recoil ($E_{out}$). The area enclosed by the loading and unloading curves is called hysteresis and represents the amount of energy dissipated through viscous processes during the cycle. The ratio $E_{out}:E_{in}$ is called the resilience and is a measure of the elastic efficiency of the structure.

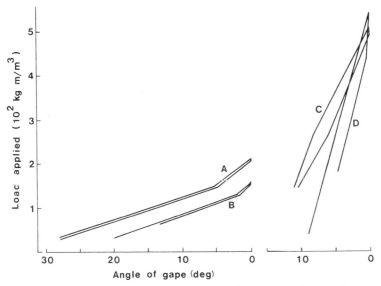

**Fig. 7.** Results of compression tests on bivalve hinge ligaments. Each curve shows one closing (loading) and opening (unloading) cycle. The ordinate gives the moment force applied, divided by the enclosed shell volume. The abscissa is the compression of the ligament, measured by decreasing angle of gape. A, *Pecten maximus;* B, *Chlamys opercularis;* C, *Spisula solidissima;* D, *Mytilus edulis.* (A, B, and D from Trueman, 1953b; C Data from Russel-Hunter and Grant, 1962.)

By comparing the hysteresis loops in Fig. 7, it appears that the hinge ligaments in the swimming Pectinidae *Chlamys opercularis* and *P. maximus* are more efficient in storing elastic energy than those in the other bivalves. Resiliences calculated from these data are quite high for *Chlamys* and *Pecten* (over 90%). In the burrowing clam *Spisula solidissima* the resilience is 74%, and it is only 63%-for the sessile mussel *Mytilus edulis*. The change in gape of the valves in response to a change in the applied load is essentially instantaneous in the scallops, but in bivalves such as *Mytilus, Spisula,* and *Anodonta* there appear to be both a sudden response and a subsequent slow change lasting several minutes (Trueman, 1953b; Labos, 1971; Kahler et al., 1976). Recall that the inner hinge ligament of the Pectinidae is not calcified and is virtually pure abductin. In contrast, the inner layer of the ligaments of other bivalves is typically from 60 to 90% calcium carbonate. Trueman (1953b) suggests that the lack of calcification in the scallop hinge is a functional adaptation for swimming that decreases internal friction and thus increases resilience. Kahler et al. (1976) found a negative correlation between calcium carbonate content and resilience in the ligament, although their values for resilience seem rather high. In the sessile bivalves it appears that calcification increases the stiffness and opening force of the hinge, at the expense of elastic efficiency (Trueman, 1953b).

Because resilience is a function of the viscous interactions within a material, the resilience should vary with the rate of deformation. The results discussed already are from "static" tests, where the gape was changed slowly over a period of several minutes. Labos (1971) showed that the resonant frequency of the shell–ligament system in *Anodonta cygnea* was 30 Hz and the damping factor, determined from the decay in amplitude of successive peaks in a free oscillation, was about 0.1. This damping corresponds to a resilience of about 55%. The resilience of the hinge ligament in this animal, when tested statically, was as high as 76% (Trueman, 1953b). Thus, as expected, resilience does decrease as the rate of deformation is increased. Alexander (1966) measured the resilience of abductin in the scallop *C. opercularis* in free oscillations at 4 Hz (i.e., about the frequency at which the valves open and close during jet swimming), and he obtained values of 90–93%. This experiment was carried out on an empty shell from which over half of the mass of the oscillating valve had been removed. Alexander found it necessary to add weights to the system in order to reduce the resonant frequency to 4 Hz, suggesting that the natural resonant frequency of the shell–ligament system is somewhat higher. It is likely that the *in vivo* resonant frequency of the system will be matched to the swimming frequency in order to minimize the energy required to drive the system. However, kinematic data that would allow us to determine this are not currently available.

In most bivalves the angle of gape attained when the adductor muscles are fully relaxed is considerably less than the maximum angle of gape, which is reached only after the muscles are excised. This means that the hinge ligament is never completely unloaded *in vivo,* but actually operates under a constant level of prestress. This is an important design feature of the bivalve shell because prestress can increase the amount of energy that is stored in the compression of the ligament. Figure 8 serves to illustrate this effect in the case of *Pecten.* The maximum angle of gape possible in scallops (i.e., when the abductin is unstressed) is about 30°. However, the valves only open to about 15° *in vivo.* Trueman (1953b) reports that when the valves are closed the abductin is compressed by about 45% (i.e., λ = 0.55). Thus, while swimming, the valves move through an angular change of about 15° and λ varies from about 0.78 to 0.55. (Note that the decrease in λ indicates

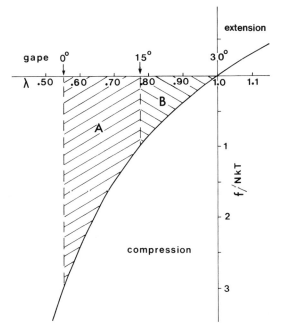

**Fig. 8.** The effect of prestress on energy storage in compression of abductin in the hinge ligament of *Pecten.* This graph is an enlarged view of the compression portion of Fig. 6. See text for detailed explanation.

compression.) The energy stored in compression of the hinge ligament during each adduction is shown by the area A in Fig. 8. If there was no prestress, the operational range of extension ratio would be from $\lambda = 1.00$ to $0.78$, and the energy stored in each cycle would be equal to area B (Fig. 8). In this case, area A is about four times greater than area B, and thus for a given change in extension ratio, the energy stored is greatly increased by prestress in the ligament. Of course the muscular force required to close the valves will increase with high levels of prestress, but this is presumably compensated for by the increased efficiency of storing larger amounts of energy in a smaller hinge. The use of prestress to increase energy storage is a principle that is used in many other structures, both biological and man-made. The walls of blood vessels provide another example of this, as we will see in the following section.

## B. Cephalopod Arteries

The functional role of the rubber-like OAE fibers is to provide the basis for long-range elasticity in the octopus aorta. Figure 9 shows circumferential stress–extension data for the artery wall. This has been constructed by using pressure–radius data from *in vitro* inflations of segments of the aorta from *O. dofleini*. The octopus aorta exhibits a nonlinear "J-shaped" stress–extension curve in which the stiffness or incremental modulus of elasticity $E_{inc}$ increases continuously with extension. A J-shaped stress–extension curve is typical of all vertebrate arteries (e.g., see Fig. 10, curve A). In fact, the stiffness of the wall of any highly extensible pressure vessel must increase with the radius in order to prevent elastic instability, which would lead to aneurysms and rupture (Burton, 1954; Gosline, 1980). This property allows the artery to be very distensible at low (diastolic) pressures, in order to provide a pulse-smoothing effect, and to become much less distensible at high pressures to prevent "blowout." In the aorta of *O. dofleini* $E_{inc}$ increases sixfold, from $1.5 \times 10^4$ to $9 \times 10^4$ N/m², over the physiological range of pressures (Fig. 9, curve A). Similarly, in the mammalian artery a sixfold increase in $E_{inc}$ is also observed over the physiological range, although the stress and modulus values are higher than in the octopus aorta.

What then is the structural basis for this functionally important increase in stiffness that occurs on inflation of the artery wall? We know that the slope of the stress–extension curve for elastin increases with $\lambda$ because the non-Gaussian properties cause this protein rubber to become stiffer as it is extended (Aaron and Gosline, 1981). However, by

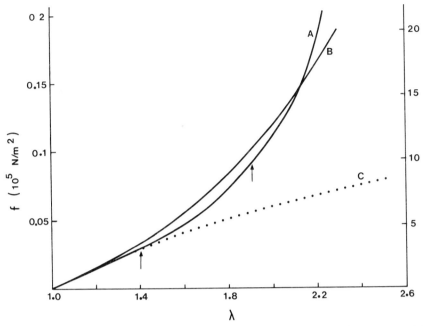

**Fig. 9.**   Curves of stress ($f$) vs. extension ratio ($\lambda$) for octopus aorta (**A**) and Octopus arterial elastomer (**B**). The upper and lower limits of extension under normal physiological blood pressures are indicated by the arrows on curve A. Over this range the incremental elastic modulus of the artery wall (see Cox, 1976) increases from $1.5 \times 10^4$ to $9 \times 10^4$ N/m². The stress scales have been arbitrarily adjusted so that the initial portions of the two curves coincide. **C** is the Gaussian curve. The left ordinate is for the aorta (**A**); right ordinate is for the OAE (**B**) and the rubber (**C**).

comparing the stress–extension curves for elastin fibers and the mammalian artery (Fig. 10), it is apparent that the change in slope of the elastin curve is not nearly sufficient to explain the sharp increase in stiffness seen for the artery wall. (In Fig. 10 we have adjusted the stress scales arbitrarily so that the initial portions of the curves coincide.) The shape of the stress–extension curve for vertebrate arteries actually arises from a complex arrangement of elastin and collagen fibers. The initial low-modulus region of the curve is dominated by elastin, whereas throughout the physiological range of extensions the modulus increases rapidly because of the recruitment of more elastin fibers and also the transfer of load to the collagen network (Roach and Burton, 1957; Wolinsky and Glagov, 1964).

The non-Gaussian properties of the OAE fibers appear to be very important in determining the shape of the stress–extension curve for the

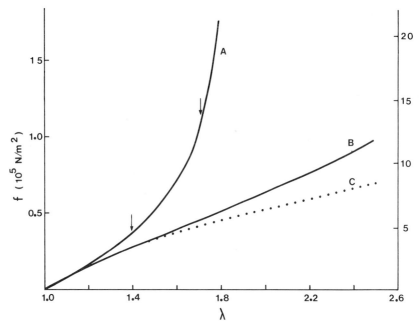

**Fig. 10.** Curves of stress ($f$) vs. extension ratio ($\lambda$) for a typical mammalian artery (**A**) and for elastin fibers (**B**). The upper and lower limits of extension under normal physiological blood pressures are indicated by the arrows on curve **A**. Over this range the incremental elastic modulus of the artery wall increases from $2 \times 10^5$ to $12 \times 10^5$ N/m². The stress scales have been adjusted so that the initial portions of the curves (up to $\lambda = 1.2$) coincide. **C** is the Gaussian curve. (**A** from Cox, 1976; **B** from Aaron and Gosline, 1981.) The left ordinate is for the artery (**A**); the right ordinate is for elastin (**B**) and the rubber (**C**).

octopus aorta. Like elastin, the OAE fibers become stiffer with extension. However, the molecular chains in the OAE protein are less flexible than those in elastin, and as a result the increase in stiffness of the OAE fibers is sufficient to provide the observed change in $E_{\text{inc}}$ for the artery wall (see Fig. 9). Here, as in Fig. 10, the stress scales have been adjusted so that the initial parts of the curves coincide, but in this case we find that the curves are almost identical up to extensions of $\lambda = 2.2$. Thus the octopus arterial elastomer has mechanical properties that are perfectly matched to the requirements of the artery wall. That is, the elastic fibers have adequate extensibility, as well as non-Gaussian properties that give rise to an increase in stiffness with extension that is sufficient to prevent elastic instability in the artery. Compared to elastin, the OAE fibers show a greater deviation from the Gaussian curve because the number of random links per molecular chain is lower. This is probably a result of the high proportion of large amino acid residues in the OAE protein.

A quantitative analysis has shown that the OAE fiber network comprises about 3% of the volume of the wall of the octopus aorta (Shadwick and Gosline, 1981), and we estimate that at least one-third of the fibers—that is, about 1% of the wall—are oriented to support load in the circumferential direction when the aorta is inflated. Figure 9 shows that up to $\lambda = 2.2$, the stress required to extend the elastic fibers is about 100 times greater than the circumferential stress in the whole artery wall. Therefore it appears that the OAE fibers not only have the correct shape of stress–extension curve but also are present in sufficient quantity to account fully for the elastic properties of the vessel, up to and including the normal physiological range of extensions. Collagen, which is present as a loose adventitia, apparently is not required to bear a significant portion of the load until very large extensions.

It is interesting to note that the octopus aorta, like the mammalian artery, is prestressed by the diastolic blood pressure ($\sim 15$ mm Hg) to extensions of about 40%. The normal systolic pressure ($\sim 40$ mm Hg) expands the artery a further 50%. The amount of energy put into stretching the wall of the aorta during each cardiac cycle is equal to the area under the stress–extension curve between $\lambda = 1.4$ and 1.9. Clearly the energy storage capacity of the artery wall is enhanced by prestressing, as is the case with the bivalve hinge ligament.

Figure 11 shows a hysteresis loop for the aorta of $O.$ $dofleini,$ as well as data for two other cephalopod species. The octopus blood vessel has a resilience of about 70% when tested by very slow inflation. This is not a direct indication of the elastic efficiency of the OAE fibers themselves, because they comprise only a small fraction of the artery wall. Although the rubber-like protein has not been isolated from cephalopod species other that $O.$ $dofleini,$ histologically identified "elastic" fibers are present in much larger quantities in the aortas of the squid and nautilus. This probably contributes to the higher resilience values for the artery walls of the squid $Nototodarus$ $sloani$ (76%) and $Nautilis$ $pompilius$ (86%), as shown in Fig. 11 (Gosline and Shadwick, 1983).

The tan δ obtained for intact octopus aorta, tested dynamically over the physiological range of frequencies (i.e., 0.2–2 Hz) was found to be 0.10 at the low frequency, increasing to 0.15 at the high frequency. These values correspond to resiliences of 72 and 61%, respectively. Because the octopus artery wall is largely muscle tissue, the dynamic resilience of the intact structure is undoubtedly lower than the true resilience of the isolated OAE fibers. A similar situation occurs in vertebrate arteries, where the resilience of the intact artery wall over the physiological range of fre-

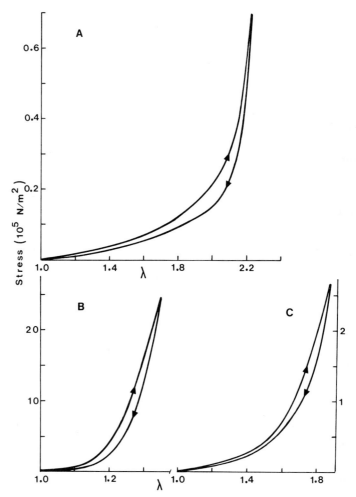

**Fig. 11.** Plots of the wall stress:extension ratio (λ) in the circumferential direction for the aortas of *Octopus dofleini* (**A**) and of *Nautilus pompilius* (**C**). Data are from *in vitro* inflations of arterial segments. Curve **B** is a uniaxial stress–extension curve for the dorsal aorta of the squid *Nototodarus sloani* extended longitudinally *in vitro*. Arrows indicate the loading and unloading phases of the cycle. Resiliences calculated from the areas under the curves are 70% for (**A**), 76% for (**B**), and 86% for (**C**).

quencies is about 65–70%, whereas the resilience of isolated elastin, tested at the same frequencies, is about 87% (Wainwright et al., 1976).

## VI. Summary

At present we have a reasonable understanding of the functional mechanical properties of two protein rubbers found in molluscan connective tissues. Abductin is present in the inner hinge ligament of the Pectinacea, and acts in compression to provide an effective antagonist to the adductor muscles. OAE fibers are found in the aortas of cephalopods and are responsible for the long-range elastic properties of these blood vessels. Both protein rubbers appear to be very efficient at storing elastic energy, although more complete data on dynamic resilience is desirable. Abductin and OAE conform well to the kinetic theory of rubber elasticity, and this suggests the presence of a kinetically free network of random-coil peptides held together by covalent cross-links. The amino acid sequence designs that allow kinetic freedom in these proteins, the cross-linking chemistry, and the biosynthetic mechanisms are interesting areas of research that certainly warrant further investigation.

## References

Aaron, B. B., and Gosline, J. M. (1980). Optical properties of single elastin fibres indicate random protein conformation. *Nature (London)* **287**, 865–866.

Aaron, B. B., and Gosline, J. M. (1981). Elastin as a random network elastomer. *Biopolymers* **20**, 1247–1260.

Alexander, R. McN. (1966). Rubber-like properties of the inner hinge ligament of Pectinidae. *J. Exp. Biol.* **44**, 119–130.

Andersen, S. O. (1963). Crosslinks in resilin. *Biochim. Biophys. Acta* **69**, 249–262.

Andersen, S. O. (1964). Crosslinks in resilin identified as di- and tri-tyrosine. *Biochim. Biophys. Acta* **93**, 213–215.

Andersen, S. O. (1967). Isolation of a new type of crosslink from the hinge ligament protein of molluscs. *Nature (London)* **216**, 1030.

Andersen, S. O. (1971). Resilin. *In* "Comprehensive Biochemistry" (M. Florkin and E. H. Stotz, eds.), Vol. 26C, pp. 633–657. Elsevier, Amsterdam.

Anthony, R. L., Caston, R. H., and Guth E. (1942). Equations of state for natural and synthetic rubber-like materials. 1. Unaccelerated natural soft rubbers. *J. Phys. Chem.* **46**, 826–837.

Barnes, H., and Gonor, J. J. (1958). Neurosecretory cells in the cirripede *Pollicipes polymerus, J. Mar. Res.* **17**, 81–102.

Beedham, G. E. (1958). Observations on the non-calcareous component of the shell of the Lamellibranchia. *Q. J. Microsc. Sci.* **99**, 341–357.

Bone, Q., Pulsford, A., and Chubb, A. C. (1981). Squid mantle muscle. *J. Mar. Biol. Assoc. U.K.* **61**, 327–342.

Bourne, G. B., and Redmond, J. R. (1976a). Hemodynamics in the pink abalone, *Haliotis corrugata*. I. Pressure relations and pressure gradients in intact animals. *J. Exp. Zool.* **200**, 9–16.

Bourne, G. B., and Redmond, J. R. (1976b). Hemodynamics in the pink abalone, *Haliotis corrugata*. II. Acute blood-flow measurements and their relationship to blood pressure. *J. Exp. Zool.* **200**, 17–22.

Bourne, G. B., Redmond, J. R., and Johansen, K. (1978). Some aspects of hemodynamics in *Nautilus pompilius*. *J. Exp. Zool.* **205**, 63–70.

Bowerbank, J. S. (1884). On the structure of the shells of molluscs and chonchiferous animals. *Trans. Microsc. Soc. London* **1**, 123–154.

Burton, A. C. (1954). Relation of structure to function of the tissue of the wall of blood vessels. *Physiol. Rev.* **34**, 619–642.

Cox, R. H. (1976). Effect of norepinephrine on the mechanics of arteries *in vitro*. *Am. J. Physiol.* **231**, 420–425.

Dahlgren, U., and Kepner, W. A. (1908). "Textbook of the Principles of Animal Histology." Macmillan, New York.

Deyl, Z., Macek, K., Adam, M., and Vančíková, J. (1980). Studies on the chemical nature of elastin fluorescence. *Biochim. Biophys. Acta* **625**, 248–254.

Dorrington, K. L., and McCrum, N. G. (1977). Elastin as a rubber. *Biopolymers* **16**, 1201–1222.

Elder, H. Y. (1966a). Organization and properties of some invertebrate collagen and elastic fibres. *Proc. R. Microsc. Soc.* **1**, 98–99.

Elder, H. Y. (1966b). The fine structure of some invertebrate fibrillar and lamellar elastica. *Proc. R. Micros. Soc.* **1**, 99–100.

Elder, H. Y. (1972). Connective tissues and body wall structure of the polychete *Polyphysia crassa* and their significance. *J. Mar. Biol. Assoc. U.K.* **52**, 747–764.

Elder, H. Y. (1973). Distribution and function of elastic fibers in the invertebrates. *Biol. Bull. (Woods Hole, Mass.)* **144**, 43–63.

Elder, H. Y., and Owen, G. (1967). Occurrence of "elastic" fibres in the invertebrates. *J. Zool.* **152**, 1–8.

Flory, P. J. (1953). "Principles of Polymer Chemistry." Cornell Univ. Press, Ithaca, New York.

Galtsoff, P. S. (1964). The american oyster, *Crassostrea virginica*. *U.S. Fish Wild. Serv. Fish. Bull.* No. 64.

Gosline, J. M. (1976). The physical properties of elastic tissue. *Int. Rev. Connect. Tissue Res.* **7**, 211–249.

Gosline, J. M. (1978a). Hydrophobic interaction and a model for the elasticity of elastin. *Biopolymers* **17**, 677–695.

Gosline, J. M. (1978b). Temperature dependent swelling of elastin. *Biopolymers* **17**, 697–707.

Gosline, J. M. (1980). The elastic properties of rubber-like proteins and highly extensible tissues. *Symp. Soc. Exp. Biol.* **34**, 331–357.

Gosline, J. M., and Shadwick, R. E. (1983). The biomechanics of cephalopod arteries. *Pac. Sci.* **36**, 283–296.

Gray, W. R., Sandberg, L. B., and Foster, J. A. (1973). Molecular model for elastin structure and function. *Nature (London)* **246**, 461–466.

Grut, N., and McCrum, N. G. (1974). Liquid drop model for elastin. *Nature (London)* **251**, 165.

Hoeve, C. A. J., and Flory, P. J. (1958). The elastic properties of elastin. *J. Am. Chem. Soc.* **80**, 6523–6526.

Hunt, S. (1970). Invertebrate structural proteins—characterization of the operculum of the gastropod *Buccinum undatum*. *Biochim. Biophys. Acta* **207**, 347–360.

Jackson, R. T. (1891). The mechanical origin of structure in pelecypods. *Am. Nat.* **25**, 11–21.

Johansen, K., and Martin, A. W. (1962). Circulation in the cephalopod *Octopus dofleini*. *Comp. Biochem. Physiol.* **5**, 161–176.

Jullien, A., Cardot, J., and Ripplinger, J. (1957). De l'éxistence de fibres élastiques dans l'appareil circulatiore des mollusques. *Ann. Sci. Univ. Besancon, Zool. Physiol.* **9**, 25–33.

Jullien, A., Cardot, J., and Ripplinger, J. (1958). Contribution à l'étude histologique des vaiseaux chez les mollusques. Sûr l'éxistence de fibres élastiques chez les pulmonés. *Ann. Sci. Univ. Besancon, Zool. Physiol.* **10**, 73–79.

Juřicová, M., and Deyl, Z. (1973). Fluoreskující složky fibrilárních bílkovin a jejich vztah K "Stařeckým pigmentům" a příčným vazbám. *Cesk. Fysiol.* **22**, 225–235.

Kahler, G. A., Fisher, F. M., Jr., and Sass, R. L. (1976). The chemical composition and mechanical properties of the hinge–ligament in bivalve molluscs. *Biol. Bull.* (*Woods Hole, Mass.*) **151**, 161–181.

Kelly, R. E., and Rice, R. V. (1967). Abductin: a rubber-like protein from the internal triangular hinge ligament of *Pecten*. *Science* **155**, 208–210.

Krumbach, T. (1935). Merostomata: Xiphosura. *In* "Handbuch der Zoologie" (W. Kukenthal, ed.), Vol. 3, pp. 47–96. de Gruyter, Berlin.

Labos, E. (1971). On the mechanical properties of the hinge ligament of *Anodonta cygnea*. *Ann. Biol. Tihany* **38**, 53–63.

Mabillot, S. (1954). Particularités histologiques de la membrane basale des mesenteron chez *Gammarus pulex*. *C. R. Hebd. Seances Acad. Sci.* **238**, 1738–1739.

Mabillot, S. (1955). Contribution a l'étude histo-physiologique de l'appareil digestif de *Gammarus pulex*. *Arch. Zool. Exp. Gen.* **92**, 20–38.

Mistrali, L., Volpin, D., Garibaldo, G., and Cifferri, A. (1971). Thermodynamics of elasticity in open systems. *J. Phys. Chem.* **75**, 142–149.

Moore, J. D., and Trueman, E. R. (1971). Swimming of the scallop *Chlamys opercularis*. *J. Exp. Mar. Biol. Ecol.* **6**, 179–185.

Morton, B. (1980). Swimming in *Amusium pleuronectes* (Bivalvia;Pectinidae). *J. Zool.* **190**, 375–404.

Nutting, W. L. (1951). A comparative anatomical study of the heart and accessory structures of the orthopteriod insects. *J. Morphol.* **89**, 501–597.

Owen, G. (1958). Shell form, pallial attachment and the ligament in the Bivalvia. *Proc. Zool. Soc. London* **131**, 637–648.

Owen, G., Trueman, E. R., and Yonge, C. M. (1953). The ligament in the Lamellibranchia. *Nature* (*London*) **171**, 73–75.

Plateau, F. (1884). Researches sur la force absolue des muscles des invertebres. *Arch. Zool. Exp. Gen.* **2**, 145–170.

Price, N. R., and Hunt, S. (1973). Studies of the crosslinking regions of whelk egg capsule proteins. *Biochem. Soc. Trans.* **1**, 158–159.

Price, N. R., and Hunt, S. (1974). Fluorescent chromophore component from the egg capsules of the gastropod mollusc, *Buccinum undatum*. *Comp. Biochem. Physiol. B* **47B**, 601–616.

Roach, M. R., and Burton, A. C. (1957). The reason for the shape of the distensibility curve of arteries. *Can. J. Physiol.* **35**, 681–690.

Russel-Hunter, W., and Grant, D. C. (1962). Mechanics of the ligament in the bivalve *Spisula solidissima* in relation to mode of life. *Biol. Bull.* (*Woods Hole, Mass.*) **122**, 369–379.

Sage, E. H., and Gray, W. R. (1977). Evolution of elastin structure. *In* "Elastin and Elastic Tissue" (L. B. Sandberg, W. R. Gray and C. Franzblau, eds.), pp. 291–312. Plenum, New York.

Sage, H. E., and Gray, W. R. (1980). Studies on the evolution of elastin. II. Histology. *Comp. Biochem. Physiol. B* **66B**, 13–24.

Shadwick, R. E. (1980). Viscoelasticity and pulse wave velocity in the aorta of *Octopus dofleini*. *Am. Zool.* **20**, 768.

Shadwick, R. E. (1982). The mechanical properties of the aorta of the Cephalopod Mollusc, Octopus dofleini (Wülker) PhD Thesis. Dep. Zool., Univ. of British Columbia, Vancouver.

Shadwick, R. E., and Gosline, J. M. (1981). Elastic arteries in invertebrates: mechanics of the octopus aorta. *Science* **213**, 759–761.

Taylor, J. D., Kennedy, J. W., and Hall, A. (1969). The shell structure and minerology of the Bivalvia. *Bull. Br. Mus. (Nat. Hist.), Zool., Suppl.* **3**.

Taylor, M. G. (1964). Wave travel in arteries and the design of the cardiovascular system. *In* "Pulsatile Blood Flow" (E. O. Attinger, ed.), McGraw-Hill, New York.

Thomas, J., Elsden, D., and Partridge, S. (1963). Partial structure of two major degradation products from crosslinkages in elastin. *Nature (London)* **200**, 651–652.

Thornhill, D. P. (1971). Abductin; locus and characteristics of a brown fluorescent chromophore. *Biochemistry* **10**, 2644–2649.

Thornhill, D. P. (1972). Elastin; locus and characteristics of a fluorophore. *Connect. Tissue Res.* **1**, 21–30.

Treloar, L. R. G. (1975). "Physics of Rubber Elasticity," 2nd ed. Oxford Univ. Press (Clarendon), London and New York.

Trueman, E. R. (1950). Observation on the ligament of *Mytilus edulis*. *Q. J. Microsc. Sci.* **91**, 225–235.

Trueman, E. R. (1951). The structure, development and operation of the hinge ligament of *Ostrea edulis*. *Q. J. Microsc. Sci.* **92**, 129–140.

Trueman, E. R. (1953a). The ligament of *Pecten*. *Q. J. Microsc. Sci.* **94**, 193–202.

Trueman, E. R. (1953b). Observations on certain mechanical properties of the ligament of *Pecten*. *J. Exp. Biol.* **30**, 453–467.

Wainwright, S. A., Biggs, W. D., Currey, J. D., and Gosline, J. M. (1976). "Mechanical Design in Organisms." Arnold, London.

Weis-Fogh, T. (1960). A rubber-like protein in insect cuticle. *J. Exp. Biol.* **37**, 880–907.

Weis-Fogh, T. (1961). Molecular interpretation of the elasticity of resilin, a rubber-like protein. *J. Mol. Biol.* **3**, 520–531.

Weis-Fogh, T., and Andersen, S. O. (1970). New molecular model for the long-range elasticity of elastin. *Nature (London)* **227**, 718–721.

Wells, M. J. (1979). The heartbeat of *Octopus vulgaris*. *J. Exp. Biol.* **78**, 87–104.

Wetekamp, F. (1915). Bindegewebe und Histologie der Gefasbahnen von *Anodonta cellensis*. *Z. Wiss. Zool.* **112**, 433–526.

Wolinsky, H., and Glagov, S. (1964). Structural basis for the static mechanical properties of the aortic media. *Circ. Res.* **14**, 400–413.

Yonge, C. M. (1936). The evolution of the swimming habit in the Lamellibranchia, *Mém. Mus. Hist. Nat. Belg.* **3**, 77–99.

Yonge, C. M. (1973). Functional morphology with particular reference to hinge and ligament in *Spondylus* and *Plicatula* and a discussion of relations within the superfamily Pectinacea. *Philos. Trans. R. Soc., London, Ser. B* **267**, 173–208.

Yonge, C. M. (1978). Significance of the ligament in the classification of the Bivalvia. *Proc. R. Soc. London, Ser. B* **202**, 231–248.

# 10

# Molecular Biomechanics of Molluscan Mucous Secretions

**MARK DENNY**[1]

Department of Zoology
University of Washington
Seattle, Washington

## I. Introduction

Molluscs' reputation for being "slimy" is easily attested to by anyone who has ever eaten a raw oyster or inadvertently stepped on a slug. Though the first cold, clammy contact with a mollusc may tend to discourage further interest in their epithelial mucous secretions, a bit of persever-

[1] Present address: Department of Biology, Stanford University, Hopkins Marine Station, Pacific Grove, California, 93950.

THE MOLLUSCA, VOL. 1
Metabolic Biochemistry
and Molecular Biomechanics

ance and closer examination reveals that these secretions, far from being simple slime, are a diverse, highly functional, and important group of bio-materials. To cite just a few examples, the gill mucus of bivalves functions both as an adhesive surface that assists in extracting food items from the water surrounding the animal and as a conveyor belt for transporting the food to the gut (Barnes, 1968). As with all animals that utilize extracellular digestion, the gut of molluscs is lined with mucus that serves to separate the digestive juices from the wall of the gut, thus allowing the organism to digest its food without digesting itself. Mucous secretions are used in vari-ous ways in reproduction, examples being the "slime threads" from which the terrestrial slug *Limax maximus* hangs during copulation (Rollo and Wellington, 1979) and the mucous coating of the eggs of the snail *As-siminea californica* (Fowler, 1980), which causes a camouflaging layer of mud to adhere to the eggs. Mucous secretions function as antipredator de-vices by rendering the animal distasteful (Simkiss and Wilbur, 1977), foul-ing the predator's feeding apparatus (Richter, 1980), or simply making the animal too slippery to pick up. Many gastropods secrete mucus onto the ventral surface of the foot, using it both as an aid in locomotion (Denny, 1979, 1980a,b, 1981; Denny and Gosline, 1980) and as a means by which the animal adheres to the surface on which it crawls (Grenon and Walker, 1980; Grenon et al., 1979). In addition, the "slime trail" of pedal mucus left behind as a gastropod crawls is often important in the animal's behav-ior, used either as a means of "homing" or for following conspecifics (Trott and Dimock, 1978; Cook, 1977).

Despite the functional diversity of molluscan mucous secretions, and their obvious importance to the physiology and ecology of these animals, little is known about the chemical and mechanical properties of these ma-terials. This chapter will first review, in general terms, what is known of the chemistry and macromolecular architecture of molluscan mucous se-cretions; this information will then be used to examine the functional me-chanical properties of these mucins. By relating macromolecular struc-ture to mechanical properties, and, where possible relating these mechanical properties to the material's function, the design of mucins may be more clearly perceived.

## I. Chemical Composition

At a superficial level of description all molluscan mucins are very much alike. The primary component of mucus is always water, forming from 90.1% (by weight) of the pedal mucus of the limpet *Patella vulgata* (Grenon and Walker, 1980) to 99.7% in the hypobranchial mucus of the whelk *Busycon canaliculatum* (Hunt, 1970). Because the mechanical properties of water are quite different from those of mucus, the 0.3–9.9% solid matter present in mucous secretions must be responsible for the mechanical properties of the material. This solid matter is formed partially of inorganic salts, which in themselves do little to affect the properties of water, and a high molecular weight complex of protein and polysaccharide, the chemistry of which, as we will see, determines the mucin's mechanical properties. The precise chemical composition of the protein–polysaccharide complex differs from one mucin to the next and has been the primary focus of studies on these materials (see Table I).

Traditionally mucous protein–polysaccharide complexes have been divided into two catagories: the mucopolysaccharides and the glycoproteins. Mucopolysaccharides consist predominantly of carbohydrate in the form of long, high molecular weight, usually linear chains. The protein component of the complex, though present, contributes little to the overall properties of the material. The term mucopolysaccharide is generally interchangeable with the term glycosaminoglycan (Hunt, 1970). Glycoproteins, on the other hand, are characterized by short, often branched carbohydrate chains (oligosaccharides) bound to a protein component that forms a substantial proportion of the material and is important in determining the mucin's properties. Although these categories are distinct when applied to vertebrate mucins (Hunt, 1970), the distinction becomes blurred when the large variety of invertebrate mucins is included. It seems likely that these terms describe the endpoints of a continuum and they will be used here in a descriptive rather than a generic sense.

## A. Polysaccharides

The polysaccharide chains of molluscan protein–polysaccharide complexes are formed from a number of monosaccharide subunits falling into three general categories: acidic sugars, amino sugars, and neutral sugars. As a general rule, the polysaccharides of molluscan mucins contain acidic monosaccharides, though there are exceptions (see Table I). Amino and neutral sugars may be either present in varying amounts or absent entirely. The acidic monosaccharides consist of an acid group (either sulfate or carboxyl) covalently bound to a hexose. Consequently each subunit carries a negative charge at physiological pH. Carboxylated hexoses are generally present as glucuronic or galacturonic acids, whereas sulfates are usually bound to glucose or galactosamine. In general the carboxylated hexoses are more prevalent in terrestrial molluscs, whereas sulfated hexoses are predominantly found in marine species (Hunt, 1970), and Hunt suggests that this may be correlated with the relative availability of sul-

## TABLE I

### Chemical Composition of the Polysaccharide of Molluscan Mucins[a]

| Species | Glandular origin | Type | Amino sugars | | | | Neutral sugars | | | | Acidic sugars | | | |
|---|---|---|---|---|---|---|---|---|---|---|---|---|---|---|
| | | | Glucosamine | N-Acetylglucosamine | Galactosamine | N-Acetylgalactosamine | Glucose | Galactose | Mannose | Fucose | Glucuronic acid | Galacturonic acid | Sulfated hexose | Sialic acid |
| Otella lactea[b] | Pedal | Mucopolysaccharide | + | + | | | + | | | + | | + | + | |
| Helix laeda[c] | Pedal | n.r.[h] | | + | | | | + | | | | | + | |
| | ? | n.r. | | + | | | | + | | | + | | | |
| Buccinum undatum[d] | Hypobranchial | Glycoprotein | + | | | | + | + | + | + | | | + | |
| | Hypobranchial | Mucopolysaccharide | | | + | | + | | | | | | + | |
| Charonia lampas[e] | Hypobranchial | Mucopolysaccharide | | | | | + | | | | | | + | |
| Neptunea antiqua[f] | Hypobranchial | Mucopolysaccharide | + | | + | | + | | | + | | | + | |
| Ariolimax columbianus[g] | Pedal | Glycoprotein | + | | + | | + | + | + | + | + | + | | |

[a] Results from histochemical studies have not been included here.
[b] Pancake and Karnovsky (1971).
[c] Suzuki (1941).
[d] Hunt (1970).
[e] Egami and Takahashi (1962).
[f] Doyle (1964).
[g] Denny (1979).
[h] n.r., Not reported.

fate. Sialic acid, which is common in vertebrate mucins, is generally absent in invertebrate mucous secretions (Warren, 1963). Among the molluscs, it has only definitely been identified in hydrolysates from the digestive gland of the whelk *Charonia lampas* (Inoue, 1965) and has never been found in a molluscan epidermal mucous secretion. The ratio of acidic to neutral subunits varies greatly among molluscan mucins. In a mucopolysaccharide such as that found in the hypobranchial mucus of the whelks *C. lampas* and *Buccinum undatum* the polysaccharide consists almost exclusively of sulfated glucose. Every monosaccharide thus carries an acidic group, and the molecule is highly negatively charged (Hunt, 1970). The polysaccharides of glycoproteins are usually less highly charged; for example, only approximately one-third of the monosaccharides found in the pedal mucus of the slug *Ariolimax columbianus* carry negative charges (Denny, 1979).

Amino sugars, usually glucosamine and galactosamine, are commonly found in molluscan mucus. In those cases where the possibility has been examined these amino sugars are present in the N-acetylated form. The generality of the acetylation of the amino sugars in molluscan mucins has not been determined though N-acetylated amino sugars are common in vertebrate mucous secretions (Pain, 1980).

A variety of neutral sugars are found in molluscan mucins, among them glucose, galactose, fucose, and mannose. In general, neutral sugars are characteristic of glycoproteins and are absent from mucopolysaccharides, but this is by no means universal (see Table I). The functional properties of neutral sugars in the polysaccharide chains of mucins are not well understood, but it has been proposed (Morris and Rees, 1978) that the nonpolar hydroxyl of fucose may allow for hydrophobic interactions between monosaccharides in aqueous solution.

The bonds by which the monosaccharides of molluscan mucins are polymerized have not been extensively studied and have been characterized only for the mucopolysaccharide fraction of the hypobranchial mucins of whelks (Hunt, 1970). In these molecules $\beta$-1,4-glycoside linkages predominate, but $\alpha$-1,4 and $\alpha$-1,6 are also present. Few measurements have been made of molecular weight for the polysaccharide fraction of molluscan mucous secretions, and again those made are restricted to the mucopolysaccharides of hypobranchial mucins. These measurements vary from a molecular weight of $1.7 \times 10^5$ for *B. undatum* (Hunt and Jevons, 1966) to $2.5 \times 10^6$ for *B. canaliculatum* (Kwart and Shashoua, 1957), or, assuming an average molecular weight per monosaccharide of 260 (glucose sulfate), approximately 650–9600 monosaccharides per chain. It seems likely that the polysaccharide chains of molluscan glycoproteins are considerably shorter than those of mucopolysaccharides; for example, the polysaccharide chains of vertebrate glycoproteins are usually only 2–18 monosaccharides long (Pain, 1980), but this has not been measured for molluscan mucins.

## B. Proteins

The protein component of the protein–polysaccharide complexes of molluscs are generally unremarkable. The amino acid composition for several such proteins is shown in Table II. Three points may be noted from these compositions: First, serine and threonine are present in relatively high proportions. In vertebrate mucins the polysaccharide fraction is often attached to the protein through $O$-glycoside bonds to either serine or threonine (Neuberger *et al.*, 1977). The presence of these bonds is detected through their alkali lability (Neuberger *et al.*, 1977), and they have tentatively been identified in the pedal mucus of the terrestrial slug *A. columbianus* (Denny, 1979). In contrast, however, the protein–polysaccharide bonds in the mucopolysaccharide fraction of *B. undatum* hypobranchial mucus are apparently alkali stable, and because these chains have been shown to be covalently linked to protein, other linkage sites must be present (Hunt, 1970). Kwart and Shashoua (1957) report that the protein of *B. canaliculatum* is bound to polysaccharide chains through electrovalent linkages with calcium. Obviously a full explanation of protein–polysaccharide linkages in molluscan mucins awaits further research. Second, cysteine (i.e., half-cystine) is present in these proteins, and due to the propensity of this amino acid to form disulfide bonds, presents the possibility of S—S linkages between protein chains. The significance of these crosslinks will be discussed later. Third, in general the acidic amino acids (glutamic and aspartic) outnumber the basic amino acids and, like the polysaccharide chains discussed already, the proteins may have a net negative charge at physiological pH. This conclusion is tentative, however, because in the intact protein aspartic and glutamic acids may be present as asparagine and glutamine, as suspected by Hunt and Jevons (1965) for the hypobranchial mucin of *B. undatum*.

The molecular weight of the protein component of molluscan mucins has only been determined for the pedal mucus of the limpet *P. vulgata* (Grenon and Walker, 1980), where eight fractions were found ranging from 23,000 to 195,000 MW (approximately 230–1950 amino acids).

In summary, molluscan mucous secretions are formed primarily of water with a small amount of dissolved salts and a protein–polysaccharide complex. The polysaccharide component is usually a mixture of acidic subunits with either or both neutral and amino sugars; and at physiological pH this component is negatively charged. The polysaccharide is bound, often covalently, to the protein component of the complex.

**TABLE II**

**Amino Acid Composition of the Protein of Molluscan Mucins** [a]

| Amino acid | Species and mucin type | | | | |
|---|---|---|---|---|---|
| | *Ariolimax columbianus* [b] pedal mucin | *Buccinum undatum* [c] hypobranchial mucin mucopolysaccharide | *B. undatum* [c] hypobranchial mucin glycoprotein | *Busycon canaliculatum* [d] hypobranchial mucin | *Patella vulgata* [e] pedal mucin |
| Aspartic acid | 9.0 | 8.15 | 11.10 | 10.0 | 9.43 |
| Threonine | 11.1 | 6.77 | 6.45 | 5.9 | 7.56 |
| Serine | 12.6 | 11.58 | 6.27 | 5.0 | 6.84 |
| Glutamic acid | 8.4 | 16.16 | 11.34 | 12.0 | 9.74 |
| Proline | 8.6 | 3.14 | 5.19 | 2.70 | 5.64 |
| Glycine | 8.6 | 13.26 | 5.19 | 7.12 | 7.14 |
| Alanine | 7.4 | 12.75 | 7.67 | 7.23 | 6.45 |
| Cysteic acid | 3.0 | n.r. [f] | 1.73 | 1.8 | 0.87 |
| Valine | 5.2 | 6.73 | 6.91 | 6.00 | 5.54 |
| Methionine | 0.2 | 3.79 | 0.31 | n.r. | 0.36 |
| Isoleucine | 4.5 | 1.61 | 4.20 | 4.39 | 4.24 |
| Leucine | 5.5 | 2.84 | 8.16 | 8.13 | 6.26 |
| Tyrosine | 1.8 | 0.86 | 2.85 | 4.04 | 2.17 |
| Phenylalanine | 2.3 | 1.42 | 3.66 | 1.05 | 3.00 |
| Lysine | 0.9 | 3.23 | 7.29 | 3.41 | 2.67 |
| Histidine | 3.4 | 3.07 | 2.45 | 2.28 | 1.29 |
| Arginine | 3.9 | 3.18 | 5.54 | 1.42 | 3.83 |
| Ornithine | n.r. | 2.46 | n.r. | 0.30 | — |
| Tryptophan | n.r. | n.r. | n.r. | n.r. | 1.98 |

[a] Composition given in residues per 100 residues.

[b] Denny (1979).

[c] Hunt and Jevons (1965); Hunt (1970).

[d] Kwart and Shashoua (1957).

[e] Grenon and Walker (1980).

[f] n.r., Not reported.

## III. Macromolecular Structure and Physical Properties—Theory

Having described the chemical composition of molluscan mucins, it is now reasonable to ask how this small amount of organic material controls the vastly larger amount of water in which it is dissolved to yield the mechanical properties of mucins. One may also wonder which aspects of mucins' chemistry account for the variation in mechanical properties among mucins. Before addressing these questions to specific examples, it will be useful to review briefly the terms and theories applicable to the physical properties of polymeric solutions.

## A. Polymer Chain Configurations

Regardless of their precise chemical composition and pattern of polymerization, the protein–polysaccharide complexes of molluscan mucins are large, flexible polymers and behave similarly to all other such macromolecules. Although the precise configuration of macromolecules in solution is an exceedingly complex subject (see Tanford, 1961; Oosawa, 1971), the general behavior of these molecules is quite intuitive. A polymeric molecule is formed of a number $N$ of subunits, each having a length $L$, and each bound to the next along the polymer chain. Each subunit thus forms one "link" in the chain. The bonds between links (in the case of mucous polysaccharides, glycoside bonds; and for proteins, peptide bonds) are flexible, allowing subunits to swivel relatively freely with respect to each other. When the polymeric chain is placed in solution at physiological temperatures, its thermal kinetic energy causes it to change configurations constantly and randomly. As a consequence it is impossible to specify the shape of the molecule or the volume it occupies, except in a statistical sense. One such statistical description of a randomly arranged molecule is its radius of gyration, a measure of the average distance of links from the center of the molecule's mass. For a truly random configuration the radius of gyration is proportional to the length of the links $L$, and the square root of $N$, the number of links in the chain (Tanford, 1961); the effective volume occupied by the molecule is thus proportional to $LN^{3/2}$. For protein–polysaccharide molecules where the length of the links (amino acids or monosaccharides) does not vary appreciably from one chain to the next, the effective average volume occupied by the molecule is controlled by the number of links in the chain; the longer the chain the larger the volume.

There are several factors, however, that can influence the configuration of large polymeric molecules in solution and thus their volume. For example, glycoside and peptide bonds, although flexible, do not allow perfectly free rotation between subunits; and real protein and polysaccharide polymers are thus less flexible than their ideal counterparts. A correction in the statistics of chain configuration is thus necessary when dealing with these molecules. This is usually accomplished by redefining the length of chain representing a link such that instead of containing one amino acid or monosaccharide, the link contains several of these basic subunits. The combination of the several partially restricted bonds of the amino acids or monosaccharides thus yields an effectively freely rotating bond for the newly defined link. The overall chain thus contains a smaller number of longer links, and the statistics for a truly random chain may be applied. A real chain of this sort has a larger radius of gyration than its ideal counterpart.

The configuration of a polymeric chain can also be affected by charges present on the chain. If, as is the case for the polysaccharides present in molluscan mucins, the individual chain links are similarly charged (in this case the charge is negative due to carboxyl and sulfate groups), the segments tend to repel each other. Segments that, if uncharged, could approach quite close to each other in the course of thermal agitation are now prevented from doing so by the forces of electrostatic repulsion. As a consequence the chains are less flexible and the effective volume occupied by the molecule increases. The degree of expansion depends on a number of factors, among these being the charge density of the chains (the more charges per length the greater the expansion) and the presence in solution of counterions. Counterions are ions of opposite charge to those bound to the chains. Through ionic bonding to charges on the polymeric chains, they reduce electrostatic repulsion between chains and result in a reduction in volume occupied. For example, a polyanion such as a mucous protein–polysaccharide complex is highly expanded in distilled water and much less expanded in a 1 $M$ NaCl solution where $Na^+$ ions mask the polymer's negative charges. For a more detailed account of such charge effects see any standard text on polyelectrolytes (see, e.g., Oosawa, 1971; Tanford, 1961). The statistics of chain configuration are also dependent on whether the chain is linear or branched (Tanford, 1961); for example, a branched chain of a given weight has a slightly smaller radius of gyration than a linear chain of the same weight, but the general behavior of the two types of chains is similar.

## B. Viscosity

For the purposes of this chapter, the effective volume and configuration of large molecules are important in that they can affect the bulk mechanical properties of a mucous solution. The simplest example of such an effect concerns the solution's viscosity. Viscosity is a measure of a fluid's resistance to flow. To understand this concept, imagine a volume of liquid sandwiched between two plates as shown in Fig. 1a. If one of the plates is moved parallel to the other the liquid is sheared; liquid directly next to each plate moves at the same velocity as that plate and a velocity gradient is established in the fluid. The fluid may be thought of as consisting of a

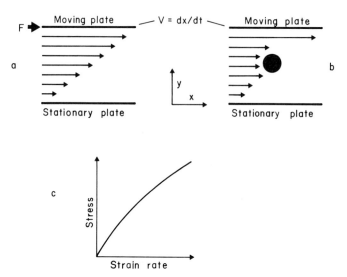

**Fig. 1.** Viscosity. (**a**) Fluid in contact with each plate moves at the same velocity as that plate, and consequently a velocity gradient ($V = dx/dt$) is established. (**b**) Rigid particles decrease the volume fraction occupied by the fluid and affect the velocity gradient. (**c**) For a purely viscous liquid, stress $\sigma$ is strictly proportional to strain rate, $\dot{\gamma}$. Strain, $\gamma = dx/dy$; strain rate, $\dot{\gamma} = d\gamma/dt$.

stack of infinitesimally thin layers, each layer moving slightly faster than the next layer below it and slightly slower than the next layer above it. "Friction" between layers opposes the establishment of a velocity gradient and consequently a certain force is required to deform the liquid at a certain rate. This force $F$ is related to the dimensions of the system and the velocity of the plates by the *viscosity* $\eta$

$$F/A = \eta \, (dx/dy)/dt \tag{1}$$

where $A$ is the area of the sample over which the force is applied, and the term $F/A$, the force normalized to the dimensions of the system, is given the name *stress*, $\sigma$. The term $dx/dy$ is the deformation of the sample, again normalized to the dimensions of the system, and is called the *shear strain*, $\gamma$. The change in strain per time $d\gamma/dt$ is the strain rate, $\dot{\gamma}$. Eq. (1) can thus be rewritten as

$$\sigma = \eta \, \dot{\gamma} \tag{2}$$

The most commonly used unit of viscosity is the poise, equivalent to g/cm sec.

Viscosity is a tangible characteristic of liquids. For example, a viscous liquid like glycerin ($\eta = 15$ poise at 20°C) requires a considerably larger force to be deformed at a certain rate than does a low-viscosity liquid such as water ($\eta = 0.01$ poise at 20°C). The force required to maintain a given strain rate can also be affected by the inclusion in the liquid of rigid particles, as shown in Fig. 1b. Here the total volume between the plates is the same as before, but now a certain fraction of the volume is occupied by nondeformable particles, and any movement of the plates must be accomplished by a deformation in the proportionately reduced volume of fluid. The particles also change the velocity gradient by tending to "tie together" fluid layers locally. As a consequence the force required to move the plate at a given velocity is greater, the greater the volume fraction of rigid particles. As a result the observed viscosity of the solution (fluid + particles) is increased. In real systems the viscosity is affected not only by the volume fraction of rigid particles but also by the shape of the particles, an effect first accurately explained by Simha (1940). Any deviation from a spherical shape tends to increase the viscosity of the system still further, and very extended particles (e.g., DNA) may have a considerable effect on solution viscosity.

In sufficiently dilute solution, the protein–polysaccharide molecules found in molluscan mucins may be thought of as "rigid" particles affecting the viscosity of the solution. The viscosity of these solutions thus depends on the volume fraction of the protein–polysaccharide molecules, and on their shape. To be more precise, the viscosity depends on the effective volume fraction of the molecules, a distinction that must be made because these molecules are porous. As discussed earlier, each protein–polysaccharide molecule is a three-dimensionally arranged, randomly configured chain, and it tends to enclose within its coils a volume of fluid much as a sponge encloses water. This fluid to some extent travels with the protein–polysaccharide molecule and thus must be included in the rigid-particle volume fraction. In general, the more expanded the molecule, the greater its hydrodynamic volume. The quantification of this parameter—that is, the precise determination of the effective hydrodynamic volume of a polymeric molecule—is a complicated subject and lies well outside the realm of this chapter. Further information can be found in Tanford (1961). Due to their random configuration, the large protein–polysaccharide molecules of molluscan mucins are generally spherical in shape (e.g., *B. canaliculatum* pedal mucin, Kwart and Shashoua, 1957). Only if the molecule is linear and very highly expanded will its shape differ appreciably from that of a sphere.

Although thorough description of the configuration of polymers in solution is not appropriate here, it is accurate to generalize to the following extent: The overall volume occupied by a protein–polysaccharide molecule depends both on its molecular weight and the polyelectrolyte nature of its chains. The larger the molecular weight and the greater the charge density, the more expanded the molecule, the larger its effective volume, and thus the greater its effect on the viscosity of a solution.

This generalization is valid only if the molecules are present in a solution so dilute that the likelihood of interaction between solute molecules can be ignored (very roughly, less than 0.1–0.5 g organic matter/100 g water). As the solute concentration is increased the protein–polysaccharide molecules begin to interact more frequently. These interactions—for example, short-term bonding between molecules—will increase the "friction" between fluid layers, and thus greatly increase the solution's viscosity. In these intermediate concentrations (approximately 0.1–1.0 g organic matter/100 g water), protein–polysaccharide solutions act as very viscous fluids. Note that the concentration ranges cited here are only rough approximations, and will vary from one mucin to the next.

## C. Elasticity

If the volume fraction of protein–polysaccharide molecules is increased beyond the range just described, at some concentration the solution ceases to behave solely as a viscous liquid. If grasped with forceps and pulled, the material forms long strings; if released it may snap back a bit, tending to return to its original shape. These new properties are those of an elastic solid. As shown previously, the force required to deform a viscous liquid is proportional to the rate of deformation, and even liquids as viscous as treacle and tar slowly flow under the constant influence of gravity. In contrast, solids exhibit the property of elasticity, where the force required to deform the material is proportional to the amount of deformation rather than the rate of deformation. For a purely elastic solid, deformed as shown in Fig. 2a, the force of deformation $F$ is

$$F/A = G \, (dx/dt) \tag{3}$$

or
$$\sigma = G \, \gamma \tag{4}$$

where $G$, the *shear modulus*, is a characteristic of the solid, quantifying its stiffness. As a result of this relationship between force and deformation, if the stress is removed from a deformed elastic solid the material returns to its original shape. A further consequence of this relationship is that the energy (force × distance) used in deforming the material is stored, and can be regained as the material returns to its original shape.

As with viscosity, the elasticity shown by concentrated solutions of protein–polysaccharide complexes is related to the configurational statistics of these molecules. As the effective volume fraction of protein–poly-

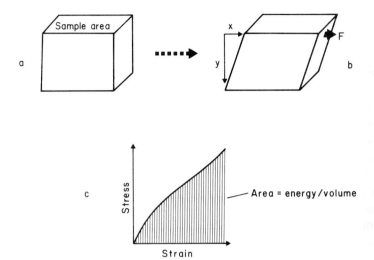

**Fig. 2.** Elasticity. (a) and (b) A given stress (force/sample area) causes a specific strain (dx/dy) in a purely elastic solid. (c) The area under a stress–strain curve is the product of stress times strain, or (force/area) × (distance/distance) = energy/volume.

saccharide molecules is increased (either by adding more molecules to a set volume of fluid, or by causing the molecules present to expand), inter-action between molecules becomes more likely. At some concentration the effective volume occupied by the protein–polysaccharide molecules is equal to the volume of the solution, and each protein–polysaccharide molecule is pressed up against its neighbors. This is not to say that water is excluded from the system; far from it. As explained previously, the neg-atively charged protein–polysaccharide chains are highly expanded, in-corporating within their effective volume large amounts of water. In fact the effective volume of pig gastric mucus glycoprotein is thought to equal the solution volume at a glycoprotein concentration as low as 2.5% (Pain, 1980). Once the concentration is such that adjacent molecules are pressed against each other, they must interact. These interactions may take sev-eral forms. For example, the chains of adjacent molecules, under the in-fluence of thermal agitation, may become entangled much as a bundle of rope becomes tangled if tossed about. If two molecules thus entangled are rapidly pulled apart, the entanglements do not have sufficient time to be-come disentwined, and the molecules over at least short times are effec-tively cross-linked. Under constantly applied stress, however, the contin-ual thermal shifting of the molecules allows them to disentangle, and a material cross-linked in this fashion will gradually flow. Cross-links of a more permanent nature may also be formed. For instance, adjacent chains of protein or polysaccharide can be bound to each other by hydrogen bonds or hydrophobic interactions. Individual hydrogen bonds or hydrophobic interactions are quite weak and it seems likely that they would continually be formed and broken as the protein–polysaccharide molecules rearranged. If, however, an appropriate arrangement of adjacent chains occurs, several bonds may form in close proximity to each other and these cooperative bonds may be quite strong (as for example in silk and collagen), and can thus serve as long-term cross-links. Finally, covalent linkages between chains are also possible, usually in the form of S—S bonds between cysteine molecules of two protein chains, and, as with cooperative hydrogen or hydrophobic bonds, form essentially permanent cross-links.

The effect of intermolecular cross-links is to tie the entire mucous solution together to form a *gel network*. If a cross-linked polymeric material is deformed, the network within it must also be deformed; and because individual molecules, being cross-linked, are no longer free to slide past each other, the average shape of individual protein–polysaccharide chains must change. Imposing this change in shape on what was a randomly arranged molecule in effect imposes some measure of order on the molecule and decreases its entropy. It can be shown through thermodynamic theory that a force is required to bring about this decrease in entropy; the larger the deformation, the larger the force. This force thus acts as an elastic restoring force; once the force is removed each molecule again becomes randomly configured, and in the process the material as a whole returns to its original shape. Thus materials formed of cross-linked, randomly configured chains have the characteristics of elastic solids: Stress is proportional to strain, and energy is stored in deformation. The statistics and thermodynamics of this process have been well documented and are expressed as the theory of rubber elasticity after the material in which they were first explained (see, e.g., Flory, 1953).

The stiffness $G$ of a cross-linked material is proportional to the number of chains per volume, the length of chains between cross-links, and the temperature. It can be shown theoretically (Flory, 1953; Alexander, 1968) that

or

$$G = \rho RT/M \tag{5}$$

$$M = \rho RT/G \tag{6}$$

where $\rho$ is the density of the protein–polysaccharide component of the mucus (in g/ml); $R$ is the universal gas constant ($8.3 \times 10^7$ erg/deg mol); $T$ is absolute temperature; and $M$ is molecular weight between cross-links. Thus from knowledge of the shear modulus and the weight proportion of protein–polysaccharide, it is possible to estimate an average weight between cross-links. As will be shown later, various values may be obtained for the stiffness of a mucin depending on how the material is tested. The value of $G$ used when estimating $M$ for a mucin is thus somewhat arbitrary, but is usually that shown by the material when it is initially deformed, before cross-links due to entanglements can disentwine.

## D. Testing Procedures

Although the cross-linking of protein–polysaccharide chains in a mucus gel gives the material some measure of elasticity, it does not in reality completely banish the viscous nature of the material. Consequently molluscan mucins simultaneously exhibit properties of both liquids and solids, and are described as being viscoelastic. Thus, depending on their chemistry and concentration, protein–polysaccharide solutions may fall anywhere on a viscoelastic continuum stretching from dilute viscous solutions with only a slight elastic stiffness, to concentrated, stiff gels with very little viscosity.

Due to their somewhat viscous nature, molluscan mucins exhibit properties that are time dependent, and these properties must be very carefully measured and described if they are to be accurately used in examining the material's biological function. For example, a material that appears quite solid if tested for short periods of time might flow over long periods. A short-term test thus might be very misleading if used as a basis for speculating on the material's possible use as a permanent structural element. Fortunately a number of standard tests have been devised to quantify the properties of viscoelastic materials.

### 1. Stress–Strain Tests

A stress–strain test is performed by deforming a mucus sample in shear as shown in Figs. 1a and 2, the stress required to deform the material being measured as a function of strain. If the mucus were a purely elastic solid, the stress–strain curve would be independent of shear rate; but for a real mucus, which is viscoelastic, the greater the strain rate the greater the stiffness $G$. Thus a plot of shear modulus versus strain rate is one method for indicating the relative importance of the viscous and elastic components of a material. A primarily elastic material has a high $G$ at low strain rates, and $G$ increases little with increased strain rate; a primarily viscous material shows the converse behavior (see Fig. 3). The same information may be obtained by cyclically deforming the sample. For example, if a purely elastic material is deformed and then allowed to return to its original shape, the force required to hold the material at any deformation is the same regardless of whether the strain is increasing or decreas-

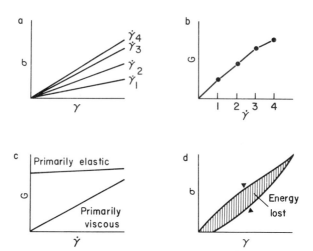

**Fig. 3.** Interpretation of stress–strain results. (**a**) The precise stress–strain curve shown by a viscoelastic material depends on the applied strain rate $\dot{\gamma}$; the higher $\dot{\gamma}$ the greater the slope of the curve. The slope $= \sigma/\gamma = G$, the shear modulus. (**b**) Graph (**a**) replotted as $G$ versus $\dot{\gamma}$. (**c**) The shape of the $G$ versus $\dot{\gamma}$ plot indicates the relative contributions of elastic and viscous processes to the overall properties of the material. (**d**) Energy is lost as heat in the cyclic extension of a viscoelastic material. The energy lost, expressed as a percentage of the total energy used in extending the material, is the material's hysteresis.

ing. Further, the energy put into the material on deformation (expressed as the area beneath the stress–strain curve) is returned as the strain is decreased. In contrast, in deforming a viscoelastic material some energy goes to deforming the viscous component; this energy is lost as heat and cannot be returned when deformation is decreased. Thus when cyclically deformed, a viscoelastic material exhibits some hysteresis (Fig. 3d). The more viscous the material in proportion to its elastic stiffness, the greater the hysteresis.

## 2. Stress-Relaxation Tests

A stress-relaxation test is performed by rapidly deforming a sample to a set strain and measuring the stress required to maintain this deformation as a function of time. A purely elastic solid does not stress-relax because by definition the stress required to maintain a set strain is independent of time. If, however, the material has a viscous component, the stress required initially to rapidly deform the material is partially a result of the material's viscosity; the stress decreases with time and the material flows or "relaxes" into its new shape. The rate and extent to which a material stress-relaxes is a measure of the relative contributions of the viscous and elastic components to the material's properties. For example, a viscoelastic material in which the cross-links between chains are due solely to entanglements will for a period of time (the initial few seconds of the test) have an appreciable elastic modulus. However, as the entanglements gradually disentwine, the material flows, and over long periods of time the material behaves as a fluid and $G$ goes to zero (see Fig. 4). The more viscous the material in proportion to its elastic stiffness, the more quickly the stress relaxes. If, on the other hand, permanent cross-links are present, the material stress-relaxes but after a certain time reaches an equilibrium modulus $G_{eq}$. Stress-relaxation tests are thus useful in quantifying the reaction of a material to long-term stresses (i.e., those applied for periods of seconds to many hours).

### 3. Creep Tests

A creep test is the converse of a stress-relaxation test, performed by subjecting a sample to a constant stress and measuring strain as a function of time. The measured variable here is the material's compliance, $\gamma/\sigma = J$. $J$ thus $= 1/G$. A purely elastic material does not creep because by definition strain is proportional to stress and independent of time. However, a viscoelastic material does creep as the viscous component of the material gradually flows. If the cross-links of the material are not permanent, strain increases indefinitely, and over long periods the material flows at a constant rate in response to the constant stress, thus behaving as a viscous fluid (see Fig. 5). In contrast, a material with permanent

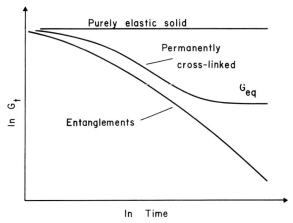

**Fig. 4.** Stress-relaxation test results. The shape of the stress-relaxation curve provides information about material properties over long time periods, and indicates whether a material is permanently cross-linked or not. $G_{eq}$ is the equilibrium shear modulus.

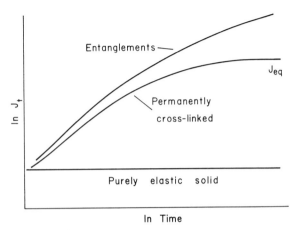

**Fig. 5.** Creep test results. The shape of the creep curve provides information about material properties over long time periods and indicates whether a material is permanently cross-linked or not. $J_{eq}$ is the equilibrium shear compliance.

cross-links reaches an equilibrium compliance and thus behaves as a solid even over long periods of time. To a first approximation, the compliance at a certain time after the start of a creep test $J_t = 1/G_t$. More precise methods of estimating $J_t$ from $G_t$ (or vice versa) are available (see Ferry, 1980).

### 4. Dynamic Tests

Although stress-relaxation and creep tests are useful in quantifying the properties of materials over long periods of time, they are of little use over short periods. This is for purely practical reasons. In a stress-relaxation test, for instance, it takes some period of time to deform the material to its initial strain. During this time the material may be relaxing, but because the material is also being deformed this relaxation cannot be measured. Thus for describing material properties over periods of less than approximately 1 sec, stress-relaxation and creep tests are useless. There are many biological situations where the short-term properties of mucus are important for their functioning (gastropod locomotor mucus, ciliary transport, etc.), and it is fortunate that another form of testing, dynamic testing, has been developed to measure material properties at short times. This form of testing is based on the fact that if a material is strained such that the strain changes sinusoidally with time, the maximum strain is separated from the maximum rate of strain by some period of time (see Fig. 6). Expressed mathematically,

$$\gamma = \sin{(\omega t)} \tag{7}$$

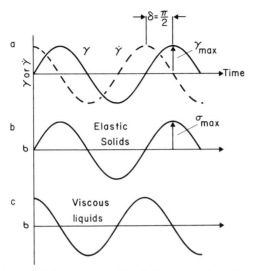

**Fig. 6.** Dynamic tests. (a) The stain rate $\dot\gamma$ leads the strain $\gamma$ by a phase angle $\delta$ of $\pi/2$ radians. (b) For a purely elastic solid, the stress $\sigma$ is in phase with strain. (c) For a purely viscous liquid, stress is in phase with the strain rate. $G^*$, the complex modulus $= \sigma_{max}/\gamma_{max}$.

where $\omega$ is the angular velocity (radian/sec) and $t$ is the time in seconds. By reference to Eq. (4), for a pure elastic solid

$$\sigma = G\gamma = G \sin (\omega t) \tag{8}$$

and the stress is in phase with the strain. In contrast, by reference to Eq. (2), for a pure viscous liquid

$$\sigma = \eta \, d\gamma/dt \tag{9}$$

$$\sigma = \eta \, d(\sin \omega t)/dt = \eta\omega \cos (\omega t) \tag{10}$$

and the stress is out of phase with the strain; it leads it by an angle $\delta$ of $\pi/2$ radians. Viscoelastic materials, being intermediate between purely elastic solids and purely viscous liquids, exhibit a phase shift somewhere between 0 and $\pi/2$ radians, and thus by measuring the phase shift the relative values of elasticity and viscosity can be determined. At the same time that the phase shift is measured, it is possible to measure the peak amplitude of both stress and strain and to express their ratio as a stiffness or modulus; this modulus is known as the complex modulus $G^*$. The in-phase or elastic contribution to $G^*$ is $G^*(\cos \delta)$, and is called the storage modulus, $G'$. As its name implies, the storage modulus is a measure of the energy stored during each deformation cycle. $G' = G^*$ when the phase shift is zero, and $G' = 0$ when the phase shift is $\pi/2$. Conversely, the out-of-phase or viscous contribution to the complex modulus is $G^*(\sin \delta)$ and is called the loss modulus, $G''$. The ratio of $G''$ to $G'$ is $\tan \delta$, a measure that at low values can be related to the hysteresis seen in the simpler cyclic stress–strain tests described previously. Dynamic tests can be conducted at high frequencies (for mucus samples usually up to 600 radians/sec), corresponding to testing the material at short time periods. Information gained from dynamic tests is related to that of creep and stress relaxation tests by considering that time $t$ in a static test is equivalent to $1/\omega$, the period in a dynamic test. Approximate methods for converting data between dynamic and static tests are given by Ferry (1980).

Various sorts of apparatus are used to conduct the tests described here. When dealing with soft solids such as mucus it is usually most convenient to confine the material as a thin layer between two plates. One plate may then be rotated or translated to impose a quantifiable deformation on the material, and the resulting force placed on the other plate can be measured with an appropriate transducer. Examples of such testing apparatuses include the commercially available Weissenburg rheogoniometer, and the home-built analog used by Denny (1980a; Denny and Gosline, 1980), shown in Fig. 7. For a full discussion of tests and testing machines for soft solids see Ferry (1980) or Dorrington (1980).

## IV. Mechanical Properties of Molluscan Mucins

With these theories and testing procedures in mind, it is now possible to examine the mechanical properties of various molluscan mucous secretions. This examination will be fragmentary, however, for at present only isolated examples of molluscan mucins have been studied. Whereas a quick perusal of the literature reveals numerous qualitative statements that suggest that molluscan mucins indeed span the viscoelastic continuum, quantitative data concerning material properties are available only for the examples cited here.

### A. Hypobranchial Mucus

The biological function of the hypobranchial gland mucin of whelks has not been extensively studied, but this secretion is reported to serve in clearing the pallial cavity of debris (Hunt, 1970; Yonge, 1947). Although the hypobranchial mucus of whelks is chemically the best characterized molluscan mucus secretion, little work has been done to determine the material's *in vivo* mechanical properties and how they relate to the mucin's function.

Mucus collected from excised *B. undatum* hypobranchial glands ex-

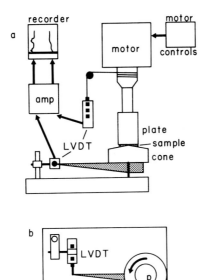

**Fig. 7.** Schematic diagram of a cone and plate testing machine for soft solids. (**a**) Side view. (**b**) Top view. The mucus sample is held between a fixed, small-angle cone (c) and a rotatable plate (p). Rotation of the plate by the electric motor places a strain on the sample. As the plate rotates, a string supporting the core of a linearly variable differential transformer (LVDT) is wound onto a capstan. The voltage output from the LVDT is proportional to the core displacement, and is thus a measure of strain. The strain rate is set with the motor controls. The cone is supported by a torsion bar. Stress placed on the sample causes this bar to twist, displacing the core of a second LVDT. The amount of twist is proportional to stress. The voltage output from this second LVDT is thus a measure of stress. For more detailed information, see Denny and Gosline (1980). (Redrawn from Denny and Gosline, 1980, courtesy of the Journal of Experimental Biology.)

hibits viscoelastic properties (Hunt, 1970). If stirred with a rotating rod, the mucus elasticity recoils when the stirring is stopped, and the mucus has sufficient tensile strength to self-siphon out of a beaker. However, the mucus is readily dispersed if stirred in a 0.6 $M$ NaCl solution, indicating that the gel network responsible for the short-term elasticity of the mucus is not permanently cross-linked. Ronkin (1955) showed that the hypobranchial mucus of *B. canaliculatum* lowers the surface tension of seawater by up to 30%, and suggests that this factor functions as an advantage in separating and ingesting detritus particles. The viscosity of fresh *B. canaliculatum* mucus measured by Ronkin was quite low, $9 \pm 1.6$ centipoise (i.e., approximately nine times as viscous as water), whereas Kwart and Shashoua (1957) report a figure for the same mucus of 60–100 centipoise.

The difference in values is presumably due to differences in the concentration of the mucus solution tested. These mucins lie toward the viscous end of the viscoelastic continuum, and further study of their mechanical properties would be informative. To date the stress–strain, stress-relaxation, creep, and dynamic characteristics of hypobranchial mucus have not been determined.

## B. *Helix pomatia* Pedal Mucus

As with most gastropods, the snail *Helix pomatia* produces a number of different types of mucous secretions serving various functions (Simkiss and Wilbur, 1977). Among these are the mucins that coat the animal's dorsal surface, serving to keep the skin moist; and the pedal mucus, which functions both as a glue allowing the snail to adhere to the surface on which it crawls, and as an aid in locomotion. The mechanical properties of these two mucins have been examined by Simkiss and Wilbur (1977).

The animal was stimulated to secrete mucus from its dorsal epithelium by being shocked with an electric current (20 V, 50–100 shocks/min). The mucus produced under these circumstances was clear and watery, with a viscosity only 25% greater than that of water, and no elastic characteristics were noted. Further characterization of this dorsal epithelial mucus was not attempted.

The pedal mucus, along with the mucus coating the side of the foot, was collected using a glass rod to roll the mucus off the epithelium. Mucus thus collected was 91.8% water (w/w). Samples from several animals were pooled and their mechanical properties tested dynamically in a Weissenburg rheogoniometer. The results of these tests are shown in Fig. 8a. At low frequencies the mucus behaves as an elastic solid; that is, the storage modulus $(G')$ is many times greater than the loss modulus $(G'')$. Although behaving as a solid, the mucus is not very stiff—approximately 1000 times less stiff than rubber. The storage modulus at low frequencies can be used to calculate an approximate molecular weight between cross-links of $4 \times 10^5$. At frequencies above 0.6 radians/sec the loss modulus rises dramatically until at approximately 40 radians/sec it surpasses in magnitude $G'$. An explanation for this unusual behavior is not immediately apparent. At sufficiently high frequencies the $G''$ for any cross-linked network rises in this manner as the rate of deformation due to the imposed strain overtakes the rate at which the polymer molecules are thermally rearranging (see Ferry, 1980). However, the frequency at which this occurs is usually on the order of 1000 radians/sec or higher, and it is difficult to see how this mechanism could operate at the low frequencies used by Simkiss and Wilbur (1977). The properties found for freshly collected

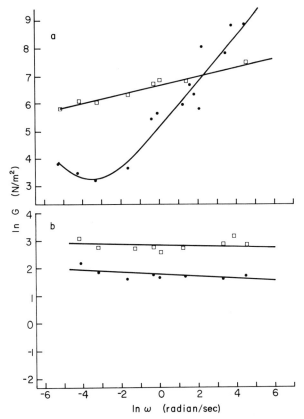

**Fig. 8.** Dynamic mechanical properties of the pedal mucus of the snail *Helix pomatia*. (**a**) Fresh mucus. (**b**) Mucus refrigerated overnight. Open squares, storage modulus (G′); closed circles, loss modulus (G″). (Redrawn from Simkiss and Wilbur, 1977, courtesy of the Zoological Society of London.)

mucus change drastically if the mucus is allowed to sit refrigerated overnight or an anionic detergent is added to the secretion (Fig. 8b). In these cases the complex modulus (G*) is lower by a factor of approximately 20 and the material behaves as a fairly viscous solid throughout the range of frequencies tested. Simkiss and Wilbur point out that their data should be interpreted with care because the mucus samples tested were pooled and heterogeneous. In any case, for at least certain frequencies of deformation the mucus behaves as a viscoelastic solid, indicating the presence of a cross-linked, kinetically free gel network. The swelling behavior of this network and the nature of its cross-links have not been examined, nor has the function of these mechanical properties in adhesion and locomotion been studied.

## C. *Patella vulgata* Pedal Mucus

In a manner similar to *H. pomatia,* the limpet *P. vulgata* uses pedal mucus as an adhesive and an aid in locomotion. These intertidal animals are frequently subjected to strong forces tending to pull them off the substratum, resulting either from water flow accompanying breaking waves or from the attacks of predators; and it is advantageous to the animal for the pedal mucus to be able to resist these forces. In addition the pedal mucus must allow the animal to move about to forage. The mechanical properties of this mucin have been examined in some detail by Grenon and Walker (1980). Pedal mucus was collected from several animals by rolling the mucus off the foot with a glass rod; samples were pooled and kept frozen until their mechanical properties were measured. The mucus was found to be 90.1% water (w/w), the remainder being composed of 6.8% organic matter and 3.1% salts. The solubility of the mucus in various solutions was tested and the only compounds found effective in dissolving the material (without severely altering its chemical structure) were 4 $M$ NaOH and 10% alkaline sodium sulfide. Grenon and Walker interpret these results as evidence for electrovalent and possibly disulfide bonds acting as cross-links in the mucus gel network. $N$-Acetylcysteine, a compound that breaks S—S bonds, was not found to dissolve the pedal mucus, although the 30-min period allowed for dissolution may have been too short to discount definitely the presence of S—S bonds as cross-links. For example, the pedal mucus of the terrestrial slug *A. columbianus* does dissolve in 1% 2-mercaptoethanol (another S—S bond cleaver), but the effect is only observed after 24 h at 40°C.

The mechanical properties of *P. vulgata* pedal mucus were tested with a creep test, and the results are shown in Fig. 9, along with these results reinterpreted and expressed as stress-relaxation data. At short times the material behaves as an elastic solid with a modulus of approximately 400 N/m$^2$, corresponding to an approximate molecular weight between cross-links of $4.1 \times 10^5$, compared to the molecular weights for the protein portion of the molecule (determined from SDS gel electrophoresis) of from $2.3 \times 10^4$ to $1.95 \times 10^5$.

At long times (i.e., low frequencies), the pedal mucus continues to flow (or stress-relax), indicating that the cross-links of the gel network, though stable for short times, are not permanent. The viscosity of the material during flow over long periods is $1.74 \times 10^7$ poise, this high figure presumably being due to considerable interaction among protein–polysaccharide

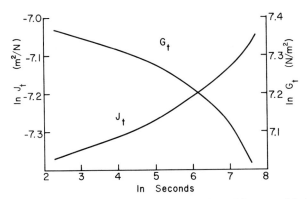

**Fig. 9.** Creep ($J_t$) and stress-relaxation ($G_t$) properties of the pedal mucus of the limpet *Patella vulgata*. There is no evidence of an equilibrium compliance or modulus, indicating that the material is not permanently cross-linked. (Redrawn from Grenon and Walker, 1980, courtesy of Comparative Biochemistry and Physiology and Pergamon Press.)

complexes. Tests on the mechanical properties of this mucus were apparently only conducted on previously frozen material, and what effect this might have had on the results obtained is not known.

Grenon and Walker (1980) point out that the short-term elasticity of this mucus would allow it to function as an effective glue, resisting the rapidly applied forces of waves. However, the manner in which the pedal mucus properties function in locomotion has not been examined.

### D. *Ariolimax columbianus* Pedal Mucus

As with the two examples just discussed, the terrestrial slug *A. columbianus* uses pedal mucus both as an adhesive and as an aid in locomotion. The manner in which the mechanical properties of this mucus allow this single-footed animal to walk on glue have been examined by Denny (1979, 1980a,b, 1981) and Denny and Gosline (1980).

Pedal mucus is collected from a slug by allowing the animal to crawl on a glass rod. If the rod is slowly rotated the animal must continually climb up the rod, and a sizeable sample (0.1–0.3 ml) of mucus may be collected. This mucus was found to consist of 96–98% water (w/w), the remainder being salts and a glycoprotein (Denny, 1980a,b). At 25°C no effective nondegradative means of dissolving the mucus was found, but at 40 or 55°C three compounds were found effective. The first of these, 1% 2-mercaptoethanol, as described previously, breaks S—S bonds, indicating the possibility of this sort of covalent cross-linking in *A. columbianus* pedal mucus. The two other compounds, 8 *M* urea and 8 *M* guanidinium HCl, both act by disrupting hydrogen or hydrophobic bonds, or both, indicating that a second form of cross-link occurs in the mucus, in this case of a noncovalent and presumably weaker variety.

The swelling behavior of *A. columbianus* pedal mucus was examined as described by Denny (1979). Samples were collected from a number of slugs and each sample immediately placed in 25 ml of a test solution. After 6–7 h the swollen (or shrunken) samples were grasped with a forceps, removed from the solution, and weighed. Samples were then dialyzed against repeated changes of distilled water to remove solute molecules of the test solution, dried, and reweighed. The degree to which the sample is swollen in each solution is then expressed as the grams of water present in the swollen sample per gram of nondialyzable solid matter. The results are shown in Table III, and confirm the polyelectrolyte nature of the gel network. In distilled water the mucus glycoprotein network contains almost 1 liter of water/g glycoprotein. As counterions ($Na^+$) are added, the glycoprotein molecules are less affected by the negative charges present on their polysaccharide chains; they assume a more compact structure, and the gel network contracts until in a 1 $M$ NaCl solution only 105 ml of water is contained per gram of glycoprotein. If the pH is lowered considerably below the p$K$ of the polysaccharide carboxyl groups (about 3.75), the glycoprotein ceases to be negatively charged and the molecules contract still further. These results are similar to those found in vertebrate mucins (Pain, 1980).

The mechanical properties of freshly collected pedal mucus were examined with a variety of tests. At low strains and short times the material behaves essentially as an elastic solid. In a stress–strain test, stress increases in proportion to strain (with a modulus of about 200 N/m²) and the viscous nature of the material is evidenced only by the slight hysteresis observed in cyclic tests (Fig. 10). The molecular weight between cross-links calculated using a $G$ of 200 N/m² is $4.2 \times 10^5$. Dynamic tests at frequencies of 0.6–600 radian/sec tell much the same story (Fig. 11): In

### TABLE III
### The Swelling Behavior of Slug Pedal Mucus in Various Ionic Solutions

| Solvent | Equilibrium hydration (g $H_2O$/g dry mucin) |
|---|---|
| Distilled water (pH 6) | $954.7 \pm 432.3$ |
| 0.1 $M$ NaCl | $206.1 \pm 38.7$ |
| 1.0 $M$ NaCl | $138.8 \pm 24.4$ |
| pH 2.1 (HCl) | $59.7 \pm 12.6$ |

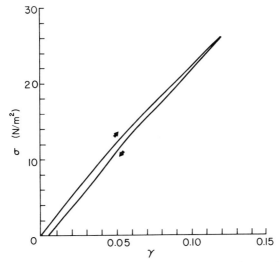

**Fig. 10.** A stress–strain curve for the pedal mucus of the slug *Ariolimax columbianus* at low strains. The viscosity of the material is apparent only in the slight hysteresis (6.9%). The strain rate for the test was 0.048/s. (Reprinted from Denny and Gosline, 1980, courtesy of the Journal of Experimental Biology.)

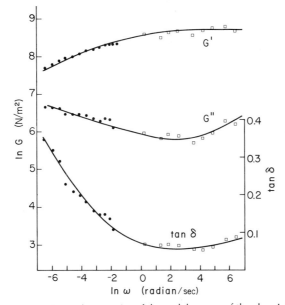

**Fig. 11.** Dynamic mechanical properties of the pedal mucus of the slug *Ariolimax columbianus*. Open squares are data derived directly from dynamic tests. Closed circles are transformed from the stress-relaxation data shown in Fig. 12 using methods described in Ferry (1980). Lines fitted by eye.

this range of frequencies $G'$ is much greater than $G''$. The dramatic increase in $G''$ at frequencies above 1 radian/sec seen in *H. pomatia* pedal mucus were not observed for this pedal mucus. The behavior of *A. columbianus* pedal mucus over long periods of time was examined with stress-relaxation tests, and was found to be similar to that of *P. vulgata* (Denny and Gosline, 1980; see also Fig. 12): A cross-linked elastic network is present over short times but the cross-links are not permanent and the material relaxes (or creeps) without reaching equilibrium. This is reflected in the decrease in $G'$ and increase in $G''$ at very low frequencies (see Fig. 11). Thus for small strains *A. columbianus* pedal mucus appears to be similar to that of *P. vulgata*.

Though useful in determining the network structure of pedal mucus and the material properties at small strains, the tests just described fail to give an adequate picture of how the mucus might behave in its normal biological function. For example, the pedal mucus of *A. columbianus* is sandwiched between the ventral surface of the animal's foot and the substratum in a layer approximately 10–20 $\mu$m thick. The animal moves by passing a series of muscular waves along the foot (Denny, 1981; see also Volume 5, Chapter 8), and each point on the foot is moved forward about 1 mm with the passing of a wave. Thus the mucus beneath the foot may be repeatedly subjected to strains of 50–100.

The behavior of *A. columbianus* pedal mucus at these high strains has been examined (Denny, 1980a, 1981; Denny and Gosline, 1980) and found to be quite different from the properties shown at low strains (Fig. 13). In Fig. 13 the upper vertical axis is a measure of stress as with previous figures; but the horizontal axis, rather than being a measure of strain, is a

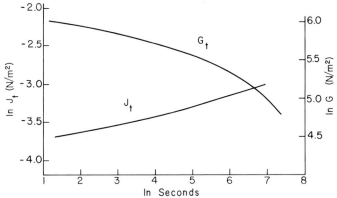

**Fig. 12.** Creep and stress-relaxation curves for the pedal mucus of the slug *Ariolimax columbianus*. There is no evidence of an equilibrium compliance or modulus, indicating that the material is not permanently cross-linked. (Redrawn from Denny and Gosline, 1980, courtesy of the Journal of Experimental Biology.)

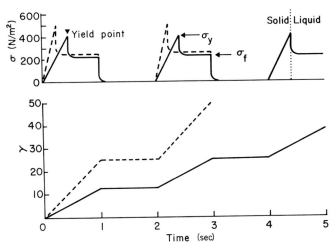

**Fig. 13.** Mechanical properties of slug pedal mucus at high strains: stress ($\sigma$) and strain ($\gamma$) as a function of time. The material behaves as a solid up to a strain of 5–6 and then yields to form a liquid. After the material is allowed to "heal" for a second, its solidity is recovered. The yield stress $\sigma_y$ and the flow stress $\sigma_f$ are strain-rate dependent. (Redrawn from Denny and Gosline, 1980, courtesy of the Journal of Experimental Biology.)

measure of time where for part of the time the mucus sample is deformed at a constant strain rate, and for part of the time the mucus is held unstressed. When the mucus is first strained it behaves as an elastic solid as discussed before, the stress increasing with strain. However, at a strain of between 5 and 6 the material abruptly yields and with further strain shows a constant stress. It will be remembered from the definition of viscosity that a constant stress at a constant strain *rate* is characteristic of a viscous liquid. Thus the pedal mucus at the point of yielding has been converted from an elastic solid to a viscous liquid, presumably by a disruption of the cross-links in the gel network. The viscosity of the mucus in this liquid form is 30–50 poise (Denny and Gosline, 1980). The strain at which the mucus yields is independent of strain rate, but the yield stress and the stress required to cause the resulting flow at that strain rate both increase as the strain rate is increased (Fig. 14). These properties, though interesting, are not unique. Any soft gel network if sufficiently deformed will "break" and flow, as can easily be observed if one stresses a piece of ordinary gelatin. However, pedal mucus is highly unusual in that once disrupted, its gel network can quickly reform. For example, in the test shown in Fig. 13, if after the mucus has yielded and shows the properties of a fluid, the strain rate is abruptly brought to zero, the stress drops to zero. This behavior is characteristic of fluids at zero $\dot{\gamma}$. The material is

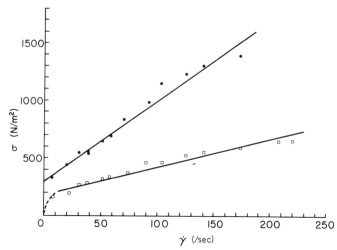

**Fig. 14.** The dependence of yield stress (closed circles, $\sigma_y$) and flow stress (open squares, $\sigma_f$) on strain rate. (Redrawn from Denny and Gosline, 1980, courtesy of the Journal of Experimental Biology.)

then allowed to "rest" for 1 sec and a constant strain rate is subsequently applied. If the material were still in a fluid state the stress should immediately rise to a constant value and remain there. However, this is not the case; the stress rises in proportion to the strain until at a strain of 5–6 the material again yields. Apparently in the 1 sec of "rest" the gel network has "healed," and this "yield–heal" cycle can be repeated 20–30 times before the mucus gel shows signs of failing to recover its solidity during the rest period. The macromolecular mechanism accounting for the yield–heal characteristics of this mucin are not known, though it may be speculated that a mechanism as discussed later (Section V) may be operating here.

The ability of A. *columbianus* pedal mucus to switch reversibly between fluid and solid properties aids in the animal's locomotion. At any one moment during movement, parts of the foot are stationary relative to the ground, while other parts are moved forward in pedal waves. It is advantageous for the mucus beneath the stationary portion of the foot to be an elastic solid, capable of resisting the forces required to slide the foot forward. Similarly, the forces of forward movement are minimized if the moving portions of the foot slide over mucus in its fluid form. Because any point on the foot is alternately stationary relative to the ground and moving forward as pedal waves pass along the foot, the pedal mucus can be most functional only if its properties can alternate between those of a solid and those of a fluid. For a more complete explanation of this phenomenon, see Denny (1980a,b, 1981) and Denny and Gosline (1980).

The mucous secretions for which mechanical properties are known form only a very small and somewhat biased group. All of the mucins discussed here for which detailed information is known are pedal mucins, primarily because these are the mucins most easily collected under natural conditions. However, much could be learned by examining the mechanical properties of mucins serving other functions. For example, many—if not most—molluscan mucous secretions are associated with ciliary transport (e.g., the feeding mucus of mussels and clams). Litt *et al.* (1977) have shown (using ciliated frog palate as a model system) that transport by cilia requires certain properties of mucus; in particular, the mucus must have some elasticity. They found that mucus with a storage modulus of about 1 N/m$^2$ was transported most effectively by the frog palate; solutions with $G'$ either smaller or larger were transported less effectively. By this criterion the pedal mucins discussed here, all with $G'$ of around 100 N/m$^2$, would not be effective in ciliary transport. It will be interesting to see whether the mechanical properties of molluscan mucus follow this pattern. Other examples abound where a knowledge of the relevant physical properties of mucus would be useful in examining the functions of molluscs. What physicochemical factors control the "stickiness" of mucins, particularly those used in feeding? Production of pedal mucus is energetically quite costly (Denny, 1980b); however, in that the mucin is mostly water, this cost may actually be the least expensive solution to the problem of adhesive locomotion. Have animals evolved to produce mucins as a minimum-cost answer to various functional problems? Detailed knowledge of the physical properties of mucus provides insight into both the biological function of the material and the material's macromolecular architecture, and thus would be a logical focus of effort for the further examination of questions such as these concerning molluscan mucous secretions.

## V. A Model for Mucus Structure

Present knowledge of the macromolecular architecture of molluscan mucus is not sufficiently precise to allow a reliable model to be drawn. However, many of the properties of the pedal mucins discussed previously are similar to those of well-described vertebrate mucins, and a description here of one model for vertebrate mucins may be helpful. This model is that of Allen *et al.* (1976) for pig gastric mucus, and is shown schematically in Fig. 15.

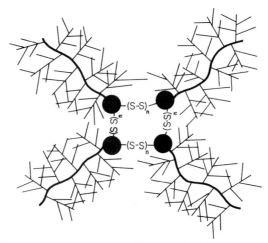

**Fig. 15.** A diagrammatic representation of the basic "building-block" molecule of pig gastric mucus. Four units, each with a molecular weight of $5 \times 10^5$, are held together by disulfide bonds to form a large, globular composite. These composites interact in solution to form a gel network. Similar building-block molecules may occur in molluscan mucins. Thick lines, protein core; thin lines, oligosaccharide chains. In the actual molecules, many more oligosaccharide chains are present per length of protein, and the molecules are contorted in three dimensions. (Redrawn from Pain, 1980, courtesy of the Society for Experimental Biology.)

The basic building block of this mucus is a glycoprotein of molecular weight $5 \times 10^5$, consisting of a protein core to which many short oligosaccharide chains are covalently bound. One end of the protein core is bare (i.e., has no attached oligosaccharides) and contains half-cystine residues that can form disulfide bonds with other similar glycoprotein molecules. When drawn in this idealized schematic fashion, this basic glycoprotein molecule resembles a bottle brush, but it should be remembered that in solution the real molecule will be continually changing shape. The oligosaccharides of this glycoprotein are negatively charged at physiological pH (due, in this case, to sulfates), and in dilute saline solutions the molecule is highly expanded. In pig gastric mucus four of these glycoprotein molecules are bound together by S—S bonds between their bare ends to form large, globular composite molecules, the outer surface of which is primarily composed of the expanded oligosaccharide chains. At glycoprotein concentrations above 2.5%, the effective volume of these globular composites is equal to the solution volume. Consequently at this and higher glycoprotein concentrations the composites are forced to interact, and they become cross-linked. The precise nature of the interaction is unclear, but Pain (1980) speculates that the cross-links are due to a lowering of the water content within the molecule below the point where the oligosaccharides are fully hydrated, causing chains to interact in order to reduce the enthalpy of the system. Regardless of their precise nature, the cross-links tie the glycoproteins together to form a gel network, and, as a consequence, the mucus behaves as a viscolelastic solid.

Although the specifics of this model apply only to pig gastric mucus, the idea that similar "building-block" molecules may occur in other mucins is useful when considering mucus macromolecular architecture in general. For example, changing the molecular weight of the individual glycoproteins or the number of glycoproteins bound into a large composite molecule should still result in viscoelastic mucus, though the storage and loss moduli will be affected. Similarly, changing the charge density of the molecule will affect its swelling behavior, but the overall material will still be recognizable as a mucin. In this sense it is possible that this model or one similar to it can be related to many different mucins, both in vertebrates and in invertebrates.

If a structure similar to that found in pig gastric mucus is present in molluscan pedal mucins, a number of their properties can be explained. For example, this model could account for the presence of two sorts of cross-links (S—S and "weak bonds"), as seen in *A. columbianus* and probably in *P. vulgata* pedal mucus. It is also possible that such a model could explain the yield–heal cycle of *A. columbianus* pedal mucin: At low strains the weak bonds between composite molecules would have sufficient strength to keep the gel network intact. At a certain strain these weak bonds would break and the material would flow as globular composites were sheared past each other. Even when sheared, the composites would have an effective volume equal to the solution volume; the composites would be always in contact, and the transient interactions between composites would account for the high viscosity of the mucus in its fluid form. As long as the material was continually sheared, cross-links of sufficient strength to allow the gel network to set up would not have time to form. However, as soon as the strain rate becomes zero, sufficiently strong cross-links could form and the mucus would quickly "heal."

At present the possibility of such a structure in pedal mucus remains pure speculation, and further research is needed before the possible occurrence in the molluscs of mucins formed from a blueprint similar to pig gastric mucus can be shown.

## VI. Summary

The mechanical properties of molluscan mucins result from the presence in solution of a small amount of a protein–polysaccharide complex. These protein–polysaccharide molecules are highly expanded and even at low concentrations may effectively enclose the entire solution volume. At very low concentrations mucous secretions act as viscous liquids. At higher concentrations cross-links between individual protein–polysaccharide molecules cause the formation of a gel network that can account for the viscoelastic properties of these mucins. These properties differ among mucins and in the cases examined to date have been shown to play an important role in the biological function of each mucous secretion.

## References

Alexander, R. McN. (1968). "Animal Mechanics." Univ. of Washington Press, Seattle.

Allen, A., Pain, R. H., and Robson, T. R. (1976). Model for the structure of the gastric mucus gel. *Nature (London)* **264**, 88–89.

Barnes, R. D. (1968). "Invertebrate Zoology," 2nd ed. Saunders, Philadelphia, Pennsylvania.

Cook, A. (1977). Mucus trail following by the slug *Limax grossui*. *Anim. Behav.* **25**, 774–781.

Denny, M. W. (1979). The role of mucus in the locomotion and adhesion of the pulmonate slug, *Ariolimax columbianus*. Ph.D. Thesis, Univ. of British Columbia, Vancouver.

Denny, M. W. (1980a). The role of gastropod pedal mucus in locomotion. *Nature (London)* **285**, 160–161.

Denny, M. W. (1980b). Locomotion: the cost of gastropod crawling. *Science* **208**, 1288–1290.

Denny, M. W. (1981). A quantitative model for the adhesive locomotion of the terrestrial slug, *Ariolimax columbianus*. *J. Exp. Biol.* **91**, 195–217.

Denny, M. W., and Gosline, J. M. (1980). The physical properties of the pedal mucus of the terrestrial slug, *Ariolimax columbianus*. *J. Exp. Biol.* **88**, 375–393.

Dorrington, K. L. (1980). The theory of viscoelasticity in biomaterials. *Symp. Soc. Exp. Biol.* **34**, 289–314.

Doyle, J. (1964). The chemical nature of the mucin secreted by the hypobranchial gland of *Neptunea antiqua*. *Biochem. J.* **91**, 6P.

Egami, F., and Takahashi, N. (1962). Studies on charoninsulphuric acid and sulphatase of *Charonia lampas*. *In* "Biochemistry and Medicine of Mucopolysaccharides" (F. Egami and Y. Oshima, eds.), pp. 53–77. University of Tokyo, Japan.

Ferry, J. D. (1980). "Viscoelastic Properties of Polymers," 3d ed. Wiley, New York.

Flory, P. J. (1953). "Principles of Polymer Chemistry." Cornell Univ. Press, Ithaca, New York.

Fowler, B. H. (1980). Reproductive biology of *Assiminea californica* (Megogastropoda: Rissoacea). *Veliger* **23**, 163–166.

Grenon, J. F., and Walker, G. (1980). Biochemical and rheological properties of the pedal mucus of the limpet *Patella vulgata* L. *Comp. Biochem. Physiol. B* **66B**, 451–458.

Grenon, J. F., Elias, J., Moorcraft, J., and Crisp, D. J. (1979). A new apparatus for force measurement in marine bioadhesion. *Mar. Biol.* **53**, 381–388.

Hunt, S. (1970). "Polysaccharide–Protein Complexes in Invertebrates." Academic Press, New York.

Hunt, S., and Jevons, F. R. (1965). The hypobranchial mucin of the whelk *Buccinum undatum* L. Properties of the mucin and of the glycoprotein component. *Biochem. J.* **97**, 701–709.

Hunt, S., and Jevons, F. R. (1966). The hypobranchial mucin of the whelk *Buccinum undatum* L. The polysaccharide component. *Biochem. J.* **98**, 522–529.

Inoue, S. (1965). Isolation of new polysaccharide sulphates from *Charonia lampas*. *Biochim. Biophys. Acta* **101**, 16.

Kwart, H., and Shashoua, V. E. (1957). The structure and constitution of mucus. *Trans. N.Y. Acad. Sci.* **19**, 595–612.

Litt, M., Wolf, D. P., and Khan, M. A. (1977). Functional aspects of mucus rheology. *Adv. Exp. Med. Biol.* **89**, 191–201.

Morris, E. R., and Rees, D. A. (1978). Principles of polymer gelation. *Br. Med. Bull.* **34**, 49–54.

Neuberger, A., Gottschalk, A., Marshall, R. D., and Spiro, A. G. (1977). Carbohydrate–peptide linkages in glycoproteins and methods for their elucidation. *In* "Glycoproteins" (A. Gottschalk, ed.), pp. 450–490. Elsevier, Amsterdam.

Oosawa, F. (1971). "Polyelectrolytes." Dekker, New York.

Pain, R. H. (1980). The viscolelasticity of mucus, a molecular model. *Symp. Soc. Exp. Biol.* **34**, 359–376.

Pancake, S. J., and Karnovsky, M. L. (1971). The isolation and characterization of a mucopolysaccharide secreted by the snail *Otella lactea*. *J. Biol. Chem.* **246**, 253–262.

Richter, K. O. (1980). Movement, reproduction, defense and nutrition as a function of the caudal mucus plug in *Ariolimax columbianus*. *Veliger* **23**, 43–47.

Rollo, C. D., and Wellington, W. C. (1977). Why slugs squabble. *Nat. Hist.* **86**, 46–51.

Ronkin, R. R. (1955). Some physicochemical properties of mucus. *Arch. Biochem. Biophys.* **56**, 76–89.

Simha, R. (1940). The influence of brownian movement on the viscosity of solutions. *J. Phys. Chem.* **44**, 25–34.

Simkiss, K., and Wilbur, K. (1977). The molluscan epidermis and its secretions. *Symp. Zool. Soc. London* **39**, 35–76.

Suzuki, H. (1941). Biochemical studies on carbohydrates LXII: prosthetic groups of snail mucoproteins. *J. Biochem. (Tokyo)* **33**, 377–383.

Tanford, C. (1961). "Physical Chemistry of Macromolecules." Wiley, New York.

Trott, T. J., and Dimock, R. V. (1978). Intraspecific trail following by the mud snail *Ilyanassa obseleta*. *Mar. Behav. Physiol.* **5**, 91–102.

Warren, L. (1963). The distribution of sialic acid in nature. *Comp. Biochem. Physiol.* **10**, 153–171.

Yonge, C. M. (1947). The pallial organs in the aspidobranch gastropods and their evolution through the Mollusca. *Philos. Trans. R. Soc. London, Ser. B* **232**, 443–518.

# 11

# Quinone-Tanned Scleroproteins

## J. H. WAITE[1]

Department of Biochemistry
University of Connecticut
Farmington, Connecticut

## I. Introduction

Despite their abundance in nature, scleroproteins are among the least understood of all proteins. Their mystery is in part attributable to their diverse composition and uncanny chemical stability. For the purposes of this chapter I shall define a scleroprotein functionally as a protein contributing mechanical strength to supporting structures in animals. This rather broad definition embraces collagens, silk fibroins, keratins, fibrins, resilin, and elastins as well as others. Some of these proteins form fibrils with distinct structural periodicities whereas others have more random structures.

[1] Present address: Orthopaedics Research Laboratory, University of Connecticut, Farmington, Connecticut, 06032.

THE MOLLUSCA, VOL. 1
Metabolic Biochemistry
and Molecular Biomechanics

The most persistent properties of the versatile scleroproteins are their insolubility *in vitro,* and their resistance to proteinases (trypsin and pepsin) and various hydrolytic solvents. Linderström-Lang and Duspiva (1936) attributed this stability to two factors: (1) special arrangement of amino acids (conformation) and (2) covalent cross-linking of the primary valence chains. In the years since Linderström-Lang's insightful predictions, the conformations and cross-links contributing to the stability of several scleroproteins have been elaborated (Table I). The cross-links have recently been extensively reviewed (Waite and Tanzer, 1982).

In molluscs, in addition to collagens, there occur a variety of scleroproteins presumed to be stabilized by quinone tanning (Brown, 1950a, 1975). Quinone tanning is an unfortunate term that has little or no demonstrated chemical significance with regard to cross-links or protein conformation. It was adopted by Pryor (1940) from the industrial practice of tanning hides with *o*-quinones (see Gustavson, 1966) to explain the sclerotization of cockroach ootheca by oxidized protocatechuic acid. Because the effect of quinones on the chemical structure of leathers and hides is to this day not fully understood, comparison of this reaction with those occurring in nature provides at best only an empirical model. Pryor (1940) was able to show that protocatechuic acid is oxidized *in vivo* by an enzyme to an *o*-quinone, and that the latter when added to ootheca led to a tanning and hardening of the structure. These, then, are the bare essentials for identi-

**TABLE I**
**Scleroproteins and Mechanisms of Stabilization**

| Scleroprotein | Precursor | Conformation | Cross-links | Reference |
|---|---|---|---|---|
| Collagen | Procollagen | Collagen helix | Aldimines Aldol condensations Pyridinoline | Eyre (1980) |
| Elastin | Elastin | Restricted random coil | Desmosines Aldimines | Bressan and Prockop (1977) Rucker and Murray (1978) |
| Fibrin | Fibrinogen | α Helix | ε-(γ-glutamyl)lysine | Folk and Finlayson (1977) |
| Keratin | Prekeratin | α Helix or β-pleated sheet | Disulfides ε-(γ-glutamyl)lysine | Steinert and Idler (1979) Baden and Goldsmith (1972) |
| Resilin | ? | Restricted random coil | Dityrosine Trityrosine | Andersen (1971) |
| Silk fibroin | ? | β-Pleated sheet | Dityrosine or none | Raven et al. (1971) Lucas et al. (1960) |

fying quinone tanning: There are quinones and a quinone-generating system, namely o-diphenols and phenoloxidase or peroxidase; the structure must be tanned and hardened following exposure to quinones. In molluscan scleroprotein structures, some of these conditions are only now being recognized through research.

The complete insolubility of quinone-tanned scleroproteins has compelled investigators to focus on precursors rather than the finished product. With this in mind, I shall discuss briefly what is known about the formation of scleroproteins such as collagen, elastin, and fibrin from precursors. With slight modifications, all of these systems share basic features: the compartmentalization of precursors, the cascade activation of components on secretion, or both (Fig. 1). In a typical case, one compartment (cellular or subcellular) contains prescleroprotein, which is incapable of polymerization (e.g., procollagen, fibrinogen, prekeratin); on secretion, this precursor is "activated" by specific proteolytic cleavage(s). Activated scleroprotein then undergoes self-assembly or aggregation. This is best illustrated by the formation of banded fibrils by collagen and fibrin, and relies chiefly on the orientation of molecules through secondary ionic, hydrogen, and hydrophobic interactions. These aggregates are then attacked by cross-linking enzymes such as lysyl oxidase (collagen and elastin), transglutaminase (fibrin and keratin), and peroxidase (resilin), which themselves frequently require proteolytic activation. The introduction of cross-links by these enzymes between the primary valence chain aggregates leads ultimately to a stable scleroprotein polymer.

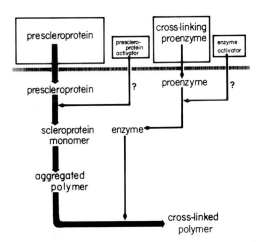

**Fig. 1.** Generalized scheme of scleroprotein formation. The question marks indicate that insufficient information exists for such a pathway in several of the proteins. The pathways do exist in fibrin polymerization, however.

Compartmentalization of precursors and cascade activation are essential features of scleroprotein formation. The reasons for this are obvious. First, the cross-linking reactions involve dangerous cytotoxic intermediates such as aldehydes, semiquinones, and activated oxygen. Second, by virtue of their stability most scleroproteins are dead-end products; that is, they are usually produced to serve a permanent or static function in the organism. They are not easily degraded or resorbed. Premature formation of scleroprotein could well spell death for the cell or organism.

Molluscan scleroprotein-containing structures are among the strongest mechanically but most inert chemically in the animal kingdom. They include the periostracum and insoluble shell matrix (Gordon and Carriker, 1980) of mollusc shells; the radula of polyplacophora, gastropods (Ducros, 1967), and cephalopods (Ducros, 1966); the hinge ligament (Chapter 8, this volume) and byssus of bivalves; the egg capsule (Hunt, 1971) and operculum (Hunt, 1976) of gastropods; and the sucker disk, beak, and pen of cephalopods (Hunt and Nixon, 1981). The purpose of this chapter is to discuss the properties and formation of two structures containing quinone-tanned scleroproteins: byssus and periostracum.

## II. Byssus

### A. Background

The byssus is an organ evolved by the bivalves for the specific function of attachment to a substratum. Hunt (1976) suggests its evolution from the gastropod operculum. Although a byssus seems universally present in bivalve larvae (Yonge, 1962; Sigurdsson et al., 1976), its occurrence in adult forms is less prevalent. In many bivalves, the byssus consists of a bundle of threads emerging from the animal at the base of the foot, extending through the two valves of the shell, and attached by a disk at the distal end of a substratum; in other species, the byssus consists of a single calcified stalk. A well-developed byssus is exemplified in species of *Arca, Mytilus, Pinna,* and *Anomia,* for example; an extensive catalog of the occurrence of byssus can be found in articles by Yonge (1962) and Stanley (1972). Other reviews of the byssus include Mercer (1972), Vovelle (1974), Brown (1975), and Tamarin (1977).

In all known instances, the byssus is secreted and assembled by specialized regions of the foot. The cytochemistry of byssus formation has been examined in detail in *Pinna* (Pujol, 1967), *Arca* (Bolognani-Fantin et al., 1973), *Anadara* (Lim, 1965), *Chlamys* (Gruffydd, 1978), *Perna* (Banu et al., 1979), and especially *Mytilus* (Brown, 1952; Smyth, 1954;

Gerzeli, 1961; Pujol, 1967; Tamarin and Keller, 1972; Ravindranath and Ramalingam, 1972; Lane and Nott, 1975; Bairati and Vitellaro-Zuccarello, 1976). *Mytilus* represents a high point in byssal development in that the byssus serves not only to anchor the animal to a substrate but also to absorb the shock of waves, and to resist periodic drying and wetting. This development has allowed *Mytilus* to inhabit the highly turbulent intertidal zone.

## 1. Byssus Secretion

The byssus in *Mytilus* consists of three parts as defined by Brown (1952): the root, the stem, and the threads (Fig. 2). The root is deeply embedded in the base of the foot and resembles an artist's paintbrush on dissection. Attached to it is a tanned acellular stem having stacked ringlike segments; from each segment of the stem emerges a thread (Tamarin, 1975). Each thread can be subdivided into four regions: a ring for attachment to the stem, a stretchy upper corrugated portion, a stiff lower portion, and a disk for adhesion to the substratum. When a new thread is produced, the animal extends its foot from the shell with the foot tip just touching the surface with which a bond is to be made (Maheo, 1970). During this quiescent period, components of the thread are secreted into the byssal groove and there molded by muscular contractions into a thread. After a brief time in this position, the animal lifts its foot, allowing a new thread to emerge; the thread, which is prestrained at 10% (Smeathers and Vincent, 1979), is initially cream colored and turns to a brownish yellow in time. The chemical components present in the completed byssal thread are derived from various exocrine glands in the foot (Fig. 2). These have been named (*a*) white or collagen gland, (*b*) purple or phenol gland, (*c*) enzyme or accessory gland, and (*d*) mucous gland. This terminology is based originally on Brown (1952), with subsequent modifications by Smyth (1954), Pujol (1967), and Allen et al. (1976). From a functional point of view, the collagen gland secretes the major fibrous components of the thread (Vitellaro-Zuccarello, 1980; Pujol et al., 1972), the phenol gland produces the adhesive substance for the disk (Tamarin et al., 1976), and the mucous gland secretes a substance that may promote the formation of a colloidal gel when mixed with the other components (Tamarin et al., 1976). The contribution of the enzyme or accessory gland is still controversial. Initially, Brown (1952) proposed this gland to be an extension of the phenol gland because both stained positively for the presence of *o*-diphenols. Smyth (1954) later rejected this hypothesis, claiming instead that the accessory gland secreted an oxidase (phenoloxidase) that was responsible for tanning or cross-linking the other proteins. Although

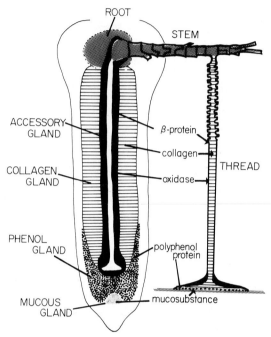

**Fig. 2.** Byssus formation in *Mytilus edulis*. Ventral surface of the foot shows approximate location of each of the exocrine glands known to secrete a substance present in byssus. For simplicity, only one thread is fully drawn (in longitudinal cross section).

there is no reason to doubt Smyth's version, there is additional evidence supporting Brown's contention that the accessory gland does secrete an $o$-diphenolic substance in addition to the enzyme (Pujol, 1967). Staining of foot sections with procedures sensitive to $o$-diphenols, but not monophenols, demonstrates that the accessory gland contains $o$-diphenols from the disk depression at the tip of the foot to the insertion of the root at the base. More will be said of this later.

## B. Physical Properties

The byssi of *Mytilus* and *Pinna* have long been shown to be largely resistant to the proteases trypsin and pepsin (Stary and Andratschke, 1925; Brown, 1952; Pikkarainen *et al.*, 1968), and to various hydrolytic solvents and protein denaturants (Brown, 1952). More recently, the adhesive-bonding properties of the disk (Allen et al., 1976) and tensile properties of the thread (Glaus, 1968; Smeathers and Vincent, 1979) have been examined in *Mytilus*. Some of these are listed in Table II. The physical properties of byssal threads in *Mytilus* vary depending on

**TABLE II**

**Some Mechanical Properties of *Mytilus* Byssal Threads and Other Scleroprotein-Containing Structures**

| Structure | Elastic modulus (N/m$^2$) | Ultimate tensile stress (N/m$^2$) | Ultimate tensile strain | Resilience (%) | Reference |
|---|---|---|---|---|---|
| Byssus | | | | | |
| Dry | $1.8 \times 10^9$ | $2 \times 10^8$ | 0.21 | — | Smeathers and |
| Wet | $8.5 \times 10^7$ | $2 \times 10^7$ | 0.44 | 60 | Vincent (1979) |
| Collagen, tendon | $6.0 \times 10^8$ | $8.2 \times 10^7$ | 0.10 | 62 | Gordon (1978) |
| Elastin, ligamentum nuchae | $6 \times 10^5$ | $3 \times 10^6$ | 1.60 | 76 | Yamada (1970) |
| Keratin, wool | $1.2 \times 10^9$ | $2-4 \times 10^8$ | 0.3–0.4 | | Danilatos and Feughelman (1979) |
| Resilin, insect wing hinge | $6.3 \times 10^5$ | $3 \times 10^6$ | 3.00 | 94 | Weis-Fogh (1961) |
| Silk fibroin, spider web frame | $3.0 \times 10^9$ | $1 \times 10^9$ | 1.25 | 53 | Denny (1976) |

the degree of humidity present (Smeathers and Vincent, 1979). In elastic modulus and tensile stress, dry byssus resembles hair keratin, whereas the wet thread is more similar to tendon collagen. Like spiders' silk and tendon collagen, wet byssus threads have low resilience; this is well suited to damping shock loads. The ultimate strain (or percentage extension) of dry byssus is one-half that of wet, and the latter is twice that of tendon collagen. As illustrated in Fig. 2, there is structural diversity within each byssal thread. The wet, corrugated upper portion of the thread is twice as extensible as the smooth distal portion (Smeathers and Vincent, 1979). Dry byssal threads have an elastic modulus 20 times greater than the wet threads (Table II). These properties serve the shock-absorbing function of byssus threads, which in *Mytilus* must withstand the sustained action of waves and tides. Moreover, during exposure at low tide, *Mytilus* is susceptible to the prying and tugging attentions of gulls and other predators. The stronger, more rigid, dry byssus may be better suited than the wet to resist such attack.

Allen et al. (1976) have reported the breaking modulus of the byssal disk–substratum bond to depend on the nature of the substratum. In disks attached to shell with periostracum, failure occurred by avulsion of periostracum from shell; when disks were attached to shell lacking periostracum, failure was caused by cohesive failure in the disk. Mean breaking load (g mm$^{-2}$) for disks attached to shells with and without periostracum was about 56 and 82, respectively. For a typical animal with 50 threads, anchorage strength would be estimated at 9–10 N normal to the substratum (Smeathers and Vincent, 1979). This agrees suitably with field test values of 5–17 N (Glaus, 1968). Fifty disks (1 mm$^2$ each) would require an ultimate breaking load of between 28 and 48 N. Clearly the threads are designed to fail before the disks in attached *Mytilus*. This would not seem to be the case in the byssus of *Arca zebra:* Here the threads reportedly have a breaking modulus seven times greater than the adhesive joints (Bowen, 1973).

## C. Chemical Properties

### 1. X-Ray Diffraction

Collagen was early discovered by X-ray diffraction to be present in *Mytilus* byssal threads (see Rudall, 1955). Curiously, collagen is not present in threads from nonmytilid species. Clearest collagen-diffraction patterns are afforded by the distal portion of the thread (Mercer, 1952); the fibrous elements of the proximal portion appear to be poorly oriented. Rudall (1955) was persuaded from his diffraction patterns that in addition to collagen there was another protein in the thread with the configuration of a $\beta$ keratin, and that the collagen constituted as much as 50% of the thread by weight. Rudall (1955) has likened the protein in the core of the stem to a fully extended $\beta$ protein. This protein demonstrates a number of crystallographic similarities with the byssus of *Pinna nobilis*. In *Subitopinna madida,* the structure of the threads was described by Mercer (1952) as "very poorly oriented."

### 2. Amino Acid Composition

Chemical analysis of *Mytilus* and *Pinna* byssal threads by Stary and Andratschke (1925) showed that the material was a "sclerotized protein." More recently, the amino acid compositions of the byssi of various species have been reported (Pujol et al., 1970a,b; Pikkarainen *et al.,* 1968; Gruffydd, 1978; Bowen, 1973; Cook, 1970); as these results have not previously been compiled, they are shown in Table III. Two main features are apparent from these data: Byssi from members of the family Mytilidae (*Mytilus* and *Modiolus*) appear to contain collagen as suggested by the high glycine content and presence of hydroxyproline; otherwise, compositions of byssi are highly variable, although several of the species do contain high cysteine (31–102 residues/1000), suggesting a cross-linked keratin-like composition.

The chemistry of *Mytilus* byssus has been studied the most. Table IV

**TABLE III**

**Amino Acid Composition of Byssus from Nine Bivalve Species[a]**

| Amino acid | Arca zebra[b] | Chlamys islandica[c] | Congeria cochleata[d] | Modiolus barbatus[d] | Mytilus edulis[e] | Pinctada alba[d] | Pinna nobilis[d] | Venerupis pullastra[d] | Anomia ephippium[f] |
|---|---|---|---|---|---|---|---|---|---|
| 3-Hyp | n.d.[g] | n.d. | n.d. | n.d. | <0.5 | n.d. | n.d. | n.d. | n.d. |
| 4-Hyp | 0 | Trace | 0 | 28 | 41 | 0 | 0 | 0 | 0 |
| Asp | 85 | 121 | 205 | 58 | 80 | 77 | 95 | 147 | 124 |
| Thr | 25 | 89 | 82 | 50 | 46 | 40 | 43 | 89 | 71 |
| Ser | 41 | 61 | 75 | 50 | 64 | 55 | 87 | 94 | 102 |
| Glu | 30 | 61 | 81 | 66 | 63 | 43 | 54 | 87 | 112 |
| Pro | 75 | 71 | 44 | 63 | 72 | 88 | 125 | 45 | 48 |
| Gly | 260 | 155 | 93 | 286 | 239 | 242 | 128 | 79 | 103 |
| Ala | 61 | 58 | 29 | 155 | 99 | 49 | 40 | 33 | 53 |
| Cys/2 | 102 | 34 | 44 | 8 | 23 | 46 | 31 | 102 | 0 |
| Val | 49 | 82 | 100 | 46 | 43 | 94 | 76 | 46 | 73 |
| Met | 7 | 3 | 18 | 8 | 9 | 22 | 16 | 20 | 33 |
| Ile | 24 | 53 | 71 | 18 | 28 | 31 | 31 | 39 | 36 |
| Leu | 20 | 32 | 44 | 20 | 44 | 40 | 47 | 36 | 52 |
| DOPA | n.d. | n.d. | n.d. | n.d. | 2[h] | n.d. | n.d. | n.d. | n.d. |
| Tyr | 97 | 18 | 22 | 16 | 20 | 8 | 43 | 31 | 27 |
| Phe | 16 | 24 | 20 | 14 | 22 | 46 | 20 | 25 | 36 |
| Hyl | n.d. | n.d. | n.d. | n.d. | 1 | n.d. | 0 | n.d. | n.d. |
| His | 28 | 18 | 5 | 21 | 19 | 39 | 18 | Trace | 17 |
| Lys | 48 | 71 | 40 | 49 | 43 | 51 | 103 | 88 | 49 |
| Arg | 33 | 32 | 25 | 45 | 43 | 29 | 43 | 39 | 57 |

[a] Given in residues per 1000.
[b] Averaged from Bowen (1973).
[c] Ribbon; Gruffydd (1977).
[d] Pujol et al. (1970b).
[e] Pikkarainen et al. (1968).
[f] Pujol et al. (1970a).
[g] n.d., Not determined.
[h] Detected by Degens and Spencer (1966).

TABLE IV
Amino Acid Composition of the Different Parts of *Mytilus* Byssus
and Its Probable Precursors[a]

| Amino acid | Stem[b] | Thread[c] | Disk[c] | Collagen[c] | β Protein[c] | Polyphenolic protein[d] |
|---|---|---|---|---|---|---|
| 3-Hyp | 0 | 0 | 2 | 0 | 0 | 30 |
| 4-Hyp | 16 | 33 | 26 | 64 | 0 | 102 |
| Asp | 91 | 68 | 94 | 68 | 107 | 23 |
| Thr | 63 | 39 | 34 | 27 | 49 | 117 |
| Ser | 74 | 66 | 70 | 49 | 85 | 102 |
| Glu | 68 | 61 | 47 | 67 | 84 | 9 |
| Pro | 90 | 64 | 44 | 47 | 27 | 81 |
| Gly | 166 | 295 | 193 | 314 | 117 | 32 |
| Ala | 79 | 129 | 80 | 150 | 55 | 81 |
| Cys/2 | 26 | 11 | 30 | 1 | 45 | 0 |
| Val | 52 | 38 | 37 | 29 | 38 | 8 |
| Met | 14 | 9 | 6 | 8 | 16 | 1 |
| Ile | 39 | 21 | 23 | 14 | 37 | 8 |
| Leu | 57 | 36 | 47 | 40 | 72 | 11 |
| DOPA | n.d. | 1 | 15 | 0 | 18 | 106 |
| Tyr | 20 | 13 | 52 | 14 | 50 | 66 |
| Phe | 21 | 16 | 35 | 17 | 40 | 1 |
| Hyl | — | Trace | 1 | — | 0 | 0 |
| His | 14 | 21 | 48 | 26 | 43 | 8 |
| Lys | 53 | 40 | 53 | 27 | 51 | 211 |
| Arg | 57 | 40 | 65 | 35 | 66 | 7 |

[a] Given in residues per 1000.
[b] Pujol et al. (1970b).
[c] Waite (unpublished data).
[d] Waite and Tanzer (1981a).

indicates that different parts of the byssus (e.g., stem, thread, disk) have significantly different amino acid compositions (Pujol et al., 1970a; Bdolah and Keller, 1976). X-Ray studies corroborate that the most collagen-like composition was obtained from the thread. This portion contained the highest glycine and hydroxyproline levels; lower glycine and higher cysteine values were obtained for the stem and disk. The disk was distinct from other parts in containing about 15 residues/1000 of 3,4-dihydroxyphenylalanine (DOPA) and a significant amount of 3-hydroxyproline, which was not detectable elsewhere in the byssus (Table IV). Traces of soluble collagen have been extracted from the byssal threads with boiling water (Pikkarainen et al., 1968), hot trichloroacetic acid (Andersen, 1968, cited in Brown, 1975), and acetic acid with pepsin (Chandrakasan *et al.,* 1977). Only the last-mentioned authors claim to have liberated a type I-like collagen from the threads.

## 3. Precursors

To circumvent the very low yields of extractable protein from byssus and possible degradation of the protein due to the severity of extraction, alternative procedures for investigating the composition of byssus need to be developed. One promising technique suggested by Tamarin et al. (1976) is to induce byssal secretion by injecting a solution of KCl into the base of the foot. The secretants thus induced can then be promptly collected and denatured before cross-linking has set in. Figure 3A illustrates the proteins present in this secretion following separation by acid gel electrophoresis. Note that three of these proteins stain strongly for $o$-diphenols. The major disadvantage of this approach is the unproven assumption that KCl-induced secretion resembles normal byssal secretion. Another strategy has been to dissect glands from the foot and extract their contents.

Bdolah and Keller (1976), Pujol et al. (1976), and J. H. Waite and M. L. Tanzer (unpublished observations) have attempted to liberate collagen from the collagen gland granules. Although this collagen has yet to be characterized, a method for isolating the granules has been devised (Bdolah and Keller, 1976), and sodium dodecyl sulfate–gel electrophoresis of a collagen gland extract suggests the presence of several collagenous proteins, the largest one of which has a molecular weight of 40,000–50,000 (Pujol et al., 1976). Waite and Tanzer (unpublished observations) have purified a collagenous protein contaminating extracts of the phenol gland (Fig. 3B; Table IV). This collagen resembles a component of the KCl-induced byssus secretion in its mobility and metachromasy on staining with coomassie blue R-250 (Fig. 3A and B). Its apparent molecular weight determined by cetylpyridinium chloride electrophoresis is about 55,000, agreeing in essence with Pujol's estimate. These estimates, however, are considerably lower than those (MW 95,000) for interstital mammalian collagens (Eyre, 1980). *Ascaris* cuticle collagen has the closest known molecular weight to byssal collagen, but unlike byssal collagen, cutical collagen has high cystine levels (Evans et al., 1976).

The major ingredient of the phenol gland is the so-called polyphenolic substance (Waite and Tanzer, 1980, 1981a). This protein is distinguished by its high content of DOPA and its insolubility in the detergent sodium dodecyl sulfate. The former property greatly facilitates assay of the polyphenolic substance by using a sensitive colorimetric method (Arnow, 1937; Waite and Tanzer, 1981b). Purified polyphenolic substance from the phenol gland always consists of two proteins that have similar mobility on acidic and detergent gel-electrophoretic sys-

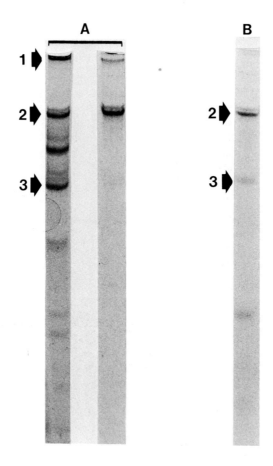

**Fig. 3.** Separation of byssal disk precursors by gel electrophoresis at acid pH. (**A**) KCl-induced secretion of byssal disks, stained for protein (left) and for →-diphenols (right), according to Waite and Tanzer (1981b). (**B**) Acid extract of neutral salt-insoluble proteins from the phenol gland (stained for protein). βProtein, polyphenolic substance, and collagen are indicated by the numbered arrows 1, 2, and 3, respectively. The collagen band exhibits metachromasy with the stain coomassie blue R-250.

tems. Estimated molecular weights of these are 135,000 and 125,000. Presumably one of these is derived from the other by a limited proteolysis because their relative proportions vary extensively from one preparation to another, and because in KCl-induced secretions left to stand, only one band remains (J. H. Waite, unpublished observations). The amino acid composition (Table IV) shows an abundance of lysine, serine, threonine, 3- and 4-hydroxyproline, and, of course, DOPA. The high lysine, serine, and threonine contents resemble the compositions of other invertebrate adhesive secretions: silkworm sericin (Sprague, 1975) and spider pyriform gland secretion (Andersen, 1970). The hydroxyproline content of the polyphenolic substance is puzzling because with rather low glycine levels, the protein can hardly be considered a normal collagen, where every third residue is glycine. One explanation is that hydroxyproline occurs in regions having a collagen helix. Other proteins reportedly have such collagenous domains (Porter and Reid, 1978; Bhattacharyya and Lynn, 1980; Rosenberry and Richardson, 1977). In the case of the polyphenolic substance, a collagenous domain could provide a specific site for grafting the adhesive to the collagen fibers in the thread (Fig. 2). Because 3-hydroxyproline is lacking in the byssal collagen, it serves as a unique marker for the amount of polyphenolic substance in the disk: about 6–7% (by weight). Another calculation, based on the 4-hydroxyproline in the disk minus that contributed by the polyphenolic substance, suggests that collagen constitutes about 26% (by weight) of the disk; in the thread, collagen may be as much as 75%. By simple deduction the remaining protein should be rich in glycine, aspartate, and cystine; it may also contain DOPA.

Chemical confirmation of the existence of Rudall's $\beta$ protein (1955) is very difficult, because crystallographic information is not readily reducible into amino acid stoichiometries. Assuming that Rudall's comparison of the noncollagenous component of byssal threads to a $\beta$ keratin *can* be extended to the composition of known $\beta$ keratins (feathers, scales, etc.), one could expect a composition rich in acidic residues (asparate and glutamate), amino acids with small side chains (glycine, serine, alanine), and cystine (Spearman, 1977). In byssal threads, the $\beta$ protein presumably is the varnish cortex seen in cross sections of the thread (Bairati and Vitellaro-Zuccarello, 1974). A putative precursor of the $\beta$ protein that has been extracted from the accessory gland (Fig. 2) is rich in aspartate, glycine, serine, and cysteine, and also contains low but significant amounts of DOPA (Table IV). There is no hydroxyproline in the protein. This composition is quite similar to those of the $\beta$ keratins; the DOPA is presumably added to facilitate quinone tanning. The $\beta$-protein precursor is soluble at neutral pH, and, in its unreduced state, migrates only slightly into acidic gels during electrophoresis. The molecular weight of this component has not yet been determined.

Another component of the accessory gland often cited is the enzyme polyphenoloxidase (Smyth, 1954; Brown, 1952; Pujol, 1967). The enzyme has not been isolated, and nothing is known about its chemical properties. Smyth (1954) and Pujol (1967) have demonstrated that in fixed sections of the gland the enzyme readily dehydrogenates the substrate catechol. In addition, Brown (1952) detected enzyme activity in newly secreted byssal threads, and Ravera (1950) and Roche et al. (1960) have confirmed the presence of extensive oxidizing activity in the threads. Using 4-methyl-catechol or DOPA as a substrate, I have assayed neutral salt extracts of the accessory gland for polyphenoloxidase. Significant activity exists, but only after brief treatment with chymotrypsin. This is reminiscent of the enzyme in the periostracum and extrapallial fluid of some bivalves (Waite and Wilbur, 1976; Misogianes, 1979).

Nothing is known about the chemistry of the mucosubstance secreted near the tip of the foot nor of the protein(s) in the stem. Tamarin et al. (1976) are of the opinion that the mucosubstance serves to form a sticky colloidal suspension when mixed with the polyphenolic substance.

## III. Periostracum

### A. Background

The periostracum of molluscs has been the subject of previous reviews. Wilbur and Simkiss (1968) and Hunt (1970) discuss the amino acid composition. More recently Vovelle (1974) and Brown (1975) deal with sclerotization of periostracum. The monograph on periostracum in *Buccinum undatum* by Hunt and Oates (1978) includes an extensive discussion of the comparative biochemistry of the structure. I do not intend to duplicate these efforts here; instead, I will focus on those aspects of the periostracum not treated at length previously. Saleuddin (Volume 4, Chapter 5) reviews the cytological events in periostracum formation; this section is intended to complement Saleuddin's chapter and is limited to biochemical events.

The periostracum may be universally present on the outer surface of mollusc shells. Presumably it was evolved to provide a cuticular covering for the naked epithelium of the animal (Morton and Yonge, 1964; Hunt and Oates, 1978). At first this cuticle was merely a composite of mineralized spines or spicules in an organic matrix (Carter and Aller, 1978). Such organization is still apparent in the primitive Aplacophora (Beedham and Trueman, 1968). As animals developed the capacity for quinone tanning, tanned scleroproteins gradually displaced the more unwieldy composite cuticles. A tanned layer is present in the periostracum of certain Polyplacophora (Beedham and Trueman, 1967); a quinone-tanned periostracum seems to be the rule for gastropods and bivalves (Brown, 1975; Vovelle, 1974).

Periostracum clearly plays a critical role in the lives of molluscs. Not only is it found in most or all species of the phylum, but it is also the first scleroprotein formed in the ontogeny of this group. In *Mytilus edulis* and *Haliotis discus* larvae, for example, it occurs only 18–24 h after fertilization and precedes initiation of shell formation (Humphreys, 1969; Iwata, 1980). Damage to intact periostracum along the shell margin is expeditiously repaired by the animal (Bubel, 1973b; Dunachie, 1963; Meenakshi et al., 1973). There has been considerable speculation about the function of the periostracum; much of this has been reviewed by Clark (1976). At present, however, there is little experimental information illustrating the function(s) of periostracum despite the commonly held opinion that periostracum plays an important role in the mineralization of the shell (Kessel, 1944; Taylor and Kennedy, 1969; Clark, 1976; Petit et al., 1979).

## 1. Periostracum Secretion

The cell biology of the secretion of periostracum is covered in the Volume 4 of this treatise. A simplified schematic version of the cell types believed to contribute to the formation of bivalve periostracum is shown in Fig. 4. This illustrates some components of the strategy of compartmentalization mentioned earlier. Four cell types are shown; all of these are located between the outer and middle folds of the mantle comprising the periostracal groove. The initial membrane of periostracum usually called the pellicle is generated by the basal cell. The pellicle is proteinaceous and may contain polyphenols (Bubel, 1973a). On the inner surface of the outer fold are granular epithelial cells and gob-

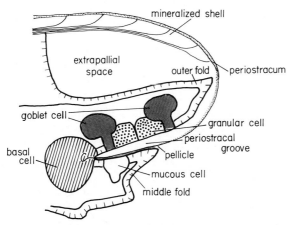

**Fig. 4.** Schematic illustration of the cells in the mantle epithelium presumed to participate in periostracum formation in bivalves. Each cell type is grossly exaggerated to emphasize its role.

let-shaped cells. The former contain DOPA-oxidase activity (Hillman, 1961); the latter, inclusions with $o$-diphenols (Bubel, 1973a). On the outer surface of the middle fold are a large number of mucosal cells. Note that substate ($o$-diphenol) and enzyme (DOPA-oxidase) are stored and produced in different cells. Mucosal cells may function to lubricate and hold the pellicle in place (Dunachie, 1963).

## B. Physical Properties

The properties of periostracum have not been well studied. Most of the early observations about periostracum are limited to mention of its stability in various hydrolytic solvents. Periostracum of *Mytilus* and *Pinna* resists digestion by pepsin and trypsin (Stary and Andratschke, 1925; Brown, 1952); that of *Buccinum* resists cold concentrated HCl, as well as NaOH, formic acid, mercaptoethanol, tetrahydrofuran, 8 $M$ urea, guanidine hydrochloride, and performic acid; it does not withstand the action of sodium hypochlorite (Hunt, 1971). Beedham (1958) found the periostraca of *Mytilus, Anodonta,* and *Ostrea* to resist thioglycolate, sodium sulfide, concentrated HCl (20°C), and HCl (55°C), but were labile to hot concentrated KOH and hypochlorite. As early as 1860, Voit observed the periostracum of *Anodonta* to be differentially soluble in alkali: In young shell it was soluble, whereas in older shell it was insoluble. This was again observed in *Geukensia* (*Modiolus*) *demissa:* Protein extracted from the margin with 8 $M$ urea and 0.5% sodium dodecyl sulfate was seven times more abundant than that liberated from the older periostracum (Waite, 1977). These results suggest that sclerotization occurs during the aging of the periostracal protein. The broad solvent stability of old periostracum indicates that sclerotization is not merely a result of hydrogen bonding, ionic bonding, or disulfide bonding, but rather of covalent bonding. Solubility in hypochlorite is empirically taken by some to suggest "quinone tanning" (Brown, 1950a).

Almost no research into the mechanical and physical properties of periostracum has been reported in the literature. Work by Digby (1968) is one exception. Using a pH indicator dye, Digby demonstrated that the pH of the exterior surface of the periostracum in *M. edulis* is consistently maintained at 1.5 units more acidic than the extrapallial side of the periostracum. This protongenic effect is attributed to the semiconducting properties of quinone-tanned proteins. Indeed, because synthetic polymers containing quinones or hydroquinones have unusual electron- and proton-conducting capabilities (Degrand and Miller, 1980), further research into the semiconducting properties of periostracum could provide critical insight into the mechanism of shell mineralization.

## C. Chemical Properties

### 1. Composition and Conformation

The amino acid composition of various periostraca has often been reported (Roche et al., 1951; Meenakshi et al., 1969; Degens et al., 1967; Hunt, 1971; Waite, 1977). The reader is referred to these articles for details. Although protein is generally accepted as the major constituent of periostracum, there are reports of other minor components such as metals (iron and manganese) (Blanchard and Chasteen, 1976; Swinehart and Smith, 1979), chitin (Goffinet and Jeuniaux, 1979), carbohydrates (Degens et al., 1967; Hunt and Oates, 1978), semiquinone-containing pigments (Wetzel, 1900; Blanchard and Chasteen, 1976), lipids (Meenakshi et al., 1969; Beedham, 1958), and polyphenols (Brown, 1952; Beedham, 1958; Bubel 1973c; Waite and Andersen, 1978).

The protein component of periostraca from bivalves and terrestrial and freshwater gastropods is characterized by rather high glycine levels (30–60% of the residues) (Degens et al., 1967; Meenakshi et al., 1969). Cystine levels are usually low in periostracum (1–22%) (Degens et al., 1967). Marine gastropods and cephalopods have, in contrast, lower glycine (usually less than 10%) but abundant serine and acidic residues (aspartate and glutamate total around 25%). Although a few gastropod and *Nautilus* periostraca contain 4-hydroxyproline, suggesting the presence of collagen, the majority of these structures appear amorphous. Only two periostraca, those of *Littorina littorea* and *B. undatum* (Bevelander and Nakahara, 1970; Hunt and Oates, 1978), so far reveal a fibrous structure with regular periodicities, that is, 30 nm in *Littorina* and 32 nm in *Buccinum*. Hunt and Oates (1978) found the *Buccinum* periostracum to contain coiled fibrous ribbons with an α-helical structure "less tightly coiled than α keratin." Nothing, unfortunately, is known about the conformations of proteins in other periostraca. The amino acid compositions, however, may allow some intelligent interim guessing. High-glycine (30–60%) proteins lacking abundant proline or hydroxyproline often assume β-pleated sheet conformations such as that in silk fibroin (Miller, 1979). At lower glycine concentrations, proteins can assume α-helical or globular conformations. Although the high glycine levels of bivalve and of freshwater and terrestrial gastropod periostraca suggests a β conformation, there has been no direct evidence to sustain this idea.

## 2. Hydrophobicity

In recent years much has been written about the hydrophobic effect in stabilizing protein structure (Tanford, 1978). There is substantial experimental evidence that much of the fibrous protein elastin, for example, is stabilized by hydrophobic interactions (Gosline et al., 1975; Sage and Gray, 1980). Degens et al. (1967) were quick to recognize that periostracum was rich in hydrophobic amino acids. I (Waite, 1976b) attempted to quantify this hydrophobicity in various periostraca using an index derived from the change in free energy in the transfer of each amino acid from ethanol to water (Nozaki and Tanford, 1971; Levitt, 1976): The more positive this value is, the more hydrophobic is the amino acid, and the more unfavorable is the transfer to water. Hydrophobicity indexes for various periostraca and scleroproteins are shown in Table V. It is important to mention the assumptions made when applying hydrophobicity to amino acid compositions of periostracum: (i) Protein conformation and heterogeneity are ignored and (ii) because asparagine and glutamine are inevitably modified by hydrolysis to their respective acids, the native protein is assumed to contain equal amounts of amides and acids, which would lead to an overestimation of hydrophobicity if, in fact, the protein had more acidic residues than amides.

Clearly, bivalve periostraca with the exception of those of *Pitar morrhuanus* are hydrophobic like the elastin-containing ligamentum nuchae and feather keratin. Among the gastropods, however, there are a significant number of hydrophilic periostraca in addition to the highly hydrophobic ones (Table V). The significance of hydrophobicity in mol-

**TABLE V**
Hydrophobicity Index (HI) of Molluscan Periostraca and Other Scleroproteins as Derived from Amino Acid Compositions[a]

| Source of scleroprotein | HI | Reference |
|---|---|---|
| Bivalves | | |
| *Artica islandica* | 171 | Degens et al. (1967) |
| *Corbicula consobrina* | 340 | Degens et al. (1967) |
| *Dosinia discus* | 547 | Meenakshi et al. (1969) |
| *Geukensia demissa* | 605 | Waite (1977) |
| *Mercenaria mercenaria* | 212 | Meenakshi et al. (1969) |
| *Mulina lateralis* | 159 | Degens et al. (1967) |
| *Mytilus californianus* | 230 | Meenakshi et al. (1969) |
| *Mytilus edulis* | 262 | Waite (unpublished data) |
| *Mytilus viridis* | 397 | Meenakshi et al. (1969) |
| *Pitar morrhuanus* | −8 | Degens et al. (1967) |
| *Quadrula quadrula* | 553 | Meenakshi et al. (1969) |

**TABLE V** (*Cont.*)

| Source of scleroprotein | HI | Reference |
|---|---|---|
| *Rangia cuneata* | 533 | Meenakshi et al. (1969) |
| *Solemya agassizii* | 336 | Meenakshi et al. (1969) |
| *Solemya velum* | 323 | Degens et al. (1967) |
| *Tagelus divisus* | 376 | Degens et al. (1967) |
| *Yoldia limatula* | 103 | Degens et al. (1967) |
| Gastropods | | |
| *Achatina* | −10 | Meenakshi et al. (1969) |
| *Astrea undosa* | 57 | Meenakshi et al. (1969) |
| *Buccinum undatum* | −90 | Hunt and Oates (1978) |
| *Cerithium lutosum* | 227 | Degens et al. (1967) |
| *Dollabella scapula* | 117 | Meenakshi et al. (1969) |
| *Haliotis cracherodii* | 4 | Meenakshi et al. (1969) |
| *Helix aspersa* | 391 | Meenakshi et al. (1969) |
| *Hydatina physis* | 32 | Degens et al. (1967) |
| *Lampusia aquatalis* | 66 | Meenakshi et al. (1969) |
| *Nassarius obsoletus* | 505 | Meenakshi et al. (1969) |
| *Nerita bernhardi* | 298 | Meenakshi et al. (1969) |
| *Neritina usnea* | 477 | Meenakshi et al. (1969) |
| *Oxytremata catenara* | 480 | Meenakshi et al. (1969) |
| *Pila virens* | 535 | Meenakshi et al. (1969) |
| *Septaria borbonica* | 410 | Meenakshi et al. (1969) |
| *Strombus gibberulus* | −427 | Meenakshi et al. (1969) |
| *Tridopsis albalorbis* | 439 | Meenakshi et al. (1969) |
| *Viviparus georgianus* | 355 | Meenakshi et al. (1969) |
| Other | | |
| Collagen (rat tendon) | 59 | Brown (1975) |
| Elastin (ligamentum nuchae) | 895 | Brown (1975) |
| Fibrinogen | 36 | Hunt and Oates (1978) |
| β Keratin (feather) | 298 | Spearman (1977) |
| α Keratin (skin) | −29 | Baden and Goldsmith (1972) |
| Paramyosin (scallop) | −491 | Hunt and Oates (1978) |
| Resilin (locust) | 17 | Andersen (1971) |
| Silk (*Bombyx mori*) | 137 | Brown (1975) |

[a] HI = $\Sigma\ n_i s_i$, where $n_i$ is the given concentration of an amino acid in a protein in residues per 1000, and $s_i$ is the free energy change (kcal/mol) of that amino acid in being transferred from ethanol to water at 25.1°C. The $s$ values used here are taken from Nozaki and Tanford (1971) and Levitt (1975): tryptophan (3.4), phenylalanine (2.5), tyrosine (2.3), DOPA (1.8), leucine (1.8), isoleucine (1.8), valine (1.5), proline (1.4), methionine (1.3), cystine (1.0), hydroxyproline (0.6), alanine (0.5), histidine (0.5), threonine (0.4), glycine (0.0), asparagine (−0.2), glutamine (−0.2), serine (−0.3), glutamate (−2.5), aspartate (−2.5), lysine (−3.0), and arginine (−3.0).

luscan periostraca can be viewed in two ways. First, periostracum is initially secreted as a soluble though highly viscous substance (Dunachie, 1963); by being hydrophobic, the periostracal precursors will tend to aggregate to avoid mixing with the water. This aggregation may be analogous to the formation of organic liquid crystals along the shell edge in *Buccinum* as proposed by Hunt and Oates (1978), although Table V suggests that, in *Buccinum*, hydrophobicity is unlikely to exert such an influence on the aggregation of periostracum. The initial aggregation brings the molecules into proximal positions for the cross-linking that presumably follows (Saleuddin, 1976). Second, periostracum in bivalves, as mentioned earlier, is a barrier to water especially along the shell margin. A hydrophobic protein aggregate would seem to provide an excellent barrier to water and the salts suspended in it.

### 3. Heterogeneity

Most past research on the chemical composition of periostracum has treated the structure as if it were uniformly homogeneous. From a physical, ultrastructural, and chemical standpoint, this interpretation is evidently too simple. Periostracum along the shell margin is more soluble and less pigmented than shell surface periostracum (Waite, 1977; Waite and Andersen, 1980). Many periostraca consist of complex layers (Kessel, 1944; Dunachie, 1963; Saleuddin, 1974; Kniprath, 1972; Waite, 1977) that are clearly different in structure. Waite and Andersen (1980) have shown that the abundance of the amino acid DOPA varies drastically with the age and solubility of periostracum, occurring at highest concentrations along the margin and decreasing to trace levels thereafter (Fig. 5). Other amino acids, with the exception of lysine, were not similarly decreased. DOPA has been detected in trace amounts (1–10 residues/1000) in many periostraca (Degens et al., 1967; Hunt, 1970; Waite, 1977; Waite and Andersen, 1978). Perhaps it is also the "polyphenol" detected histochemically (Brown, 1952; Beedham, 1958; Bubel; 1973c; Jones and Saleuddin, 1978). The variable amount of DOPA found in relation to age and solubility of periostracum in *G. demissa* and *M. edulis* suggests that this amino acid might play an active role in sclerotization.

### 4. Precursors

In *Mytilus* and *Geukensia,* the DOPA detected in periostracum is an integral part of the protein. None can be extracted by 5% trichloroacetic acid, in which free DOPA is ordinarily quite soluble. Is DOPA then a peptide-bound amino acid in the protein precursor secreted into the periostracal groove? Evidently yes. Waite et al. (1979) have isolated a

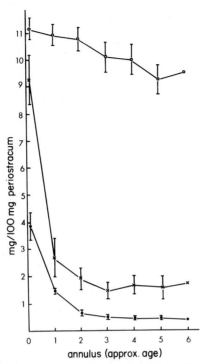

**Fig. 5.** Variation of DOPA (●), tyrosine (○), and protein solubility in formic acid (×) in the periostracum of *Mytilus edulis* with respect to age. Numbers in the abscissa denote the average age of each annulus. The portion of periostracum projecting (without underlying shell) beyond the shell margin is identified as the "O" annulus. From Waite and Andersen (1980), with permission from the publisher.

soluble DOPA-containing protein from marginal periostracum of *Mytilus* (Fig. 6; Table VI). The protein is insoluble at neutral pH and 5% sodium chloride. This DOPA protein or periostracin has an apparent molecular weight of about 20,000; however, in electrophoretic sample buffer, it rapidly undergoes degradation and polymerization (Fig. 6). The degradation is not sensitive to common protein denaturants, nor to the protease inhibitor phenylmethylsulfonylfluoride. The biological significance, if any, of the degradation *in vitro* is unclear. A similar reaction may occur in extracts of the marginal periostracum of *Geukensia* (Waite, 1977). In amino acid composition, periostracin resembles closely the marginal periostracum (Table VI). Aspartate is a notable exception; periostracin contains only about half that present in periostracum. The higher aspartate level in intact periostracum may come from

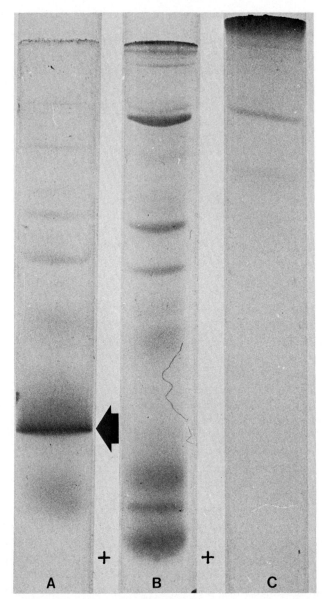

**Fig. 6.** Sodium dodecyl sulfate gel electrophoresis of formic acid-extracted periostracal proteins from marginal periostracum. (**A**) Sample run immediately after removal of formic acid by dialysis (stained for protein). Arrow marks position of periostracin. DOPA is detected by fluorescence (Yagi et al. 1960) or colorimetric (Waite and Tanzer, 1981b) methods. (**B**) Protein-stained sample 2-3 h after removal of formic acid. Note both polymerization and degradation of periostracin band. Only polymerization is inhibitable with addition of 2-mercaptoethanol. (**C**) Protein-stained sample 24 h after removal of formic acid. From Waite et al. (1979), with permission from the publisher.

TABLE VI

Comparison of the Amino Acid Compositions of
Periostracin and Marginal Periostracum of
*Mytilus edulis* [a]

| Amino acid | Periostracin | Marginal periostracum |
|---|---|---|
| Aspartate | 21 | 43 |
| Threonine [b] | 14 | 14 |
| Serine [b] | 56 | 51 |
| Glutamate | 22 | 22 |
| Proline | 19 | 21 |
| Glycine | 549 | 535 |
| Alanine | 32 | 33 |
| $\frac{1}{2}$-Cystine | 11 | 8 |
| Valine | 33 | 39 |
| Methionine | 2 | 3 |
| Isoleucine | 9 | 13 |
| Leucine | 18 | 19 |
| DOPA [b] | 21 | 12 |
| Tyrosine | 117 | 101 |
| Phenylalanine | 23 | 20 |
| Histidine | 10 | 10 |
| Lysine | 15 | 15 |
| Arginine | 35 | 34 |
| Tryptophan [b] | 9 | 7 |

[a] Given in residues per 1000.
[b] Corrected for losses during hydrolysis.

adhering shell matrix proteins, which are in fact rich in aspartate (Degens et al., 1967).

The disappearance of DOPA during aging of the periostracum may be caused by the enzyme $o$-diphenoloxidase (EC 1.14.18.1), which has been detected in the mantles of various species (Hillman, 1961; Tsujii, 1962; Timmermans, 1969; Jones and Saleuddin, 1978), in extrapallial fluid (Misogianes and Chasteen, 1979; Misogianes, 1979), in shell (Gordon and Carriker, 1980; Samata et al., 1980), and in periostracum (Hillman, 1961; Waite and Wilbur, 1976; Waite, 1977). In *Mytilus* and *Geukensia,* the enzyme contains copper, has a molecular weight of about 80,000, and requires treatment with chymotrypsin for activation (Waite and Wilbur, 1976; Misogianes, 1979). The fact that active enzyme is often detected in histological studies of mantle (Minganti and Mancuso, 1962; Jones and Saleuddin, 1978; Hillman, 1961) may be an artifact of fixation or some other procedure in tissue preparation. In my experience, $o$-diphenoloxidase from the mantle of *Geukensia* never occurs in active form (Waite, 1976a,b). Although chymotrypsin is the best activator of the enzyme, trypsin also promoted activity somewhat, as did the detergent sodium dodecyl sulfate. Enzyme extracted from periostracum of *Geukensia*, on the other hand, contains approximately equal amounts of active and inactive forms, both having similar molecular weights (Waite, 1976a). Again, chymotrypsin treatment leads to expression of activity in the inactive form. Periostracal *o*-diphenoloxidase in *Geukensia* has a pH optimum of 8.0–8.5 (roughly the pH of seawater), lacks detectable activity toward monophenolic substrates (e.g., tyrosine, tyramine), and preferentially utilizes *o*-diphenols with hydrophobic side chains or side chains lacking carboxyl groups (Waite and Wilbur, 1976). Of the experimental substrates, tertiary butylcatechol has the lowest $K_m$ and highest $V_{max}$. Because DOPA is the only known substrate in the periostracum, the evidence suggests that a DOPA strategically located in a hydrophobic pocket of periostracin might present the enzyme with a substrate at least as favorable as the butylcatechol. This conjecture has yet to be kinetically tested with purified periostracin and enzyme.

Recently, Dogterom and Jentjens (1980) demonstrated that in *Lymnaea stagnalis* periostracum formation may be stimulated by a growth hormone from the light-green neurosecretory cells in the cerebral ganglia. This is the only evidence so far suggesting regulation of scleroprotein formation in molluscs.

## IV. The Meaning of Quinone Tanning

### Sclerotization and DOPA

As in the better studied fibrous proteins (collagen, fibrin, elastin, etc.), the stability of molluscan scleroproteins is presumably determined by both conformation and cross-linking. Conformations detected thus far include the $\beta$-keratin and collagen helix of *Mytilus* byssus (Rudall, 1955), the $\alpha$ helix in *Buccinum* periostracum (Hunt and Oates, 1978), and the antiparallel $\beta$-pleated sheet of *Buccinum* operculum (Hunt et al., 1979). Hydrophobic interactions may also contribute to scleroprotein stabilization in periostracum.

The presence of cross-links in byssus and periostracum (and other scleroprotein structures) has largely eluded direct verification. Disulfides, although abundant in the periostracum and byssus of some species, may not contribute substantially to the stability of these scleroproteins. This is suggested by the observation that reduction of disulfides does not make byssus or periostracum any more soluble

(Brown, 1975). An aromatic compound from hydrolysates of byssus was identified as dityrosine by DeVore and Gruebel (1978), but I have been unable to confirm this (Waite and Tanzer, 1980). Hunt (1978) has described 2,4-quinoline dicarboxylic acid, isolated from *Buccinum* operculum, as a possible cross-link, but its derivation as such is difficult to envision. "Quinone tanning" is as close as anyone has come to identifying the mechanism of cross-linking in molluscan scleroproteins (Trueman, 1949; Brown, 1950a,b; Smyth, 1954; Beedham, 1958; Bubel, 1973c). The term, however, is only suggestive about cross-linking because it is empirically based on the presence of $o$-diphenols, quinones, polyphenoloxidase/peroxidase, and the browning and hardening of proteins.

DOPA is an $o$-diphenol and readily forms quinones and semiquinones by photolysis, autoxidation, and enzyme catalysis (Felix and Sealy, 1981). DOPA has now been identified in hydrolysates of (*a*) periostracum (Waite and Andersen, 1978; Degens et al., 1967) and its precursor, periostracin (Waite et al., 1979); (*b*) byssus (Waite and Tanzer, 1980) and two of its precursors, polyphenolic substance (Waite and Tanzer, 1981a,b) and $\beta$ protein; (*c*) insoluble shell matrix protein (Degens et al., 1967); and (*d*) egg capsules (Hunt, 1970). The disappearance of DOPA, presumably to $o$-quinones, during the sclerotization of these proteins is further evidence that DOPA may have a direct role in this process (Waite and Andersen, 1980). In view of this, I propose that quinone tanning and sclerotization of molluscan scleroproteins revolve around the concept of the DOPA protein. The model in Fig. 7 is based on still-fragmentary results on periostracum and byssus formation, but illustrates both the compartmentalization and cascade activation of other scleroproteins as discussed in the introduction. The scheme involves the hormone-induced secretion of a pre-DOPA protein, that is, one that is less capable of aggregation, perhaps. The preprotein is activated by proteolysis to an aggregating species; the aggregated mass is then attacked by $o$-diphenoloxidase. This enzyme, which in turn requires proteolytic activation, converts the DOPA residues to quinones, semiquinones, or both, and polymerization to the scleroprotein follows. Two crucial questions arising from this model need to be addressed: (1) How does DOPA become a part of the protein? and (2) What happens to DOPA after it has been oxidized by $o$-diphenoloxidase?

## 1. Origin of DOPA in Protein

It is well known that only 20 amino acids are directly coded for in the genomes of animals. Many unusual amino acids found to occur in proteins owe their existence to posttranslational modification of the protein by en-

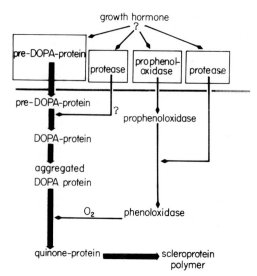

**Fig. 7.** Model illustrating the role of DOPA protein in the formation of quinone-tanned periostracum and byssus in molluscs.

zymes (Uy and Wold, 1977). Hydroxyproline and hydroxylysine in collagen are familiar examples of this. DOPA is parenthetically included among these modifications in that it can be obtained *in vitro* by the action of mushroom tyrosinase on various tyrosine-containing proteins (Nishi and Tomura, 1969; Lissitzky et al., 1962; Yasunobu et al., 1958). However, because tyrosinase catalyzes a two-step reaction consisting of (1) the hydroxylation of tyrosine to DOPA and (2) the dehydrogenation of DOPA to DOPA-quinone (Mason, 1955), DOPA occurs in these proteins only as a transient intermediate. The DOPA of periostracin and polyphenolic protein, on the other hand, is the end product of a reaction that seems to occur intracellularly and probably involves the posttranslational hydroxylation of specific tyrosyl residues in the proteins (Waite et al., 1979). An enzyme, tyrosine hydroxylase, has been characterized from the adrenal medulla of various mammals; this enzyme hydroxylates free tyrosine to form DOPA en route to the synthesis of catecholamines (Nagatsu, 1973). Thus far no one has shown this or any other enzyme to possess activity capable of hydroxylating protein tyrosyl residues.

The old model of quinone tanning in molluscs (also termed autotanning) held that during byssus or periostracum formation, tyrosine-rich proteins are secreted and cross-linked by the action of tyrosinase, which converts tyrosyl residues to quinones (Smyth, 1954; Pujol, 1967; Lissitzky et al., 1962). This model is in doubt not only because DOPA is present in scleroprotein precursors before they are secreted, but also because polyphe-noloxidase in molluscs is more accurately an o-diphenoloxidase catalyz-ing the dehydrogenation of DOPA to quinones but not the oxidation of tyrosyl residues (Waite and Wilbur, 1976; Misogianes, 1979).

Finally, the old model of autotanning is inadequate because many qui-none-tanned proteins are altogether lacking in tyrosine or contain only low levels of it, even in the most recently secreted scleroproteins.

## 2. Reactions of DOPA

**a. Nucleophilic Addition to o-Quinones.** o-Quinones such as DOPA-quinone readily condense with various nucleophiles in reactions de-scribed as Michael additions (Fig. 8) (Horspool, 1969). In DOPA-quinone the most available nucleophile happens to be the amino group on the side chain; this intramolecular condensation reaction leads to the formation of dihydroxyindoles (Mason, 1948, 1955; Nagatsu, 1973). Because intramo-lecular condensation preempts most tendencies for intermolecular con-densation reactions, molecules like DOPA are not good candidates as pro-tein cross-linkers (Pryor, 1962). Insects circumvent this problem by designing their presumptive cross-linker with a blocked amino group [i.e., N-acetyldopamine (Andersen, 1976)] or omitting the amino group alto-gether [i.e., protocatechuic acid (Pryor, 1940)]. The same concern about the efficacy of DOPA as a cross-linker in molluscs can be eliminated by making its amino group participate in peptide bonds (Fig. 8).

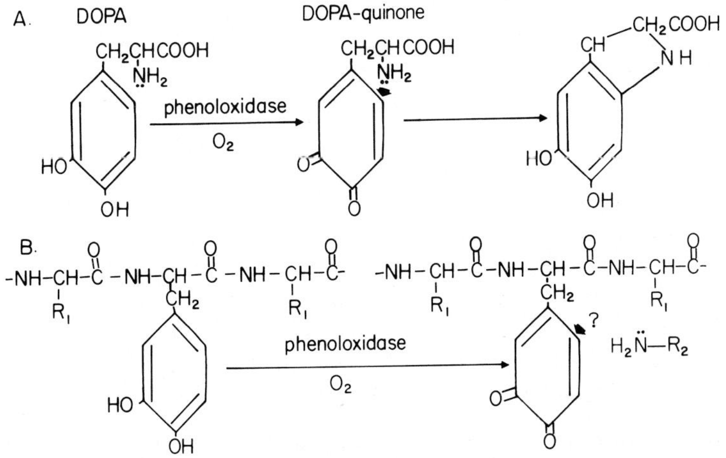

**Fig. 8.** Reactions of DOPA. **(A)** DOPA oxidation and intramolecular condensation with amino group. **(B)** DOPA-protein oxidation and potential cross-link formation.

If $o$-quinones are prevented from intramolecular condensation reactions, they can form adducts with several different nucleophiles. Primary amines attack $o$-quinones with great avidity *in vitro* at pH values near the $pk_a$ of the amines. Monosubstituted $o$-quinones form adducts with $\beta$-alanine (Peter, 1978), with various primary aliphatic amines (Davies and Frahn, 1977; Liberato et al., 1981), and with aniline (Hackman and Todd, 1953). Amino acids react primarily through their $\alpha$-amino groups, but lysine also reacts through its $\epsilon$-amino group (Pierpont, 1969a). The amino group of amino-terminal proline also reacts with $o$-quinones (Mason and Peterson, 1955).

Adducts of $o$-quinones (including DOPA-quinone) with cysteine and other sulfhydryl compounds occur readily *in vitro* and have been isolated from natural sources as well as (Agrup et al., 1976; Roston, 1960; Ito and Prota, 1977; Ito and Nicol, 1977; Burton and Stoves, 1950; Liberato et al., 1981, Mason and Peterson, 1965). DOPA also reacts with disulfides such as cystine (Ito et al., 1981). The sulfur of methionine also reportedly reacts with $o$-quinones (Gupta and Vithayathil, 1980).

Chemical evidence clearly supports the view that $o$-quinones can form covalent adducts with various nucleophiles, particularly amines and thiols, be they free or protein residues (Pierpont, 1969b). To be a crosslinker, a free quinone must react twice with nucleophiles; say, for example, the cysteine of one protein and the lysine of another. In peptide-bound DOPA-quinones, this requirement is reduced by one because the DOPA residue is already a part of one protein. Such adducts have yet to be characterized from any molluscan scleroprotein, but this may be due in part to their instability on hydrolysis.

**b. $o$-Diphenols as Metal Chelators.** $o$-Diphenols and semiquinones are capable of chelating various metals with their vicinal aromatic hydroxyls. DOPA, for example, can form stable ligands with copper(II) (Boggess and Martin, 1975), manganese(II) (Martius and Schwarzhans, 1978), zinc(II), and nickel(II) (Gergely and Kiss, 1979; Felix and Sealy, 1981). The iron-scavenging $o$-diphenol (enterobactin) of many bacteria forms iron complexes with dissociation constants of $10^{-52}$ (Raymond and Carrano, 1979). This property of $o$-diphenols may contribute to sclerotization of molluscan proteins by fostering a passive mineralization resulting from a selective sequestration of metals from seawater. It may also explain the enrichment of various heavy metals in the byssus and periostracum of *Mytilus* (Hamilton, 1980; Sturesson, 1976; Swinehart and Smith, 1979).

**c. Dehydration of Proteins by Polyphenols.** In recent years, the quinone-tanning mechanism as originally espoused by Pryor (1940) has come under attack for failing after 40 years to provide evidence for the existence of a single quinone-related cross-link. Vincent (1979) and Vincent and Hillerton (1979) suggest that the sclerotizing effect of polyphenols on insect cuticle can be explained entirely by the dehydration of cuticle proteins by polyphenols. This eliminates any need, they claim, for the assumption that covalent cross-links are formed. Indeed vegetable tannins and polyphenols do stabilize and render insoluble various globular (Haslam, 1974; McManus et al., 1981) and fibrous proteins (Grant and Alburn, 1965; Gustavson, 1966). The $o$-diphenol catechin, for example, renders a soluble type I collagen insoluble as well as resistant to the action of collagenase (Kuttan et al., 1981). McManus et al. (1981) have proposed a model for the association of proteins with polyphenols. An $o$-diphenol, or tannin, can be hydrogen-bonded to peptide bonds in the protein; specifically, hydrogen bonding takes place between the hydrogen of the peptide nitrogen and the oxygen of one phenolic OH group, and between the oxygen of the peptide carbonyl and the hydrogen of the other phenolic OH. At increasing levels of $o$-diphenols, a phenolic micelle is formed around the protein peptide backbone. Precipitation of the protein is caused when the surface of the micelle is less hydrophilic than the protein itself. Haslam (1974) has shown that tannin-precipitated serum albumin can be redissolved in water by removing tannins with organic solvents (e.g., acetone). Although protein dehydration by $o$-diphenols may play some role in stabilizing molluscan scleroproteins, these proteins are not rendered significantly more soluble by treatment with organic solvents (Brown, 1950, 1952; Hunt, 1971).

## V. Concluding Remarks

It is still premature to say exactly how DOPA-containing scleroproteins are stabilized or even whether they occur in all of the so-called quinone-tanned structures of molluscs. Among the stabilizing conformational factors are the collagen helix, $\alpha$ helix, $\beta$-pleated sheet, and hydrophobic interactions. With regard to cross-linking, I have presented evidence that (1) $o$-quinones can form covalent adducts with primary amines (i.e., $\epsilon$-amino group of lysine) and thiols (cysteine); (2) $o$-diphenols and semiquinones can chelate various metals with high affinity; and (3) $o$-diphenols dehydrate proteins. Future research will determine whether and to what extent these are present in molluscan scleroproteins.

Finally, the significance of quinone tanning in molluscs via DOPA proteins deserves comment. The insect cuticle constitutes one of the few invertebrate structures in which sclerotization has been scrutinized, albeit incompletely (Andersen, 1976). Instead of using DOPA proteins, insects mix the $o$-diphenol $N$-acetyldopamine with oxidases, cuticular proteins, and chitin. By blending these four basic ingredients in various proportions, insects have achieved an intricate capacity to modulate the mechanical properties of their cuticles, for example, the hard mandibles, resilient femurs, and extensible abdomens of locusts. Such versatility is not inherent in the DOPA protein of periostracum, for example. Periostracin, with a fixed amount of DOPA residues, is secreted by the mantle; the DOPA is oxidized, sclerotization ensues, and a highly regimented type of scleroprotein is formed. Periostracum, however, despite its apparent lack of versatility (compared to insect scleroproteins), has served molluscs well. Periostracal DOPA proteins are water-insoluble ($N$-acetyldopamine is not), rapidly form chemically inert scleroproteins, and may have contributed fundamentally to the organism's ultimate defense—the mineralized shell. The DOPA protein, too, is capable of conferring mechanical versatility on molluscan scleroproteins. This is best illustrated in *Mytilis* byssal threads, which are an amalgam of two DOPA proteins (polyphenolic protein and the $\beta$ protein) and collagen. Just how these proteins are processed and mixed by the animal to produce the properties exhibited by finished byssus threads remains a challenge for future research.

# References

Agrup, G., Hansson, C., Rorsman, H., Rosengren, A. M., and Rosengren, E. (1976). Mass spectrometric analysis of enzymatically synthesized cysteinyl DOPA isomers. *Commun. Dep. Anat., Univ. Lund, Swed.* No. 5, 1–17.

Allen, J. A., Cook, M., Jackson, D. J., Preston, S., and Worth, E. M. (1976). Observations on the rate of production and mechanical properties of the byssus threads of *Mytilus edulis* L. *J. Molluscan Stud.* **42**, 279–289.

Andersen, S. O. (1970). Amino acid composition of spider silks. *Comp. Biochem. Physiol.* **35**, 705–711.

Andersen, S. O. (1971). Resilin. *In* "Comprehensive Biochemistry" (M. Florkin and E. H. Stotz, ed.), Vol. 26C, pp. 633–657. Elsevier, Amsterdam.

Andersen, S. O. (1976). Cuticular enzymes and sclerotization in insects. *In* "The Insect Integument" (H. R. Hepburn, ed.), pp. 121–144. Elsevier, Amsterdam.

Arnow, L. E. (1937). Colorimetric determination of the components of 3,4-dihydroxyphenylalanine and tyrosine mixtures. *J. Biol. Chem.* **118**, 531–537.

Baden, H. P., and Goldsmith, L. A. (1972). The structural protein of epidermis. *J. Invest. Dermatol.* **59**, 66–76.

Bairati, A., and Vitellaro-Zuccarello, L. (1974). Ultrastructure of the byssal apparatus of *Mytilus galloprovincialis*. II. Observations by microdissection and scanning electron microscopy. *Mar. Biol.* **28**, 145–158.

Bairati, A., and Vitellaro-Zuccarello, L. (1976). Ultrastructure of the byssal apparatus of *Mytilus galloprovincialis*. IV. Observations by transmission electron microscopy. *Cell Tissue Res.* **166**, 219–234.

Banu, A., Shymasundari, K., and Rao, K. H. (1979). Foot glands in *Perna indica* and *Perna viridis* (Pelecypoda: Mytilidae). Histology and histochemistry. *Folia Histochem. Cytochem.* **17**, 395–404.

Bdolah, A., and Keller, P. J. (1976). Isolation of collagen granules from the foot of the sea mussel *Mytilus californianus. Comp. Biochem. Physiol. B* **55B**, 171–174.

Beedham, G. E. (1958). Observations on the non-calcareous component of the shell of lamellibranchia. *Q. J. Microsc. Sci.* **99**, 341–357.

Beedham, G. E., and Trueman, E. R. (1958). The utilization of $I^{131}$ by certain lamellibranchs, with particular reference to shell secretion. *Q. J. Microsc. Sci.* **99**, 199–204.

Beedham, G. E., and Trueman, E. R. (1967). Relationship of the mantle and shell of the polyplacophora in comparison with that of other mollusca. *J. Zool.* **151**, 215–231.

Beedham, G. E., and Trueman, E. R. (1968). The cuticle of the Aplacophora and its evolutionary significance in Mollusca. *J. Zool.* **151**, 215–231.

Bevelander, G., and Nakahara, H. (1970). An electron microscopic study of the formation and structure of the periostracum of a gastropod, *Littorina littorea. Calcif. Tissue Res.* **5**, 1–12.

Bhattacharyya, S. N., and Lynn, W. S. (1980). Characterization of the collagen-and non-collagen-like regions present in a glycoprotein isolated from alveoli of patients with alveolar proteinosis. *Biochem. Biophys. Acta* **625**, 343–355.

Blanchard, S. C., and Chasteen, N. D. (1976). EPR spectrum of a sea shell, *Mytilus edulis. J. Phys. Chem.* **80**, 1362–1367.

Boggess, R. K., and Martin, R. B. (1975). Copper chelation by DOPA, epinephrine and other catechols. *J. Am. Soc. Chem.* **97**, 3076–3081.

Bolognani-Fantin, A., Pujol, J. P., Bouillon, J., Bocquet, J., and Gervaso, M. V. (1973). Étude histochemique du complexe byssogène d'Arca (Mollusque bivalve). *Bull. Soc. Linn. Normandie* **104**, 194–203.

Bowen, H. J. (1973). Potential dental cement from the marine mussel *Arca. In* "Dental Adhesive Materials" (H. D. Moskowitz, G. T. Ward, and E. D. Woolridge, eds.), pp. 82–93. Institute of Dental Research, U.S. Dept of Health, Education and Welfare, Bethesda, Maryland.

Bressan, G. M., and Prockop, D. J. (1977). Synthesis of elastin in aortas from chick embryos. *Biochemistry* **16**, 1406–1412.

Brown, C. H. (1950a). A review of the methods available for the determination of forces stabilising structural proteins in animals. *Q. J. Microsc. Sci.* **91**, 331–339.

Brown, C. H. (1950b). Quinone tanning in the animal kingdom. *Nature (London)* **165**, 275.

Brown, C. H. (1952). Some structural proteins of *Mytilus edulis. Q. J. Microsc. Sci.* **93**, 487–502.

Brown, C. H. (1975). "Structural Materials in Animals." Wiley, New York.

Bubel, A. (1973a). An electron microscope study of periostracum formation in some marine bivalves. I. Origin of the periostracum. *Mar. Biol.* **20**, 213–221.

Bubel, A. (1973b). An electron microscope study of periostracum repair in *Mytilus edulis. Mar. Biol.* **20**, 235–244.

Bubel, A. (1973c). An electron microscope investigation into the distribution of polyphenols in the periostracum and cells of the inner face of the outer fold of *Mytilus edulis. Mar. Biol.* **23**, 2–15.

Burton, H., and Stoves, J. L. (1950). Quinone tanning in the animal kingdom. *Nature (London)* **165**, 569–570.

Carter, J. G., and Aller, R. C. (1975). Calcification in the bivalve periostracum. *Lethaia* **8**, 315–320.

Chandrakasan, G., Geetha, B., Krishnan, G., and Joseph, K. T. (1977). Studies on invertebrate collagens: I. Nature of the collagenous protein of byssus threads of *Mytilus edulis. Indian J. Biochem. Biophys.* **14,** 132–137.

Clark, G. R. (1976). Shell growth in the marine environment: Approaches to the problem of marginal calcification. *Am. Zool.* **16,** 617–626.

Cook, M. (1970). Composition of mussel and barnacle deposits at the attachment interface. *In* "Adhesion in Biological Systems" (R. S. Manly, ed.), pp. 139–150. Academic Press, New York.

Danilatos, G., and Feughelman, M. (1979). Dynamic mechanical properties of α-keratin fibers during extension. *J. Macromol. Sci., Phys.* **B16,** 581–602.

Davies, R., and Frahn, J. L. (1977). Addition of primary aliphatic amines to 1,2-benzoquinone. The absence of reaction between a secondary amine and 1,2-benzoquinone. *J. Chem. Soc.,* Perkin Trans. 1, 2295–2297.

Degens, E. T., and Spencer, D. W. (1966). "Data File on Amino Acid Distribution in Calcified Tissues," Tech. Rep. No. 66–27. Woods Hole Oceanogr. Inst., Woods Hole, Massachusetts.

Degens, E. T., Spencer, D. W., and Parker, R. H. (1967). Paleobiochemistry of molluscan shell proteins. *Comp. Biochem. Physiol.* **20,** 553–579.

Degrand, C., and Miller, L. L. (1980). An electrode modified with polymer-bound dopamine which catalyzes NADH oxidation. *J. Am. Chem. Soc.* **102,** 5728–5732.

Denny, M. (1976). The physical properties of spider's silk and their role in the design of orb webs. *J. Exp. Biol.* **65,** 483–505.

DeVore, D. P., and Gruebel, R. J. (1978). Dityrosine in adhesive formed by the sea mussel, *Mytilus edulis. Biochem. Biophys. Res. Commun.* **80,** 993–999.

Digby, P. S. B. (1968). Mechanism of calcification in the molluscan shell. *Symp. Zool. Soc. London* **22,** 93–107.

Dogterom, A. A., and Jentjens, T. (1980). The effect of the growth hormone of the pond snail *Lymnaea stagnalis* on periostracum formation. *Comp. Biochem. Physiol. A* **66A,** 687–690.

Ducros, C. (1966). Tannage quinonique du bec, de la plume et des dents radulaires chez le Calmar *Loligo vulgaris. Bull. Soc. Zool. Fr.* **91,** 331–332.

Ducros, C. (1967). Contribution à l'étude du tannage de la radula chez les Gasteropods. *Ann. Histochim.* **12,** 243–272.

Dunachie, J. F. (1963). The periostracum of *Mytilus edulis. Trans. R. Soc Edinburgh* **65,** 383–411.

Evans, H. J., Sullivan, C. E., and Piez, K. A. (1976). The resolution of *Ascaris* cuticle collagen into three chain types. Biochemistry **15,** 1435–1439.

Eyre, D. R. (1980). Collagen: Molecular diversity in the body's protein scaffold. *Science* **207,** 1315–1322.

Felix, C. C., and Sealy, R. C. (1981). Electron spin resonance characterization of radicals from 3,4-dihydroxyphenylalanine: semiquinone anions and their metal chelates. *J. Am. Chem. Soc.* **103,** 2831–2835.

Folk, J. B., and Finlayson, J. S. (1977). ε-(γ-glutamyl) lysine. *Adv. Protein Chem.* **31,** 1–133.

Gergely, A., and Kiss, T. (1979). Coordination chemistry of L-DOPA and related ligands. *Met. Ions Biol. Syst.* **9,** 143–172.

Gerzeli, G. (1961). Ricerche istomorfologiche e istochimiche sulla formazione del bisso in *Mytilus galloprovincialis. Pubbl. Stn. Zool. Napoli* **32,** 88–103.

Glaus, K. J. (1968). Factors influencing the production of byssal threads in *Mytilus edulis. Biol. Bull. (Woods Hole, Mass.)* **135,** 420.

Goffinet, G., and Jeuniaux, C. (1979). Distribution and quantitative importance of chitin in mollusk shells. *Cah. Biol. Mar.* **20,** 341–349.

Gordon, J., and Carriker, M. R. (1980). Sclerotized protein in the shell matrix of a bivalve mollusc. *Mar. Biol.* **57**, 251–260.

Gordon, J. E. (1978). "Structures, or Why Things Don't Fall Down." Penguin Books, London.

Gosline, J. M., Yew, F. F., and Weis-Fogh, T. (1975). Reversible structural changes in a hydrophobic protein, elastin, as indicated by fluorescent probe analysis. *Biopolymers* **14**, 1811–1826.

Grant, N. H., and Alburn, H. E. (1965). Enhancement of collagen aggregation by catecholamines and related polyhydric phenols. *Biochemistry* **4**, 1271–1276.

Gruffydd, L. D. (1978). Byssus and byssus glands in *Chlamys islandica* and other scallops (Lamellibranchia). *Zool. Scr.* **7**, 277–285.

Gupta, M. N., and Vithayathil, P. J. (1980). Chemical modification of methionines of ribonuclease A with o-benzoquinones. *Int. J. Pept. Protein Res.* **15**, 236–242.

Gustavson, K. H. (1966). The function of the basic groups of collagen in its reaction with vegetable tannins. *J. Soc. Leather Trades' Chem.* **50**, 144–160.

Hackman, R. H., and Todd, A. R. (1953). Some observations on the reaction of catechol derivatives with amines and amino acids in the presence of oxidizing agents. *Biochem. J.* **55**, 631–637.

Hamilton, E. I. (1980). Concentration and distribution of uranium in *Mytilus edulis* and associated materials. Mar. Ecol.: Prog. Ser. **2**, 61–73.

Haslam, E. (1974). Polyphenol–protein interactions. *Biochem. J.* **134**, 285–292.

Hillman, R. (1961). Formation of the periostracum in *Mercenaria mercenaria*. *Science* **134**, 1754–1755.

Horspool, W. M. (1969). Synthetic 1,2-quinones: Synthesis and thermal reactions. *Q. Rev. Chem. Soc.* **23**, 204–235.

Humphreys, W. J. (1969). Initiation of shell formation in the bivalve *Mytilus edulis*. *Proc. — Annu. Meet., Electron Microsc. Soc. Am.* **27**, 272–273.

Hunt, S. (1970). "Polysaccharide–Protein Complexes in Invertebrates." Academic Press, New York.

Hunt, S. (1971). Comparison of three extracellular structural proteins in the gastropod mollusc, *Buccinum undatum* L.: The periostracum, egg capsule, and operculum. *Comp. Biochem. Physiol. B* **40B**, 37–46.

Hunt, S. (1976). The gastropod operculum: A comparative study of the composition of gastropod opercular proteins. *J. Molluscan Stud.* **42**, 251–260.

Hunt, S. (1978). Operculin, a molluscan scleroprotein: 2,4-quinoline dicarboxylic acid, a possible cross-link. *Comp. Biochem. Physiol. B* **62B**, 55–60.

Hunt, S., and Nixon, M. (1981). A comparative study of protein composition in the chitin protein complexes of the beak, pen, sucker disc, radula, and esophageal cuticle of cephalopods. *Comp. Biochem. Physiol. B* **68B**, 535–546.

Hunt, S., and Oates, K. (1978). Fine structure and molecular organization of the periostracum in a gastropod mollusc *Buccinum undatum* L. and its relation to similar structural proteins systems in other invertebrates. *Philos. Trans. R. Soc. London, Ser. B* **283**, 417–463.

Hunt, S., Fraser, R. D. B., MacRae, T. P., and Suzuki, E. (1979). Molecular organization in molluscan operculae. *J. Mol. Biol.* **129**, 149–153.

Ito, S., and Nicol, J. A. C. (1977). A new amino acid 3-(2,5-S,S-dicysteinyl-3,4-dihydroxyphenyl)-alanine, from the tapetum lucidum of the gar (Lepisoteidae) and its enzymatic synthesis. *Biochem. J.* **161**, 499–507.

Ito, S., and Prota, G. (1977). Facile one-step synthesis of cysteinyl-dopas using mushroom tyrosinase. *Experientia* **33**, 1118–1119.

Ito, S., Inoue, S., Yamamoto, Y., and Fujita, K. (1981). Synthesis and antitumor activity of cysteinyl-3,4-dihydroxyphenylalanines and related compounds. *J. Med. Chem.* **24**, 673–677.

Iwata, K. (1980). Mineralization and architecture of the larval shell of *Haliotis discus hannai* Ino (Archaeogastropoda). *J. Fac. Sci., Hokkaido Univ., Ser. 4* **19**, 305–320.

Jones, G. M., and Saleuddin, A. S. M. (1978). Cellular mechanisms of periostracum formation in *Physa* spp. (Mollusca: Pulmonata). *Can. J. Zool.* **56**, 2299–2311.

Kessel, E. (1944). Über der Periostracum-Bildung. *Z. Morphol. Oekol. Tiere* **40**, 348–360.

Kniprath, E. (1972). Formation and structure of the periostracum of *Lymnaea stagnalis*. *Calcif. Tissue Res.* **9**, 260–271.

Kuttan, R., Donnelly, P. V., and Differante, N. (1981). Collagen treated with (+)-catechin becomes resistant to the action of mammalian collagenase. *Experientia* **37**, 221–223.

Lane, D. J. W., and Nott, J. A. (1975). A study of the morphology, fine structure and histochemistry of the foot of the pediveliger of *Mytilus edulis* L. *J. Mar. Biol. Assoc. U.K.* **55**, 477–495.

Levitt, M. (1976). A simplified representation of protein conformations for rapid simulation of protein folding. *J. Mol. Biol.* **104**, 59–108.

Liberato, D. J., Byers, V. S., Dennick, R. G., and Castagnoli, N. (1981). Regiospecific attack of nitrogen and sulfur nucleophiles on quinones derived from poison oak/ivy catechols and analogues as models for urushiol–protein conjugate formation. *J. Med. Chem.* **24**, 28–33.

Lim, C. F. (1965). Functional morphology of the byssal and associated glands in the bivalve genus *Anadara*. *J. Anim. Morphol. Physiol.* **12**, 113–131.

Linderström-Lang, K., and Duspiva, F. (1936). Studies in enzymatic histochemistry. XVI. The digestion of keratin by larvae of the clothes moth (*Tineola biselliela*. Humm.). *C. R. Trav. Lab. Carlsberg, Ser. Chim.* **21**, 1–82.

Lissitzky, S., Rolland, M., Reynaud, J., and Lasry, S. (1962). Propriétés des protéines oxydées et des certaines DOPA-protéines. *Biochem. Biophys. Acta* **65**, 481–494.

Lucas, F., Shaw, J. T. B., and Smith, S. G. (1960). Comparative studies of fibroins. *J. Mol. Biol.* **2**, 339–349.

McManus, J. P., Davis, K. G., Lilley, T. H., and Haslam, E. (1981). The association of proteins with polyphenols. *J.C.S. Chem. Commun.* **7**, 309–311.

Maheo, R. (1970). Étude de la pose et de l'activité de sécrétion du byssus de *Mytilus edulis* L. *Cah. Biol. Mar.* **11**, 475–483.

Martius, K. V., and Schwarzhans, K. E. (1978). Preparation of the Mn(II) complex with 3,4-dihydroxyphenylalanine (DOPA). *Z. Naturforsch., 33B*, 124.

Mason, H. S. (1948). Chemistry of melanin: III. Mechanism of oxidation of dihydroxyphenylalanine by tyrosinase. *J. Biol. Chem.* **172**, 83–99.

Mason, H. S. (1955). Comparative biochemistry of the phenolase complex. *Adv. Enzymol. Relat. Subj. Biochem.* **16**, 105–185.

Mason, H. S., and Peterson, E. W. (1955). The reaction of quinones with protamine and nucleoprotamine: N-terminal proline. *J. Biol. Chem.* **212**, 485–490.

Mason, H. S., and Peterson, E. W. (1965). Reactions between enzyme generated quinones and amino acids. *Biochim. Biophys. Acta* **111**, 134–146.

Meenakshi, V. R., Hare, P. E., Watabe, N., and Wilbur, K. M. (1969). The chemical composition of the periostracum of the molluscan shell. *Comp. Biochem. Physiol.* **29**, 611–620.

Meenakshi, V. R., Blackwelder, P. L., and Wilbur, K. M. (1973). An ultrastructural study of shell regeneration in *Mytilus edulis*. *J. Zool.* **171**, 475–484.

Mercer, E. H. (1952). Observations on the molecular structure of byssus fibres. *Aust. J. Mar. Freshwater Res.* **3**, 199–204.

Mercer, E. H. (1972). Byssus fiber—Mollusca. In "Chemical Zoology" (M. Florkin and B. T. Scheer, eds.), Vol. 7, pp. 147–154. Academic Press, New York.

Miller, A. (1979). Structure and function of fibrous proteins. Int. Rev. Biochem. 24, 171–210.

Minganti, A., and Mancuso, R. (1962). Tyrosinase activity in embryos of Physa fontinalis. Acta Embryol. Morphol. Exp. 5, 199–205.

Misogianes, M. J. (1979). A physicochemical and spectral characterization of the extrapallial fluid of Mytilus edulis. Ph.D. Thesis, University of New Hampshire, Durham, New Hampshire.

Misogianes, M. J., and Chasteen, N. D. (1979). A chemical and spectral characterization of the extrapallial fluid of Mytilus edulis. Anal. Biochem. 100, 324–334.

Morton, J. E., and Yonge, C. M. (1964). Classification and structure of the mollusca. In "Physiology of the Mollusca" (K. M. Wilbur and C. M. Yonge, eds.), Vol. 1, pp. 1–58. Academic Press, New York.

Nagatsu, T. (1973). "Biochemistry of Catecholamines." Univ. Park Press, Baltimore, Maryland.

Nishi, H., and Tomura, R. (1969). Effects of tyrosinase on sericin. Nippon Sanshigaku Zasshi 38, 117–122.

Nozaki, Y., and Tanford, C. (1971). The solubility of amino acids and two glycyl peptides in aqueous ethanol and dioxane solutions. J. Biol. Chem. 246, 2211–2217.

Peter, M. G. (1978). Reaktionen von naszierenden Chinonen mit β-Alanin methyl ester in Essigsäure und in wässriger Lösungen. Z. Naturforsch., C: Biosci. 33C, 912–918.

Petit, H., Davis, W. L., and Jones, R. (1979). Morphological studies on the periostracum of the freshwater mussel Amblena (Unionidae). Tissue Cell 11, 633–642.

Pierpont, W. S. (1969a). o-Quinones formed in plant extracts: Their reactions with amino acids and peptides. Biochem. J. 112, 609–618.

Pierpont, W. S. (1969b). o-Quinones formed in plant extracts: Their reaction with bovine serum albumin. Biochem. J. 112, 619–629.

Pikkarainen, J., Rantanen, J., Vastamaki, M., Lampiaho, K., Kari, A., and Kulonen, E. (1968). On collagens of invertebrates with special reference to Mytilus edulis. Eur. J. Biochem. 4, 555–560.

Porter, R. R., and Reid, K. B. M. (1978). The biochemistry of complement. Nature (London) 275, 699–704.

Pryor, M. G. M. (1940). On the hardening of the ootheca of Blatta orientalis. Proc. R. Soc., London, Ser. B 128, 378–392.

Pryor, M. G. M. (1962). Sclerotization. In "Comparative Biochemistry" (M. Florkin and H. S. Mason, eds.), Vol. 4, pp. 371–397. Academic Press, New York.

Pujol, J. P. (1967). Le complexe byssogène des Mollusques bivalves histochimie comparée des sécrétions chez Mytilus edulis L. et Pinna nobilis L. Bull. Soc. Linn. Normandie 8, 308–332.

Pujol, J. P., Bocquet, J., Tiffon, Y., and Rolland, M. (1970a). Analyse biochimique du byssus calcifié d'Anomia ephippium L. (Mollusque bivalve). Calcif. Tissue Res. 5, 317–326.

Pujol, J. P., Rolland, M., Lasry, S., and Vinet, S. (1970b). Comparative study of the amino acid composition of the byssus in some common bivalve molluscs. Comp. Biochem. Physiol. 34, 193–201.

Pujol, J. P., Houvenaghel, G., and Bouillon, J. (1972). Le collagene du byssus de Mytilus edulis L. Arch. Zool. Exp. Gen. 113, 251–264.

Pujol, J. P., Bocquet, J., and Borel, J. P. (1976). Le byssus de Mytilus: étude électrophorétique de fractions protéiques riches en hydroxyproline extraites de la "glande du colla-gene." *C. R. Hebd. Seances Acad. Sci., Ser.* D **283**, 555–558.

Raven, D. J., Earland, C., and Little, M. (1971). The occurrence of dityrosine in Tussah silk fibroin and keratin. *Biochim. Biophys. Acta* **251**, 96–99.

Ravera, O. (1950). Ricerche sul bisso e sulla sua secrezione. *Pubbl. Staz. Zool. Napoli* **22**, 95–105.

Ravindranath, M. H., and Ramalingam, K. (1972). Histochemical identification of Dopa, Dopamine and catechol in phenol gland and mode of tanning of byssus threads of *My-tilus edulis. Acta Histochem.* **42**, 87–190.

Raymond, K. N., and Carrano, C. J. (1979). Coordination chemistry and iron transport. *Acc. Chem. Res.* **12**, 183–190.

Roche, J., Ranson, G., and Eysseric-Lafon, M. (1951). Sur la composition des sclero-protéines des coquilles des Mollusques (Conchiolines). *C. R. Seances Soc. Biol. Ses Fil.* **145**, 1474–1476.

Roche, J., André, S., and Covelli, I. (1960). Sur la fixation de l'iode par la Moule (*Mytilus galloprovincialis* L.) et la nature des combinaisons iodées elaborées. *C. R. Seances Soc. Biol. Ses Fil.* **154**, 2201–2206.

Rosenberry, T. L., and Richardson, J. M. (1977). Structure of 18S and 14S acetylcholines-terase: identification of collagen-like subunits that are linked by disulfide bonds to catalytic subunits. *Biochemistry* **16**, 3550–3559.

Roston, S. (1960). Reaction of the sulfhydryl group with an oxidation product of $\beta$-3,4-dihy-droxyphenylalanine. *J. Biol. Chem.* **235**, 1002–1004.

Rucker, R. B., and Murray, J. (1978). Crosslinking amino acids in collagen and elastin. *Am. J. Clin. Nutr.* **31**, 1221–1236.

Rudall, K. M. (1955). The distribution of collagen and chitin. *Symp. Soc. Exp. Biol.* **9**, 49–71.

Sage, H., and Gray, W. R. (1980). Studies on the evolution of elastin. III. The ancestral protein. *Comp. Biochem. Physiol. B.***68B**, 472–480.

Saleuddin, A. S. M. (1974). An electron microscopic study of the formation and structure of periostracum in *Astarte. Can. J. Zool.* **52**, 1463–1471.

Saleuddin, A. S. M. (1976). Ultrastructural studies of the formation of periostracum in *Helix aspersa. Calcif. Tissue Res.* **22**, 49–65.

Samata, T., Sanguansri, P., Cazaux, C., Hamm, M., Engels, J., and Krampitz, G. (1980). Biochemical studies on components of Mollusc shells. *In* "Mechanisms of Biomineralization in Animals and Plants" (M. Omori and N. Watabe, eds.), pp. 37–47. Tokai Univ. Press, Tokyo.

Sigurdsson, J. B., Titman, C. W., and Davies, P. A. (1976). The dispersal of young post-larval molluscs by byssus threads. *Nature (London)* **262**, 386.

Smeathers, J. E., and Vincent, J. F. V. (1979). Mechanical properties of mussel (*Mytilus edulis*) byssus threads. *J. Molluscan Stud.* **45**, 219–230.

Smyth, J. D. (1954). A technique for the histochemical demonstration of polyphenoloxidase and its application to egg shell formation in helminths and byssus formation in *Mytilus. Q. J. Microsc. Sci.* **95**, 139–152.

Spearman, R. I. C. (1977). Keratins and keratinization. *Symp. Zool. Soc. London* **39**, 335–352.

Sprague, K. U. (1975). The *Bombyx mori* silk proteins: characterization of large polypep-tides. *Biochemistry* **14**, 925–930.

Stanley, S. M. (1972). Functional morphology and evolution of byssally attached bivalve mollusks. *J. Paleontol.* **46**, 165–212.

Stary, Z., and Andratschke, I. (1925). Bertrage zur Kenntniss einiger Skleroproteine. *Hoppe-Seyler's Z. Physiol. Chem.* **148**, 83–98.

Steinert, P. M., and Idler, W. W. (1979). Postsynthetic modifications of mammalian epidermal α-keratin. *Biochemistry* **18**, 5664–5669.

Sturesson, U. (1976). Lead enrichment in the shells of *Mytilus edulis. Ambio* **5**, 253–256.

Swinehart, J. H., and Smith, K. W. (1979). Iron and manganese deposition in the periostraca of several bivalve molluscs. *Biol. Bull. (Woods Hole, Mass.)* **156**, 369–381.

Tamarin, A. (1975). An ultrastructural study of byssus stem formation in *Mytilus californianus. J. Morphol.* **145**, 151–177.

Tamarin, A. (1977). How mussels get attached. *Nat. Hist.* **86**, 42–47.

Tamarin, A., and Keller, P. J. (1972). An ultrastructural study of the byssal thread forming system in *Mytilus. J. Ultrastruct. Res.* **40**, 401–416.

Tamarin, A., Lewis, P., and Askey, J. (1976). The structure and formation of the byssus attachment plaque in *Mytilus. J. Morphol.* **149**, 199–221.

Tanford, C. (1978). The hydrophobic effect and the organization of living matter. *Science* **200**, 1012–1018.

Taylor, J. D., and Kennedy, W. J. (1969). The influence of the periostracum on the shell structure of Bivalve mollusks. *Calcif. Tissue Res.* **3**, 274–283.

Timmermans, L. P. M. (1969). Studies on shell formation in molluscs. *Neth. J. Zool.* **19**, 417–523.

Trueman, E. R. (1949). Quinone-tanning in the Mollusca. *Nature (London)* **165**, 397–398.

Tsujii, T. (1962). Studies on the mechanism of shell- and pearl-formation VIII. On the tyrosinase in the mantle. *Mie-Kenritsu Daigaku Suisangakubu Kiyo* **5**, 378–383.

Uy, R., and Wold, F., (1977). Post-translational covalent modification of proteins. *Science* **198**, 890–896.

Vincent, J. F. V. (1978). Cuticle under attack. *Nature (London)* **273**, 339–340.

Vincent, J. F. V., and Hillerton, J. E. (1979). The tanning of insect cuticle: A critical review and a revised mechanism. *J. Insect. Physiol.* **25**, 653–658.

Vitellaro-Zuccarello, L. (1980). The collagen gland of *Mytilus galloprovincialis:* an ultrastructural and cytochemical study on secretory granules. *J. Ultrastruct. Res.* **73**, 135–147.

Voit, C. (1860). Anhaltspunkte für die Physiologie der Perlenmuschel. *Z. Wiss. Zool.* **10**, 470–498.

Vovelle, J. (1974). Sclerotisation et mineralisation des structures squellettiques chez les Mollusques. *Haliotis* **2**, 133–165.

Waite, J. H. (1976a). Rosewood polyphenols alter phenoloxidase activity from the mantle of the marine bivalve mollusc, *Modiolus demissus* Dillwyn. *Pestic. Biochem. Physiol.* **6**, 239–242.

Waite, J. H. (1976b). The function and properties of phenoloxidase in the periostracum and mantle of the bivalve mollusc *Geukensia demissa* (Dillwyn). Ph.D. Thesis, Duke University, Durham, North Carolina.

Waite, J. H. (1977). Evidence for the mode of sclerotization in a molluscan periostracum. *Comp. Biochem. Physiol. B* **58B**, 157–162.

Waite, J. H., and Andersen, S. O. (1978). 3,4-Dihydroxyphenylalanine in an insoluble shell protein of *Mytilus edulis. Biochim. Biophys. Acta* **541**, 107–114.

Waite, J. H., and Andersen, S. O. (1980). 3,4-Dihydroxyphenylalanine and sclerotization of periostracum in *Mytilus edulis* L. *Biol. Bull. (Woods Hole, Mass.)* **158**, 164–173.

Waite, J. H., and Tanzer, M. L. (1980). The bioadhesive of *Mytilus* byssus: A protein containing L-DOPA. *Biochem. Biophys. Res. Commun.* **96**, 1554–1561.

Waite, J. H., and Tanzer, M. L. (1981a). Polyphenolic substance of *Mytilus edulis:* Novel adhesive containing L-DOPA and hydroxyproline. *Science,* **212**, 1038–1040.

Waite, J. H., and Tanzer, M. L. (1981b). Specific colorimetric detection of o-diphenols and DOPA-containing peptides. *Anal. Biochem.* **111**, 131–136.

Waite, J. H., and Tanzer, M. L. (1982). Macromolecular cross-links in proteins and DNA. *In* "Handbook on Biochemistry of Aging" (J. L. Florini, ed.), p. 195–220. CRC Press, Boca Raton, Florida.

Waite, J. H., and Wilbur, K. M. (1976). Phenoloxidase in the periostracum of the marine bivalve *Modiolus demissus,* Dillwyn. *J. Exp. Zool.* **195,** 359–368.

Waite, J. H., Saleuddin, A. S. M., and Andersen, S. O. (1979). Periostracin: a soluble precursor of sclerotized periostracum in *Mytilus edulis* L. *J. Comp. Physiol.* **130,** 301–307.

Weis-Fogh, T. (1961). The thermodynamic properties of resilin, a rubber-like protein. *J. Mol. Biol.* **3,** 520–531.

Wetzel, G. (1900). Die organischen Substanzen der Schaalen von *Mytilus* und *Pinna. Hoppe-Seyler's Z. Physiol. Chem.* **29,** 386–410.

Wilbur, K. M., and Simkiss, K. (1968). Calcified shells. *In* "Comprehensive Biochemistry" (M. Florkin and E. H. Stotz, eds.), Vol. 26A, pp. 229–295. Elsevier, Amsterdam.

Yagi, K., Nagatsu, T., and Nagatsu-Ishibashi, I. (1960). Condensation reaction of Dopa with ethylene diamine. *J. Biochem. (Tokyo)* **48,** 617–620.

Yamada, H. (1970). "Strength of Biological Materials." Williams & Wilkins, Baltimore, Maryland.

Yasunobu, K. T., Peterson, E. W., and Mason, H. S. (1958). The oxidation of tyrosine containing peptides by tyrosinase. *J. Biol. Chem.* **234,** 3291–3295.

Yonge, C. M. (1962). On the primative significance of the byssus in the bivalvia and its effects in evolution. *J. Mar. Biol. Assoc. U.K.* **42,** 113–125.

# Index